THE RISE
OF
ANIMALS

THE RISE of ANIMALS
EVOLUTION *and*
DIVERSIFICATION *of the*
KINGDOM ANIMALIA

Mikhail A. Fedonkin

James G. Gehling

Kathleen Grey

Guy M. Narbonne

Patricia Vickers-Rich

Foreword by Arthur C. Clarke

THE JOHNS HOPKINS UNIVERSITY PRESS, Baltimore

The authors are grateful for support from the UNESCO International Geological Correlation Program, Project 493: The Rise and Fall of the Vendian Biota, coordinated by Mikhail Fedonkin, Patricia Vickers-Rich and Jim Gehling (2003–2007).

The Johns Hopkins University Press
2715 North Charles Street
Baltimore, Maryland 21218-4363
www.press.jhu.edu

Library of Congress Cataloging-in-Publication Data

The rise of animals: evolution and diversification of the kingdom animalia / Mikhail A.
    Fedonkin . . . [et al.].
        p. cm.
    Includes bibliographical references and index.
    ISBN-13: 978-0-8018-8679-9 (hc : alk. paper)
    ISBN-10: 0-8018-8679-1 (hc : alk. paper)
    1. Geology, Stratigraphic—Proterozoic. 2. Paleontology—Proterozoic. 3. Biodiversity.
    4. Organisms. 5. Life—Origin. 6. Evolution. I. Fedonkin, M. A. (Mikhail Aleksandrovich)

QE653.R567 2007
560'.1715—dc22   2007061351

Frontispiece: The Rawnsley Quartzite, top unit in the Pound Subgroup and latest Ediacaran, Flinders Ranges, South Australia (P. Trusler).

Reconstruction Art Work by:
    Peter Trusler, Honorary Research Associate, School of Geosciences, Monash University, Melbourne,
    Victoria, Australia

Major Editorial Assistance:
    Patricia Komarower, School of Geosciences, Monash University, Melbourne, Victoria, Australia

Major Research Assistance:
    Patricia Komarower, School of Geosciences, Monash University, Melbourne, Victoria, Australia
    Maxim Leonov and **Andrey Ivantsov**, Paleontological Institute, Russian Academy of Science, Moscow, Russia

Major Photography by authors, and:
    Francesco Coffa and **Steven Morton**, Monash University, Melbourne, Victoria, Australia
    Maxim Leonov and **Alexander Mazin**, Paleontological Institute, Russian Academy of Science, Moscow, Russia

Image Editors: **Draga Gelt** and **Steven Morton**, Monash University, Melbourne, Victoria, Australia

Book Design: **Draga Gelt** and **Patricia Vickers-Rich,** Monash University, Melbourne, Victoria, Australia

This book was prepared in Australia using Adobe InDesign. Design and layout were done by Draga Gelt, in collaboration
with the authors.

To Boris Sokolov, Reg Sprigg, and Mary Wade

And with sincere gratitude to
Tatiana and Timofei Burych
Barbara Boehm-Erni and Bruno Boehm
Marg and Doug Sprigg
Jane and Ross Fargher
Nathan Hunt and his family

CONTENTS

Opposite page: Ediacarans, reconstructed for Australia Post Creatures
of the Slime *stamp issue, 2005. This material has been reproduced with
permission of the Australian Postal Corporation. The original work is held
in the National Philatelic Collection (art by Peter Trusler).*

FOREWORD

ARTHUR C. CLARKE

Like many, my fascination with the Earth's past really started with dinosaurs. I was about 10 years old, and I can still see the event clearly in my mind's eye, though perhaps it's a false memory. I was riding in a pony-cart with my father not far from our home in Somerset when he gave me a cigarette postcard from the pack he'd just opened. It showed a weird monster with a row of plates protruding from its back. It was a *Stegosaurus*. I don't know why cigarette makers in the 1930's were fascinated by dinosaurs, but around the same time another company put out a beautiful collection of 3D photographs, so life-like that they could have been mistaken for reality. A folded-down viewer went with them, and I must have spent hours peering through it at these strange beasts of the past. How ironic that I owe a debt to the drug that killed my father, and which I have never inhaled even once!

In 1980, Luis Alvarez and his geologist son Walter advanced the theory that the dinosaur extinction *circa* 65 million years ago was caused by the impact of an asteroid or comet, with catastrophic consequences to life on Earth. As is commonly known now, other mass extinctions before and after that of the dinosaurs have also been attributed to such extraterrestrial "invaders." Early in the Precambrian this must have been the case – at a time when volcanoes were still belching out the ingredients that were to eventually make up our own protective shield of atmosphere, and at a time when the galaxy was filled with even more galactic bullets than we see today. During this most ancient of times, the mysterious Precambrian – spanning some 4 billion years and ending a mere 542 million ago – there simply must have been interruptions along life's path at the hands of cosmic collisions. The odds against it are just too small. Indeed, I opened my 1972 novel *Rendezvous with Rama* with these words:

"Sooner or later, it was bound to happen. On 30 June 1908, Moscow escaped destruction by three hours and four thousand kilometers. On February 12, 1947, another Russian city had a still narrower escape, when the second great meteorite of the twentieth century detonated less than four hundred kilometers from Vladivostok. . . . The meteorites of 1908 and 1947 struck only uninhabited wilderness, but by the end of the 21st century there was no region left on Earth that could be used for celestial target practice. And so, inevitably at

0946 GMT on the morning of 11 September in the exceptionally beautiful summer of the year 2077, most of the inhabitants of Europe saw a dazzling fireball appear in the eastern sky . . ." I then go on to destroy much of Europe; but why, 30 years ago, did I choose the infamous date 11 September? This still gives me a rather spooky feeling." The novel went on to suggest that a warning system, Spaceguard, should be set up to protect the Earth, and I am happy to say this is exactly what has now happened. Our forbears in the Precambrian, however, had no such system. And yet life did survive, in some cases, and evolve into the many forms we see today.

Imagine with me, if you will, that time long before the dinosaurs where there existed a world so unfamiliar to us that it was like a scene from a strange story, perhaps even one of my own. That world is revealed for all in *The Rise of Animals*. It is a book about the first animals, long before dinosaurs – animals that lived in the Precambrian, a time when life was just beginning to generate its diversity, eventually resulting in more and more complex forms and even those with skeletons.

Study of this Precambrian time is a real scientific frontier. New discoveries of weird and wonderful life forms pop up in scientific journals every week – forms that look like strange mollusks (snails, clams and their kin), or bizarre spiny-skinned animals related to starfish or maybe the ancestors of the backboned animals, of which we are one. Others are so weird that it's hard to tell just what they were. There is much debate about just what these animals were – and scientific argument is rife and fun.

Written by paleontologists, *The Rise of Animals* reveals lifetimes spent searching rocks in remote areas of the world looking for the clues that allow these paleontologists to imagine what this ancient world was like. The story they tell deals with the excitement of discovery and the scientific debates about interpretation. But the riot of photography and reconstructions of these ancient assemblages is the end result that will impress on the brain the wondrous nature of earth in deep, deep time. Such images will inspire the reader to realize there is something from long ago beyond dinosaurs that sheds light on an age long faded. And when we finally reach Europa, with its ice-covered seas, we are more likely to meet the animals in this book than we are to encounter anything resembling a dinosaur.

Opposite page: Rocks on the Avalon Peninsula of Newfoundland have produced some of the oldest Ediacarans. Near Mistaken Point (P. Trusler).

ACKNOWLEDGMENTS

So many people have given of their time and information to make this book possible that it is difficult to know where to begin and where to finish. The extensive bibliography gives some idea of those who have freely given of their scientific knowledge and the use of their graphics.

For the privilege of using figures and ideas of others, we thank them each in the captions to illustrations in this book.

Sincere thanks are due the following for reading in depth and commenting on various stages of this manuscript: Soren Jensen, Dan Condon, Maxim Leonov, Jim Valentine, Andrey Ivantsov, Ekatarina Serezhnikova, Xiao Shu-hai, Anthony Martin, Tom Rich, Marc Laflamme, John Latannzio, Mary Walters, Richard Jenkins, Rob Hengeveld, Gabi Schneider, Li Chia-wei, Hans Hofmann, Enrico Savazzi and Mary Lee Macdonald.

We would also like to sincerely thank many people who helped us gather material, photograph specimens, wrestle with computers, curate fossils, find fossils, carry out fieldwork and with many other tasks: Maxim Leonov, Andrey Ivantsov, Dan Condon, Sam Bowring, Paul Eriksson, Rob Creaser, Ekaterina Sereznikova, Alexei Rozanov, Dimitri Grazhdankin, Nathan Hunt, Ellis Yochelson, Jo Bain, Dennis Rice, Natalie Schroeder, John Gribben, Tim Flannery, Karl Heinz (Charlie) Hoffmann and his family, Marion Anderson, Yana Malakhovskaya, Lena Sossi, Cecilia Hewlett, Meaghan McDougall, Gail O'Toole, Florita Henricus, Sue Turner, Corrie Williams, Jo Monaghan, Rick Squire, Bruce Schaeffer, Peter Betts, Andrew Prentice, Jeff Weekes, Mario Juric, Richard Gott, Margaret Fuller, Hung-Jen Wu, J. L. Crowle, Bob Goodale (and the Naragebup, Rockingham Regional Environment Centre), Boyd Wykes, Karen Majer, Gail and Alan Curl, Michael Snow, Mary Droser and family, Mary Hoban, Marg Towt, Belinda Marshall, Peggy Cole and the Melbourne Aquarium, the Manly Aquarium staff, Andrew Constantine, Father Peter Golden, Paul Strother, Felicity O'Brien, Mary and Jerome Devereaux, Robert Franceschini and Dolf Seilacher and many others that we have ashamedly forgotten to mention. Guy Narbonne dedicates his chapter on the Canadian NW to Hans Hofmann, a geological pioneer and his postdoctoral mentor in Precambrian paleobiology, without whom there would have been no Windermere story to tell.

Funding to underwrite this project has come from many sources, and we thank each and every one of them immensely: the UNESCO International Geological Correlation Program (for IGCP493 and IGCP478 support, from both the international committee and the Australian committee), especially Margarete Patzak; Australia Post and the Federal Government of Australia for its tax incentive scheme which allows some fund return for artwork donated to public institutions; Monash University with special thanks to the Monash University Publications Committee for providing some funds to allow significant color work in this book; to Jim Cull and Ian Cartwright, heads of the School of Geosciences, who have significantly underwritten the drafting and photographic work that makes this book stand out; to the Geological Survey of Namibia and the Paleontological Institute of the Russian Academy of Sciences for both field and laboratory support that made work on this book possible; and to all of our respective institutions for the support given to all of us throughout this project – Monash University, the Paleontological Institute, the South Australian Museum, Queens University and the Geological Survey of Western Australia; the Russian Fund for Basic Research (Grant 05-05-64825), the Grant 1790.2004.5 of the President of the Russian Federation in support of leading scientific schools, the National Geographic Society, which has supported field research on the Vendian fauna of Russia and the field research on material from the Belt Series of Montana; Program 25 "Origin and Evolution of Biosphere" of the Russian Academy of Sciences; and the Australian Research Council for work on the Australian Edicaran fauna.

We owe a great debt of gratitude to our ever patient editor, Vince Burke of Johns Hopkins University Press, who has been part of this project from the very beginning. He has only mildly chided us at missed deadlines, helped us pick an engaging title, and significantly improved the text in many ways.

Fundamental thanks must go to Patricia Komarower who has edited, and edited, and edited as well as served as a research assistant throughout the several years of this project, and to Draga Gelt, who has produced the graphics and systematized the photo data base for this publication, and Maxim Leonov, who has cross-checked aspects of the Atlas and photographs fundamentally as well as provided a multitude of photographs, assisted with the bibliography and cross checked measurements. The reconstruction artwork of Peter Trusler has involved him in significant field and museum work and the reconstructions that he produced have involved him deeply in the research on those metazoans he has rendered. His concepts of how to present such enigmatic organisms has added fundamental new insights. Likewise, the very fine photographs provided by Francesco Coffa, Steven Morton, Alexander Mazin and Enrico Savazzi are much appreciated and admired. Anastasia Besedina and Iya Tokareva rendered many of the black-and-white reconstructions of Vendian fauna from Russia.

Opposite page: Exposed soles of the Rawnsley Quartzite that are covered with Ediacarans, south of the Ediacara Hills, Flinders Ranges, South Australia under study by J. Gehling, M. Drosser and their teams (F. Coffa).

We would also like to sincerely thank our work collegues, families and friends, who have put up with us and cheerfully (well, most of the time) and tolerated our obsession with this project since 2002. Perhaps now we can have that long-awaited vacation and come out of our respective "writers' caves"! Especial thanks to Prof. Terufumi Ohno of the Kyoto University Museum and to Kyoto University, who made it possible for two of us (MAF and PVR) to spend months in Kyoto unburdened with the normal pressures of life. Most of the final work on this book was accomplished there – the quiet thinking time was essential to the completion of this project.

The book is dedicated to three people, Boris Sokolov, Reg Sprigg and Mary Wade – all of whom have made extraordinary contributions to the understanding of early metazoan evolution. There are many others who have contributed greatly in these areas, but these three individuals often do not receive the accolades that many others do – thus we heartily salute them here and spotlight their fundamental gifts to us all. We also recognize Tatiana and Timofei Burych for their gift of friendship and logistic support to the Russian teams working along the White Sea, to Marg and Doug Sprigg for the contribution the Sprigg family has made to first discovering and later establishing the true nature of the Ediacara fauna of South Australia. We also thank two families of landowners who have opened their properties for detailed exploration of Ediacaran territory – Jan and Ross Fargher in South Australia and Barbara Boehm-Erni and Bruno Boehm in Namibia. Nathan Hunt and his family have greatly facilitated work of our international teams in Russia and have provided significant financial assistance in the curation and storage of the Vendian metazoans of Russia and other areas of the Eastern European Platform. To all of these altruistic people we give our sincere thanks by dedicating this book to each and every one of them.

Kathleen Grey publishes with the permission of the director of the Geological Survey of Western Australia. Patricia Vickers-Rich is grateful to Monash University for the Long Service Leave, which made it possible to complete this project.

Below: Ediacarans, reconstructed for Australia Post Creatures of the Slime *stamp issue, 2005. This material has been reproduced with permission of the Australian Postal Corporation. The original work is held in the National Philatelic Collection (art by Peter Trusler).*

INTRODUCTION

Over the past few decades an unprecedented interest in rocks older than 542 million years has led to a vast literature on what has been called the Precambrian. It was during this time that the Solar System came into being and the Earth solidified into the unique planet that it is – a planet with a narrow temperature range allowing liquid water to exist, large enough to have gravity entrap a permanent atmosphere and dynamic enough to produce and maintain one in the first place. The Earth is also graced with life, the only planet yet known to have such a quality – life that appeared relatively early in its history.

The main aim of this book is to highlight one part of the immense sweep of time called the Precambrian – the Proterozoic – and, in fact, only a part of that Eon – the time when the first animals appeared – in a wide variety of places on Earth – Australia, Newfoundland, the White Sea, Siberia and the Urals of Russia, the Ukraine, Namibia, the Northwest Territories of Canada, Montana and more. It was a time that may range back as far as 1.6-1.8 billion years until around 542 million years ago, when a veritable explosion of life forms with hard skeletons appeared on the scene.

Brief chapters are devoted to the time before the Proterozoic (the Hadean and Archean), and the time right after (the Cambrian), but so much literature has been generated covering those times, it would be repetitious and an impossible job to cover, in this work, the detail. Such publications as that by J. William Schopf and Cornelius Klein (1992) extensively document the whole of the Proterozoic.

Four authors of this book, Mikhail Fedonkin, James Gehling, Kathleen Grey and Guy Narbonne, have spent their lives working on this time slice, in the Soviet Union, Australia and Canada. The fifth author, Patricia Vickers-Rich, is a geologist and biostratigrapher by trade, working until recently in the Phanerozoic, drawn to the Proterozoic in search of ancestors. All authors have been able to work in many parts of the world on sites of similar age to their home bases with a variety of colleagues from around the globe. They bring their field and research experience to the pages which follow and highlight the work of many others besides themselves – not only the scientific research and results but also the human side of the story – the field experiences, the debates, the differences of opinion and, sometimes, resolutions that result from long-term thinking and discussion about a group of most enigmatic fossils.

The text of this book sets out to characterize the physical environments in which multicelled organisms developed, prospered and faced innumerable crises, from global Greenhouse to ice ages. It then proceeds to flesh out many species that developed during this period, leaving impressions of their soft bodies or their traces.

Besides text, the book hosts a wealth of high-quality photography by Francesco Coffa, Steve Morton (who also rendered backgrounds clean and made sure scales were inserted), Alexander Mazin and Maxim Leonov, along with the beautiful and meticulous reconstructions rendered by anatomist, *cum* artist Peter Trusler. *The Rise of Animals* traces the history and current state of this rapidly advancing field of science, changing with each new fossil find that adds detail, highlighting input from geochemistry, geochronology, molecular biology and sedimentology, which provides new insights about the strange world of the Proterozoic and the biota that populated it.

The first animals will always be of profound interest to scientist and layperson alike. Some may be our very ancient ancestors, whose origins prepared the way for the modern world we inhabit, and us.

PART 1: THE BACKGROUND
THE ARCHEAN AND PROTEROZOIC EONS

CHAPTER 1

The Background. The Archean
(4.5 Billion to 2500 Million Years Ago)

P. Vickers-Rich and M.A. Fedonkin

Visions of the biblical Hell would not be far removed from the pictures painted of earliest Earth by rocks and the theoretical constructs of geologists and astrophysicists. Reconstructions of the first moments of a solid Earth conjure up a most inhospitable place. There was no atmosphere. Round and round this early Earth circled a huge, reddish Moon, closer than it is today. A dim Sun, the Earth's nearest star and source of great energy, hung in a black, starry sky. Slicing through that darkness were hundreds, thousands of extraterrestrial visitors, meteorites that pummeled the Earth's dark surface, their impacts bright, fiery flashes that glowed, then dimmed into blackness. Enormous clouds of dust shot up from these impacts, but fell back quickly because there was no atmosphere to suspend particles. All was silent – for with no atmosphere, no air, there was no sound transmission. Only ground vibrations recorded the strength and intensity of this bombardment from space. And at first it was intolerably cold – some have suggested the phrase "Hell Froze Over." But that soon changed. The Earth warmed up, then soon began to cool towards the present, as the chemistry of tiny zircon crystals reflect. Certainly by 2.5 billion years ago, life had originated and fundamentally changed the atmosphere and oceans towards modernity. The internal heat of the planet, its size and its distance from the Sun ensured the presence of liquid water, seasonality and the presence and dynamism of an atmosphere as well as a mobile and ever-changing surface crust. The Earth remains unique in the Solar System and perhaps rare within the Universe.

The Setting: Earth's Origin

Physicists and astronomers have proposed a number of possibilities for how the Universe and Earth formed. Literature on this subject is vast. Thomas Gold, Herman Bondi and Fred Hoyle preferred a Universe that was always there – no beginning, no end. The more accepted model, however, is the Big Bang Theory, eloquently discussed by Steven Weinberg in his book *The First Three Minutes* published in 1977 – a name given by Fred Hoyle in derision of a theory with which he could not agree. In fact, Hoyle rather ungraciously noted that the Big Bang idea was about as elegant as "a party girl jumping out of a cake"! (Microsoft, Encarta 2002). Hoyle and his colleagues saw no need for a Big Bang; most others disagreed.

Today, most cosmologists suggest that our Universe began somewhere between 12.5 and 14 billion years ago (Davies, 2004; http://wmap.gsfc.nasa.gov/results). They suggest that it began with an explosion, which occurred simultaneously, everywhere, filling all space. Every particle of matter rushed apart from every other particle (Weinberg, 1977). Was this Universe finite? Did it have edges or ends? Or, was it infinite, endless? Einstein simply explained this with his concept of warped space-time – there was no space outside of the Universe. Space formed with the Big Bang. So, no edge problem. No infinity.

During the next three minutes, temperature began excessively high - around a hundred thousand million degrees Centigrade (10^{11} – 10 times itself 11 times!). The very building blocks of matter (atoms, molecules) could not stick together when it was so hot. Only elementary particles existed, such as electrons, those negatively charged entities that move through wires as electric current, those same particles that orbit the nuclei of atoms. Others of these elementary particles were positively charged with the same mass as electrons, positrons, today found only in particle accelerators, like synchrotrons, in cosmic rays or associated with some radioactive elements. Hanging around too were the ghostly neutrinos – entities with no charge and almost no mass. Add to those light, which was everywhere – made up of photons (particles with no mass and no electrical charge, essentially little packets of energy) – and you have the principal players in this early drama of the Universe.

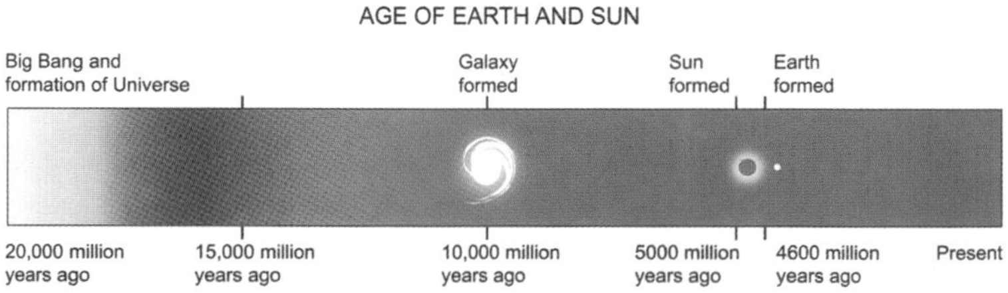

AGE OF EARTH AND SUN

Figure 1. Big Bang Sequence of Events. A chart of the events proposed for the "Big Bang" hypothesis (modified from Moore, 1990). Simulations of how galaxies are born and evolve can be accessed on line at nationalgeographic.com/ngm/0302 based on an article by Ron Cowen in National Geographic Magazine in 2003 (D. Gelt).

Such elementary particles appeared, had short lives and disappeared. Numbers of these particles remained balanced. But then, as the explosion ran its course, things began to cool, after those first three minutes. Only then did protons and

3

neutrons appear and form nuclei – the first being hydrogen, then heavy hydrogen (deuterium) with a single proton and one neutron. Next to form was tritium (even heavier hydrogen, with two neutrons and one proton) and finally the isotopes of helium with two protons and one or two neutrons.

It would take several hundred thousand years, however, according to predictions of the Big Bang Theory, before things were cool enough for atoms to form. This required nuclei to capture free electrons. The first atoms were hydrogen (H) and helium (He). This initial Universe they formed "looked like a very bright fog, similar to the present solar photosphere" (Emiliani, 1988). Electrons were scooped up, elements formed as gases and gravitation concentrated them into stars and galaxies. Only then, did "the Universe become transparent" (Emiliani, 1988). So the theory goes – but it is not entirely satisfying, because the first tiny fraction of a second (maybe the first ten-thousandth of a second or even one microsecond) is a great unknown.

A. WITHOUT MOTION

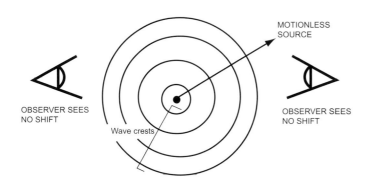

B. WITH MOTION

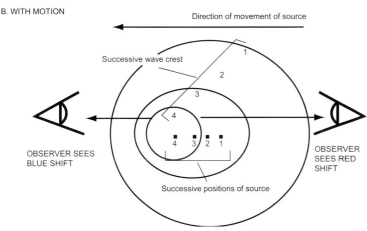

Figure 2. Red Shift and Blue Shift in the viewed spectra colors depends on whether an object is approaching (blue shifted) or moving away (red shifted) from the viewer (modified from Lurquin, 2003) (D. Gelt).

Interesting hypothesis – but one could ask just how this view of the Universe has been constructed by physicists and astronomers, those who seek to understand the vastness of space and the Earth's place in it. What is the evidence supporting such an idea? There are three basic observations satisfyingly explained by the Big Bang Theory but not its current competitors: 1) the Universe is expanding; 2) there

is a uniform background microwave radiation present everywhere in the Universe; and 3) the Big Bang conditions predict the observed amounts of the most abundant elements, hydrogen and helium.

The behavior of light suggests that the Universe is indeed expanding and also gives some idea of what the Universe is actually made of. Composition of the Universe, in particular stars or any luminous bodies, can be determined by the nature of the light they emit. Isaac Newton first noticed this in the 17th century. When light passes through glass, such as a prism, white light is split into a rainbow of colors, a spectrum, because different wavelengths of light are bent at different angles – blue and violet bend more than orange and red. If the luminous body is hot and dense, be it solid, liquid or gas, it will produce a rainbow spectrum, and if its density is very low, it will yield no rainbow at all. When the light from a luminous body passes through a prism, there will be a few bright lines in the spectrum that are separated from one another. Each represents some element present in the luminous body. As Patrick Moore points out in his *The Universe for the Under Tens* (1990): "Suppose we throw some salt into a flame? Salt contains what we call 'sodium', and this produces two distinct lines in the yellow part of the spectrum. See these lines, and you may be sure that they are due to the presence of sodium." The light from our Sun produces such spectral lines, so it is clear that one of the elements in the Sun is sodium. The makeup of other stars and galaxies outside our own solar system can be determined by studying their light spectra. To understand what the lines in these spectra mean, their patterns can be checked out here on Earth.

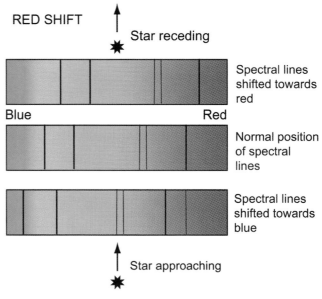

Figure 3. Red Shift and Blue Shift in spectral lines in response to a star approaching the viewer or receding (modified from Moore, 1990) (D. Gelt).

Light also allows determination of whether a star is moving towards the Earth or away from it. Light behaves in a wave-like fashion – a light source produces light waves, and the distance between the wave crests (or the wave troughs) is called the wavelength. Sound behaves likewise. If a person is standing in one place and sound is coming towards the person

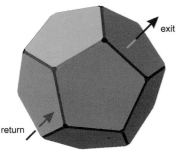

Figure 4. The Shape of Space. A model proposed by Luminet, Weeks, Riazuelo, LeHoucq & Uzan (2003). Their article in Nature suggested that the Universe was infinite and flat, constantly and forever expanding. They used a dodecahedron as a model for the shape. Jeff Weeks further notes (pers. com.) "contrary to appearances, Figure 3a in the Nature paper is not the starting point for the dodecahedral space. Instead, one starts with an ordinary (flat-faced) dodecahedron." "The 3D volume within the dodecahedron represents space, so that's where the galaxies go. At first glance this space has boundaries (namely its 12 pentagonal faces) but – this is the fun part – the faces get 'glued' to make a wraparound Universe, so that the traveller leaving the dodecahedron across one face returns from the opposite face. An analogous 2D construction is to take a piece of paper and 'glue' its left and right edges to make a cylinder, so 2D travellers on the cylinder can travel to the West, go all the way around and return to their starting point from the east. Of course, one's line of sight also wraps around, leading to some interesting visual effects. The Shape of Space video explains this far better…." (The Shape of Space video can be ordered from: http:www. keypress.com/catalog/products/supplementals/Prod_ ShapeOfSpace.html (courtesy of Jeff Weeks, http:// www.geometrygames.org).

from a truck moving towards that person, the sound waves are compressed. Their wavelength is shortened. Likewise, if the truck is moving away, the waves of sound are stretched – that is the wavelength is increased. End result – the truck sounds

are higher pitched when it is barreling towards a person and lower when the truck races away, provided the person is still standing and hearing! As the truck passes, you can hear the distinct change in sound from higher to lower pitch. Exactly the same thing happens with light. We cannot hear it, but we can see its effects in its spectrum. Johann Christian Doppler first described this in 1840, and it bears his name, the *Doppler Effect*. So, if the lines produced by the presence of certain elements (such as sodium) are shifted from where they normally appear on Earth into the red part of the spectrum (called *Red Shift*), then this is evidence that the star is moving away from Earth. Its wavelengths in parts of the spectrum have been stretched and thus the position of the spectral lines shifted. On the contrary, if lines have been shifted towards the blue part of the spectrum, the wavelengths are compressed, then that star is moving towards the Earth. The greater the displacement of these familiar spectral lines, the greater the speed (and velocity) of the star, or the galaxy, which is simply a huge collection of stars and gas. Based on observation of such spectral shifts, it is clear that the Universe is expanding.

The Big Bang Theory is also supported by observations made by Arno Penzias and Robert Wilson concerning the microwave background radiation (CMB) – the same wavelength used to cook in microwave ovens or communicate globally. Such was the leftover echo of the initial, universal explosion – the Big Bang.

So, what is the future of this Universe of which the Earth is a part? One possibility is that it may just keep on expanding forever – the outward expansion being greater than the force of gravity. In a second scenario, strong gravity could stop and actually reverse the expansion – so, once the Universe reaches its maximum size, it might collapse – a backwards Big Bang (the "Big Crunch"). This would produce a closed Universe – rather like a rubber band being stretched, then collapsing, and then quite possibly being stretched again (Lightman, 1991). So far, the issue is unresolved.

Luminet *et al.* (2003) suggest that the Universe has a rigid, and quite specific, shape – a dodecahedron (space with 12 sides) and that we may indeed be able to see all the way around it. They suggest that the Universe acts like an expanding bubble, and that the rate of expansion seems to be accelerating (J. Lattanzio, *pers. com.*, 2005). One can actually view this model of space by accessing a website (http://www.keypress.com/catalog/productssupplementals/Prod_ShapeOfSpace.html). Of course, there are other models. The search goes on.

After the initial formation of the Universe, it was some time before the star we call the Sun formed. Cosmologists speculate on the course of evolution – which leads from Big Bang to the formation of galaxies with stars consolidated from clouds of cosmic gases, and then to planets, some with their encircling moon or moons. Best estimates at present are that first atoms formed, then later began to clump, forming clouds of matter. The more matter clumped, the greater the attraction of matter to other matter. Local gravitational attraction became greater than

Following pages: Figure 5. Modified from a map of the Universe proposed by Richard Gott and Mario Juric (Princeton University [http://www.astro. princeton.edu/~mjuric/universe]) and originally published in New Scientist (22 November 2003). With permission of Richard Gott, Mario Juric and NASA/JPL-Caltech (D. Gelt).

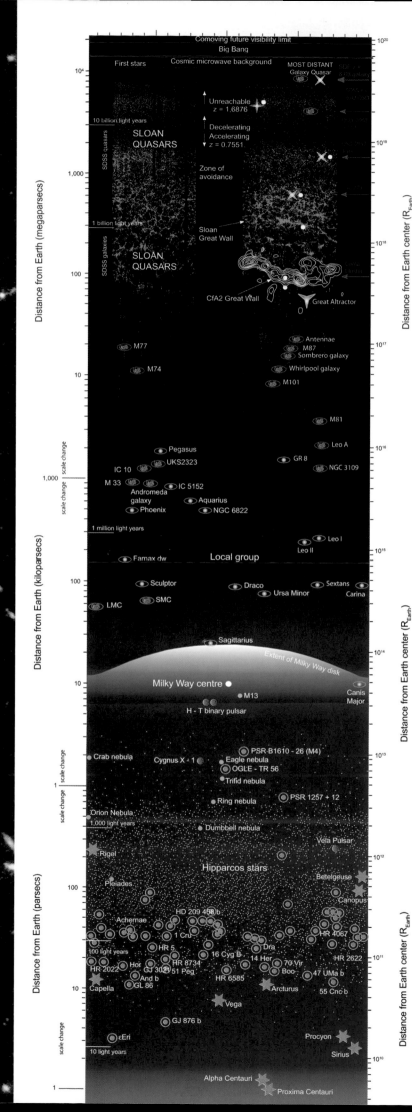

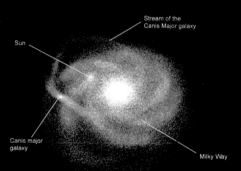

The cosmic microwave background is the afterglow of the big bang. Tiny temperature differences within it tell us the universe is flat and 13.7 billion years old. Only 4 per cent of it consists of ordinary matter that makes up stars. The rest is mysterious stuff known as "dark energy" and "dark matter".

The Sombrero galaxy is 5 million light years from Earth. Even further out towards the edge of space, galaxies cluster into a sponge-like structure filled with enormous voids separated by giant walls. This structure reflects the spread of matter shortly after the big bang.

Canis Major Dwarf: a new closest galaxy.

Gas pillars of creation. The Eagle nebula is one of the most famous star nurseries ever photographed. It is the same distance from us as the Crab nebula, the site of a supernova seen in 1054, and Cygnus X-1, a star system containing a black hole.

Over 118,000 of the Milky Way stars had their position pinpointed by the Hipparcos satellite. The Milky Way is a member of the Local Group of galaxies. Another is Andromeda, which is the most distant object visible to the naked eye.

Andromeda Galaxy

Comet Hale-Bopp originated in the Oort cloud, a chilly reservoir full of comets. We only see these icy snowballs when they are nudged out of the cloud by the gravitational tug of a passing star or some other object.

Voyager 1 is the most distant man-made object in space. Launched in 1977, it has now reached the heliopause, the edge of the solar system. On its journey it dodged the asteroid belt, visited Saturn and Jupiter and raced through the Kuiper belt, a swarm of icy rocks left over from planet formation.

Astronaut collecting samples on the Moon.

the overall "pull" of the expanding Universe – and so galaxies formed. It was fortunate that the gas cloud generated by the Big Bang was not perfectly uniform in the distribution of helium and hydrogen or there would be no galaxies (Emiliani, 1988). Such irregularity, where there was a denser part, attracted other matter and things began to clump on a megascale.

In these early galaxies, where gravitational force was greatest, stars switched on. In some parts of young galaxies, cosmologists predict that the density became so great that super massive entities formed – black holes – where the gravitational attraction is so great that light cannot escape. One would think that in these places, because light was pouring in, all would be black. Instead, because of superheating, such places flash brightly, forming quasars – places of great violence. It is from here that GRB's, gamma ray busters, emanate. These are immense explosions of only a few seconds in duration – but perhaps linked to some of the major extinctions on Earth.

In other places where the mass density was slightly greater than elsewhere, mass attracted more mass, and then

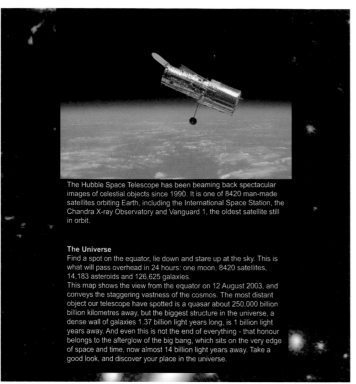

The Hubble Space Telescope has been beaming back spectacular images of celestial objects since 1990. It is one of 8420 man-made satellites orbiting Earth, including the International Space Station, the Chandra X-ray Observatory and Vanguard 1, the oldest satellite still in orbit.

The Universe
Find a spot on the equator, lie down and stare up at the sky. This is what will pass overhead in 24 hours: one moon, 8420 satellites, 14,183 asteroids and 126,625 galaxies.
This map shows the view from the equator on 12 August 2003, and conveys the staggering vastness of the cosmos. The most distant object our telescope have spotted is a quasar about 250,000 billion billion kilometres away, but the biggest structure in the universe, a dense wall of galaxies 1.37 billion light years long, is 1 billion light years away. And even this is not the end of everything - that honour belongs to the afterglow of the big bang, which sits on the very edge of space and time, now almost 14 billion light years away. Take a good look, and discover your place in the universe.

Figure 5. (Continued)

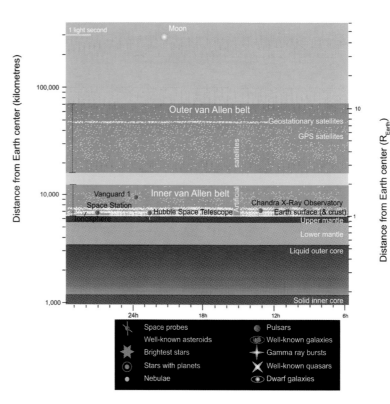

The Earth's Moon

The Moon has not always been where it is today. It was once much closer, now retreating from Earth at approximately 4 centimeters per year. Such a precise measurement has been made using lasers on Earth bouncing off reflectors left on the Moon by astronauts on the Apollo missions. The Moon's gravity deforms the Earth, causes tides in the oceans, and even affects the atmosphere. When the Moon was closer to Earth, such tidal effects were much greater. Tidalites – repeating sequences of sediments – indicate that 900 million years ago, at or near the time animals appeared on Earth, days may have been 18 hours long and years as long as 481 days. At the very beginning of Earth history days would have been even shorter – the Earth was spinning faster than now. The Moon helped stabilize Earth's axis of rotation – which varies as little as 2-3 degrees. By contrast, the rotational axis of Mars has varied as much as 25-35 degrees over 50,000 years.

There are a few competing models that suggest how the Moon was formed. One idea was proposed by Canup and Asphaug (2001) – a large planetoid the size of Mars plowed into the Earth about 4.51-4.55 billion years ago (Albarede, 2003; Albarede and Erikkson, 2004). The energy of impact instantaneously liquefied Earth's surface and ejected a huge bolide into space. Some of this material was lost, some pulled back to Earth, some forming a debris ring around Earth. Gravitation consolidated this space debris into the Moon. This mighty impact gave the Earth its tilt, which brings about seasons of the year and its day-night cycle as well as its tides. This tilt also led to distribution of the heat energy earthwide and thus habitability of this planet (Nisbet & Sleep, 2001).

The Moon is rather different from the Earth. With a radius of around 1700 km and gravity about 17% of that on Earth, the Moon is not able to hold onto an atmosphere. Based on data sent back from such satellites as Clementine, it appears that the Moon's crust is between 60 and 100 km thick – thicker than Continental Crust on Earth. The Moon has no magnetic field – and so apparently no significant metallic Core – and its surface is "depleted in volatile elements and extremely reduced" (Albarede, 2003). Unlike Earth, the Moon is silent now, with no volcanic activity recorded for more than 3.2 billion years. Before that, however, from 4.5-3.2 billion years ago it was tectonically active, thoroughly battered by meteorites until about 3.5 billion years ago that left enormous basins, which subsequently filled with liquid basalts, which solidified forming the Maria (or "seas"). Then, tectonics appears to have shut off and left the Moon we see today with only a few other minor alterations by chunks of rocks from space (McSween, 1999).

began the battle of survival between attraction (gravitation) and heating, which caused expansion. Where the amount of material was too great, supernova explosions took place and the stars tore themselves apart, spreading gas and dust outwards, which then was destined to be recycled into other stars. It was in such clumping, star growth, explosion and dispersal where elements which make up the Earth's Sun were formed. Thus, the scene was set for the origin and evolution of the Earth's Solar System.

The Earth's star, the Sun, is not an old star, probably no older than 4.6 billion years or so. Stars in our galaxy, the Milky Way, also appear to not exceed 14 billion years. Recent CMB observations give an upper age limit. The Sun is a Yellow Dwarf, burning at a surface temperature of around 6000° K. (0 degrees Kelvin equals -273° C or -460° F or Absolute Zero). It is a young star in the Universe and appears to be made of recycled material – stardust that has come from the life and death of older stars, now long burned out. It is made primarily of hydrogen, the rest being helium (about 29% by mass) and minor other elements (about 1.5% of such components as oxygen, carbon and neon). It generates energy by nuclear fusion – again mainly of hydrogen to helium. Other elements in the Solar System have been generated by processes not occurring in the Earth's Sun. For example, once a star runs out of hydrogen fuel, it can swell to become a Red Giant. Then helium ignites to form carbon and oxygen. In stars that burn more intensely, heavier elements such as magnesium, sodium, silicon and iron will be formed – but the formation of elements more dense and massive than iron are at present difficult to understand. Greater amounts of energy must be injected into the system to form them, such as when Red stars explode in supernovae. Such blistering giants eventually exhaust their nuclear fuel and collapse inwards. Material falling into the collapsed core then rebounds, causing the star to eventually explode and the result includes such elements as silver, tin, gold, lead and uranium (McSween, 1999).

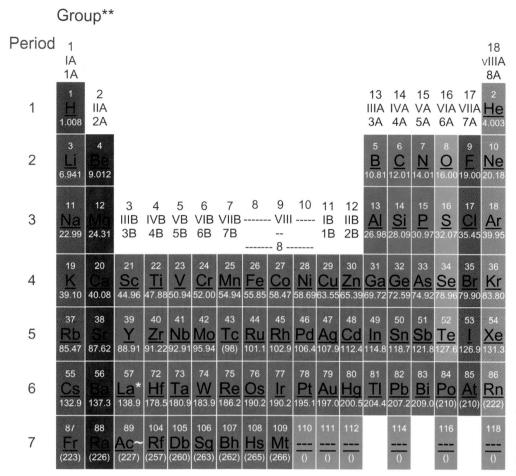

Group**

Period	1 IA 1A	2 IIA 2A	3 IIIB 3B	4 IVB 4B	5 VB 5B	6 VIB 6B	7 VIIB 7B	8	9 ------- VIII ----- -- ------- 8 -------	10	11 IB 1B	12 IIB 2B	13 IIIA 3A	14 IVA 4A	15 VA 5A	16 VIA 6A	17 VIIA 7A	18 vIIIA 8A
1	1 H 1.008																	2 He 4.003
2	3 Li 6.941	4 Be 9.012											5 B 10.81	6 C 12.01	7 N 14.01	8 O 16.00	9 F 19.00	10 Ne 20.18
3	11 Na 22.99	12 Mg 24.31											13 Al 26.98	14 Si 28.09	15 P 30.97	16 S 32.07	17 Cl 35.45	18 Ar 39.95
4	19 K 39.10	20 Ca 40.08	21 Sc 44.96	22 Ti 47.88	23 V 50.94	24 Cr 52.00	25 Mn 54.94	26 Fe 55.85	27 Co 58.47	28 Ni 58.69	29 Cu 63.55	30 Zn 65.39	31 Ga 69.72	32 Ge 72.59	33 As 74.92	34 Se 78.96	35 Br 79.90	36 Kr 83.80
5	37 Rb 85.47	38 Sr 87.62	39 Y 88.91	40 Zr 91.22	41 Nb 92.91	42 Mo 95.94	43 Tc (98)	44 Ru 101.1	45 Rh 102.9	46 Pd 106.4	47 Ag 107.9	48 Cd 112.4	49 In 114.8	50 Sn 118.7	51 Sb 121.8	52 Te 127.6	53 I 126.9	54 Xe 131.3
6	55 Cs 132.9	56 Ba 137.3	57 La* 138.9	72 Hf 178.5	73 Ta 180.9	74 W 183.9	75 Re 186.2	76 Os 190.2	77 Ir 190.2	78 Pt 195.1	79 Au 197.0	80 Hg 200.5	81 Tl 204.4	82 Pb 207.2	83 Bi 209.0	84 Po (210)	85 At (210)	86 Rn (222)
7	87 Fr (223)	88 Ra (226)	89 Ac~ (227)	104 Rf (257)	105 Db (260)	106 Sg (263)	107 Bh (262)	108 Hs (265)	109 Mt (266)	110 --- ()	111 --- ()	112 --- ()		114 --- ()		116 --- ()		118 --- ()

Lanthanide Series*	58 Ce 140.1	59 Pr 140.9	60 Nd 144.2	61 Pm (147)	62 Sm 150.4	63 Eu 152.0	64 Gd 157.3	65 Tb 158.9	66 Dy 162.5	67 Ho 164.9	68 Er 167.3	69 Tm 168.9	70 Yb 173.0	71 Lu 175.0
Actinide Series~	90 Th	91 Pa	92 U	93 Np	94 Pu	95 Am	96 Cm	97 Bk	98 Cf	99 Es	100 Fm	101 Md	102 No	103 Lr

Figure 6. Periodic Table of Elements. Courtesy of Los Alamos National Laboratory, Chemistry Division, New Mexico. Note that O, C, H, N, Ca, P, S, K and Mg make up 99.76% of the general weight of the living biomass on Earth, the first three elements accounting for 98.5%.

After being spun off into space from such events, elements tended to condense as solid mineral grains. Some of the atoms of carbon, hydrogen, oxygen and nitrogen combine to form organic compounds in dense and cold molecular clouds. At such low temperatures gases freeze and bind themselves onto the surface of solid material. With this combination and the exposure to ionizing radiation, more complex compounds can form – with carbon-hydrogen bonds – called polycyclic aromatic hydrocarbons. The next step in this process is that once this mixture melts and combines with water, even more complex organic molecules can be generated – kerogens – which have been detected in comets and asteroids. These organic compounds may well have played a role in the development of life on Earth.

After the Earth's Sun formed from a spinning gas and dust cloud, a number of other clumps in this system consolidated into the 10 planets, plus the asteroid belt between Mars and Jupiter. Earth formed as the third planet out from the Sun at around 93,000,000 miles (around 150,000,000 kilometers). Four of the planets closest to the Sun, the Inner Planets (Mercury, Venus, Earth and Mars) are formed of solid materials (the two Outermost Pluto and Charon, as are possibly captured asteroids and not true planets), but most of those beyond Mars (Jupiter, Saturn, Uranus, Neptune) are made up primarily of gas and lack solid surfaces.

Of the four Inner Planets, Earth is unique. It is dynamic and has been for its more than 4 billion years. It is just the right distance from the Sun to support liquid water – just the right amount of energy reaches the surface of the Earth so that water can remain liquid, not frozen or "boiled off." The Earth is of a size allowing for a source of internal heating produced by radioactive decay, leading to differentiation of elements, such as iron, into layers – thus a solid inner Core. Internal heating led to movement of material from the Mantle below towards the Crust above. Thus, the outermost solid part of the Earth is divided into a number of large, brittle plates, and these have moved and still move relative to one another, leading to massive recycling, building of mountains, movement of continents, massive earthquakes and volcanic activity. This also led to the formation of continents and ocean basins, and their ever-changing shape and position – all in a state of perpetual movement.

Of the remaining three Inner Planets, as well as the Earth's Moon, history has been different, and from life's point of view rather unkind. The surfaces of both Mercury and the Moon belie a static history – preserving a record of heavy meteorite bombardment during the early phases of the Solar System formation. At this time the leftover debris that did not clump into the planets and moons pummeled the surfaces of such solid bodies, and left a clearly visible, pock-marked

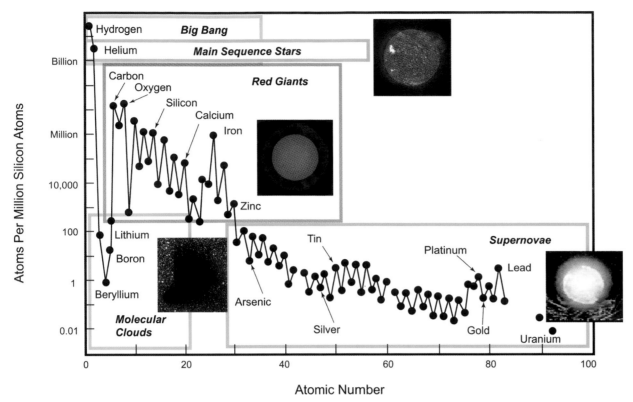

Figure 7. Cosmic abundance of elements (images courtesy of the National Solar Observatory/Sacramento Peak, NASA, and modified from McSween, 1999) (D.Gelt).

planetary surface little changed for more than 3.5 billion years. The Moon, unlike Earth, was too small to sustain internal heat to any great degree. Their early volcanoes, which formed the massive lava plains (pock-marked by impacts), shut off, never to reactivate. Their small size meant that it was impossible to retain either an atmosphere or a hydrosphere (water). They stand as mirrors or "fossils" of the early history of the Solar System and the Earth.

Mars is somewhat larger than Mercury and had more internal heating in the past. Its surface reflects previous massive volcanic eruptions (Olympus Mons), earthquake rifting (Valles Marineris) and modification by wind and water in times past. Visible crater impacts are nowhere as numerous and well preserved as on Mercury and the Moon. The smaller size of Mars, however, limited to the past the dynamic nature of Mars, which does not continue into the present. Small size also made it impossible for Mars to retain a significant atmosphere.

Venus, closer to the Sun than Earth and larger than Mars, has quite a different nature. It retains an internal heat engine, which moves and rearranges the crust of the planet continuously. Volcanoes still erupt on Venus, producing a thick, carbon dioxide atmosphere – a runaway Greenhouse atmosphere. Temperatures on the surface of Venus reach upwards of 500°C – hot enough to melt lead – and the atmospheric pressure is nearly 100 times that of Earth. Venus appears to have no water, no oceans and certainly no ice caps. It lacks any concentration of craters – its surface is young and dynamic.

The Earth stands alone, a beautiful blue speck in the blackness of space, sprinkled with stars. Although other planets of similar size are undoubtedly present elsewhere, well beyond the Solar System, perhaps with conditions similar to those on Earth, so far, this third planet from a rather ordinary Yellow Dwarf star remains unique – perhaps even ostentatious in its provision of the right conditions for life to prosper.

The Setting: Earth's Structure

It is not easy to reconstruct the earliest history of Earth. As Evan Nisbet (2000) quotes Rachel Carson, "Beginnings are apt to be shadowy." The very first period of Earth's history, for which we have little or no direct geological evidence, is called the Hadean Eon (Cloud, 1976; Schopf, 1983) or the Archean (Gradstein *et al.*, 2004), more than 2 billion years long. The dating of Earth's beginnings is based on the age of meteorites – a maximum of about 4.6 billion years (Harland *et al.*, 1989). Two major events took place during this time – initial accretion of the Earth from the solar nebula followed by the stabilization of the young planet into a solid mass.

Figure 8. Murchison meteorite, fell near Murchison, 100 km to the east of Melbourne, Victoria, Australia in 1969. It is a carbonaceous chondrite containing water and complex organic molecules, such as amino acids. Although controversial, some researchers have suggested its contents, like others of its kind, suggests that complex organic compounds exist in outer space (W. Birch, Museum Victoria).

In the beginning, when Earth was actively "assembling" from dust and all sorts of space matter, its surface likely resembled that of today's Moon. It was dark and covered with a layer of fine-grained rock fragments constantly pummeled, melted and reworked by meteorites. In the absence of the air, meteorites did not burn or leave spectacular fiery flourishes in the dark sky, and only ground vibrations recorded contact for there was no substantial atmosphere to transmit sound. Even at mid-day stars were clear in an inky black sky, for without atmosphere, the Sun's rays did not dull these bright points of light.

During its early days, the Sun was about 30% less luminous than it is today, 1981; Hoyt & Shatten, 1997), with solar luminosity increasing to present, where the Sun today loses 4,290,000 tons of its mass every second (T. Rich, J. Monaghan, *pers. com.*, 2005), decreasing its gravitational pull, so that the orbit of Earth has increased in diameter. Earth now is farther from the Sun than in times past.

Because of the lower luminosity of the young Sun, some astrophysicists have predicted surface temperatures of early Earth were lower than present, but then again the Earth may well have been closer to the Sun and thus the total energy reaching its surface could have been equivalent to or even greater than at present (T. Rich, *pers. com.*). Zahnle (2001) describes the early Archean as a time when impacts ruled the Earth and goes further to say "one imagines [the Earth] with craters and volcanoes; crater oceans and crater lakes, a scene of mountain rings and island arcs and red lava falling into a steaming sea under an ash-laden sky." "The climate, under a faint sun and with little CO_2 to warm, would likely have been in the median extremely cold, barring the intervention of biogenic greenhouse gases (such as methane)." Nisbet (2000) suggests that the Earth prior to 4 billion years steered "a bumpy course between brief periods of hot inferno after meteorite impacts, and long episodes of Norse ice-hell." But evidence presented by Valley (2005), based on study of zircon crystals, suggests that conditions may not have been so extreme.

Heat on Earth's surface was (and is) not generated solely by the Sun. Soon after consolidation, the Earth's surface warmed because of the release of kinetic energy due to at least three factors: gravitation of consolidation, meteorite bombardment and the celestial collision that formed the Moon. In addition, warmth came from below. The decay of radioactive elements into more stable elements released heat. And, some radioactive isotopes, such as uranium 235, were far more abundant 4 billion years ago than today. This could have contributed significantly to surface warming. So, after a cold beginning, things began to heat up.

The interior of the early Earth partially melted. Under these conditions dense iron (Fe) along with nickel (Ni) and other heavy elements sank to the Center, forming a Core. The lighter, silicate-rich melt rose, forming a molten ocean on the Earth's surface birthing a similarly hot atmosphere above it. Geochemist Francis Albarede (2003) suggested that this Core segregation took place within a few tens of millions of years after the Earth developed into a solid body, affecting areas of the planet closer to the surface and ultimately the developing atmosphere (Galimov, 2005). This process generated enormous amounts of kinetic energy, melted the iron-nickel alloy as well as a major part of the overlying Mantle (the part of Earth between Core and overlying Crust), removing iron and other siderophilic elements (such as nickel, chromium, rhenium [Re], osmium [Os] and iridium [Ir]). Segregation was followed by intense meteorite bombardment, so replenishing the depleted Mantle with iron and other siderophiles (again Re, Os, Ni), explaining an anomaly that had bothered geologists – why these elements were in the Mantle when segregation that formed the Core took them away! Such segregation and the energy generated by it switched on the great convection gyres circulating material from the Mantle and lower Crust upwards – the powerhouse of plate tectonics.

With cooling of the molten Earth surface, a layer of solid, basaltic crust formed – somewhat like the basement underlying oceans today, true Oceanic Crust, composed of sodium/calcium-rich aluminum silicates (plagioclase) and iron/magnesium-rich silicates (olivines and amphiboles), normally quite dark in color. Only later did continents form, made up of a different kind of material, Continental Crust, consisting of granites and related rocks. These were composed primarily of sodium/calcium-rich aluminum silicates (plagioclase), quartz (a silicate mineral with a tetrahedral crystal structure) and potassium/rich silicates (K-feldspars), depleted in the iron-magnesium rich minerals, and differing from Oceanic Crust. This thin, initial crust was recycled by tectonic processes and melted again and again by meteorite impacts until about 3.8 billion years ago, when sustained bombardment finally ceased. The surface Crust then cooled and thickened enough so as not to continuously remelt. Stable cratons formed, and plate tectonics switched on. Heat flowing upwards from the Mantle below would have produced many hotspots, likely leading to rifts (faults), many small individual plates and even subduction zones, where one slab of Crust dived under another. Finally, due to interaction of water from volcanic eruptions and possibly occasional incoming comets, Continental Crust began to differentiate and true, albeit small, continents formed.

With differentiation of the iron Core, both the inner solid and the outer liquid, Earth developed its magnetosphere, a "cocoon" that shielded it and protected developing life from cosmic rays and high-energy solar radiation. Such energy can impart serious, even fatal, genetic damage to living systems. Other planets of the Solar System – Mercury, Jupiter, Saturn, Uranus and Neptune – also have significant magnetospheres, but our closest neighbors, Venus and Mars, do not. Mars has a magnetic field of less than 0.0003 times that of Earth (Carr, 1981), and Venus even less (0.00001).

The Setting: Early Earth (The Archean: 4.6 to 2.5 Billion Years Ago)

The first record of solid rock on Earth is around 4.4 to 4.5 billion years old – zircons from Western Australia (Harrison *et al.*, 2005). These, and the oldest known crustal rocks, aged at around 4.04 billion years, are from the Archean Eon, which lasted until 2.5 billion years ago. It was during this time that the first evidence of life appeared on Earth, a period dominated by bacteria. The ancient environments that fostered this first life were very different from those of today, where oxygen is present in some abundance in most places. Archean environments favored the formation of major ore deposits dominated by lead, zinc and iron – all critical to civilization as we know it. Some of these ancient deposits are unique to this time, no longer forming in quantity – examples being the banded iron formations (BIF's) of Western Australia, southern Africa and northern North America.

the planets at a glance

relative sizes of the planets in the solar system

	MERCURY	VENUS	EARTH	MARS	JUPITER	SATURN	URANUS	NEPTUNE	PLUTO
AVERAGE DISTANCE FROM SUN (kilometers)	57.9 million	108.2 million	149.6 million	227.94 million	778.3 million	1,429.4 million	2,871 million	4,504.3 million	5,913.5 million
EQUATORIAL DIAMETER (kilometers)	4,879	12,103.6	12,756.28	6,794.4	142,984	120,536	51,118	49,492	2,274
MASS (kilograms)	3.3×10^{23}	4.9×10^{24}	6.0×10^{24}	6.4×10^{23}	1.9×10^{27}	5.7×10^{26}	8.7×10^{25}	1.0×10^{26}	1.3×10^{22}
DENSITY (grams per cubic centimeter)	5.41	5.25	5.52	3.9	1.3	0.7	1.3	1.6	2.05
LENGTH OF DAY (relative to Earth)	58.6 days	243.0 days	23.9 hours	24.62 hours	9.92 hours	10.2 hours	17.9 hours	16.1 hours	6.39 days
LENGTH OF YEAR (relative to Earth)	87.97 days	224.7 days	365.2 days	686.98 days	11.86 years	29.46 years	84 years	164.8 years	248.5 years
NUMBER OF KNOWN MOONS	0	0	1	2	16	At least 18	At least 16	8	1
ATMOSPHERIC COMPOSITION	Traces of sodium, helium and oxygen	96% carbon dioxide, 3.5% nitrogen	78% nitrogen, 21% oxygen, 0.9% argon	95% carbon dioxide, 3% nitrogen, 1.6% argon	90% hydrogen, 10% helium, traces of methane	97% hydrogen, 3% helium, traces of methane	83% hydrogen, 15% helium, 2% methane	85% hydrogen, 13% helium, 2% methane	Probably methane, possibly nitrogen and carbon monoxide

NEW LIGHT ON THE SOLAR SYSTEM

Figure 9. Characteristics of the planets of the Solar System (modified from illustration in Scientific American Special Edition *vol. 13, no. 3, 2003 and with permission of* Scientific American *and artist Laurie Grace) (D. Gelt).*

The oldest known crustal rocks on Earth, the Acasta gneisses from northwest Canada, have been dated at 3.8-4.05 billion years. They are rich in silica, suggesting that plate tectonics was already in operation (Nisbet & Sleep, 2001). Slightly younger volcanics and sediments, including banded ironstones, from southwest Greenland in the Isua Greenstone Belt (Cas, Beresford & Appel, 2002), range from 3.7 to 3.85 billion years (Gradstein *et al.*, 2004). After initial formation, these rocks were later subjected to heating and folding (metamorphosed), at least twice. They include pillow basalts, formed when lava poured out into ocean waters, and appear to have been deposited under quiet, deep water conditions. Despite a concerted effort by geologists in this part of Greenland, no evidence of nearby

continents has yet been discovered. The presence of the BIF's (Banded Iron Formations) and the now disputed biomarkers have been used to suggest that living organisms capable of photosynthesis were present and producing oxygen in some quantities. But, another option could have been that in the absence of an ozone screen, powerful UV radiation penetrated surface waters and oxidized iron photochemically (Knoll, 2003), thus giving rise to the BIF's.

Even older are the recycled grains of a mineral called zircon from the Jack Hills on the Yilgarn Craton of Western Australia. SHRIMP analyses (Sensitive High Resolution Ion Micro Probe using a mass spectrometer) measure the ratio of uranium to lead in the tiny crystals. One was dated at 4.404 billion years (Wilde *et al.*, 2001) and most recently an even

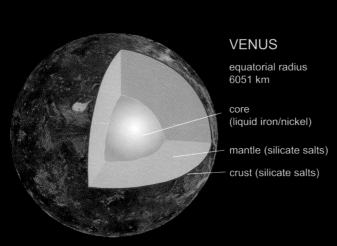

VENUS

equatorial radius
6051 km

core
(liquid iron/nickel)

mantle (silicate salts)

crust (silicate salts)

Venus is the second planet from the Sun. It rotates
in a retrograde direction (opposite to its direction of orbit
around the Sun) in about 243 days and travels around the
Sun every 224 days. Its atmosphere is made of highly reflective
sulphuric acid clouds that completely hide its surface, an
atmosphere that is 96% carbon dioxide.

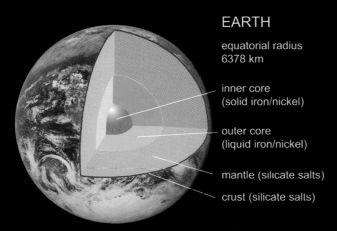

EARTH

equatorial radius
6378 km

inner core
(solid iron/nickel)

outer core
(liquid iron/nickel)

mantle (silicate salts)

crust (silicate salts)

Earth is the third planet from the Sun and the only planet known
to support life. Water is liquid on its surface. Earth rotates every
24 hours (one Earth day), and orbits the Sun every 365 days.
Earth has a single moon, which is about a quarter of Earth's size.

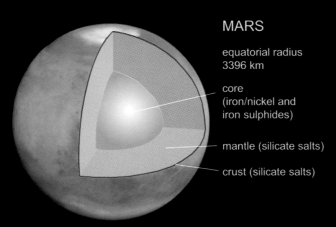

MARS

equatorial radius
3396 km

core
(iron/nickel and
iron sulphides)

mantle (silicate salts)

crust (silicate salts)

Mars is the fourth planet from the Sun. It has the climate
that most closely resembles Earth's. It rotates in about
25 hours and travels around the Sun in around 687 days.
Like Earth, Mars has a tilted rotation axis and so has four
seasons. Its surface has significant topographic relief, including
the magnificent volcano Olympus Mons, 24 kilometers high,
the tallest peak in the Solar System. It is clear that at some
point in its history, liquid water flowed on Mars, leaving its
mark on the surface of the planet.

older date of 4.4-4.5 billion has been determined based on
isotopes of the element hafnium (Harrison *et al.*, 2005). Zircon
crystals (zirconium silicate [$ZrSiO_2$]) originally formed when
molten rock solidified, in an "environment where liquid water
interacted with pre-existing continental crust to produce a
new rock with a composition similar to that of granite" (Wilde,
interview comment, 2001). Somewhat later, the crystal grain
was swept away and deposited along with sediments of a
somewhat younger age, the Jack Hills metaconglomerate (part
of the Narryer Gneiss), then found by geologists and analyzed.

Also of interest is that when analyzed for its oxygen
isotope composition, results suggested that the zircon had
formed when liquid water was present on Earth, further
implying that temperatures were lower than previously
predicted – high enough not to freeze and low enough not to
boil water. Conditions, even at 4.4 billion years, could have
been tolerated by life – but this conclusion by Wilde and his
colleagues is not without some challenge (Pidgeon, Nemchin
& Williams, 2001). And finally, additional chemical analysis of
the crystal implied that it formed as part of a granite – the first
evidence of true Continental Crust.

The Setting: First Continents and Ocean Basins, Plate Tectonics Begins

Plate tectonics switched on early in Earth history (Leitch,
2001), perhaps even as early as 4.4 to 4.5 billion years ago
(Harrison *et al.*, 2005). This led to the formation of ocean basins
that then filled with water. As this dynamic process continued, less
dense rocks, such as granites, formed in association with tectonic
trenches in greenstone belts – areas where one great crustal
plate dived under another. Unstable volcanic islands formed
along the spreading zones, the mid-ocean ridges.

The very oldest evidence of the Continental Crust,
mentioned above, is around 4.4 to 4.5 billion years old,
preserved in the Jack Hills of Western Australia (Harrison
et al., 2005: Valley, 2005). What has survived from that
remote past makes just a small fraction of the present-day
continents, which cover about 41% (Eriksson *et al.*, 2004) of
the Earth's surface. Continents grew by accretion of smaller
fragments around the edges of larger landmasses. Lowe
suggested that about 5% of the Continental Crust was formed
by 3.3-3.1 billion years ago, around 58% was formed later
between 2.7 and 2.3 billion years ago and another 33% of
crust was likely formed between 2.1 and 1.6 billion years ago.
So, about 96% of the Continental Crust was likely in place
by the Middle Proterozoic – though these percentages are
estimates only and still open to lively debate. Only 10 small
pieces of the ancient Archean Continental Crust (cratons) has
been preserved (de Wit *et al.*, 1992), the rest younger.

The first real evidence of land that lay above sea level
dates from around 3.5 billion years in the Pilbara region
of Western Australia. The Warrawoona Group consists of
metamorphosed volcanics and sediments of the sort today
formed around continents and named greenstones because of
the prevailing color of the weathered rocks. These Warrawoona
rocks were invaded (intruded) by granite, which formed from a
hot, molten rock magma that rose into these ancient volcanic

*Figure 10. Internal structure of some planets in the
Solar System (modified from illustration in* Newton
magazine, November-December 2000) (D. Gelt).

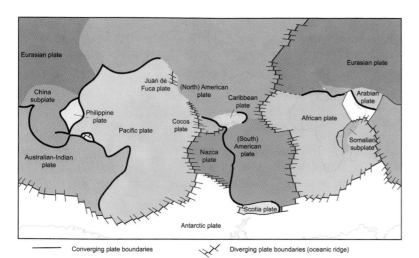

Figure 11. Major subdivisions (lithospheric plates) of the Earth's surface. It is the movement of these rigid plates relative to one another that is responsible for the "drift of continents" over the last billions of years of Earth history. The boundaries between these plates are marked by ridges (heavy lines), trenches or subduction zones (light lines) and transform faults, which displace both ridges and trenches laterally. Along the ridges new molten material is added from below and the plates spread apart. Along the trenches one plate dives beneath the other (modified from Vickers-Rich & Rich, 1999, with permission) (D. Gelt).

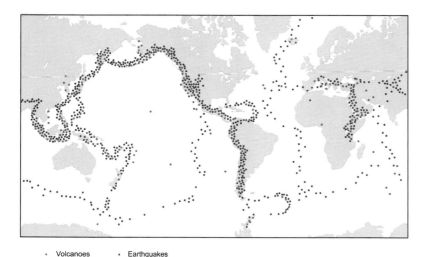

· Volcanoes   · Earthquakes

Figure 12. Global distribution of tectonic features – earthquakes and volcanoes – which define the edges of the Earth's lithospheric plates (D. Gelt).

rocks. Afterwards, the entire sequence was eroded and deposited as well rounded quartz-chert grains as sandstones and conglomerates. These sediments were laid down on a surface, which had been cut into and planed down while it stood above sea level. The Warrawoona sediments themselves were evidently deposited in shallow marine water, overlying submarine lavas.

Uplift of continents markedly changed the face of Earth. Composed of the lighter rocks, they moved (and still move) over the heavier Oceanic Crust like ice slabs over water. A modern example is the west coast of South America, where the eastern Pacific Ocean Crust dives under Continental Crust of Chile, Ecuador and Peru building the Andes. With continental uplift came the formation of continental shelves and

The Way the Earth Works

Today's Earth is dynamic. The *theory of plate tectonics* best explains its dynamism. This theory suggests that the outer few kilometers of the Earth (upwards of 100 km in continental areas) is made up of a series of thin, cold, rigid plates, which fracture if stressed too much rather than bend or flow. Geologists call these lithosphere plates. They meet each other along one of three kinds of boundaries – trenches (or subduction zones), ridges and transform faults. The plates are internally relatively stable – earthquakes and volcanic activity are mainly restricted to their edges.

The Mid-Atlantic Ridge is a ridge boundary. It is a hot region where volcanoes add new material and shallow earthquakes frequently occur. Earthquake faulting is generally tensional – indicating that the rocks are being stretched apart in this region. Valleys form atop this mid-ocean, undersea ridge – down dropped valleys like those in the Red Sea and East African Rift Valley on land.

The Marianas Trench and the Japan Trench are examples of a trench boundary, also called subduction zones. Here earthquakes begin shallow and deepen away from the undersea topographic trench. Volcanoes occur some distance away from the oceanic trench, one of the most famous examples being the holiest mountain in Japan, Mt Fuji, to the west of the Japan Trench. Such earthquake activity and volcanic eruptions are related to the heating up of the massive crustal slab, which plunges into the Japan Trench, underneath the Asian continent to the west. Once heated to a certain point, rock melts and rises towards the surface, producing volcanoes (Mt Fuji). In contrast, rocks near the topographic trench on the ocean floor are quite cool – after all, they are the oldest rocks in the ocean basin and farthest from the hot oceanic ridge. But, as these rocks descend and pass below the overlying crustal plate, frictional heating occurs, eventually melting the overlying rocks and giving rise to volcanism and the building of mountains as the molten rock rises.

Transform faults form the third type of plate boundary. These appear as great offsets of ridges and trenches in the ocean basins, such as the Kangaroo Fracture Zone, which displaces the mid-ocean ridge south of Australia. Along these fractures some volcanism occurs, forming submarine volcanic islands along with shallow earthquakes. Fault movement is sideways, not vertical. Transform faults are special faults because they have a two-part history – first they move to relieve tension along a ridge or trench boundary, but then subsequent movement is due to the continuous addition of material along ridges (spreading) or the continuous contraction along trenches or subduction zones.

The dynamics along all these plate boundaries brings about the continual rearrangement of the continents, which ride on the backs of the crustal plates. Convection in the asthenosphere (part of the inner-Earth layer known as the Mantle) transfers heat from the interior to the surface. While the asthenosphere is rock, over millions of years it acts like a liquid, slowly flowing beneath the outer, rocky, lithosphere. Movement of the lithosphere (the outermost layer of the Earth which includes the Crust) displaces the continent and ocean floors at a rate of 1-10 cm per year, about a rate at which fingernails grow!

The restless nature of the Earth's Interior ensures a dynamic planetary surface. It is the reason that supercontinents like the great southern continent Gondwana and the Precambrian Rodinia existed and by the same token no longer exist. Such dynamism allows explanation of often now disjunct distributions of animals and plants, and why volcanoes and earthquakes and even ocean basins and continents occur where they do and have not always been in the same place forever. Plate tectonics theory also provides a basis for understanding how life might have originated and why it is maintained today on Earth – in a system that recycles materials like no other planet in the solar system, and like no other heavenly body yet known in the Universe.

environments that today host the highest marine biodiversity. In this sort of environment the Carravia Dolomite of the eastern Pilbara region of Western Australia was deposited.

Once there was land above sea level and plate tectonics was in action, continents began to interact. Some collided with one another, and these were sites of major mountain belts. But unlike today topographic relief was not so great. The world

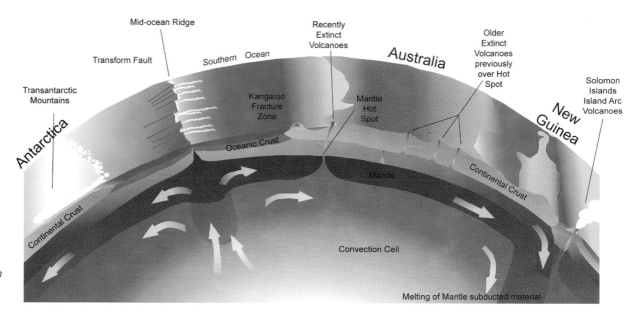

Figure 13. Cross-section structure of Earth (modified from Vickers-Rich and Rich,1999, with permission) (D. Gelt).

was much flatter. Half a billion years ago continents were isolated blocks scattered around the Equator, not recognizable as the continents of today.

The Late Archean and Early Proterozoic (2.6-2.2 billion years ago) was a quiet period of Earth history. There would have been little in the way of mountain-building activity and accumulation of the oceanic sediments was slow. However, between 2.5 and 1.7 billion years ago there were two major episodes of intensive rifting accompanied by massive deposition of sediments along the edges of some continents. This is also a time of maximum accumulation of the banded-iron formations. Continent building must have had an enormous impact on life. Vast, stable continents trapped metals and other elements (including carbon) bound up in the sediments that made them up – thus removing them from recycling by the biosphere. This led to a global glaciation between 2.4 and 2.2 billion years ago, a time called "Snowball Earth" (Hoffman & Schrag, 2000).

The Setting: Oceans Past and Present

Life is animated by water. All living things need some liquid water. Water forms 50% to more than 90% of total weight in most living organisms. In the protoplasm of each living cell, water acts as an agent that facilitates a multitude of chemical reactions. Water provides much of the internal environment of a cell. Water transports nutrients and oxygen and removes the waste products (which can also be oxygen!) of metabolism. Water provides the environment for reproduction and transportation of seeds, gametes and larvae of aquatic organisms. Water is the major means of transport of substances important for life activity, both inside and outside of organisms. It makes possible such important biological processes as osmosis and turgor. Water plays the major role in the process of thermoregulation. All these qualities of water are related to its unique molecular structure and composition. Without water, life as we know it would not exist.

Water is also one of the best-known ionizing agents. Because most substances are somewhat soluble in water, water is the universal solvent. It reacts with metal oxides to form acids. It acts as a catalyst in many critical chemical reactions. Water provides an active environment in which the most intimate biochemical reactions take place inside the living cell at an

extraordinary rate even at room temperature. So, clearly, it is of critical importance just when water first appeared on Earth.

The oldest water-deposited sediments are about 3.8 billion years old – from southwest Greenland (Nisbet & Sleep, 2001). Heinrich Holland and James Kasting (1991) noted that these rocks proved that liquid water was present. And based on the amount of volatiles predicted to have been released in the first few hundred million years of Earth history, there should have been enough water to form oceans by then – perhaps earlier. By 3.5 billion years ago, the presence of the mineral gypsum in the Warrawoona succession of Western Australia indicates that ocean temperatures could not have been much more than 57°C. Above that temperature gypsum turns into another mineral, anhydrite, and there is no anhydrite in this sequence. However, this has been questioned, some suggesting that what was thought to be gypsum is actually barite. Further work is in order to resolve this issue.

Today's oceans are not good models of these first Archean oceans. Water in this primary ocean likely came from two sources, outgassing from volcanic eruptions on Earth and comets from outer space. Comets probably did not contribute more than about 15% of the water based on the ratio of deuterium to hydrogen in comets (like Halley's Comet) (Albarede, 2003). At the time of formation, because of the presence of such gases as HCl (hydrochloric acid), high levels of carbon dioxide (CO_2) as well as sulfur dioxide (SO_2), seas were likely quite acidic (with a pH of around 2-4) (Williams & Frausto da Silva, 1996). Chloride, which combined with sodium to form salt in these seas, may have been present from the beginning, derived from HCl. Acidity likely decreased with the weathering of basic rocks and interaction of the first water with Oceanic Crust. Early oceans were also reductive in nature due to high levels of carbon dioxide, sulfides and hydrogen and the low levels of oxygen.

These ancient oceans were strongly and permanently stratified, made up of two main layers. The lower and thicker was anoxic due to injection of dissolved gases and minerals from numerous hydrothermal vents on the sea floor. Surface layers of the ocean were affected by the overlying atmosphere, eventually becoming enriched in oxygen. Cyanobacteria provided oxygen as a byproduct of photosynthesis, even

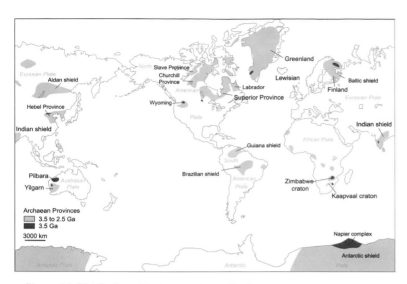

Figure 14. Distribution of Archean rocks on Earth (modified from Condie, 1993, courtesy of the Geological Society of London, from their Special Publication 199) (D. Gelt).

during the Archean. Such layering persisted for some time because Archean oceans did not overturn. Upward movement, or upwelling, of deeper oceanic water masses toward the surface along the edge of the continents is commonplace in today's oceans, but not so in the Archean. Upwelling today brings nutrients to the surface waters – phosphorus, for example, required in the life cycles of most phytoplankton. Mixing today also occurs because of the sinking of cold, dense and oxygen-enriched waters from the polar regions down to oceanic depths. This forces warmer, metal-enriched bottom waters upwards. Both mechanisms maintain a permanent overturn of oceans and ensure the recycling of vitally important nutrients needed by the biosphere from bottom to top in the ocean basins. Due to this lack of overturn, surface waters of Archean oceans were nutrient-starved deserts, unlike water at depth. Such a permanently stratified ocean likely existed until very latest Archean or Early Proterozoic times, between 2.7 and 2.5 billion years ago, and then, with formation of substantial continents, upwelling began, promoting recycling nutrients, profoundly affecting the biota.

The Setting: Atmospheres Past and Present

Early in Earth history, volcanic activity formed the primitive atmosphere (Holland & Kasting, 1992). This early atmosphere likely contained significant carbon dioxide (CO_2) at a high pressure (10 times the atmospheric pressure of today), nitrogen (N_2), carbon monoxide (CO), water vapor (H_2O), methane (CH_4) and smaller amounts of hydrogen sulfide (H_2S), hydrogen cyanide (HCN) and ammonia (NH_3). Oxygen was present in small amounts, but just how much is contentious. Rocks known from the Pilbara region of Western Australia, apparently laid down in river systems between 3.25 and 2.75 billion years ago, contain detrital uraninite and pyrite, both not oxidized (Rasmussen & Buick, 1999). Grains of these minerals show that they have been transported for some distance, and if there had been any significant amount of oxygen in the environment, they would not have survived. Light gases, such as hydrogen and helium, escaped Earth's grip, although they must have been part of the Earth's very first atmosphere, and there is

now a suggestion that some hydrogen remained (Tian *et al.*, 2005). For a few hundred millions of years this atmosphere remained anoxic and chemically aggressive, smelled like rotten eggs (H_2S) and looked like visions painted of a biblical Hell, full of acrid hydrogen sulfide and unbreathable by most modern surface-dwelling organisms. Radiation from the Sun would have bathed the planet with deadly UV, for without oxygen present there was no protective ozone layer. That same UV, however, initiated chemical reactions within the primal atmosphere, which in turn gave rise to organic compounds and life, which assisted in the formation of the ozone layer as well (Williams & Frausto da Silva, 1996).

This early atmosphere was affected by the heat generated in the Earth's interior as well as by the Sun. This initiated temperature gradients, vertical and horizontal, producing a dynamic atmosphere. For the first time there was wind. Movements within this atmospheric system together with water and other compounds which accumulated within it were factors maintaining the Earth's surface temperatures close to 25°C for much of its history after initial cooling, although with some variations such as glacial and greenhouse periods (Williams & Frausto da Silva, 1996).

With these initial conditions, oxygen then played a major role in the generation of the Earth's modern atmosphere. As D. E. Canfield (2005) notes: "The evolution of oxygen-producing cyanobacteria was arguably the most significant event in the history of life after the evolution of life itself." Oxygen present in some quantity changed the surface chemistry of Earth, significantly increased primary production and vastly increased the possibilities for biotic diversity (Canfield, 2005). Two models chart the course of oxygen buildup from the Archean to present, the Dimroth-Ohmoto Model and the Cloud-Walker-Holland-Kasting Model (Eriksson *et al.*, 2004). The Dimroth-Ohmoto Model suggests that oxygen levels reached 50% of present-day levels as long ago as 4 billion years. In contrast the Cloud-Walker-Holland-Kasting Model proposes that the early atmosphere was quite reducing with a small rise at about 3.0 to 2.8 billion years and a major rise at 2.2 billion years ago, the Great Oxidation event, that spurred the emergence of the eukaryotes.

From the perspective of today, the rise in oxygen seems beneficial, but as Nick Lane (2002) points out this increase in oxygen level was environmental pollution without parallel in the history of life. "For the tiny single-celled organisms that lived on the early Earth …. oxygen was anything but life-giving. It was a poison that could kill, even at trace levels." The first small single-celled organisms were "oxygen haters" – their living relatives are bacteria that thrive in stagnant swamps, in our guts, or near the mid-ocean ridges where black smokers churn out hot or heavy metal-enriched waters where oxygen levels are less than 0.1 % of present levels elsewhere.

Where did oxygen come from? How did life contend with it when it first occurred in increasing amounts, since it was clearly a toxin to the "oxygen haters" of the world? The most significant source is photosynthesis, a chemical reaction where energy from sunlight fuels the splitting of water, which in turn produces O_2 as a waste product and binds carbon to form 'sugars, fats, proteins and nucleic acids that make up organic matter' (Lane 2002). Another source is the splitting of water by cosmic rays into hydrogen and oxygen, hydrogen being lost into space, leaving O_2 behind.

Figure 15. Banded Iron Formation (BIF) in southern Namibia (M. Fedonkin).

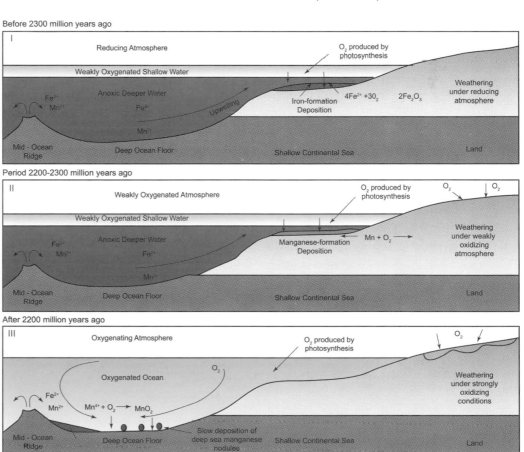

Before 2300 million years ago

I

Reducing Atmosphere

O_2 produced by photosynthesis

Weakly Oxygenated Shallow Water

Anoxic Deeper Water

Fe^{2+}
Mn^{2+}

Fe^{2+}

Upwelling

Iron-formation Deposition

$4Fe^{2+} + 3O_2$ $2Fe_2O_3$

Weathering under reducing atmosphere

Mn^{2+}

Mid - Ocean Ridge

Deep Ocean Floor

Shallow Continental Sea

Land

Period 2200-2300 million years ago

II

Weakly Oxygenated Atmosphere

O_2 produced by photosynthesis

O_2 O_2

Weakly Oxygenated Shallow Water

Anoxic Deeper Water

Fe^{2+}
Mn^{2+}

Fe^{2+}

Manganese-formation Deposition

$Mn + O_2$

Weathering under weakly oxidizing atmosphere

Mn^{2+}

Mid - Ocean Ridge

Deep Ocean Floor

Shallow Continental Sea

Land

After 2200 million years ago

III

Oxygenating Atmosphere

O_2 produced by photosynthesis

O_2

Oxygenated Ocean

O_2

Fe^{2+}
Mn^{2+}

$Mn^{4+} + O_2$ MnO_2

Weathering under strongly oxidizing conditions

Slow deposition of deep sea manganese nodules

Mid - Ocean Ridge

Deep Ocean Floor

Shallow Continental Sea

Land

Figure 16. Model for Banded Iron Formation depositions (modified from Beukes et al., 2002) (D. Gelt).

Figure 17. The Archean Earth (courtesy of the Smithsonian Institution, Washington).

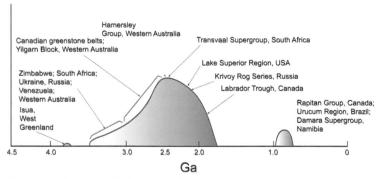

Figure 18. Periods in Earth history when Banded Iron Formations were deposited (modified from Schopf and Klein, 1992) (D. Gelt).

Today, and for much of Earth time, oxygen is used to fuel metabolism of animals, some bacteria and fungi. They in turn release CO_2 and water (H_2O) back into the atmosphere, but about 0.01% of organic matter gets buried in the sediments and thus goes "out of circulation" – unable to take up oxygen because of its burial. Such buried organics become coal, oil and gas, as well as vast, less concentrated reserves preserved in the pores of sandstones and other sediments. With these reserves locked up, the complete consumption and recycling of oxygen by the consumers is prevented because some lie buried in sediments. Oxygen begins to accumulate in the atmosphere. So, this treasured commodity in the world today is likely due to a "3-billion-year

mismatch between the amount of oxygen generated by the primary producers and the amount used up by consumers" (Lane, 2002).

So, oxygen may have accumulated at an increasing rate for some time, driven by photosynthesis. Just how the "oxygen haters" dealt with this is the second question posed above. Was life beset with massive extinctions, or did it already have the inbuilt ability to cope with this changing world? Early organisms may have been able to cope with the changing concentrations of oxygen because they were able to switch methods of respiration, from one to another in response to changing environmental conditions (Lane, 2002). The living bacterium *Thiosphaera pantotropha*, for example, prefers to feed on excrement. If there is a ready supply of oxygen, *Thiosphaera* gains its energy using aerobic (oxygen-supported) respiration. However, when conditions are starved of oxygen, it derives energy from sulfides by using nitrogen oxides in place of oxygen. Perhaps those organisms that survived the buildup of oxygen, which could have been sporadic at best in the beginning, might have used the same duel strategy (Lane, 2002).

Chemistry to Life: From Black Smokers to Simple Cells

Sometime, during Earth's very early history, life appeared, certainly long before the first structures built by life, stromatolites, which occur in the rock record 3.5 billion years ago. Paleontologists agree that stromatolites were constructed by once living organisms – they are still being formed in places like Shark Bay in Western Australia. Biomarkers and actual cellular structures provide evidence for an early presence of life on Earth detailed below. The

Banded Iron Formations

Banded Iron Formations are heavy rocks with alternating red bands of iron-rich, sometimes almost black layers rich in amorphous silica, with up to 30% total iron. They polish up nicely and are often fashioned into jewelry. BIF's are no longer forming and have not been forming since @ 900 million years ago. Most BIF's range from around 3.5 to about 2.7 billion years in age (Western Australia, southern Africa, Russia and northern North America) with minor deposition in West Greenland around 3.8 billion years ago and a major spike at 1.8-1.9 billion globally. The iron used to make steel comes from this non-renewable resource, albeit available in tremendous quantities – so we are not likely to run out.

How these formed and why they are no longer forming is a great enigma. BIF's seem to have formed in oceans full of dissolved Fe_{+2} iron (Holland, 2004) and low levels of oxygen. Iron is used by many microorganisms in their metabolic cycles. Both "oxygen-loving" and "oxygen-hating" photosynthesizing and chemosynthesizing organisms use iron. Jelte Harnmeijer (*pers. com.*, 2003) has proposed that the environment in which BIF's formed was a likely place for the emergence of life itself. He suggested that perhaps the best setting for the switch from non-life to life were hydrothermal vents on the ocean floor (Russell & Hall, 1997) where the presence of clay mineral surfaces could have played a catalytic role (Cairns-Smith, 1982) in a hot and mineral-rich environment, where phosphatic crusts could have served as the first formed membranes separating life from the surrounding environment.

The association of BIF's with evidence of nearby hydrothermal activity, a diverse mineralogy, hydrous clay minerals and early biomarkers, followed in slightly younger sediments by actual fossils in cherts gives Harnmeijer's theory strength. There are many types of BIF's – the most abundant being oxide-rich BIF's, silicate-rich BIF's and carbonate-rich BIF's. Oxide-rich BIF's are made up of alternating bands of hematite, with or without magnetite. Siderite and iron silicate can also be present. Silicate-rich BIF's contain several different hydrous iron silicates, carbonates and cherts. Carbonate-rich BIF's generally have about the same proportions of chert and ankerite (sometime with siderite, sometimes without) and are often very finely bedded.

Two major sorts of regional deposits have been noted: Algoma-type and Superior-type. Algoma-type BIF's are generally limited in extent (under 10 km in lateral extent and no more than around 100 m in thickness) and associated with volcanics – likely deposited around spreading centers, back-arc basins and rift zones on cratons (Veizer, 1983; Gross, 1983; Bjerrum & Canfield, 2002). The Superior-type BIF's are much larger, both in areal extent and thickness, and are associated with sediments. They seem to have formed in shallow oceans (Simonson & Hassler, 1996) and are some of the youngest, dating at around 2.0 to 1.9 billion years, not strictly BIF's but granulated iron formations, as are the similar deposits in the Earaheedy Group of western Australia (Kath Grey, *pers. com.*, 2005).

literature on early life and the transition from chemical soup to organism, from the RNA world to the DNA world, is vast (see Schopf & Klein, 1992; Schopf, 1999; Bengtson, 1993; and Seckback, 2004 for excellent summaries) and is constantly expanding. It will be discussed only briefly here as the background which led to metazoans.

What is life and how does it differ from non-life? Life is an organized system of chemical reactions, which take place in a confined space, usually a cell. Life is a self-perpetuating chemical reaction, a self-assembling dynamic system. Reactions "extract energy from the nutrients, produce structural material from basic building blocks of amino acids, sugars and fats" – a process called metabolism (Ainsworth, 2003). The cell is separated from the environment by a membrane, which allows reactions to take place independent of surrounds. It provides structural integrity. The cell membrane also allows communication with the outside world, but regulates what comes in and what goes out. Additionally, living systems must store information in the form of a double helical molecule called DNA – and it must be able to

"repair this molecule when damaged, translate it into proteins and copy it, to reproduce" itself so as to have a future (Ainsworth, 2003). Life, as we know it, also requires the presence of liquid water.

Where and when did life begin? Beginnings may have been in extreme environments, very hot environments such as hydrothermal vents associated with ocean floor ridges. Some of the most primitive living forms are hyperthermophiles, the Archaea and Bacteria. Some prosper in temperatures between $150°$ and $350°C$ and pressures up to 265 atmospheres. As Nisbet and Sleep (2001) point out: "Life on Earth dates from before about 3,800 million years ago, and is likely to have gone through one or more hot-ocean 'bottlenecks.' Only hyperthermophiles (organisms optimally living in water at $80°$-$110°$ C) would have survived." In such hot regions, both present and past, the contrast between ocean water chemistry and the magmatic fluids coming from the Earth's interior is and would have been considerable. The redox conditions (that is, whether the environment was reducing, *i.e.* gaining electrons, or oxidizing, *i.e.* losing electrons) would have encouraged energy flow. With the development of photosynthesis both in non-oxygenated and then oxygenated environments, life could have "escaped" the hydrothermal environments.

There are other ideas about how life developed. Sherwood Chang (ms., 1992), based on work of the Space Science Division of the NASA Ames Research Center suggested that a "slightly acid ocean surface film [could] have benefited chemical evolution. Diverse processes acting on and within the top layers of the mixed zone would have made it a complex, physically and chemically active environment perhaps well suited for chemical evolution" (Lerman & Teng, 2004). The ocean-atmosphere interface would have been a dynamic and energy-filled environment. It would have had input from many sources – oceanic, continental, volcanic, even extraterrestrial. Energy would have come from wave and wind, solar radiation, coronal discharges, lightning, etc. Volcanoes would have provided chemical input, such as airborne polyphosphates – the base for phosphorylation reactions. And, in addition the ocean surface is a place where bubbles form as a result of wind and wave action. Bubbles can act as concentrators of material produced near the atmosphere-water interface, then eject such materials as they break and form and break over and over again. Perhaps condensation reactions and synthesis of organic phosphates were initiated under such conditions. The persistence of the surface "monolayer" could have served as an ordering device – encouraging preferred molecular orientations and thus encouraging molecular self-assembly.

Other researchers have suggested an extraterrestrial origin of life, panspermia, but that just moves the question about beginnings to another place in the Universe. There may not be just one solution to the origin of life problem. Perhaps life has multiple origins. As Antonio Lazcano notes: "In the past decade considerable progress has been achieved in our understanding the emergence and early evolution of living systems, but we are still haunted by major uncertainties, the magnitude of which is matched only by our ignorance." ...."we are still very far from understanding the origin and nature of the first living beings. These are still unsolved problems – but they are not completely shrouded in mystery, and this is no minor scientific achievement" (Bengston, 1993).

Figure 19. Ancient zircon crystal from rocks more than 4 billion years old in Western Australia (courtesy of S. Bowring and J. Crowley, from Crowley et al., 2005).

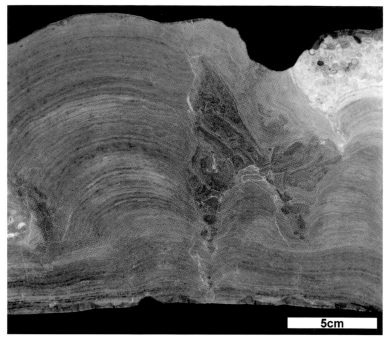

Figure 20. Fossil stromatolite colony. Earaheedia kuleliensis, *Kulele limestone, Earaheedy Group, about 1.8 billion years old. Proterozoic, Western Australia (M. Fedonkin).*

How to Date a Rock: Radiometrically!

Today many different radiometric dating techniques are used, and they rely on the principle that radioactive decay processes proceed at a constant rate. Temperature, pressure and chemical conditions typical of the Earth's surface do not affect the rate of decay. One technique that has been used extensively for dating very old rocks is the potassium-argon method – rocks ranging in age from 1 million years to as much as 4000 million years. This technique involves the minerals biotite, hornblende and sanidine. Another method measures the uranium-lead ratios derived from studies mainly on zircon crystals found in volcanic rocks or in sediments that have been derived from the weathering of such rocks. Each technique has its good points and bad, but if it is possible to use one to check another, then more reliable dates can often be obtained (Bowring, Schmitz & Condon, *pers. com.*).

Radiometric techniques rely on the fundamental principle that after a known period of time, called the "half-life," the amount of original radioactive material remaining is reduced by one-half. The radioactive element decays into another, stable element. Decay continues halving radioactive material, at the same rate over time, until the amount remaining is infinitesimally small, undetectable because of background radiation from outer space. As an example the half-life of one isotope of uranium (U238), the "parent," is around 4500 million years. In that time it loses half of its mass through alpha particle decay. This converts uranium (U238) into stable lead (Pb206), the "daughter product." Another commonly used technique involves the decay of uranium 235 (U235) into lead 207 (Pb207). By calculating the ratio of U238 to Pb206 (or U235 to Pb207) the age of the rock can be worked out in years. Most recently, these techniques have been superseded by the SHRIMP uranium-lead and uranium-uranium dating of zircons, but this technique uses similar principles.

In the rocks containing some of the oldest animals on Earth, the most reliable of dates have used air-fall tuffs, ash-flows and lavas. Ashes are of particular value. When a volcano erupts explosively, it ejects debris that may reach the stratosphere and can be spread widely around the world – a single, local event acts as a global time line.

Fission track dating is yet another way of aging rocks in numerical terms. This technique depends on the spontaneous splitting (fission) of an atom of uranium 238. The large, heavy nucleus of U238 splits into two pieces that separate at high speed. These fragments move through the enclosing zircon crystal structure, leaving a trail of damage. So, the number of trails left in a crystal where U238 has decayed and left its trails of damage can be counted. In order to count these trails, crystals must be chemically etched after being finely sliced. The prepared sample is then viewed under a high-powered microscope and the etched holes, the "fission tracks," tallied. By estimating the approximate amount of original uranium in the sample and knowing the fission half-life of U238, the track count yields an approximate age of the zircon. Other minerals used besides zircons are apatite and some volcanic glasses, such as obsidian. These all provide closed systems lacking outside interference. Zircon in particular is hard and quite resistant to weathering, and it protects the "radioactive record" stored in its crystal structure, despite long transit after weathering out of its magmatic coffin and even being transported by streams and wind.

How might have the first cell formed? William Martin and Michael Russell (2003) suggest that an iron monosulfide crust might have formed the first functional membrane. This was likely deposited in the region of hydrothermal mounds where there exists a redox, pH and temperature gradient between sulfide-rich hydrothermal fluid and iron (II)-containing waters on the ocean floor. Really quite a nasty environment – acidic, smelling of rotten eggs and filled with dark minerals. Here metal-rich seawater comes into contact with water from the Earth's molten interior through tiny holes in cooling basalts and forms small, inorganic compartments, physically separated from the surrounding seawater. The alkaline water is trapped inside these compartments with walls formed of iron sulfides (FeS) and nickel sulfides (NiS), both compounds which act as catalysts in the formation of acetylmethylsulfide from carbon monoxide and methylsulfide and which have originated from the hydrothermal fluids. Once these compounds are produced, they act as barriers for rapid loss by diffusion of the contents of the tiny compartment.

Martin and Russell suggest that once this process took place the "chemistry of what is known as the RNA-world could have taken place within these naturally forming, catalytic-walled compartments to give rise to replicating systems." Energy flow initiated for the charge on either side of this sulfide crust would have been different, just as in a battery – low energy redox (reducing/oxidizing) reactions – basic to the metabolic processes that define life. RNA is a very versatile molecule with the potential to be involved in many of the processes that define life – energy conversion, replication, catalysis. The small "cells" formed by sulfides in the depths of growing oceans may have nurtured the first RNA's and given life a cradle in which to develop.

Despite the uncertainties of just how and where life got started, biomarkers give some idea of when life first appeared on Earth and just how complex it was early in Earth history. Some of the very oldest biomarkers were recorded by Bill Schopf and his colleagues (2002), carbonaceous residues of kerogens that signal the carbon isotopic record of photoautotrophic carbon fixation about 3.5 billion years ago. Additional "molecular fossils" have been recovered from the Mount Bruce Supergroup of the Pilbara region of Western Australia (Brocks *et al.*, 2003). Here, in shales that have been little altered since deposition in shallow seas between 2.45 and 2.78 billion years ago, are hydrocarbons, the leftovers of life that lived at the time. These shales were recovered from several drill cores in two different mines in the Hamersley Basin near Wittenoom. From these cores complex polycyclic biomarkers have been extracted, examples being hopanes and steranes (with a wonderful variety of nearly unpronounceable names like end-branched monomethylalkanes, w-syslohexylalkanes, diamondoids, tri- to pentacyclic terpanes, polyaromatic hydrocarbons, aromatic steroids, etc.). These compounds suggest that bacteria, in particular photosynthesizing cyanobacteria (Canfield, 2005), were present, metabolizing as they do today and producing a variety of waste product "biomarkers" – they mark the past presence of bacterial life. These do not appear to be modern contaminants, for such "biomarkers" as those produced by plants or hydrocarbons characteristic of the last 550 million years (Phanerozoic times) are not present.

In every case, when samples of such ancient times are analyzed, a thorough evaluation as to their syngenicity (that they represent compounds deposited at the same time sediments now forming the rocks were deposited and were not introduced later) must be established. Sampling, storage of the samples and analysis must be carried out under strict guidelines; sterile conditions, to avoid contamination. Jochen Brocks and his colleagues (2003) outline in detail how such sampling and

Bacterial Communities at Depth (The Subsurface Biosphere)
Subsurface biosphere is a term referring to organisms that prosper without sunlight, and frequently without oxygen. Biosphere was originally defined as space on the surface of the planet where the energetics of life are directly or indirectly dependent on the sunlight that powers photosynthesis. Most familiar life forms depend on photosynthetically generated organic carbon and free oxygen. Recent discoveries, however, clearly demonstrate that the habitable zone for life extends to depths of thousands of meters, both beneath the continents and below the ocean floor (Pederson, 1993).

The subsurface biosphere is occupied by Bacteria and Archaea and estimated to form a total biomass of microorganisms probably equivalent to that of all life in the surface and near-surface of the planet. Bacteria and archaeans inhabit not only permeable and porous sediments but also such solid rocks as basalts, where they can be found in microscopic pores and cracks (Fisk at al., 1998). Drilling into the ocean floor, particularly the basalts of the Atlantic, Pacific and Indian oceans, has provided evidence that bacteria inhabit a significant part of the upper oceanic crust. The volume of this habitat is estimated to be upwards of 10^{18} cubic meters, which is, in fact, the largest habitat on Earth.

What do they do in the darkness of the Earth's interior, at great depths, under such very high pressure? Most of these microorganisms are chemolithoautotrophs. They consume gases and minerals to fuel their metabolism (Liu *et al.*, 1997). The chemical reactions that fuel their metabolism do not require input of light energy. High temperature is critical for their survival.

Discovery of such abundant subsurface life offers opportunities of marked importance to industry and agriculture. Microbiologists have proposed possible uses to include recycling of human and animal effluent, purification of the toxic industrial wastes, confining and locking up dangerous nuclear byproducts, increasing efficiency of the mining and oil-gas extraction and purifying drinking water. In order to utilize this realm of invisible life, it is critical to understand how it originated and evolved over time. Studies of the Precambrian, the early history of the Earth itself, is one place to begin.

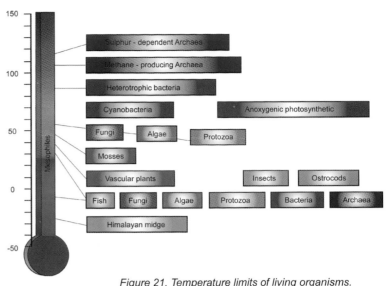

Figure 21. Temperature limits of living organisms. The lowest and highest temperature tolerated by each major group is given: Archaea (red), Bacteria (Blue), Algae (light green), Fungi (brown), Protozoa (yellow), plants (dark green) and Animalia (purple) (from Rothschild & Mancinelli, 2001, and with their permission and that of Nature) (D. Gelt).

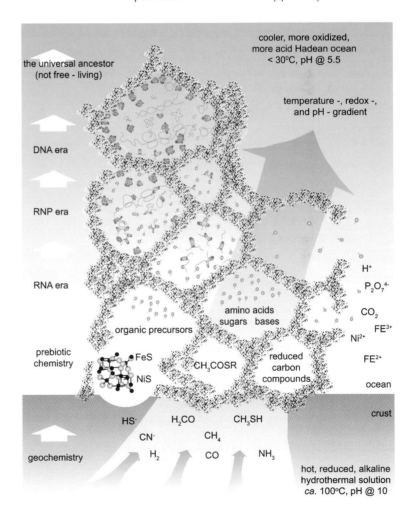

Figure 22. A model for the origin of life at a redox, pH and temperature gradient typical of a modern-day submarine hydrothermal vent. The first "membranes" to support early environments that eventually developed into organic cells (which separate outside chemistry from that inside the cell) may have been inorganic crusts deposited in such an environment (from Martin & Russell, 2003, with their permission and that of the Royal Society) (D. Gelt).

Figure 23. Underwater view of modern
stromatolites, Shark Bay, Western Australia
(N. Kawazoe).

analysis should be done in order to make sure that contamination has not given the wrong idea of when the first life appeared.

So, certainly by 3.5 billion years ago the byproducts of metabolic processes characteristic of eukaryotes left their marks in the rock record. Real fossils also left their marks. Stromatolite fossils signal the activities of life, probably that of cyanobacteria, and these organosedimentary structures occur in rocks at least 3.5 billion years ago. In the somewhat younger Apex Chert of northwestern Western Australia (@3.46 billion years) and the Kromberg Formation of South Africa (@3.37 billion years) fossilized remains of prokaryotes are preserved, with kerogen biomarker signals indicating true biogenicity (Schopf *et al.*, 2002). Bill Schopf and many others (Westall *et al.*, 2005) present a case for biogenicity of these remains, but this has been challenged by others (Brasier *et al.*, 2002). Roger Buick (1992) suggested that the lacustrine stromatolites known from sediments in Australia dated at about 2.7 billion years were of a type built today only by photosynthesizing organisms, as there are no signs in these structures of reduced sulfide or iron (Canfield, 2005). This is the most reliable first record of photosynthesizing cyanobacteria. It is the association of

Some Definitions
  Oxidation – the relative loss of electrons in a chemical reaction, including removal to form an ion or sharing with another substance. Often associated with the liberation of energy (Lane [2002] compares this process to "paint-stripping")
  Reduction – the addition of electrons by chemical reaction. (Lane compares this to applying the paint)
  Redox Reaction – a chemical reaction where one reactant becomes oxidized and the other reduced (Purves *et al.*, 2003).

biomarkers with these early fossils that give added strength to Schopf and his associates' observations (Alterman & Corcoran, 2002).

Certainly studies on the protein and nucleotide sequences of the chromosomes in the living relatives of ancient lineages, act as genomic clocks, and support the ancientness of the fossil and biomarker data. By observing differences in the genotypes of living prokaryotes and taking into account known rates of change in genetic structures, a differentiation of cyanobacteria is estimated, by some, as long ago as 4 billion years (Hedges & Kumar, 2003). This certainly needs further support, as it is far older than the first fossil or biomarker data.

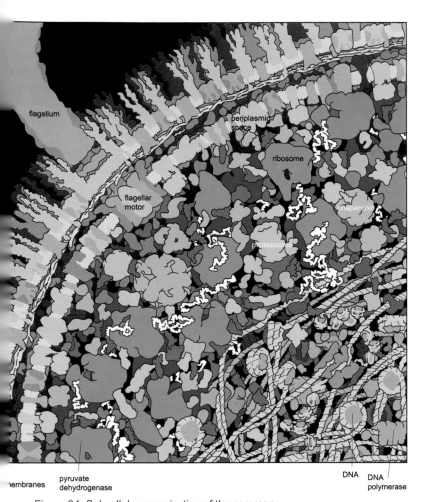

flagellum

periplasmic space

ribosome

flagellar motor

chaperonin

proteasome

pyruvate dehydrogenase

membranes

DNA

DNA polymerase

Figure 24. Subcellular organization of the common bacteria Escherichia coli. Such prokaryotes are frequently described as "bags of enzymes" but a closer look reveals several levels of subcellular organization. This cell is encased by two membranes that contain a compartment that stores and sorts nutrients and waste products. In the center of the cell is the DNA, tightly packed, and storing the genetic data – but as in all prokaryotes, it is not enclosed by a nuclear membrane. The rest of the cell is occupied by cytoplasm, filled with ribosomes and many different sorts of enzymes. Multiprotein complexes carry out many different tasks – one the "motor" of a flagellum which whips about, moving the cell (from Hoppert & Mayer, 1999 and by artist David S. Goodsell; courtesy of Scripps Research Institute) (D. Gelt).

Noncellulosic cell wall

Cell membrane

Nucleoid

Small ribosomes

Golgi body

Large ribosomes

Mitochondria

Plastid

Thylakoids

Plastid inner membrane

Plastid outer membrane

Nucleus

Nucleolus

Nuclear membrane

Cell membrane

Cell wall (cellulose or chitin)

Flagellum

Endoplasmic reticulum

Chromatin

Kinetosome (9 + 0)

Kinetochores

Undulipodium (9 + 2)

Cell membrane

PROKARYOTE

EUKARYOTE

Figure 25. Organization of a prokaryote and a eukaryote cell (modified from Margulis & Schwartz, 1988; courtesy of L. Margulis) (D. Gelt).

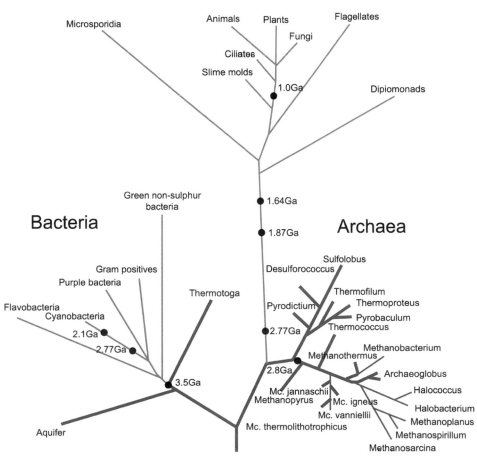

Eucarya

Microsporidia

Animals

Plants

Flagellates

Fungi

Ciliates

Slime molds

1.0Ga

Dipiomonads

Green non-sulphur bacteria

1.64Ga

1.87Ga

Bacteria

Archaea

Sulfolobus

Desulforococcus

Gram positives

Purple bacteria

Thermofilum

Thermoproteus

Thermotoga

Pyrodictium

Pyrobaculum

Thermococcus

Flavobacteria

Cyanobacteria

2.77Ga

Methanobacterium

Methanothermus

2.1Ga

Archaeoglobus

2.77Ga

2.8Ga

Halococcus

3.5Ga

Mc. jannaschii

Methanopyrus

Mc. igneus

Mc. vanniellii

Halobacterium

Methanoplanus

Methanospirillum

Aquifer

Mc. thermolithotrophicus

Methanosarcina

Figure 26. Minimum ages for the appearance of groups of Bacteria, Archaea and Eucarya. Red indicates organisms that can tolerate very high temperatures based on molecular studies (modified from Brocks et al., 2003) (D.Gelt).

23

Some Definitions

Archaea – Microorganisms of a size similar in appearance to bacteria but distinct in their molecular organization. In classifications that divide the biological world, they form one of three major divisions, the other two being Bacteria and Eukarya. This differs from the five kingdoms proposed by Margulis and Schwartz (1988). The membranes of archaeal cells have a single lipid layer, whereas the Bacteria and Eukarya have two. The chemical composition of archaeal membranes also differs from that of Bacteria and Eukarya.

Archaebacteria - A subkingdom of prokaryotes (Margulis & Schwartz, 1988) that includes methanogenic (producing methane as a byproduct of metabolism), halophilic (adapted to high salinity) and thermoacidophilic (adapted to high temperature and acidic conditions). Archaebacteria differ from Eubacteria in the shape of their ribosomes, in having a distinctive cell wall structure (lacking a compound called peptidoglycan), having a distinctive sequence of nucleotides – thus a distinctive genetic sequence on their RNA and a distinctive structure of their lipids. The environments in which these microorganisms prosper are typical of those that characterize the Archean Earth surface – in hot springs, undersea vents that inject sulfide-rich liquids and a variety of other extreme environments.

Cytoplasm (from the Greek *kytos*, hollow vessel; plasma, anything molded). Excluding the nucleus, the living matter in a cell.

Eubacteria – All bacteria that are not archaebacteria. The vast majority of bacteria, diverse in their shape and structure and in the metabolic pathways used to produce energy and process nutrients. Mostly single-celled but some multicelled forms exist. Groups within the eubacteria (three) are recognized by their unique cell wall structure, which in turn is different from that of archaebacteria.

Eukaryote (Eucaryote). Name derived from the Greek *eu*, good + *karyon*, kernel. A cell with a nucleus and specialized organelles bound by a membrane and thus separated from the cytoplasm of the rest of the cell. These cells have chromosomes bound by a membrane in the nucleus.

Prokaryote (Procaryote). Name derived from the Greek *pro*, before + *karyon*, kernel. Unicellular life forms lacking a membrane-bound nucleus, structured chromosomes and complex internal organization. They lack any sort of membrane-bound internal structures, but the differential structure of water inside the prokaryotic cell may form microenvironments that act like the more structured compartmentalization in eukaryotes (Hoppert & Mayar, 1999). Prokaryotes lack the firm structural support found in nucleated eukaryotes, the cytoskeleton. Prokaryotes must depend for the most part on their rigid cell wall to keep their shape. Some, the xenophyophores, can reach large sizes – *Stannophyllum* can reach up to 25 cm in diameter, but most are small.

Prokaryotes are invisible to the naked eye, but are essential components of the Earth's biota (= living organisms). They catalyze unique and indispensable reactions in the biogeochemical cycles of the biosphere, produce important components of the Earth's atmosphere and represent a large portion of life's genetic diversity and biomass. They function with "stunning efficiency. Anyone who has ever had a badly infected cut can appreciate just how rapidly bacteria multiply" (Hoppert & Mayer, 1999).

Prokaryotes contain much of the cellular carbon on Earth. Whitman and colleagues (1998) at the University of Georgia have made some estimates of these unseen life forms: 1) the total amount of prokaryotic carbon is around 60 to 100% of the estimated total carbon in plants, and inclusion of prokaryotic carbon in global models will almost double estimates of the amount of carbon stored in living organisms; 2) the Earth's prokaryotes contain about 10 times as much nitrogen and phosphorus as do all the Earth's plants and represent the largest pool of these nutrients in living organisms; 3) most of the Earth's prokaryotes occur in the open ocean, in soil, and in oceanic and terrestrial subsurfaces, really out of sight; 4) the cellular production rate for all prokaryotes on Earth is highest in the open oceans. Whitman suggests that the large population size and rapid growth of prokaryotes provide enormous genetic diversity, which has great potential for future research in medicine and many related fields (Whitman *et al.*, 1998).

Ribosome – A molecular combination of protein and a particular kind of RNA that manages protein formation – the most complex aggregation in the cell. There are three different kinds of RNA molecules, each with a different job: mRNA, tRNA and rRNA. rRNA is an essential structural and functional unit of ribosomes. By ensuring the binding of mRNA and tRNA it guarantees the correct ordering of amino acids in protein synthesis.

RNA World – RNA or ribonucleic acid in living organisms is involved with DNA (deoxyribonucleic acid) in replicating and storing genetic information. In the early history of life, RNA may have stood on its own, in the absence of DNA. RNA is an amazing set of molecules that not only carries genetic data but can act as enzymes. Thus the RNA World was a time when RNA was the sole carrier of genetic information. All four bases as building blocks of RNA (adenine, cytosine, guanine and uracil) are present in the abiotic world, but not so the 5th amino acid, thymine, which replaces uracil to produce DNA. In vivo thymine is synthesized from uracil precursors. What is also interesting is that the universal molecule used by all cells to produce energy, ATP, is formed from RNA precursors (Barbieri, 2003).

Sex (sexual reproduction). The fusion of two sex cells during meiosis with a copy of DNA from each cell combined into a new organism. This leads to an entirely new combination of genetic material, not an exact replica of either "parent." Besides novelty this method of replication gives certain advantages to the offspring by the possibility of removing harmful mutations, more opportunities for improved fitness, flexibility to adapt to new environments. Unlike asexual reproduction which is primarily a "copy and divide" strategy.

Stromatolites. Laminated structures that commonly form mounds or columns or even flat-lying layers, built over long time periods by a succession of organic mats, often by cyanobacteria that have entrapped sedimentary material or brought about the deposition of carbonate precipitates. These structures are convex upwards and often have fossil cells associated. They are synsedimentary – that is, deposited at the same time as the surrounding sediments. Such structures are still forming in some areas of the world today.

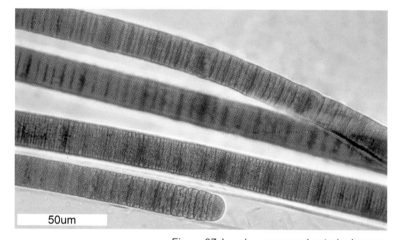

Figure 27. Lyngbya, *a cyanobacteria, Laguna Mamona, Mexico (courtesy of W. Schopf).*

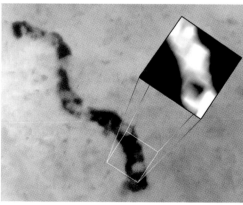

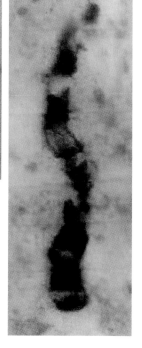

Figure 28. Primaevifilum amoenum, *Apex Chert, Western Australia, which has been dated at around 3.46 billion years. Medium diameter of filaments 2-5 microns (courtesy of W. Schopf).*

Figure 29. Primaevifilum amoenum, *Apex Chert, Western Australia (courtesy of W. Schopf). Medium diameter of filaments 2-5 microns.*

The geological history of surface ore formation

Era (billion years before present)	Geological and chemical features	Major ores formed
Early Archean (3.8 - 3.0)	Submarine trench formation; basic magma flows give *primary greenstones**	Fe, Ni, Cu sulphides, Au
Late Archean (3.0 - 2.5)	Recycling of primary greenstones, hydrothermal processes	Cu, Zn hydrothermal sulphides
Early Proterozoic (2.5 - 1.7)	Uplifted crust erodes O_2 produced by photosynthesis, oxidation of Fe^{2+}	Au deposits Banded Fe formations
Mid-late Proterozoic (1.7 - 0.7)	Thick continental crust forms Atmospheric O_2 increases: active sulphur redox chemistry	Ti, Cr oxides, Fe sulphide. Pt metals Co, Cu, U deposits
Phanerozoic (0.7 - present)	Extensive crustal recycling Tropical weathering conditions Secondary enrichment	Hydrothermal Cu, Zn, Mo, Sn, Pb Al, Fe resistates+ Co, Ni, Cu minerals

*Greenstones — named after deposits in South Africa
+Resistates — rocks resisting weathering

Figure 30. *Timing of the formation of major ore accumulations (modified from Williams & Frausto da Silva, 2001, with their permission) (D.Gelt).*

Hard Rock and Heavy Metal

Early in the history of Earth, developing oceans were rich in certain elements, in particular metals, heavy metals, that nowadays are not there – they are instead locked up in sediments, mined for the continuation of civilization as we know it. Such elements as iron, chromium, zinc and manganese were produced in abundance by volcanoes fueling developing atmosphere and ocean water composition. Living organisms had but to absorb their surrounds to gain substances needed for metabolism.

These needed elements form "extinct sediments" – uranium and gold-bearing conglomerates, banded iron formations, laminated copper deposits, lead and zinc ore-bearing shales and carbonates, sedimentary manganese ores and phosphorites (Schopf & Klein, 1992) that are abundant in sequences older than 2.5 billion. Oceans were stripped of those elements fundamental to metabolism of many organisms with the increase in oxygen. Frequent association of such sedimentary ores with carbonaceous deposits, characteristic biominerals and microfossils (mineral pseudomorphs of once living bacterial cells) all indicate a significant role of the biota and biogenic organic matter in the formation of such ore deposits.

Biochemical reactions in living cells are sped up by enzymes. Composed of polymers of amino acids, these large proteins act as catalysts in regulating and speeding up rates of the chemical reactions of metabolism in cells, sometimes up to 1000 times. Over 700 enzymes have now been identified, and the majority of their active catalytic sites involve metals, essential for the functioning of enzymes. With such metals becoming scarce in the environment, one way to assure availability for living was for organisms to sequester and concentrate these substances within their structure.

Some exceptions to this pattern, however, do exist. One such is molybdenum. This heavy metal is associated with the polymerization of some proteins, with biological breakdown of sulfur and carbon compounds and with oxygen transport in animals. In the anoxic biosphere of early Earth molybdenum was simply unavailable. Its function was carried out by other metals, such as tungsten, vanadium or even iron. With the oxygenation of the atmosphere, many metals became scarce, and molybdenum likely took their place as an enzyme cofactor.

Impoverishment of many of the heavy metals from circulation in ocean waters, related to and coupled with the increase in oxygen from 2.3 billion years to present (Rasmussen & Buick, 1999; Beukes *et al.*, 2002; Yang & Holland, 2003), essentially forced organisms either to adapt or to perish – it forced the sequestering of essential metals within living systems or required a change in their metabolic pathways. Rather than causing catastrophe for many, this may have spurred innovation – apparently forcing a combination of prokaryotic cells into a symbiotic relationship which led to eukaryotes – a solution for the surrounding geochemical starvation and oxygen build–up. One organism depended on the waste products of another and both were protected from the outside environments.

Beginnings of a Modern Earth

The boundary between the Archean Eon and the Proterozoic is significant. On one side of the boundary, the Archean Earth is an unfamiliar place. On the side of the Proterozoic, the world is more familiar, though atmosphere composition may have been an unusual mix until late in the Neoproterozoic, with CO^2 levels quite high for some time (Kaufman and Xiao, 2003). During the Archean, plate tectonics switched on, oxygen levels rose, familiar sediments began accumulating; both marine carbonates, and those typical of rivers and lakes (Awramik & Buchheim, 2001). Sometime during the Archean continents emerged above sea level. The rock record becomes a richer tapestry above this boundary and the geological history much easier to reconstruct. Not only is the record better, but also there is less alteration of rocks by metamorphism. The Proterozoic record is more true to original form. The stage is set for the rise of complex life, the metazoans, and the development of a modern Earth.

Figure 31. Microbial-bound sands that typify marine environments of the Precambrian when cyanobaterial mats were widespread, Pound Quartzite, Flinders Ranges, South Australia (P. Vickers-Rich).

Figure 32. Mildred Fenton inspecting the microbial mats and stromatolite structures in Shark Bay, Western Australia, one of the few parts of the marine environment today which still hosts biotas reflective of the Precambrian (M. Fenton).

Billions of years ago

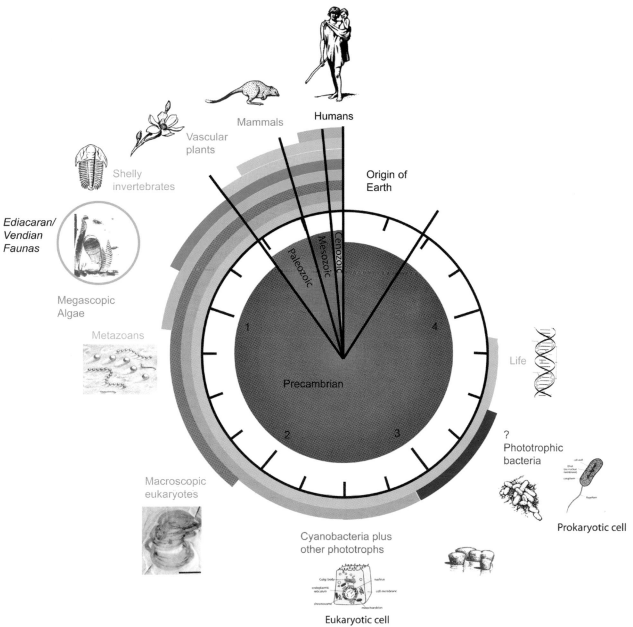

Humans

Mammals

Vascular plants

Shelly invertebrates

Origin of Earth

Ediacaran/ Vendian Faunas

Paleozoic

Mesozoic

Cenozoic

Megascopic Algae

Metazoans

Precambrian

Life

1

4

2

3

Macroscopic eukaryotes

?
Phototrophic bacteria

Prokaryotic cell

Cyanobacteria plus other phototrophs

Eukaryotic cell

Figure 33. History of life through time (modified from Des Marais, 2000) (D. Gelt).

CHAPTER 2

The Background. The Proterozoic
(2.5 Billion to 542 Million Years Ago)

P. Vickers-Rich and M.A. Fedonkin

When did the first multicellular animals appear on Earth? And what sorts of environments gave birth to these first metazoans? Atmospheric oxygen concentrations may supply the answer. It may have provided "encouragement" for the development of complexity – in order to survive one of the greatest pollutants of all time. In some ways oxygen may have been more a curse than a blessing in the short term. Increased oxygen levels may have chemically impoverished both oceans and atmosphere, removing certain elements that otherwise had been critical to organisms whose metabolism prospered in anoxic conditions, the "oxygen haters." The reaction of organisms to oxygen "poisoning" was to sequester needed resources, thus preserving an archaic biochemistry, adding structural and metabolic "blocks" to neutralize oxidation, and in doing so they became more complex and isolated themselves from environmental toxins – first developing into a new kind of cell (eukaryote) and then banding together to form multicellular organisms and eventually metazoans.

A New Eon

By 2.5 billion years ago, the beginning of the Proterozoic Eon, Earth had begun to modernize. Oxygen was produced in increasing amounts by photosynthesizing organisms, as reflected by the cessation of formation of Banded Iron Formations between 1.6 and 1.8 billion years ago (with one additional spike of local formation between 600-800 million years). The ozone blanket formed as oxygen responded to light energy, providing genetic protection for living systems. Ocean chemistry was changing, as was the whole physical nature of the oceans themselves. Proterozoic oceans on Earth were modestly oxic at the surface, but much less oxygen-filled and likely sulphic at depth (Anbar & Knoll, 2002). These were oceans intermediate between the oxygen-starved (anoxic) Archean seas and the well-oxygenated oceans of today, which have persisted for more than the last billion years. Levels of CO_2 and methane were likely decidedly higher and the Sun not so bright. Yet, despite this, the Earth began to cool towards the first major global glaciation.

Sometime after oxygen began to build up in quantity, the first eukaryotes (organisms with a defined nucleus) appeared.

Opposite page: The late Neoproterozoic was a time of intense cold, referred to by some as "Snowball Earth." This photo in the Transantarctic Mountains of Antarctica conjures up an image that could have been global in Ediacaran times (T. Rich).

Then followed the first multicelled organisms, including metazoans. Bacteria still dominated the warm, tropical seas and the deeper, less oxygenated ocean depths. During this time large continental masses emerged, producing the first fully terrestrial environments. Plate tectonics, certainly underway by this time, changed continental positions through time, opening and closing ocean basins, cutting off or altering the pathways of ocean currents, creating and destroying continents.

Figure 34. Glacial conditions, Transantarctic Mountains, Antarctica (T. H. Rich).

Biological activity and tectonic plate movement affected sedimentation patterns and rates in ocean basins. The building of mountain chains locked heavy metals such as manganese, iron, zinc, copper and chromium, as well as carbon stored in the buried organic remains, for lengthy periods. This in turn had an effect on climate. Wide continental shelves developed for the first time on Earth, providing the first shallow, sunlit seas that today nurture the greatest biodiversity in the oceanic realm. The oceans, for the first time, began to mix, with the nutrient-rich bottom waters welling up from below, injecting energy-rich nutrients into those shallow, clear seas, stimulating innovation. Such upwelling was fueled by water masses colliding with the newly formed high-standing continents.

It was during these dynamic times on Earth that metazoans appeared. Their world was one without eyes and ears, and these early, many-celled life forms depended entirely on touch and chemical signals to evaluate their surrounds, just as the jellyfish and worms do today. To us humans, with our emphasis on visual acuity, this seems bizarre, almost unimaginable, only truly appreciated by those who cannot see or hear.

The Continents and Ocean Basins. Rodinia or Not?

Just before and during the time that the first metazoans appeared in the fossil record, a supercontinent seems to have formed and remained stable from about 1000 to 750 million years ago – Rodinia. There are several different paleogeographic reconstructions proposed for Rodinia (Dalziel, 1997; Weil *et al.*, 1998; Dalziel *et al.*, 2000; Murphy *et al*, 2001 for an animated viewing; Condie, 2001; Cordani *et al.*, 2003; Pesonen *et al.*, 2003; Bowring & Schmitz, 2006), and these may change in the future with further studies that strive for more precision. At present, the scenario proposed by Condie (2003) seems the most plausible:

Rodinia
- 1300 and 950 mya (million years ago) Rodinia formed
- 850-600 mya Rodinia broke up
- 680-550 mya Pangaea-Gondwana formed

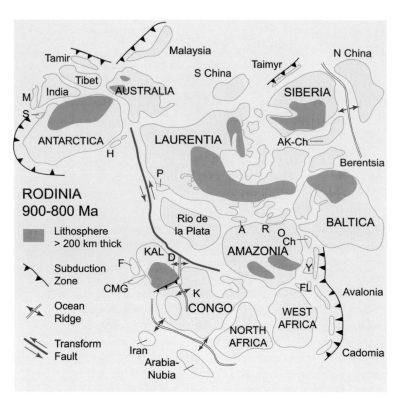

Figure 36. The supercontinent Rodinia 900-800 million years ago (modified from Condie, 2001) (D. Gelt).

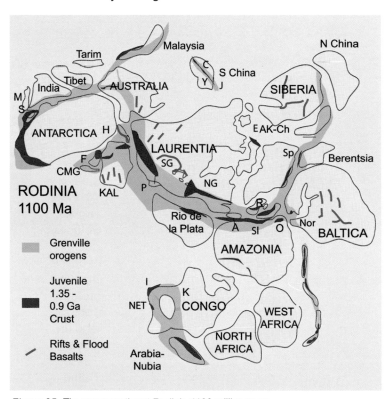

Figure 35. The supercontinent Rodinia 1100 million years ago (modified from Condie, 2001) (D. Gelt).

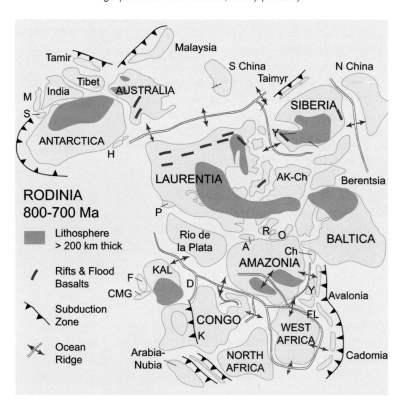

Figure 37 The supercontinent Rodinia 800-700 million years ago (modified from Condie, 2001) (D. Gelt).

The result of this megacontinent formation and breakup was the building of an enormous mountain range between around 650 and 515 million years ago, upwards of 8000 km in length – the Transgondwanan Mountain Chain. Over time, this range was eroded, which in turn gave rise to enormous volumes of quartz-rich sediments – the Gondwana Super-fan system (Squire *et al.*, 2006). Perhaps this mountain

chain, with its primarily east-west orientation, initiated cooling because of cloud formation, and thus enhanced the albedo of the Earth reflecting incoming solar energy. In part such a mountain system may have been responsible for some of the severe glacial events (but not the earliest at around 700 mya) in the Neoproterozoic.

As continents interacted due to plate tectonic motions, the mountain range was reoriented in the very latest Neoproterozoic and Early Cambrian. This may have initiated

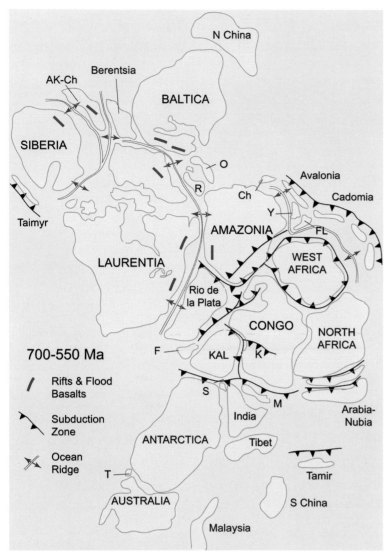

Figure 38. The supercontinent Rodinia 700-550 million years ago (modified from Condie, 2001) (D. Gelt).

Figure 39. The supercontinent Pannotia-Gondwana, 580 million years ago (modified from Condie, 2001) (D. Gelt).

a monsoonal system, triggering the final meltdown of glacial conditions, the Cambrian explosion of life, the acquisition of hard skeletons and deposition of carbonates that became widespread during the mid to Late Cambrian (Squire *et al.*, 2006). Between 544 and around 515 million years ago, seawater concentrations of calcium increased about threefold. This may have been linked to uplift and erosion of the Transgondwanan Supermountains and associated rise in hydrothermal activity along spreading ridges, itself altering seawater composition (Brennen *et al.*, 2004).

From Single Cell to Many: The Proterozoic Fossil Record

Eukaryotic cells (and cyanobacteria) had appeared sometime before 2.7 billion years, as indicated by hydrocarbon biomarkers preserved in late Archean rocks of Western Australia (Brocks *et al.*, 1999), at least a billion years after life first appeared on Earth. Eukaryotes are complex organisms with numerous genes, which are required for generating the cytoskeleton, compartmentalization, cell-cycle control, proteolysis (important in digestion), protein phosphorylation (important for preventing the loss of metabolites from the cell and in energy control) and RNA splicing (important in

reproduction) to develop, all the things that make a cell work.

How the eukaryotic cell originated is a debated topic. Symbiosis is one answer supported by many (Margulis, 1981). Nisbet and Fowler (1999) suggested that prokaryotes living in close association within or on bacterial mats, common in the oceans of early Earth, might have had a "tight physical association," which led to a permanent relationship. One example of such symbiosis is the heterotrophic proteobacteria. These are able to produce molecular H^2 by anaerobic fermentation with anaerobic Archaea, which depend on hydrogen (Martin & Muller, 1998). Another suggested evolutionary pathway to the eukaryotic cell is the association between anaerobic archaebacteria that developed the ability to survive in oxygenated environments and respiring proteobacteria (Vellai & Vida, 1999).

Maria Rivera and James Lake (2004) have proposed another idea, that of genome fusion and "horizontal" gene transfer, which could have led to the first eukaryotes. The fusion of genetic material from two diverse groups of prokaryotes may have led to more complex eukaryotes. What Rivera and Lake noted was that "informational genes (genes involved in transcription, translation and other related processes) are most closely related to archaeal genes, whereas operational genes (genes involved in cellular

million years ago (Porter & Knoll, 2000; Butterfield, 2004) and acritarchs (sphaeromorphic) as old as 1.7 billion years or more (Butterfield & Rainbird, 1998). Acritarchs with complex processes on their outer surfaces have been interpreted by some as fungi, as old as 1.4 billion years old (Butterfield, 2000). Andy Knoll (1994) suggests that diversity levels of certain eukaryotes, the protistans (single-celled eukaryotes and some of their immediate multicellular derivatives) increased significantly around a billion years ago. Molecular phylogenies (that is, relationship trees based on comparison of the gene sequences on DNA or RNA [ribosomal RNA] molecules of organisms) predict that this was about the right time for a broad radiation of "higher" eukaryotic groups (phyla) (Budin & Philippe, 1998). Data gained from the study of fossils and from genetics certainly support each other.

Throughout the Proterozoic, the fossil record clearly reflects increasing abundance and diversity of megascopic

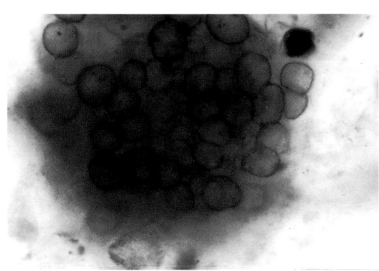

Figure 40. Microfossils preserved in the 820 million year old Bitter Springs Chert, Macdonnell Ranges, Northern Territory, Australia. Individual cells are 8-12 microns in diameter (courtesy of Jim Warren and Ian Stewart).

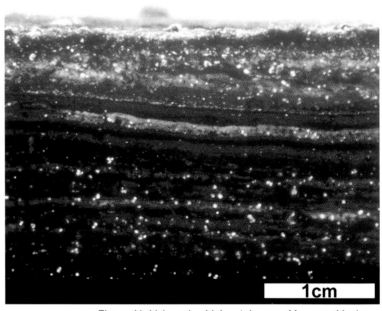

Figure 41. Living microbial mat, Laguna Momona, Mexico (courtesy of W. Schopf).

metabolic processes such as amino acid biosynthesis, cell envelope and lipid synthesis, and so on) are most closely related to eubacterial genes."

After the appearance of the first cellular structure preserved in the fossil record before 3.5 billion years, there was an increase in the size of cells preserved in progressively younger and younger rocks. The earliest known cells are very tiny, ranging from 0.005 to 0.06 mm across. The upper size limit for most living prokaryotic cells is about 0.06 mm (5-60 microns), and significantly, this size was exceeded about 1.9 million years ago (Schopf & Klein, 1992). The principle of larger size translating into a cell being eukaryotic instead of prokaryotic, however, is not clear-cut. There are certainly some examples of living, giant bacterial cells and also some very small eukaryotic cells. The lower size limit of eukaryotic cells is about 20µm (that is, 0.02 mm), so their real history may have begun before 2.7 billion years ago.

In rocks 1.7 to 1.0 billion years old, true eukaryotic microfossils are widespread and well preserved, but their diversity during this time is low, their evolutionary change slow.

Early eukaryotes which left significant fossil remains include rhodophytes (red algae) over 1.2 billion years ago, chlorophytes (green algae) in sediments approaching 1.0 billion years, brown algae at about 1 billion (Knoll, 1994), with ciliates and testate amoebae appearing around 750

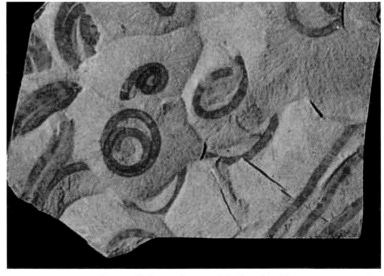

Figure 42. Grypania, *possibly the oldest known megascopic eukaryote – perhaps an alga from 1.8 billion year old rocks of Michigan, USA (courtesy of Sun Weiguo). Some researchers interpret it as a possible colony of cyanobacteria (Schopf, 1999). Block width 5 cm.*

fossils – remains of organisms that you can see with your naked eye, without the aid of a microscope. Most of these fossils consist of carbonaceous material and resemble "dried and compressed hot-dogs or raisins" (Hofmann, 1994). They certainly reflect the rise of multicellularity, which likely developed many times independently in different kingdoms and in different places. Buss (1987) suggested that multicellularity could have risen at least six times: in animals, in several algal groups and in fungi. A multicelled state could have been forced by advantages of larger body size and efficiencies of certain cells having specialized tasks in an environment that was changing, and not all that predictable – rising oxygen levels, impoverishment of many mineral nutrients that had previously been more abundant – a symbiosis that could ensure availability of "nutrients" otherwise rare in the environment. The waste products of one cell, or group of cells, might provide energy for another of differing construction.

The oldest megascopic carbonaceous fossils of uncertain nature appear in rocks around 1.8 billion years old. Some may be flattened remains of prokaryotic colonies or fragments of the bacterial mats, whereas some may indeed be true eukaryotes. The oldest reported megascopic eukaryotic alga is *Grypania* from the Early Proterozoic iron-bearing rocks of Michigan (Han & Runnegar, 1992). Originally, the age of these rocks hosting *Grypania* was estimated to be 2.1 billion years, but was recently revised to around 1.85 billion years (Porter & Knoll, 2000). Schopf (1999) has suggested another interpretation – that it represents a colonial cyanobacteria, in the fashion of the "living" *Nostoc*. Only details of the cellular construction and possibly biomarkers may lead to a better understanding of these fossils.

Fossils of cellular structures are extremely rare. Morphologically more complex carbonaceous fossils with some anatomical details preserved appear in rocks of about 1.4 billion years, but more complex and diverse forms first appear only about 800 to 900 million years ago. Most of these seem to be eukaryotic algae, but a few may be some sort of colonial prokaryotes, fungi or even metazoans.

Figure 44. Outcrops of Appekunny Formation, Belt Series in Glacier National Park, Montana (courtesy of E. Yochelson and the National Geographic Society).

An unexpectedly complex life form was discovered in quite ancient rocks of Middle Proterozoic age, 1.5 billion years old. The enigmatic bedding-plane impressions, resembling a necklace or string of beads, were initially described from the Belt Supergroup of Montana, USA (Horodyski, 1982) and later from the Manganese Group, Bangemall Supergroup, Western

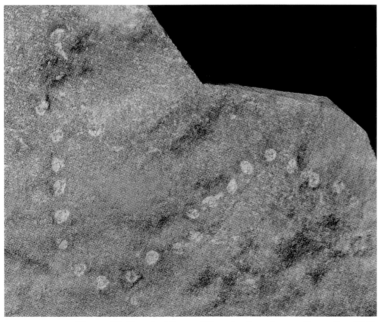

Figure 43. Horodyskia moniliformis, Appekunny Formation, Belt Series, Glacier National Park, Montana. Each bead is about 2 mm in diameter (courtesy of E. Yochelson and the National Geographic Society). and the National Geographic Society).

Figure 45. Proterozoic rock outcrop in the Bangemall Basin of Western Australia with J. Williams for scale (M. Fedonkin).

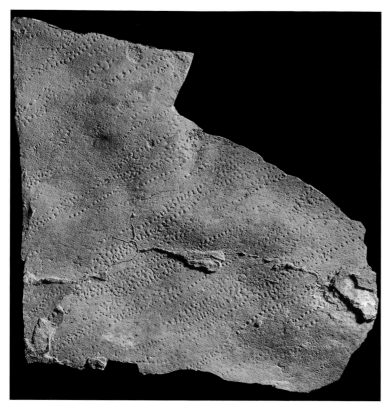

Figure 46. Horodyskia from the Bangemall Basin, Western Australia (S. Morton). Slab about 35 cm wide.

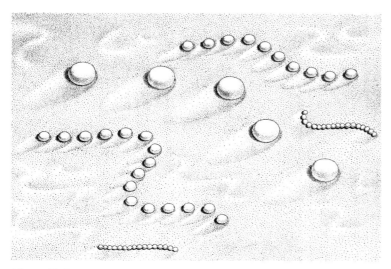

Figure 47. Reconstruction of several Horodyskia "strings of beads" as living organisms from the Appekunny Formation, Belt Series, Glacier National Park, Montana (courtesy of E. Yochelson and the National Geographic Society).

Australia (Grey & Williams, 1990). These megascopic fossils, some named *Horodyskia moniliformis* (Yochelson & Fedonkin, 2000), are both abundant and geographically widespread, but they occur in only a narrow time band, at usually only one stratigraphic level.

Horodyski was unsure about the nature of the beads, whereas Grey and Williams (1990) interpreted them as seaweed stands with float bladders. Others have suggested different origins: non-biogenic or perhaps even symbiotic Archaea and Bacteria. However, detailed studies of the taphonomy (the study of the transition of all or part of an organism and its traces from the biosphere to the rock record or lithosphere), paleoecology and overall morphology of

these fossils suggest that *Horodyskia* was a colonial, bottom-dwelling organism of tissue-grade organization, perhaps a metazoan (Fedonkin & Yochelson, 2002). Bead-like structures appear to have grown upward to a maximum of 1.0 cm, and these were evidently connected by a tube that may have allowed communication between individuals in the colony. The maximum length of the colonies was 30 cm. The size of beads relative to size of spacing remains the same for each colony. As the colony grew, the spacing increased in a regular pattern. The complex mode of growth of the colony, regular arrangement of the individual organisms, the absence of branching forms, the rigid nature of the cone wall and the strong stolon that even survived some transportation of the colony by water currents, are all features that suggest this organism had proper tissues. And if so, it means that there

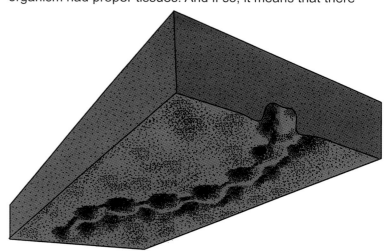

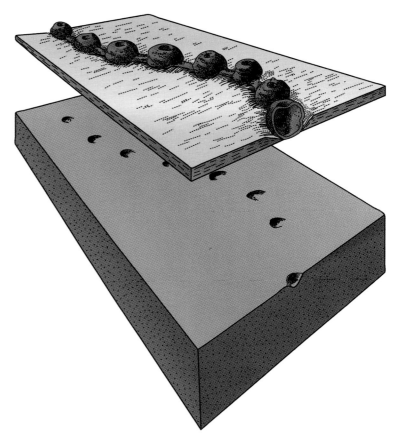

Figure 48. Reconstruction of Horodyskia fossils from the Bangemall Basin of Western Australia (courtesy of K. Grey).

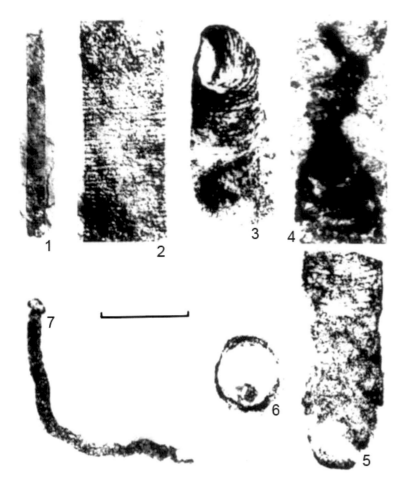

Figure 49. Tubular fossils from the early Neoproterozoic Jiuliqiao Formation, Anhui Province of China. 1-2, Sinosabellites huainanensis; 3-6, Pararenicola; and 7, Protoarenicola. Scale bar 5 mm (courtesy of Sun Weiguo).

What is a pogonophoran? Not your average worm!
Pogonophorans are called "beard worms" because they
have a set of tentacles that project from their front end – certainly resembling a beard. They are generally only a few centimeters long (but with a range of from 10 to 90 centimeters) and live attached to the sea bottom in fairly deep waters, generally more than 100 meters down. They live in vertical, chitinous tubes that can be from 0.1 to 2 mm wide and are open at both ends. Bodies of pogonophorans are divided into three parts, a short front section, a long trunk region and a short rear end. In the front there is a head, and beneath that are the tentacles, which they extend to collect organic detritus and plankton, then move by minute waving cilia into the head region. Nutrients can also be directly absorbed into the worm's body through the cilia. This group also harbors sulfide-loving bacteria that provide a further source of nutrition and energy.

Pogonophorans have no mouth, no gut and no anus, so absorption of nutrients seems to be their method of nutrition, but this appears to be a secondary adaptation (Rouse & Pleijel, 2001). They do have paired excretory ducts that open to the outside through one or more pores, so once the nutrients are in there is a specific way out for wastes! Pogonophorans do have a coelom, an internal body cavity, which extends not only into the main body of the worm but also into the tentacles, and there is a closed blood system – two vessels each per tentacle in the beard. There are male and female pogonophorans, with males releasing the sperm into the sea and fertilization taking place inside the maternal tube, where the larvae develop until ready to settle and metamorphose into sessile, attached adults where they spend the rest of their lives – in one place.

Pogonophorans have only been known since the early 20th century, and now there are more than 100 different species recognized. Many of these have been collected by the Russians in their global oceanographic expeditions, especially in waters of the western Pacific, but this group seems to have a worldwide distribution in cold, marine seas – deeper in the warmer oceans but present in shallower depths in the cold Arctic Ocean. Interestingly, some forms allied with the pogonophorans are known to prosper around deep-sea vents, along the mid-ocean ridges where temperatures are high and the water is saturated with sulfides. In these conditions they have a special association with sulfide-oxidizing chemoautotrophic bacteria from which they derive nutrition. These special "worms" have been placed by some (see Margulis & Schwartz, 1988) in their own separate phylum, Vestimentifera, as they do show significant structural differences from the pogonophorans, despite many similarities.

were well-established regulatory genes in action at this time. *Horodyskia* marks the first occurrence of complex, integrated multicellularity and such organisms, not simply a collection of independently functioning cells, became an essential unit in the evolutionary process. If the interpretation of *Horodyskia* as an early metazoan is correct, it almost doubles the time span of the Animalia. However, there is not total agreement among Precambrian paleobiologists on the interpretation of this organism, and better preserved material is needed to resolve the true identities of this early complex form.

Additional associated problematic fossils in Montana and Western Australia further suggest a greater diversity of megascopic organisms living 1.5 billion years ago in addition to *Horodyskia*. An important peculiarity of these metazoans/metaphytes is that they are preserved primarily in siliciclastic rocks in the same manner as the globally occurring Ediacara fauna in younger sediments (600-545 million years). These sediments were deposited under colder conditions, not in warmer environments where carbonates dominated.

Recently reported problematic discoidal and trace-like fossils occur in tidal sandstones of @1.2-1.8 billion years age from southwestern Australia (Rasmussen *et al.*, 2002). But they are not convincingly metazoan. More likely metazoan are the macroscopic, worm-like organisms discovered in the early Neoproterozoic deposits in Huainan District, Anhui Province, northern China (Wang, 1982). The age of these fossils is @740 million years. The Huainan fossils are carbonaceous, thin dark filaments, ranging from 15 to 20 mm in length and

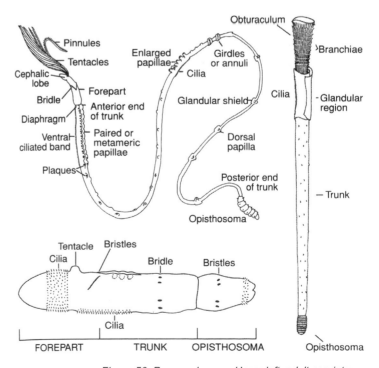

Figure 50. Pogonophorans. Upper left, adult perviate; lower left, larva of a perviate; right, generalized adult vestimentiferan (modified from Valentine, 2004).

show a very fine segmentation. Named *Protoarenicola* and *Pararenicola*, these fossils were interpreted to be annelid worms and pogonophorans. In addition to the general worm-like form, size and type of segmentation, they show a broad, circular aperture with a proboscis-like structure on one end of their slender body, presumably the anterior end, with the opposite (posterior) end rounded.

Sun Weiguo and co-authors (1986) reported even older worm-like fossils of *Sinosabellidites* in 800-850 million year old rocks from the same region of China. These had very fine and regular transverse striations but lacked any other features of complexity. Both ends of the band-like fossils were rounded, and were it not for the fine striations, these fossils closely resemble *Tawuia* and *Chuaria* – compressed, sausage-like fossils, interpreted as multicellular eukaryotic algae, Metaphyta or perhaps even colonies of prokaryotes.

Gnilovskaya (1998) published a description of the new genus *Parmia*, increasing the diversity of worm-like forms inhabiting the ocean during the Late Riphean, significantly older than the Ediacara faunas. More than 80 specimens of *Parmia* were recovered from a borehole core drilled in southern Timan, on the northeastern Russian Platform. With a maximum length up to 60 mm and a body width up to 2.5 mm, these organisms have regular segmentation (7-11 segments per millimeter of body length). *Parmia* is interpreted as a probable predecessor of annelid worms. All geological and paleontological evidence

Figure 52. Marine glacial conditions on the Davis Ice Shelf, Antarctica (N. Schroeder).

(including the associated microfossil assemblage) indicates *Parmia* lived about 1 billion years ago (Gnilovskaya *et al.*, 2000). Although the exact age of *Sinosabellidites* and *Parmia* is yet to be determined with certainty, these metazoan fossils are without doubt decidedly older than the Ediacara-type metazoans known from the late Neoproterozoic.

Thus, there is certainly evidence of complex multicellular eukaryotes older than 1 billion years living in relatively cool seas. And some of these multicellular eukaryotes have been interpreted as probable metazoans, the first evidence of the Animalia on Earth.

A Cooling Biosphere. Snowball Earth, Fact or Fiction?

Earth has known ice ages through much of its history from Archean times to the present-day glacial conditions that we are experiencing. But the glacial events for the past 540 million years, during the Phanerozoic Eon, have been fairly restricted in extent, centering around the North and South Poles and only occasionally extending "fingers" onto Equator-ward continents. It is also clear that ice ages are unusual events on Earth. For most of Earth history, the planet has been ice-free.

During the Proterozoic, however, there were "ice ages of all ice ages," including several periods during which some researchers (Hoffman *et al.*, 1998) suggest that the Earth almost or completely froze over – "Snowball Earth"! This term was coined in 1992 by Joe Kirschvink, who was the first to suggest truly global glaciation, based on the paleogeographic positioning of continents bearing Neoproterozoic glacial deposits. Later authors have been inclined to support a less extreme model "Slushball Earth" – where not the entire Earth froze over. Evidence for these glacial events is found in the sediments called diamictites, tillites, dropstones and polished rock surfaces that all suggest thick sheets of ice smoothed surfaces and left behind the cargo transported for some distance. Varves in lake sediments reflect annual freeze and thaw regimes. Not all researchers accept this theory (see Grey, 2005). However, Snowball Earth has become a popular interpretation of events in the Neoproterozoic. Hoffmann and his colleagues suggest that at these times glaciers and frozen oceans reached the Equator, such as in Australia, lasting for tens of millions of years. During these times bioproduction in

Figure 51. *Parmia, Late Riphean (about 1 billion years old) from the siliciclastic deposits, southern Timan, northeast of the Russian Platform, about 1 mm wide (courtesy of M. B. Gnilovskaya and M. A. Fedonkin, in the collections of the Institute of Geology and Geochronology of the Precambrian, Russian Academy of Sciences, Sankt Petersburg).*

Definitions. Sedimentary Evidence for Ancient Ice Ages

Abrasion Surface or Polished Pavement - the erosion or polishing of rocks over which a moving glacier passed. Often the surface is polished to a shiny finish, but on that shiny surface may be deep grooves or scratches made by rocks carried along in the glacial ice at the base of the glaciers.

Cap Carbonates - regionally persistent, thin layers of carbonate rock directly overlying glacial deposits. Abrupt lower and upper boundaries. The cause of these is debated but many geologists think they are related to the cessation of a glaciation with the influx of oxygen into an otherwise anoxic environment, resulting in the deposition of carbonate rocks.

Diamictite - a rather chaotic collection of rock fragments lacking much sorting that can be the result of glaciers dropping the load of material picked up as they scour over the countryside – and then as they melt they deposit the rocky materials held in the ice. Sometimes these sorts of deposits can also be formed by downslope movement like landslides or by flooded rivers, so geologists must be careful in making interpretations.

Dropstones - rock fragments dropped from melting icebergs onto the sea floor, where deposits are very different in both texture and nature from the dropstones. Often the dropstones disrupt and deform the deposits into which they drop.

Tillite - a rock body formed by lithification (solidification) of unsorted and unstratified glacial drift deposited directly by glacial ice on the land surface.

Varves - fine, laminated sediments consisting of paired, thin beds, one coarse, one fine. Every couplet represents a year's cycle or interval of thaw of ice followed by an interval when the water body is frozen over, such as one sees in glacial lakes.

Figure 53. The Nuccaleena Formation cap carbonate (Jim Gehling has his foot resting on it) that overlies the Elatina Formation glacial tillite at the Ediacaran GSSP site in the Flinders Ranges of South Australia. Such rocks appear to represent significant carbonate deposition as the Earth warmed rapidly after the "Snowball Earth" event (J. Gehling).

Figure 54. The Elatina Formation tillite (diamictite) at the GSSP site in the Flinders Ranges, South Australia, an indicator of glacial conditions (P. Vickers-Rich).

the oceans may have slowed considerably, photosynthesis almost ceasing, and hydrologic and weathering styles would have been severely affected, even shut down in many places (Bekker *et al.*, 2001). Oceans would have become quickly anoxic and dominated by hydrothermal input. Only when volcanic activity on the continental areas pumped gases into the atmosphere, significantly raising CO_2 levels to as much as 350 times modern levels, did the Earth escape the "Icehouse" conditions that gripped it (Hoffman *et al.*, 1998; Kaufman & Xiao, 2003). Then climate rapidly deteriorated into severe Greenhouse, with immediate precipitation of calcium carbonate from the warming seas to form distinct layers (cap carbonates) formed on top of glacial sediments.

Andrey Bekker and his colleagues (2001) have suggested that the last of these great glacial events in the late Neoproterozoic (with dates spanning the period from 580 [or perhaps even younger] to 740 million years ago) might be related to the breakup of the supercontinent Rodinia, as outlined above. The severe glaciation produced a massive "graveyard" repository of organic carbon – the very basis of life on Earth. As animals and plants died, their remains fell to the bottom of the shallow seas surrounding the continental masses formed as the supercontinent fragmented. These sediments with high organic carbon content were buried along the expanded continental margins, preventing oxidation and recycling of carbon and other elements, so lowering CO_2 in the atmosphere. This, in turn, led to global refrigeration (Kaufman, Knoll & Narbonne, 1997).

Because these glacial events tied up water in ice cover, sea level dropped perhaps as much as 250-300 meters and strongly fluctuated over the few million years of each glacial period. Most of the continental shelf would have been exposed to the atmosphere (or covered with ice), and the area of shallow-water habitats significantly reduced to a small strip along the edge of continental shelves. Life may have been restricted to these narrow shelf strips or to deeper

Figure 55. Close up of the Elatina Formation, GSSP site, Flinders Ranges, South Australia (P. Vickers-Rich).

Figure 56. Dropstone emplaced when sediment load carried by a glacier entered a marine environment, forming icebergs which transported this sediment load. As the iceberg melted, it dropped its sediment load, often introducing such "passengers" as this stone into an otherwise fine-grained sedimentary package. Neoproterozoic of southern Namibia (M. Fedonkin).

Figure 57. Portage Glacier near Anchorage, Alaska calving icebergs into an inlet of the sea, an environment where dropstones are deposited in fine sediments when the icebergs melt, dropping their sediment load (P. Vickers-Rich).

seas associated with hydrothermal vents. Along with the cold climate, vast temperature gradients, changes in atmospheric circulation and an increasing storm frequency would have been the norm. All of this would have led to a radical shrinkage of suitable benthic environments and severely affected life there and pelagic life as well. Not all researchers agree with this scenario, but discussion will continue as more data is collected and more precise dates are determined for individual physical and biologic events.

During these glacial events, geographic isolation of species would have increased because of the growing expanse of terrestrial environments, the isolation of ocean basins from one another and climatic zonation. The low input nutrients into the oceans from the frozen continents lacking

liquid water may have been compensated partly by the input of metabolites from the deeper ocean zone – heavy, cold water was sinking, forcing bottom water rich in dissolved nutrients to the surface. Surface winds, too, might have had some effect on oceanic circulation. The same mechanism caused the "ventilation" of the ocean: sinking waters brought oxygen into deeper parts of the ocean.

At the end of most of the glacial periods that affected the seas of this time, rather dramatic events occurred. Cap carbonates formed due to rather abrupt changes in ocean chemistry, perhaps with massive influx of oxygen. And it appears that rapid melting of nearly global ice sheets with a concomitant rapid rise in sea level occurred over a very short period, perhaps in as little as 2000 years (Allen & Hoffman, 2005). Structures that occur in rocks of two of these late Precambrian glacial sequences, called tepee structures, hint at high wind velocities exceeding 20 meters per second (72 km per hour) during the early stages of deglaciation for sustained periods, with dates of around 710 and 635 million years ago (Allen & Hoffman, 2005) – cyclone-like weather persistent over perhaps 100's or 1000's of years. The winds, of course, would have stirred massive waves, which in turn would have had effects on near surface water communities, particularly along shorelines.

Bernd Bodiselitsch and colleagues (2005) noted one unusual marker in these post-glacial sediments flagging that some of these glacial events could have had a significant duration – the Neoproterozoic glaciation with a U-Pb date of 635.5 +/- 1.2 recorded on the Congo Craton of Africa may have lasted from 3 to 12 million years. Their evidence is a layer of iridium at the base of a cap carbonate. They suggest that the iridium was extraterrestrial in derivation and accumulated over time from meteorite impacts on the ice cover during the glaciation. This concentration of iridium was then dumped en masse on the sea floor as the ice melted and cap carbonates formed as the world moved into Greenhouse conditions. When based on the rate of modern-day iridium concentrations, others have suggested that the iridium may have had a volcanic source and thus little time significance.

Hoffman and his colleagues (Hoffman et al., 1998; Hoffman & Schrag, 2000, 2002) claimed that such dramatically changing conditions had profound effects on life. The most negative effect of these glaciations would have been the destruction of the shallow-water habitats of the benthos on the continental shelves. In recent oceans, it is the shallow, well-lit continental shelves that normally support over 80% of the benthic biomass. In addition to the benthic ecological disaster, a large part of the phytoplankton community was shut off from sunlight by oceanic ice cover, and thus eliminated. In most cases, organisms living around hot vents in the deep ocean may have been the only survivors of these glacial periods. But the low oxygen content (if not anoxia) of the deep benthic realm may not have allowed the metazoans to recolonize these environments until the very end of the Ediacaran, or even later (Crimes & Fedonkin, 1994).

Increasing Oxygen, Saline Giants, Cold Cradles and Global Playgrounds

The Neoproterozoic glacial periods, united under the name Cryogenian, began about 750 million years ago and appear to have significantly lowered global temperatures.

These were not the first of the Precambrian glacials, which date back into the Archean. But these younger glacials and global refrigeration may have given significant advantages to eukaryotic organisms, which could have developed before in isolated cold areas near the poles when the Earth was warmer (Fedonkin's [1996] "cold cradles"). Another possibility, or even a "parallel-possibility," for such "cradles" of metazoans could have been in fresh waters or in less saline environments, the interface between fresh waters and the "salty," open marine conditions, which could have had twice the concentration of salt of modern seas (Kanuth, 2005). Both cold and less saline estuarine or fresh waters would have had another advantage, with their higher oxygen supply (Kanuth, 2005; Vickers-Rich, 2006, 2007).

Between 610 and 605 million years ago massive accumulations of salts were laid down globally, preserved today in Australia, Oman, Saudi Arabia, Iran and Pakistan (Kanuth, 2005). Before this rather dramatic removal of NaCl from the oceans, salinity could have been as high as 1.6 to 2 times what it is today – a problem for most living metazoans, but certainly not an issue for cyanobacteria (Kanuth, 2005). Increased salinity also reduces oxygen solubility, and so with decreased salinity, oxygen levels would have risen in the global oceans. Further enhancing oxygenation may have been a marked change in weathering styles, highlighted by Kennedy and colleagues (2006). With a greater input of weathered clays, enhanced perhaps by the development and expansion of an early terrestrial biota, greater volumes of carbon would have been sequestered, bumping up free oxygen levels in the environment. This terrestrial input could also have been increased by tectonic activity at this time that generated a lengthy mountain range, forcing the burial of organic carbon that occurred in the late Neoproterozoic (Des Marais et al., 1992; Squire et al., 2006).

Then, with Earth-wide refrigeration, which lowered ocean salinity and increased oxygen content (both because of the cold and the lower salinity), the isolated "cradles" could have provided the sources of metazoan biodiversity. Metazoans could have expanded into more inviting, less saline, well-oxygenated environments – the "cold playgrounds" of the global oceans (Vickers-Rich, 2006, 2007) and new ecospace that provided the chance to construct the new niches that Douglas Erwin (2005) termed the inventions that led to innovation – of body plans and diversity. This could explain why the Ediacaran fauna appeared rather abruptly and long after genetic studies predict that metazoan body plans should have arisen (Blair & Hedges, 2004). Condon and Bowring (2005) also note that there was a major perturbation in the global carbon cycle (the Shuram anomaly) at around 560 to 551 million years, unrelated to glaciation, but "consistent with progressive oxidation and re-mineralization of the organic carbon reservoir." They note that this coincides with the appearance of such macro-Ediacarans as *Kimberella*.

These early macro-metazoans are preserved in the siliciclastics laid down in marine waters, reflected by the sediments in the White Sea region of Russia, the Flinders Ranges of South Australia and the Avalon Peninsula of Newfoundland, suggestive of cool waters. Even in the carbonate-rich, siliciclastic deposits of southern Namibia, Ediacara-type macrofossils are preserved in the quartzites that may have been deposited during relatively cool episodes. In contrast, in the carbonates of Namibia, *Cloudina*,

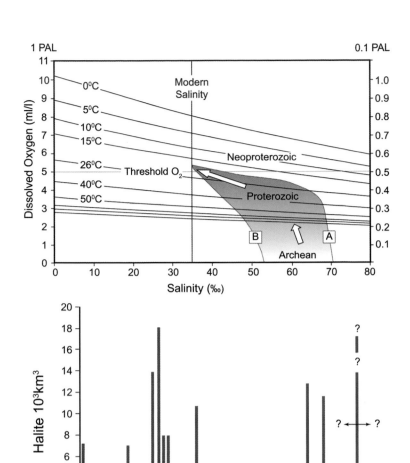

Figure 58. The amount of oxygen in seawater is related to both salinity and temperature. Decreasing salinity of the oceans in the late Proterozoic combined with a number of glacial events could well have been the key to the global spread of the Ediacara biota, which may have existed in isolated enclaves before their record became global (P. Vickers-Rich, 2006, 2007) (modified from Kanuth, 2005) (D. Gelt).

Figure 59. Cloudina, a tubular fossil, visible to the naked eye, formed massive reefs in the late Neoproterozic globally. They built their shells with carbonate. Slab about 12.5 cm wide (M. A. Fedonkin and P. Vickers-Rich, courtesy of the Geological Survey of Namibia).

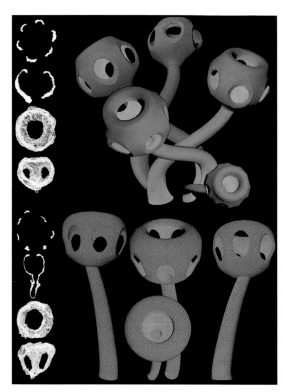

Figure 60.
Reconstruction of
Namacalathus from
the Nama Group of
Namibia, Pinacle
Reef Complex (from
Grotzinger et al.,
2000). Diameter
of bulbous "head"
about 2 cm.

Figure 61. Namacalathus hermanastes, Nama Group,
Namibia (R. Jenkins). Each spear about 2 cm in diameter.

Namacalathus and *Namapoika* are the first metazoans to
form reefs, especially *Cloudina* with its weakly calcified
shells, documented in the Nama Group of southern Africa
and elsewhere. *Cloudina* lived at a time when the oceans
may have had salinity similar to those of today because of the
formation of the rare "saline giants" around 540-550 million
years ago, which formed major salt deposits globally. Thus,
low salinity combined with the rapid degradation of a lengthy
mountain chain spanning much of Rodinia (Squire *et al.*, 2006)
may have supersaturated the global oceans with both oxygen
and the building blocks of hard parts (Brasier & Lindsay, 2001).
Higher temperatures, too, would have favored the deposition of
calcite. So, both ocean chemistry and temperature may have
played a role in the origin and expansion of the metazoans and
the first development of shells in the "ocean playgrounds of
summer" (Kanuth, 2005; Vickers-Rich, 2006, 2007).

Life, as it became more complex, may have had a
significant effect on the glacial styles of Earth. As outlined
before, during the time before 540 million years ago, ice
reached the Equator (Evans, 2003). This has not been
the Phanerozoic style, for when there were several global
glaciations, these were confined to regions near the poles.
Certainly none have sent great ice sheets to the Equator.
David Evans and his colleagues (2003) have suggested that
this may be due to biofeedback, where life acts as a check on
global climate – "a net regulatory mechanism on global climate
imposed by an increasingly complex biosphere."

The details of the process of growing complexity,
especially the development of metazoans is reflected by
the distribution of the oldest animal assemblages on Earth,
covered in detail in later chapters of this book. These
first diverse metazoan faunas are preserved in siliciclastic
sediments – sandstone, mudstone and clay, which have
accumulated in cold ocean basins of the Ediacaran Period,
from around 610 to 554 million years ago. Precambrian
carbonates yield few metazoan fossils. Ironically, today's
carbonates – limestones and dolomites that are precipitated
in warm, tropical waters of the world – host the greatest
biodiversity of metazoans – not polar oceans.

Though there is certainly evidence between major
glaciations of warm-water carbonate deposition, of stromatolite
reefs, oolitic dolomites and even evaporites, there are no
biodiverse metazoan assemblages of Ediacaran composition
known from these. Could it be that metazoans were spawned
and diversified in the colder waters away from the carbonate-
rich "tropics." Could it also be that during one or more of these
global freezes that a variety of metazoan body plans appeared
where the climate was most severe, the "weedy" environments
that even today foster innovation?

How does one explain such a paradox – cold waters
of the past providing the cradle for metazoans, but today
warmth nurturing their greatest biodiversity? It is likely linked
to availability of oxygen, for oxygen content in cold water
is significantly higher than in warm water. Perhaps higher
levels of O_2 enhanced the development of complexity and the
beginning of Animalia. Oxygen is a necessity of metazoan
metabolism and does not favor many bacterial communities,
which dominated Earth in the Archean and early Proterozoic.
The idea that increased oxygen levels on a world scale could
have precipitated multicellular diversification has certainly
been suggested before (*e.g.* Knoll, 1996). It would likely have
been quite important in the synthesis of collagen, a building
block of metazoan tissues, such as muscle (Budd & Jensen,
2000). But were there other reasons for metazoans to prefer
the cold? Perhaps lower temperatures excluded certain life
forms, such as some groups of bacteria, which themselves
were dependent on warmth. It is in these warmer basins of the
Proterozoic, preserved in ancient dolomites and limestones,
that bacteria retained dominance, but in the colder waters
with lowered bacterial and archaebacterial biodiversity,
there may have been unused ecospace for a new and naïve
organism (Fedonkin, 1996) and the possibility for bilaterian
animal diversification (Budd & Jensen, 2000). Perhaps only
in those places where bacteria were "uncomfortable," and not
able to survive or prosper because of their cold-temperature
intolerance, were the metazoan new kids on the block able
to take a foothold and evolve, around 600 million years ago.
The world's cold, and likely less saline, oceans became global

"playgrounds." (Vickers-Rich 2006, 2007). In these cooler oceans, the first metazoans had an advantage – they were consumers (heterotrophs) – thus gaining their energy primarily by eating other organisms.

The First Animals. Genes and Fossils

Just where the first animals came from is a question puzzled over by humanity from its beginnings. Charles Darwin himself was stumped by the question of animal origins, as he so clearly spelled out in his *Origin of Species*, Chapter 9 (1859): "To question why we do not find records of these vast primordial periods, I can give no satisfactory answer." Several of the most eminent geologists, with Sir R. Murchison at their head [Murchison was one of the early geologists who set up major parts of the geological time scale, such as the Permian in Russia and the Silurian in England] are convinced that we see in the organic remains of the lowest Silurian stratum the dawn of life on this planet. Other highly competent judges, such as Charles Lyell, disputed this conclusion. Darwin further commented: "The case [the lack of a Precambrian fossil record] at present must remain inexplicable; and may be truly urged as a valid argument against the views here entertained [the origin of species by natural selection]." Darwin does give one reason for this in noting: "We should not forget that only a small portion of the world is known with accuracy." Finds showcased in this book highlight the discoveries since Darwin that have helped solve his dilemma.

Darwin's problem was that in 1859, and for a significant time after that, Animalia seemed to have sprung full blown into the Cambrian record, some 542 million years ago, without many precursors. With time, fossils were found in rocks older than the Cambrian, as is documented in this and later chapters, but their exact connections, if any, with Cambrian and younger animals is yet hotly debated and will remain so for some time to come. The fossil record is sometimes difficult to read, especially when the writing on those pages is difficult to interpret. The first Animalia, which have left their traces, have no skeletons, for the most part no hard parts, to clarify the details of their complexity. Their original nature, too, has been obscured by the manner in which they are preserved. Precambrian animals have left traces, "death masks" and impressions, which yet challenge the minds of paleontologists and zoologists alike.

Studies of the detailed sequence of genes on the DNA molecule over the last few decades and development of "molecular clock" techniques (see below) have provided insights

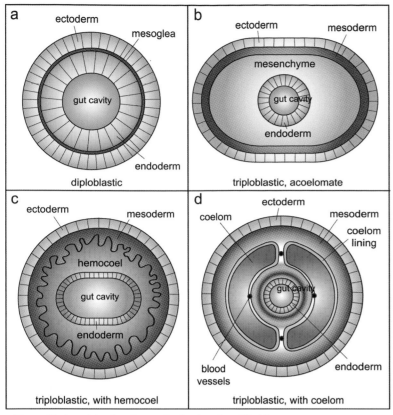

a — ectoderm, mesoglea, gut cavity, endoderm — diploblastic

b — ectoderm, mesoderm, mesenchyme, gut cavity, endoderm — triploblastic, acoelomate

c — ectoderm, mesoderm, hemocoel, gut cavity, endoderm — triploblastic, with hemocoel

d — ectoderm, coelom, mesoderm, coelom lining, gut cavity, blood vessels, endoderm — triploblastic, with coelom

Figure 62. A variety of metazoan body plans (modified from Erwin et al., 1997; courtesy of The artist D. W. Miller and American Scientist) (D. Gelt).

What is a Metazoan?
Metazoans are usually called "animals." More specifically metazoans (Animalia) are:
Multicellular – more than one cell
Mobile – ability to move at some stage in the life cycle
Heterotrophic – a consumer, cannot produce its own food or form a symbiotic relationship whereas an autotroph can produce energy by using sunlight or chemical energy
Diploid – has two sets of chromosomes in the nucleus
Diploblastic or Triploblastic – possesses two or three embryonic layers that develop in the ball of cells that form as the fertilized egg begins to multiply into many cells in embryonic development. This hollow, fluid-filled ball of cells (the blastula) can either have two embryonic layers (diploblastic – with ectoderm and endoderm only) or three layers (triploblastic – with ectoderm, mesoderm and endoderm). These embryonic germ layers develop into various adult tissues – such as the mesoderm into the heart and other muscles, the ectoderm into skin and the nervous system.
Most animals are now thought to be basically triploblastic (Raven & Johnson, 1986) except two that appear to be diploblastic – the Cnidaria (corals and medusae) and the Ctenophora (combjellies). There are, of course, even simpler multicellular metazoans, such as the sponges.

into the relationships of the Animalia and the timing of the appearance of different phyla of the Metazoa. Together with more traditional approaches, such as comparative anatomy of fossil and recent animals and embryology of living forms, a much better understanding of the family tree of animals is current.

One of the earliest predictions made by paleontologists concerning the timing of animal origins was that of J. Wyatt Durham, who extrapolated observed evolutionary rates of major Phanerozoic invertebrate groups back into the Precambrian. His conclusion was that the common ancestor of the deuterostomes should have differentiated between 800 million and 1.7 billion years ago (Durham, 1978).

Durham's model was challenged by J. John Sepkoski (1948-1999), best known for his statistical models of diversity dynamics and mass extinction periodicity. Sepkoski's studies on patterns in the history of life were based on his analysis of fossils collected by generations of paleontologists, a massive database (Sepkoski, 1978, 1979, 1981, 1984, 1986, 1989, 1992; Sepkoski et al., 1981). Tallying of the numbers of animal genera (in the name Homo sapiens, Homo is the genus, sapiens is the species) from the Ediacaran to the Cambrian clearly demonstrated a slow increase in exponential growth of diversity through the Ediacaran with an increasing rate once into the Cambrian Period.

Retrospective extrapolation of this diversification of the metazoans, especially during the Cambrian, brought Sepkoski to the conclusion that the roots of metazoan diversity must be very close to the beginning of the Ediacaran. But, analysis of the fossil record has some traps. Exponential growth of biodiversity in the Cambrian is most likely an artifact of the massive skeletonization that took place about 542 million years ago, almost simultaneously in many groups of the metazoans. As a result, these hard parts, internal and external skeletons, were preservable, and they made

the real biodiversity more visible in the fossil record. Ediacaran animals were primarily soft-bodied, not easily preservable, but despite this, though the diversity of Ediacaran metazoans is low, the range of body plans is broad and varied, certainly comparable to what we see in the early Cambrian. In the Ediacaran faunas shapes and complexity range from sponge-like to the joint-limbed arthropods, coral-like forms to snails.

Yet another attempt to date the origin of the first animals was made by James W. Valentine. Valentine, a paleobiologist, known for his detailed analyses of animal evolution (Valentine, 2004), adopted an earlier idea of J. T. Bonner (1988). Bonner, himself a prominent biologist and author of many thought-provoking books, realized that the complexity of an organism is directly correlated with the number of different cell types it possesses – the greater the number, the more complex the organism. Valentine extrapolated from a modern tally of 210 cell types (observed in recent mammals) backwards to the average of the 30 and 50 cell types that should have existed, respectively, in the animals which produced the oldest horizontal trails in the Ediacaran and the higher invertebrates, ancestors of today's snails and starfish.

Horizontal trails have a meaning beyond just marks on the sea bottom. Those who made the trails in the Ediacaran were not burrowers, but horizontal surface movers, which had "moderately large bodies capable of prolonged creeping." "These organisms were too large for purely ciliary locomotion, and at any rate created grooves in the sediments, implying locomotion involving body-wall musculature antagonized by a hydrostatic skeleton" (Valentine, 1992). Thus, by the complexity of organization, the least being the ability to develop specialized sorts of tissues, Valentine made a minimum estimate of the number of cell types necessary to support a certain kind of complexity, and put a date on it. From that and other information on the kinds of organisms that were around at a particular time, and using modern counterparts where he knew the number of cell types, he was able to extrapolate and obtained an estimate for the origin of metazoans of around 680 million years. If new data that has been collected since he made this estimate in 1992 were incorporated into his analysis, this date would likely be pushed even farther back in time.

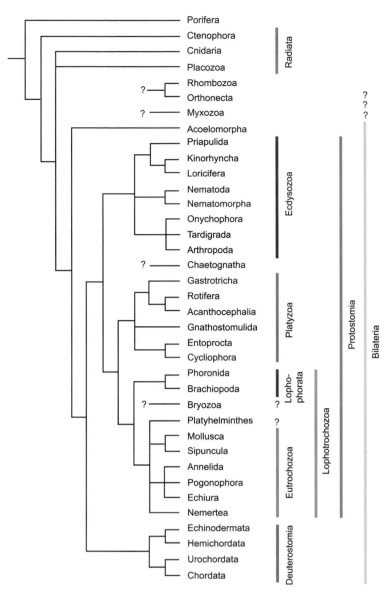

Figure 63. One hypothesis about the relationships of metazoan phyla, using information gleaned from molecular, developmental and morphological studies. Modified from Valentine, 2004 (D. Gelt).

Another indicator of when animals capable of burrowing might have evolved is bioturbation, the disruption of bedding and sediment laminations by organisms that churn through sediments in search of food or protection from predators or environmental disasters. It appears that late Neoproterozoic sequences lack such sediment disruption, whereas Early Cambrian sequences show disturbance. This signals the appearance of metazoans with the complexity of structure to support such behavior – muscles, neuromuscular control, some with appendages and shells.

Molecular Data on the Origin of Metazoans

"Of all the properties of 'living' things, heredity is perhaps the most characteristic. The ability to preserve advantageous changes was, and is, an essential aspect of life" (Raven & Johnson, 1986). The basis of passing of information from one generation to the next is a complex molecule called DNA (deoxyribonucleic acid). The DNA molecule is made up of series of what are called base pairs, single-ring compounds, made up of the elements hydrogen (H), nitrogen (N), oxygen (O) and carbon (C). Four base pairs (the purines, adenine and guanine, and the pyrimidines, cytosine and thymine) are bound together in an elegant structure called a double helix, a shape like a spiral staircase, where the steps are hydrogen bonds. Also involved in these chains are a sugar (deoxyribose, with not only carbon but also hydrogen and oxygen in its structure) and a phosphate group (PO_4). Along this double helix are segments called genes that give orders about how to produce or synthesize a particular protein. The sum total of all the genes on a DNA molecule is the genome of a particular organism. In prokaryotes (those organisms without an organized nucleus) the genome is held on a single molecule of DNA, a chromosome that forms a closed circle. Eukaryotes are more complicated. They have an organized nucleus enclosed within a nuclear membrane and have many chromosomes. During reproduction, the offspring ends up with a new combination of DNA, half of the information coming from each parent. In prokaryotes, during reproduction a duplicate copy is made of the single DNA molecule, and thus the "child"

is a carbon copy of the "parent" unless mutation occurs. Thus, the making of a novel organism from the act of reproduction is of low probability in prokaryotes, but a certainty within eukaryotes.

To attempt an understanding of metazoan relationships, study of their DNA or RNA molecules can be useful. The greater the difference in the genetic codes encrypted on their DNA molecules, the more distant their relationships and the longer ago they shared a common ancestor. Understanding the rates of genetic change, how fast mutations or recombinations of genetic material occur, allows prediction of how long ago the family tree branched. And if phenotypes of the genotypes (that is, the observable morphological differences – phenotypes - encoded by the genes as genotypes on the DNA molecule) can be recognized in the morphology of fossil animals and these animals dated in millions of years, then rates can also be calculated and used as "minutes and hours" on what is called a "molecular clock." This allows molecular biologists to extrapolate backwards, using the determined rates, and predict the time of origin of groups of animals (Lodish *et al.*, 2003).

Heckman *et al.* (2001) used this method to map out the degree of genetic differences on the chromosomes of many different organisms and suggested that fungi diverged from plants and animals roughly 1.6 billion years ago. This is the oldest molecular date placed on the appearance of animal life. Although Heckman and colleagues' conclusion was supported by several studies on the sequence and patterns of genes on the DNA molecule, there are problems with this method, for it assumes a constant rate of genetic change – the "molecular clock" is assumed to run smoothly without change of the speed at which it is ticking. This is not always the case (Ayala, 1997), and different groups seem to have different rates. Wray *et al.* (1996) suggest even older origins of metazoan phyla, taking into account the difference of these rates in vertebrates and some invertebrate groups. Other studies (Bromham *et al.*, 1998) in which several different genes on the DNA molecule of a number of metazoan groups were analyzed suggested that invertebrates (animals without backbones) had diverged from the chordates (animals with a stiffening rod called a notochord) about a billion years ago. According to this study, the protostomes diverged from chordate lineages well before the echinoderms (the spiny-skinned animals such as starfish and sea cucumbers). Such observations suggest a prolonged radiation of animal phyla and an origin of the Animalia at least 1.2 billion years ago.

The conflict between various molecular estimates on the one hand, and the traditional studies of first occurrences in the fossil record, such as the Cambrian evolutionary explosion, makes specialists and non-specialists alike question the molecular data. For instance, Ayala *et al.* (1998) analyzed 18 protein-coding genes and estimated that the protostomes (arthropods, annelids and mollusks) diverged from the deuterostomes (echinoderms and chordates) about 670 Ma ago, and chordates from the echinoderms about 600 Ma ago. Even though both dates are older than the fossil record currently reflects, they are still not so different from paleontological estimates. As the fossil record becomes better with future finds, these theoretical estimates may have fossils to back them up. Ayala and colleagues further suggested that the results obtained by Wray and co-authors (1996), suggesting divergence times of 1.2 and 1.0 billion years ago, may be too old because they extrapolated using vertebrate rates instead of those of invertebrate groups, which have proven rate differences. But if molecular studies take into account such varying rates, and use those known within related groups, molecular rates may indeed support conclusions drawn from paleontological and embryological studies.

In addition to looking at branches on the tree of life and when the branching occurred, molecular biologists (Wood *et al.*, 2002) have also looked for characteristics on the DNA molecule, specific genes, that are related to great jumps in complexity – such as the jump between single-celled and multicelled life and what brought about the variety of body forms in the late Precambrian. What they found was that very few genes seem specific to all multicellular species, and they suggest that the transition from single cells to multicellularity may not have required the appearance of many new genes. The metabolic pathways necessary for multicellular organization may have been in existence already in unicellular eukaryotes (a cell that has membrane-bound organelles, especially a nucleus in which the DNA is concentrated). It follows that the evolution from

DNA vs. RNA

How the genetic information stored in the gene sequences on the double-helix DNA moves out from the nucleus of a cell and directs the production of proteins is the job of RNA, a single-strand molecule similar in structure to DNA, specifically one form called mRNA – messenger RNA – ribonucleic acid. RNA is a little different from DNA. The sugar in RNA is ribose (instead of the deoxyribose in DNA) and one of the base pair pyrimidines is a compound called uracil (instead of the thymine in DNA). DNA transfers the genetic coding by splitting apart and encoding the single-strand RNA, which then transfers the code to a protein that in turn carries out its specific job within the cell. This is a dangerous time for the RNA molecule, as it is subject to "random cleavage" and it is probably for this reason that DNA, which likely developed from the single-strand RNA, took on the double-strand structure– a structure that "protected it from the ongoing wear and tear that is associated with cellular activity" (Raven & Johnson, 1986). After that, the more delicate RNA was employed to transfer the genetic directions when needed.

Just to complicate the story there is another kind of RNA – ribosomal RNA. This aggregates with certain proteins to form ribosome. Ribosomes are great translators – they translate the mRNA copies of genes in the production of further proteins.

a unicellular protozoan to some multicellular animal could have taken place in a very short space of time with little structural change needed. The transition from a colonial cell accumulation to a truly coordinated multicellular organism, one with tissue-grade organization, required that a part of the genome responsible for autonomous cell life was switched off or, perhaps more accurately put, unnecessary (Zavarzin, 2002). The development of tissues (that is, a group of cells organized into functioning and structurally sound units) limited the exchange of genetic information between individual cells within the organism (and between the multicellular organisms as well). This function was successfully taken on by sexual reproduction.

The Stage is Set for the Next Act: A Plethora of Body Plans

Whatever it was that spurred the development of the many different shapes and internal organizations of bodies between the time of *Horodyskia* and *Grypania* and the riot of diversity in the late Neoproterozoic and Cambrian is the subject of many papers and books (Raff, 1996; Blair & Hedges, 2004; Budd & Jensen, 2000; Carroll, 2001; Davidson *et al.*, 1995; Erwin, 1999; Erwin, Valentine & Jablonski, 1997; Valentine, 2004).

Organisms begin small and over time got larger and more complex. Sean Carroll (2001) points out that with increased size cell, diversity increases. With increased complexity there is opportunity for an organism to use ecospace in a different manner than others have used it in the past. There are three trends of development clearly evident in the fossil record: development of multicellularity, increase in size and diversification. Development of multicellularity likely happened independently many times. Once multicellularity was attained there followed larger and larger forms with a variety of new body plans and higher grades of complexity – examples are multicellular protists, animals and land plants. There were periods of rapid diversification, one example being the Ediacaran biodiversification and later the Cambrian "explosion," which was an event spread over tens of millions of years.

Some of the simplest animals to develop were the sponges (Phylum: Porifera). They possess only a very few cell types, each performimg special tasks. They "lack the sort of cell-to-cell junctions that form sheets of tissues" (Erwin, Valentine & Jablonski, 1997), which typify more complex forms. More complex than sponges are the Ctenophora (combjellies) and Cnidaria (jellyfish, anemones, corals). They have two tissue layers, ectoderm and endoderm, separated by a jelly-like material (mesoglea) and are termed diploblastic organisms because of this two-fold layering. The outer layer is protective, the inner digestive. Some Ediacarans have this type of organization.

Still more complex are organisms with three tissue layers, the triploblasts, which also possess bilateral symmetry – the Bilateria. Flatworms (Platyhelminthes) have a middle tissue layer, generally the most massive of the three layers, which develops into muscles, and some internal organs, such as complex digestive and excretory organs, nerves and much of the reproductive system. The outer ectoderm still functions to protect the organism. Flatworms have no circulatory system, so oxygen must diffuse from outer to inner tissue – thus, they remain flat to maximize surface to volume ratio. They are active predators, like most primitive metazoans. Like the anemones, they have only one opening to the gut.

Even more complex than flatworms are those organisms that left the first tracks and trails on the ocean bottoms. Erwin and colleagues (1997) note that these triploblastic animals had to have a more three-dimensional shape than flatworms. They were creepers or burrowers, sometimes leaving behind fecal pellets. Of necessity, they had to have muscles capable of producing waves of contraction and expansion so they could move. They needed a complete gut and fluid-filled spaces that could act as a hydrostatic "skeleton" to assist in the muscular action/reaction allowing movement of the animal through or on the sediments. So, from the flatworm body plan two more complex body plans arose – the protostomes and the deuterostomes. The names reflect how the mouth and anus form, but in addition to that there are other differences such as how the body cavity of each forms, one by splitting of internal tissue (protostomes) and the second by an inpocketing of tissue (deuterostomes). Typical protostomes are most of the invertebrates – the joint-legged animals or arthropods – insects; the annelid worms – earthworms; nematodes; mollusks – *Nautilus*, bivalves, snails; and many others. Deuterostomes include the spiny skinned-echinoderms such as starfish and sea urchins on the one hand and the chordates – sea squirts, fish and humans as examples.

Many of these body plans were in existence or foreshadowed in Ediacaran times, prior to the Cambrian diversification event, and in addition there may have even been some body plans unique to this time. It was upon some of these body plans that skeletons and shells developed first in the Late Ediacaran with *Cloudina* and then more fully in the Early Cambrian. These developments left a much better fossil record than soft-bodied organisms that lived during all of the billions of years before, paving the way for the modern world.

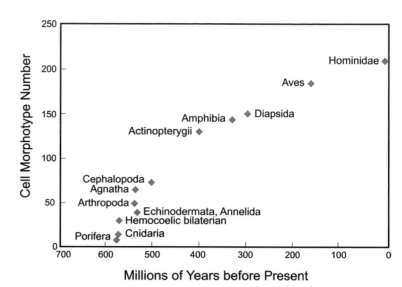

Figure 64. Numbers of different cell morphotypes for phyla plotted against time of the first appearance of phyla in the fossil record. Taxa with greatest number of cell morphotypes for each phylum is plotted. Valentine (2004) noted that such genetic complexity had the greatest rate of origination for metazoans early in their history, but further stated that despite appearances "taxa that appear later are increasingly more complex than can be represented by this simple index, the slowdown in complexity increase suggested….is misleading" (modified from Valentine, 2004) (D. Gelt).

PART 2: THE FOSSIL SITES
RARE AND EXTRAORDINARY

Introduction

P. Vickers-Rich

The road leading to the discovery of the most ancient animals has been long, oftentimes winding and serendipitous. The journey along this road has been to many parts of Earth, from the dry deserts of Australia and Namibia in West Africa to the blustery, cold coasts of northern Russia and Newfoundland. So, "if you are a dreamer, a wisher," perhaps enjoying a few embellished stories (not lies), "we have some flax-golden tales to spin" in the chapters that follow, told by some of those very explorers that have ferreted out these most ancient

of remains. So "Come in, Come in" as Shel Silverstein, New York poet, wrote in 1974 about storytelling, and follow the circuitous path that led to such discoveries. These are the very discoveries that illuminate Darwin's query of 1859: "To question why we do not find records of these vast primordial periods, I can give no satisfactory answer." There are now some answers, offered up by the Ediacarans, and the following chapters illustrate how they were found, and the questions they have posed that yet remain to be answered.

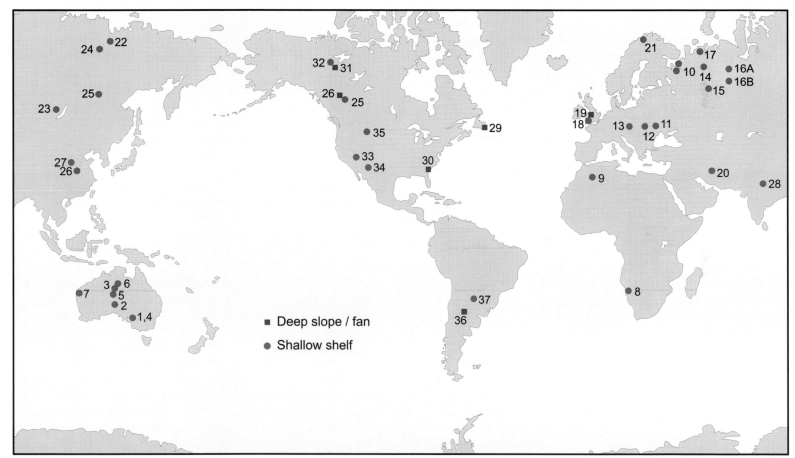

Figure 65. Locales where fossil Ediacarans (and other early possible metazoans –7, 35) have been found globally. AUSTRALIA: 1,4, Flinders Ranges, South Australia; 2, Punkerri Hills, Officer Basin, South Australia; 3, Amadeus Basin, Northern Territory; 5, Mt Skinner, Northern Territory; 6, Georgina Basin, Northern Territory; 7, Bangemall Basin, Western Australia (non-Ediacaran). AFRICA. 8, Namibia; 9, Algeria. EUROPE. 10. White Sea Region, Russia; 11, Dniester River, Podolia, Ukraine; 12, Ternopol District, Ukraine; 13, Moldavia; 14, Yarensk borehole, Russia; 15, Lorino borehole, Russia; 16a, Southern Ural Mountains, Russia; 16b, Central Ural Mountains, Russia; 17, Timan, Russia; 18, Carmathen Area, Dyfed, south Wales; 19, Chamwood Forest, England; 20, Kushk, Iran; Digermul Peninsula, Finmark, Norway. ASIA. 22, Yenisey River, Siberia; 23, Lake Baikal, Siberia; 24, Anabar and Olenek, Siberia; 25, Maya River, Yudoma, Siberia; 26, Yangtze Gorge, China; 27, Anhui Province, central China; 28, Uttar Pradesh, Lesser Himalayas, India. NORTH AMERICA. Avalon Peninsula, Newfoundland; 30, Carolina Slate Belt, North Carolina; 31, Mackenzie Mountains, Northwest Territories, Canada; 32, Wernecke Mountains, Yukon Territory, Canada; 33, Montgomery Mountains, Great Basin, southwestern USA; 34, Northwestern Sonora, Mexico; 35, Appekunny Mountains, northwestern Montana (non-Ediacaran). SOUTH AMERICA. 36, Puncoviscana outcrops, northwestern Argentina; 37, Mato Grosso do Sul, Brazil (D. Gelt).

Opposite page: Outcrop of the Nama Group, Swartpunt. Quiver tree and bushman shelter made of Nama rocks, often with Ediacaran fossils preserved on their surface. Indigenous peoples most likely would have seen these patterns on the rocks. Aus region, Namibia (P. Vickers-Rich)

The Ediacara biota contains the oldest unequivocal evidence of fossil animals on Earth, and thus is of critical significance in understanding the origin and early evolution of animals, as well as the ecosystems that they were part of and the environments that nurtured them. These were animals reaching large size and considerable complexity, which lived during the last part of the Neoproterozoic between @575 to 542 million years ago with all the truly biodiverse assemblages sandwiched between the youngest Neoproterozoic glacial sediments and the base of the Cambrian. This was a time when animals began depositing hard shells and burrowing deeply into the ocean sediments. The earliest biodiverse assemblages (575-560 million years ago) are dominated by complex, attached, suspension-feeding forms. Ocean waters contained abundant food particles (microplankton, eggs, larvae). Indeed, the Ediacarans were the very first hunters, so there was likely no incentive to "invent" sophisticated mechanisms of feeding – yet. Somewhat later, during the "boom times" of Ediacaran biodiversity (560-542 million years ago) a wide variety of body plans developed, some mobile and clearly grazing on the abundant microbial mats. The low-diversity "Twitya discs" of northwestern Canada may be somewhat older and a few Ediacarans may have survived into the Cambrian, but most show a restricted time range (Narbonne, 2005). Ediacaran fossils are for the most part impressions of soft-bodied animals, preserved on the undersides (or soles) or tops of sandstone beds or blanketed by volcanic ashes, all of which represent catastrophic, short-term events, such as sand avalanches down submarine slopes induced by storms, or volcanic ash falls (Narbonne, 1998, 2005; Gehling, 1999). Only rarely are such fossils preserved within sediments, animals buried alive where they lived or ripped up and transported in large masses of sediment.

Just how they were preserved has had an influence on the way different Ediacaran metazoans are interpreted, and in some cases this has encouraged a proliferation of names – several different names for the same animal – an example being the holdfasts for many of the frondose metazoans being interpreted originally as jellyfish. Narbonne (2005) notes at least four different styles of preservation which have recorded the Ediacarans – Flinders, Fermeuse, Conception and Nama – and notes a useful analogy that beautifully describes in general what occurred, outlined in more detail below. "It is as if event beds provided the taphonomic 'camera' capable of recording the living assemblage as fossils, but that microbial mats, sediment, and volcanic ash provided three different types of 'film' that were sensitive to very different wavelengths of light."

Flinders-style preservation produced a complex surface that resulted from microbial mats playing a major role. Muddy sea bottoms between wave base and storm wave base with such mats were buried by avalanches, mainly of sand, which instantaneously entombed the living communities of the time. The decaying mats quickly lithified (turned to stone) the bottom of the entombing sand, so preserving the impressions left by the metazoans that had been living on the surface of the mud below. So, preserved were the bases of holdfasts (such as *Inaria*) recorded as negative impressions and then as the animal decayed and the microbial mat disintegrated, thus solidifying the overlying sand into a "death mask," a depression was formed in this overlying surface, and thus the negative impression (Gehling, 1999). Where a "death mask"

did not form quickly enough, the bulbous body collapsed and was cast as a positive feature on the bed above. Frond-like forms (such as *Charnia*) were preserved as positive casts on the bases of beds for they were living on the surface, were preserved and partly collapsed before their body shapes were "molded from beneath and cast by the sand as the organisms . . . decayed" (Narbonne, 2005). Mobile animals, such as *Kimberella*, *Dickinsonia*, *Yorgia* and *Parvancorina*, as well as sessile *Tribrachidium,* were preserved as negative impressions on bed bases. Burrows were preserved as both ridges and grooves on the soles. This sort of preservation is typical of most fossils found in the Ediacara Member of the Flinders Ranges of South Australia and in the sandstone/shale sequences of the White Sea (Gehling, 1999; Grazhdankin, 2004; Narbonne, 2005), for that matter somewhere in the sequences of most Ediacara assemblage-bearing sediments around the world (J. Gehling, *pers. com.*, 2005).

Fermeuse-style preservation (Narbonne, 2005) characteristically is associated with outer shelf and slope environments typical of the Fermeuse Formation of Newfoundland (Gehling *et al.*, 2000), the Windermere Supergroup of northwestern Canada (Narbonne & Aitken, 1990), the Innerelv Member in the Finnmark region of Norway (Farmer *et al.*, 1992) and perhaps some locales in the White Sea area of Russia. Trace fossils and the bases of holdfasts typify this preservational style. Narbonne (2005) suggests that the environment that engendered this sort of preservation has a limited development of microbial mats or perhaps was more typified by heterotrophic or sulfur-oxidizing bacteria that did not feature the quick solidification of the overlying beds. Thus what was preserved primarily were metazoan parts that were partially buried.

Conception-style preservation (Narbonne, 2005) is characterized by the presence of a layer of overlying volcanic ash capping a layer strewn with Ediacarans – just as the "mummies" were in the Italian disaster of Pompeii. Characteristic fossils are the tops of holdfasts, collapsed rangeomorph shapes, impressions of the top of microbial mats – the ash instantaneously solidified the top of the mudstone layer – so preserving the organisms as they looked in life in positive relief. So, with this style of preservation one is actually viewing the top of the bed with a volcanic blanket revealing the structure below, unlike in the Flinders-type preservation where one is observing the impression left on the bed above. This sort of preservation is characteristic of the Conception Group of Newfoundland and sediments from Charnwood Forest in England. Gehling and Narbonne recently observed pyrite forming on assemblage surfaces of the ash fall beds in Newfoundland and noted variations in the preservation-style even in these volcanic crusts. Crystal tuffs were observed to preserve high relief, while thin ashes produced low. In both cases, pyrite was observed to be crystallized at the interface (Gehling, *pers. obs.*, 2005).

The fourth preservational style proposed by Narbonne (2005) is that of Nama, in which fossils are preserved as three-dimensional molds and casts of organisms caught up in or covered in downslope avalanches rather than being covered by such event beds. Examples of this type of preservation occur in the Nama Group of southwest Africa, Australia, the southwestern United States, parts of the White Sea succession of Russia and Newfoundland. Interestingly, due to partial degradation of the organisms before final preservation and the lack of microbial "death mask" formation, some internal structures can be observed, such as specimens of *Rangea* from Namibia and *Ventogyrus* from the White Sea (Dzik, 1999; Ivantsov, 2004; Narbonne, 2004, 2005).

Yet another type of preservation seems to be concretionary, an example being the exquisitely preserved rangeomorphs reported on by Narbonne in 2004 (Gehling, *pers. com.*, 2005).

This Ediacara biota is now known from around 30 different localities on five continents, with particularly abundant and diverse assemblages reported from Flinders Ranges of Australia, the White Sea region of Russia and the Ukraine of Europe, southwestern Africa, especially Namibia, and the Avalon Peninsula of Newfoundland. A variety of less biodiverse assemblages are also known from NW Canada, SW North America, North Carolina, England and Scotland, Iran, northern Siberia and the Ural Mountains of Russia, India, China, Argentina and Brazil.

Ediacarans have been known since 1872 – a form named *Aspidella terranovica* from the Avalon Peninsula of Newfoundland, described by E. Billings. As noted by Gehling and others (2000), these megascopic discoidal impressions were not accepted as organic by most researchers of the time or even much later. They were dismissed as sedimentary structures. Even as late as the mid-20th century other specimens from the Precambrian described as fossils by such workers as Gürich (1930, 1933) and Reg Sprigg (1947, 1949) from the Flinders Ranges of South Australia were dismissed as either non-metazoan or non-organic or assigned to a younger, Cambrian age. Martin Glaessner's seminal paper in 1959 rather changed those misconceptions when he coined a name for this unusual assemblage – the Ediacaran fauna. What followed in the years after, until the last two decades, was an assignment of fossil forms to modern phyla. There was a proliferation of names of metazoans allied closely to jellyfish, sponges, soft corals and segmented worms, to trilobites and other arthropods (Glaessner, 1984). But in the last 25 years these ideas have been challenged and modified by a number of researchers (Seilacher 1984, 1992; Gehling *et al.*, 1991; Retallack, 1994; McMenamin, 1998; Steiner & Reitner, 2001; Peterson *et al.*, 2003).

As Narbonne (2005) succinctly put it: "In a little more than a decade, the affinities of the Ediacaran biota went from being a well-established 'fact' to becoming one of the greatest controversies in paleontology." Significantly this stimulated a great deal of further research in cross-disciplinary fields, resulting in the present, still unsettled state of affairs. One point made quite clear by Jim Gehling (1999) was that Ediacarans were made of soft, flexible tissues that were not out of the ordinary for Phanerozoic and still living metazoans (Narbonne, 2005). It was also clear from other studies that at least some of these animals were not in need of light in order to survive, the best example being the faunas from the deep water, in faunas of Newfoundland (Wood et al., 2003) and northwest Canada (Dalrymple & Narbonne, 1996; Narbonne, 2005).

Narbonne's (2005) most recent summary of Ediacaran paleobiology points out the following affinities that most researchers would agree upon: the simplest forms are those with radial symmetry, the sponges (Phylum: Porifera) and the corals, jellyfish and sea anemones (Phylum: Cnidaria). Both fossils and biomarkers flag the presence of the poriferans (Gehling & Rigby, 1996; McCaffrey *et al.*, 1994). If indeed *Thaumaptilon walcotti*, a mid-Cambrian form from the Burgess Shale, preserves true remains of zooids (Conway Morris, 1993), and if indeed it is related to *Charniodiscus*, perhaps soft corals make up part of the Ediacaran biota – this is still not agreed upon (see Williams, 1995). Other groups that seem to have some presence among the Ediacarans include those with bilateral symmetry: the joint-legged arthropods, the nematode worms, mollusks, annelids, brachiopods, the platyhelminth worms, the spiny-skinned echinoderms, the chordates and one further group with radial symmetry, the combjellies (the ctenophorans) (Narbonne, 2005). There are still a number of Ediacarans, however, whose relationships are unclear and may represent experiments that eventually failed or were replaced by others – that is, lost architectures. These include such forms like *Yelovichnus* and *Palaeopascichnus*, once labeled trace fossils, but which may, in fact, be body fossils (Narbonne, 2005). *Tribrachidium* is another puzzler, some suggesting that it could be related to sponges (Seilacher *et al.*, 2003) or some of the Small Shelly Fossils, abundant in the Early Cambrian. *Pteridinium* and *Ernietta* along with *Swartpuntia*, so well represented in the Namibian assemblages, have defied placement and the rangeomorphs, so abundant in the Newfoundland localities, appear to have been most successful in early Ediacaran times, but represent one of the failed experiments with the arrival of the Cambrian faunas.

So what finally brought an end to the Ediacarans? Narbonne (2005) and many others (Seilacher, 1984; Gehling, 1999; Bengtson, 2002) have speculated on this. There are numerous ideas: perhaps a mass extinction at the end of the Neoproterozoic occurred related to a global anoxic event or widespread methane release, which has some credence offered by a carbon-isotope anomaly at this time. Another suggestion is that the rapid diversification of burrowing and grazing organisms at and directly after the Precambrian-Cambrian boundary not only affected food sources but also the preservational environment of the Ediacarans (Gehling, 1999; Droser *et al.*, 1999). A third possibility is the development of predators, first noted in borings in the shells of *Cloudina* from China (Bengtson & Yue, 1992; Hua *et al.*, 2003; Narbonne, 2005) in the late Neoproterozoic.

The section of the book that follows examines the biota and setting of several Ediacaran locales, their history of discovery and the wealth of information they have provided in an attempt to give an up-to-date account of just how much and how little we know about the origin of the diverse body plans and lifestyles of these first metazoans. Undoubtedly, these words will soon be out of date!

CHAPTER 3

The Misty Coasts of Newfoundland

G. M. Narbonne, J. G. Gehling, and P. Vickers-Rich

As the tourist brochures acclaim, Newfoundland is "the place where land, water and sky embrace like old friends." It is the eastern edge of North America where storms and the rugged, fog-shrouded rocky coastline has been the doom of many a seagoing craft. The southwestern part of this first landfall from Europe is the Avalon Peninsula. One particular spot, Mistaken Point, named because of the particular navigational hazard it is, is a most significant paleontological site. Preserved on the flat bedding planes of sandstone and shale that have challenged the best of navigators – to avoid, that is – are some of the very oldest complex animal fossils known on Earth.

The Avalon Peninsula

Of all the Ediacara assemblages, perhaps the most spectacularly preserved and compositionally perplexing are those from Mistaken Point and Fermeuse on the Avalon Peninsula and the Bonavista area of eastern Newfoundland. Up to 30 distinct types of impressions of soft-bodied Ediacara organisms are present, many of them bizarre fractal constructions unknown anywhere else on Earth. The fossils of Avalon include abundant discs similar to those reported elsewhere, but the simple burrows that occur in most Ediacara fossil assemblages worldwide are conspicuously absent from Newfoundland.

The Avalon Peninsula of Newfoundland contains Ediacara fossils that range through more than 3 km of stratigraphic thickness, spanning 10 million years (Narbonne & Gehling, 2003), making the Avalon the thickest and longest-ranging

Figure 66. Fortune Head, Burin Peninsula, GSSP for the base of the Cambrian Period and Phanerozoic Eon, Avalon Peninsula, Newfoundland, with Soren Jensen for scale, pointing to exact level (J. Gehling).

Opposite page: Mistaken Point, Avalon Peninsula, Newfoundland (painting by Peter Trusler).

Figure 67. Vendotaenids from the Chapel Island Formation below the Cambrian GSSP, Newfoundland. Average width of tubes about 2 mm (J. Gehling).

single record of Ediacara fossils known anywhere. Fossils are exceptionally abundant – there probably are more Ediacara-type fossils at Mistaken Point than in the combined collections of every museum on Earth! Fossils occur in nearly 100 different layers within this rock stack, allowing paleontologists to analyze the effects of time and changing environments on ecological associations of fossilized assemblages as well as observe the evolutionary pathways of many different animal lineages. This evolutionary record can be dated using U-Pb radiometric techniques carried out on volcanic ashes that cover each fossil surface. This part of the world has long been affected by volcanic activity – providing a unique opportunity to calibrate the early evolution of animals.

In contrast with the predominant preservation of Ediacara-type fossils on the bottoms of sandstone beds in Australia, the White Sea and most other Ediacaran localities worldwide, Mistaken Point organisms died where they lived when they were suddenly buried under beds of volcanic ash. Subsequent weathering away of the soft ash over the ages has exposed tennis-court-sized surfaces, each littered with hundreds or even thousands of fossil impressions that faithfully preserve everything from the morphology of the organisms to the ecological structure of their communities. Walking across a fossil surface in Newfoundland is like snorkeling over a 565 million year old sea bottom.

Such spectacular "natural exhibits" of this age are unknown from any other place on Earth. The Newfoundland fossils provide a unique natural laboratory allowing scientists to study biodiversity and ecology of these early animals (Conway Morris, 1998; Jenkins, 1992; Bendick & Pflüger, 2001; Clapham & Narbonne, 2002: Clapham *et al.*, 2003; Laflamme *et al.*, 2004; Narbonne, 2005). A few forms that occur in these Newfoundland rocks have broad distributions across the globe, examples being *Aspidella* and the frond-like *Charnia* and *Charniodiscus*. Most species, however, are limited to the Mistaken Point and Fermeuse assemblages (Waggoner, 1999, 2003). And what is quite revealing is that Ediacaran forms often quite common elsewhere in the world are not present at all in the Newfoundland rocks – *Dickinsonia, Pteridinium, Tribrachidium*.

The Oldest of Ediacarans, in Deep Water

The history of Ediacaran research in Newfoundland goes back far before the term "Ediacaran" had even been coined, to Elkanah Billings, whose enormous lifetime contributions

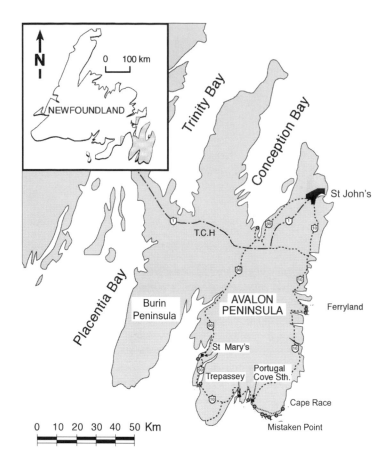

Figure 68. Map of the fossiliferous Ediacara locales on the Avalon Peninsula, Newfoundland (after Narbonne et al., 2005).

to Paleozoic paleontology eventually led the Geological Association of Canada to name the highest medal in Canadian paleontology after him. In 1872, Billings named *Aspidella terranovica* based on cm-scale discs found in rocks near the post office in downtown St John's, in sediments now called the Fermeuse Formation. This single rock outcrop still exists, and the fossil called *Aspidella* is incredibly abundant, not only in St John's but also in many other outcrops of the Fermeuse Formation throughout the Avalon Peninsula.

The naming and description of *Aspidella terranovica* was a bold move by Billings. Definite Precambrian fossils were almost unknown anywhere in the world at that time, a problem that vexed Darwin greatly. Yet the *Aspidella*-bearing rocks clearly lay below local trilobite-bearing "Primordial" strata, and furthermore these older rocks were separated from the younger trilobite-rich rocks by a marked unconformity or break in sedimentation – which means that some time had passed, and probably considerable erosion had occurred, before the trilobite-bearing sediments were deposited. This led Billings to conclude that *Aspidella* was of pre-Cambrian age and thus considerably predated any animal fossils known at the time. The use of *Aspidella* fossils as "time pieces" to link, or correlate, similar-aged rocks throughout the Avalon was also obvious to geologists (Murray, 1868, 1873; Billings, 1872; Walcott, 1899). But, despite this geologic utility of the fossils, it was another matter for paleontologists to explain the enormous numbers of fossils of apparently soft-bodied organisms in such old sedimentary rocks, and most workers at the time and throughout the succeeding 20th century regarded *Aspidella* as a "pseudofossil." A century later, Gehling, Narbonne and Anderson (2000) carried out a comprehensive study that vindicated Billings

Figure 69. Several spindle-shaped rangeomorphs underlying a thick ash that covers the E-surface at Mistaken Point. U-Pb dating of zircons in this ash provide a precise date of 565 million years. (G. Narbonne).

Figure 71. Specimens of Aspidella terranovica preserved in part and counterpart, Fermeuse Formation, Ferryland, Newfoundland (G. Narbonne).

and showed that *Aspidella* is an organic structure representing the first Ediacaran species ever named.

Time passed, and much later, in 1967, remains of even older soft-bodied animals were discovered at Mistaken Point during the M. Sc. thesis research of S. B. Misra, a student at Memorial University of Newfoundland, assisted by an undergraduate student. When described, the Mistaken Point biota was the first indication of a diverse collection of highly complex Ediacara-type fossils in the New World. It was also the first discovery anywhere on Earth of Ediacarans found in deep-water sediments. These fossils were unusual, too, in being extremely large, abundantly covering broad tracts of bedding planes representing the surface of ancient sea beds, their diversity reflecting the nature of deep-water biotas so long ago.

The initial report of the fossils from Mistaken Point appeared in the prestigious scientific journal *Nature* (Anderson & Misra, 1968), which figured a single, new fossil species. This announcement of such unexpected fossils was questioned by Goldring (1969), also in *Nature*, but Anderson and Misra's (1969) quick reply and subsequent finds of abundant and biodiverse fossil assemblages led to worldwide agreement by the scientific community that the Mistaken Point biota was unquestionably organic and represented one of the most important Proterozoic fossil assemblages known on Earth. It shed new light on the early evolution of animals and their ecology.

Since that time, Ediacara-type fossils have been found throughout most of the Avalon Peninsula (Anderson, 1978) and recently on the nearby Bonavista Peninsula (O'Brien & King, 2004), a geographic range rivaled only by the extensive fossil beds of the Flinders Ranges and Namibia. The Mistaken Point biota includes as many as 30 different forms, but until recently none of the taxa were named scientifically, and they were instead referred to using consistent but informal names such as "spindles," "bushes," "pectinates," "fractal frond," etc. (Anderson & Conway Morris, 1982). Ongoing work by G. M. Narbonne, J. G. Gehling, and colleagues at Queen's University, in conjunction with M. M. Anderson (Memorial University of Newfoundland), has sought to redress this problem. It is anticipated that this group will formally describe and name all of the major taxa in the Mistaken Point Assemblage over the next five years. New Mistaken Point and Fermeuse taxa defined thus far include the clover-leaf-shaped fossil *Triforillonia costellae* (Gehling *et al.*, 2000), the arrowhead-shaped to triangular fossil *Thectardis avalonensis* (Clapham *et al.*, 2004) and three species of Ediacaran fronds (*Charnia wardi* - Narbonne & Gehling, 2003; *Charniodiscus procerus* - Laflamme *et al.*, 2004; and *Charniodiscus spinosus* - Laflamme *et al.*, 2004).

The volcanic ash beds that directly cover each Mistaken Point surface can be dated precisely using U-Pb (uranium-

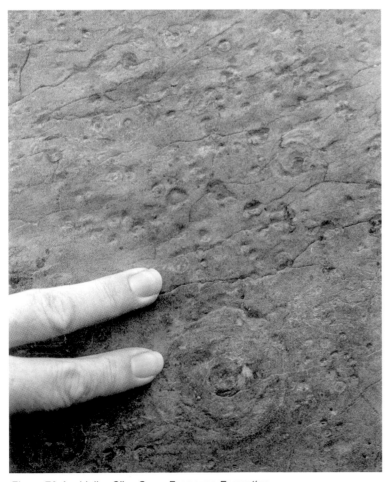

Figure 70. Aspidella, *Silos Cove, Fermeuse Formation, Mistaken Point, Newfoundland (P. Trusler).*

Figure 72. As in Australia, there have been many attempts to illegally remove fossils from Mistaken Point. Kate Ward, guardian of Portugal Cove South on the Avalon Peninsula, foiled one such attempt to remove an Ivesheadia disc (G. Narbonne).

Figure 73. Ivesheadia specimens inspected by students in the Mistaken Point Ecological Reserve, Newfoundland (J. Gehling).

lead) radiometric dating of zircons in the ash. Greg Dunning's analysis of a volcanic ash covering the main fossil horizon in the Mistaken Point Formation ("bed E") gave a date of 565 ±3 million years (Benus in Landing et al., 1988), which at the time was the only Ediacara fossil assemblage that was directly dated using high-precision geochronology. Originally this date was regarded as implausibly young to those who accepted a 570-610 million year date for the base of the Cambrian (Harland et al., 1982). Today, to the contrary, it is regarded as one of the oldest, reliable dates for the Ediacara biota, while the base of the Cambrian has been redated at 542 Ma (Grotzinger et al., 1995; Amthor et al., 2003). Lower fossil beds containing the oldest Ediacara-type fossils (Narbonne & Gehling, 2003) have now been dated as being 10 million years older (575 million years old) by Sam Bowring (Massachusetts Institute of Technology). Every one of the dozens of surfaces of Mistaken Point fossils throughout the succession is directly overlain by a volcanic ash bed, and further dates would help to constrain the rate of evolution of early Ediacarans.

Complementing the paleontological research were detailed studies of the sediments enclosing the Mistaken Point fossils in order to determine the environments in which the organisms lived and died. Ediacarans were mainly attached organisms living on the sea bottom (Jenkins, 1992; Seilacher 1992); the sediments with which they are associated lend significant insights into the exact environment that hosted these first animals. Based on such sedimentological studies, most Ediacaran biotas appear to have lived in shallow, well-lighted settings, somewhere between fair-weather wave base and storm wave base (see Narbonne, 1998). Quite the contrary, all sedimentological studies of the assemblages from Mistaken Point and other localities on the Avalon Peninsula have concluded that these represent very deep-water settings, dark environments well below the influence of waves or storms (Misra, 1971, 1981; Benus, 1988; Conway Morris, 1998; Jenkins, 1992; Myrow, 1995; Narbonne et al., 2001, 2005; Wood et al., 2003). The fossil-bearing Conception and St John's groups consist of nearly 5 kilometers of stacked turbidites, which show all the classic features of deep-sea sedimentation. Thorough search by generations of sedimentologists has failed to reveal any mud cracks, wave ripples, storm-generated sedimentary structures or any other indication of shallow water in any part of the fossiliferous succession. Strata in the Conception Group are undisturbed, but those of the overlying St John's Group show abundant evidence of downslope sliding. The overall succession is interpreted as representing a deep-sea environment that shoaled from a basin plain to a continental slope over time (Wood et al., 2003). This, most certainly, would have impacted on the types of animals that are now preserved as fossils in the Mistaken Point Assemblage.

All sedimentologists to date have agreed on the interpretation of the Mistaken Point sediments as having been deposited in deep water, well below both wave base and the photic zone. However, some paleontologists have challenged this interpretation – such as Greg Retallack (1994), who has interpreted Ediacarans as lichens, or Mark McMenamin (1998), who suggested they were an extinct group of compulsory photoautotrophs. Both of these interpretations require that the Mistaken Point fossils lived in shallow, sunlit settings, and not surprisingly both of these authors have concluded that the sedimentological interpretations must be incorrect! Based on visits to Mistaken Point in 1979 and 2001, Fedonkin (pers. com., 2006) accepted the deep-water interpretation of the strata but regarded the fossil assemblages as having been transported downslope into an environment in which they did not live, principally because they contain fossil organisms

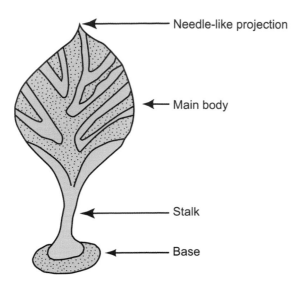

Figure 74. "Leaf-shaped" metazoan from the Cape Race area, Conception Group, Avalon Peninsula, Newfoundland – an early reconstruction (from Misra, 1969) (D. Gelt).

Labels: Needle-like projection, Main body, Stalk, Base

Figure 75. Rangeomorph module. Ediacara fossils near Spaniard's bay show exquisite, three-dimensional preservation that reveals the fine structure.
(G. Narbonne; this specimen was illustrated on the cover of Science Magazine, 20 August 2004). Length of module about 2.5 cm.

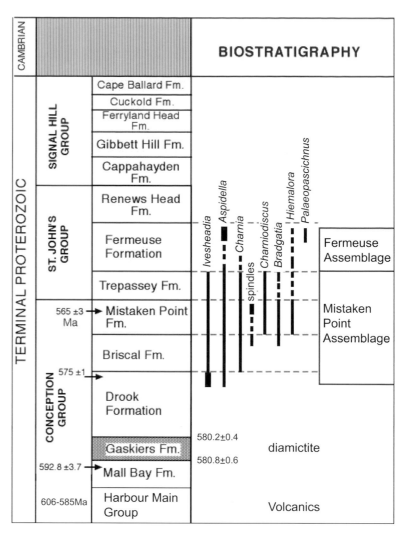

Figure 76. Stratigraphic column and fossil occurrence typical of the Avalon Peninsula, Newfoundland (from Narbonne et al., 2005).

(such as *Rangea, Charnia* and *Hiemalora*) that are known from the shallow-water deposits elsewhere and also because of unusual preservational features (*e.g.* preservation of fossils on the tops of beds, preservation of fronds and their holdfasts on the same bedding surface, etc.). However, other workers have cited the strong current alignment of tethered fronds (Seilacher, 1999; Wood *et al.*, 2003), evidence for lateral displacement during growth of the organisms (Gehling *et al*, 2000) and preservation of vertical and spatial ecological structure in the communities (Clapham *et al.*, 2003) as evidence that most or all of the fossils in the Mistaken Point Assemblage are in their original life positions.

Other deep-water occurrences of Ediacara-type fossils have been reported from Charnwood Forest in England, the Carolina Slate Belt in the eastern USA, and the Mackenzie Mountains in NW Canada (Chap. 10). The taxa described from Charnwood (*Charnia, Charniodiscus, Bradgatia* and discs such as *Ivesheadia*) are basically a subset of those in the more diverse Mistaken Point biota (Waggoner, 1999). Recent studies suggest that rangeomorphs, like those at Mistaken Point, characterized both older and deeper-water Ediacaran assemblages and were much less common in younger and shallower-water assemblages (Grazhdankin, 2004; Narbonne, 2005; O'Donoghue, 2007).

Figure 77. Detail of the rangeomorph fossils from two layers – spindles, combs and Charniodiscus, Mistaken Point, Newfoundland (P. Trusler).

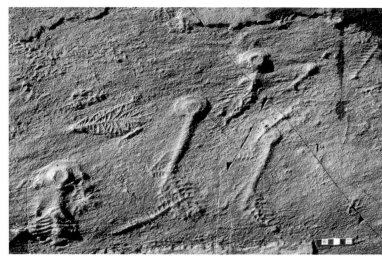

Figure 78. Ediacara fronds (Charniodiscus procerus, Charniodiscus spinosus and an unnamed rangeomorph) and spindle rangeomorphs preserved on the E-surface at Mistaken Point, Newfoundland (G. Narbonne).

At the time the Ediacarans were alive, microbial mats were common, and though not exclusive, they were often a factor in the preservation of Ediacaran megafossils. Surface textures of rock layers, known as "old elephant skin," along with carbonaceous or pyritic coatings are some of the features that have been noted commonly associated with most of these fossil assemblages (Fedonkin, 1992; Narbonne, 1998; Gehling, 1999), and they are certainly a feature of the Mistaken Point sediments (Wood et al., 2003). In the Mistaken Point Formation, there are two characteristic kinds of fossiliferous horizons. Limonitic and crinkled surfaces, associated with very thin layers of volcaniclastic silt and low-relief impressions of fossils, are likely the products of microbial mats. These mats may have been similar to the mats laid down by heterotrophic and/or sulfur-oxidizing bacteria that coat large areas of the modern deep-sea floor. On the other hand, layers with more three-dimensional fossil impressions are always associated with thin beds of crystalline tuff, representing volcanic ash falls into a nearby ocean, which must have smothered the sea floor organisms. The volcanic coating would have been quickly cooled by the ocean waters and solidified to produce a "death mask" casts of the recently deceased organisms, including the easily compressed spindle-like organisms and the more resistant fronds.

The most spectacular example of these "death mask" fossil assemblages occurs in "bed E" (Landing et al., 1988) that is exposed along more than 2.5 km of coast near Mistaken Point. The arrays of fossils that lie atop such horizons appear to be death assemblages of once living benthic communities, encased in place just as the unfortunate humans and their pets were entombed in Pompeii when Vesuvius erupted on the 23rd of August AD79, the only difference being that Pompeii was in air fall of ash, whereas the Mistaken Point Assemblage was preserved by volcanic ash that fell into the ocean and settled on the sea bottom, thus entombing a living community. Unfortunately, morphological detail of organisms is limited by the crystal size of tuff and, in most places, by alteration of the sediments later by metamorphism. Most Newfoundland specimens suffer from having been deformed

(had their length in one direction shortened during subsequent mountain-building episodes), which requires mathematical or photographic techniques to restore the fossils to their original shape (Wood et al., 2003).

In 2004, Narbonne reported what are some of the most exquisitely preserved Ediacaran fossils known anywhere. A photograph of the best-preserved specimen graced the cover of Science Magazine. Fossils from Spaniard's Bay in the northern Avalon Peninsula are undeformed and exquisitely preserve morphological details as fine as 30 μm (3/100th of a millimeter) in width, the finest Ediacara-type preservation known anywhere. This Lagerstätten provides a unique insight into the architecture of these long-extinct life forms (Narbonne, 2004), and has aptly been called "a Rosetta stone for decoding Ediacaran animal evolution" (Brasier & Antcliffe, 2004).

The task of working out what these bizarre fossils might represent has not been easy. Early studies (Misra, 1969, 1971; Anderson, 1978; Anderson & Conway Morris, 1982) regarded some or most of the fossils as representing primitive "coelenterates" (cnidarians), but also recognized that many of the taxa were difficult to compare with living forms. In 1992 Dolf Seilacher proposed that the Mistaken Point forms were "serially and fractally" quilted organisms, which belonged to his newly proposed kingdom of life forms, which he called "Vendobionta." These were supposed to be organisms with a structure rather like an inflated inner tube, which did not move under their own power, but instead simply "lazed around" absorbing their nutrients from the environment through their extensive surface. They had no mouths, no eyes, no legs, no digestive tracts. Recently Kevin Peterson, Ben Waggoner and James Hagadorn (2003) have suggested that such forms as Aspidella, Charnia and even Charniodiscus may be some sort of fungi.

Studies of the exquisitely preserved fossils at Spaniard's Bay in the northern part of the Avalon Peninsula led Narbonne (2004) to conclude that they were "rangeomorphs," complexly fractally branching constructions that represent a "failed experiment" in the Neoproterozoic history of life. This view received support from Brasier and Antcliffe (2004), who regarded all rangeomorphs as fundamentally related and

Figure 79. Ecological sudies at Mistaken Point. Matthew Clapham studying the E-surface, where the greatest abundance and diversity of fossils occurs at Mistaken Point, Avalon Peninsula, Newfoundland (G. Narbonne).

suggested that heterochronous development led to the wide array of forms present in the Mistaken Point and Charnian assemblages. Rangeomorphs show "centimeter-scale frondlets exhibiting three orders of fracticality in branching" (Narbonne, 2004). These frondlets were used as modules to construct a variety of forms outlined below (excluding *Aspidella* and *Thectardis*). There is no evidence of muscles or sensory organs. These are what have been termed the "rangeomorphs," which seem to have diversified widely in deep marine waters. They seem to have been forms that lived above the sea floor, taking on a variety of shapes and sizes.

Previous comparisons of rangeomorphs with cnidarians or ctenophores now seem unlikely. The fact that both sides of the rangeomorph structure are essentially identical, is not consistent with the rangeomorph frondlet "representing the bases of an array of open tubes that housed polyps or other filter feeding organisms." Narbonne further noted that the very small diameter of the secondary and tertiary branches would rule against the rangeomorph structure housing cnidarian polyps.

Rangeomorphs seem to have nicely partitioned the food web. Spindle-shaped rangeomorphs lay on the sea bottom, while others (bush-shaped, plume-shaped, frond-shaped, comb-shaped) stood up at various levels in the water column, filtering food from above the level utilized by the spindle-shaped forms. Such habits would have allowed a partitioning of the environment, reduced direct competition and allowed a range of species of rangeomorphs with subtly, and some not so subtly, different lifestyles to coexist. Narbonne (2005) speculated that the lesser amount of genetic information it would take to code for the fractal architecture and modular construction of rangeomorphs would have allowed them to get an early start over more traditional animals, but that this simplicity did not permit them to compete with the metazoans which gradually evolved throughout the Ediacaran. He views rangeomorphs, in sum, as a "forgotten" architecture that thrived at the beginning of the Ediacaran, but was later displaced by other metazoan groups and left no progeny in the Phanerozoic.

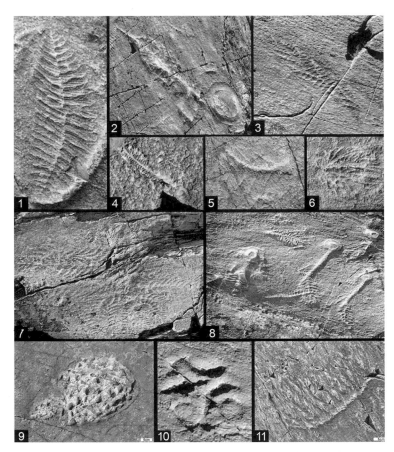

Figure 80. A variety of fossils in the Mistaken Point Assemblage, Avalon Peninsula, Newfoundland (J. Gehling). 1, spindle-shaped form, Mistaken Point Formation; 2, Charniodiscus spinosus *with discoidal holdfast from Bed-E; 3,* Charnia masoni; *4,* Charniodiscus arboreus *(ROM 36504); 5, small tree-like form with short trunk and rudimentary attachment zone; 6, bush-shaped form of* Bradgatia; *7, network form; 8,* Charniodiscus *fronds with holdfasts, aligned SSE overprinting spindles,* Charnia *composite form in top center; 9, "pizza disc" form of* Ivesheadia; *10, lobate morph of* Ivesheadia; *11, pectinate form. 1-3, 5-11 are specimens still in the field.*

Recognition that the rock surfaces upon which the fossils are preserved actually represent true populations of ash-killed organisms recently has led researchers to apply methods and models that modern-day ecologists use to study living communities (Clapham & Narbonne, 2002; Clapham *et al.*, 2003). These studies suggest that Mistaken Point communities were most similar to those of Phanerozoic and modern sessile, suspension-feeding animals in terms of their community attributes (*e.g.* richness, density, diversity, evenness) and tiering above the sea bottom. Strikingly different assemblages on surfaces throughout the succession were interpreted as reflecting an ecological succession that progressed from high-dominance communities of low-level suspension feeders to more diverse and equitable communities with complex tiering and spatial structure. However, there is no evidence of mobility, herbivory or predation among any of the Mistaken Point taxa, and these ecological innovations would have to wait for corresponding innovations in morphology (Narbonne, 2005).

Further taxonomic and ecologic studies of the Mistaken Point biota are ongoing, and promise to be every bit as exciting, illuminating and perplexing as the studies we have seen above.

Fossils of Geometric Design

Ediacara fossils on the Avalon Peninsula comprise two stratigraphically and taphonomically distinctive assemblages: the Mistaken Point Assemblage and the Fermeuse Assemblage (Narbonne *et al.*, 2001, 2005; Narbonne, 2005). Fossils of the Mistaken Point Assemblage are preserved under beds of volcanic ash as an "Ediacaran Pompeii." They first appear in the Upper Drook Formation and occur on more than 50 horizons through the Conception Group up into the Trepassey Formation of the St John's Group. However, the most dense and diverse fossil horizons are confined to a 50 meter interval in the upper part of the Mistaken Point Formation. These horizons were defined by volcanic ash or microbial mat partings within otherwise massive outcrops. The gently tilting rock benches at Mistaken Point provide an unequalled opportunity to study the rich collections of fossils that are preserved on upper bedding surfaces, making up the Mistaken Point Assemblage (Anderson & Misra, 1968; Misra, 1969; Anderson, 1978; Anderson & Conway Morris, 1982). This area has fascinated paleontologists for decades and been the site of intensive study. Fossils of the Fermeuse Assemblage are preserved on the soles (undersides) of sandstone and siltstone beds in the otherwise shale dominated Fermeuse Formation (Gehling *et al.*, 2000). *Aspidella* appears to be the only fossil that ranges between the Mistaken Point Assemblage and the Fermeuse Assemblage on the Avalon Peninsula. Ash beds occur much higher in the section on the nearby Bonavista Peninsula, and consequently the Mistaken Point Assemblage ranges much higher up there than it does on the Avalon (O'Brien, King & Hofmann, 2006). This confirms that the primary control on these two assemblages reflects preservational differences rather than age or environment.

Figure 83. *Spindle-shaped rangeomorphs, Mistaken Point, Newfoundland (J. Gehling).*

Figure 84. *Assorted rangeomorphs and* Charniodiscus, *Green Head, Spaniard's Bay, Newfoundland (P. Trusler).*

Figure 81. Charnia wardi, *Drook Formation, Mistaken Point Ecological Reserve, Newfoundland (G. Narbonne).*

Figure 85. *Thectardis "triangle," Avalon Peninsula, Newfoundland (G. Narbonne). Length of specimen 8 cm.*

Figure 82. *Mistaken Point, Newfoundland (P. Trusler)*

Figure 86. Reconstruction of Bradgatia linfordensis, *Mistaken Point, Newfoundland (P. Trusler). Width of the specimen about 10 cm.*

Mistaken Point Assemblage

The Mistaken Point Assemblage is dominated by rangeomorph "fossils exhibiting fractal-like quilting" that is unlike the construction of any known animal group, fossil or living, even at the higher levels of classification – at the phylum level (Narbonne, 2004). The most common and best preserved are the yet unnamed spindle-shaped rangeomorphs, preserved as external molds of an animal that once lived on a muddy sea bottom. Volcanic ash rained down on this organism-covered sea floor and formed a film over collapsed, three-dimensional bodies, forming one of Jim Gehling's "death masks." The spindles are made up of a bipolar array of branches emerging from a zigzag midline. Each rectangular-shaped branch consists of a close-packed bundle of rod-like elements, which in turn split toward the outside of the organism. The branches decrease in width and length from the center toward both ends. Specimens with a bent axis

have their branch-like bundles of elements compressed on the concave side and splayed or split between bundles on the convex side. Individuals may lie in contact or even partly overlap, but never entirely overlap, suggesting that there may be a dimension to the organisms that is not clearly preserved. Spindles are among the most common and diagnostic fossils in the Mistaken Point biota but have not been recorded in any Ediacara assemblages outside of Newfoundland.

Frond-like, three-dimensional forms with attachment discs are preserved on several horizons. These forms show distinct orientations, which can be used to reconstruct the currents operating during the life and burial of the organisms (Wood *et al.*, 2003). Some fronds, such as those in the classic "bed E," show downslope orientation to the SSE, implying that they were buried by an ash-rich turbidite flowing down the slope. Most fronds are oriented to the NNE, implying that they lived in a gentle current that flowed parallel to the contours of the slope and that brought food, and possibly oxygen to the organisms.

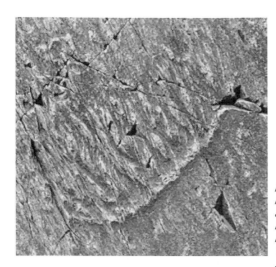

Figure 87. Pectinate rangeomorphs and frondose taxa, Mistaken Point, Newfoundland (P. Trusler). Scale in cm .

Figure 88. Mistaken Point surface, Mistaken Point, Newfoundland (P. Trusler)

Seilacher (1992) regarded bending of the frond to one side of the aligned stalk as suggesting a reverse flow or "backwash," but Laflamme and colleagues (2004) disputed this and noted that the kink in the stem was a morphological trait of this species.

An upright life position is suggested by the way the fossils are preserved. The internal and surface impressions are poorly preserved, in contrast to the detailed impressions of their holdfasts, which, of course, would have been essentially buried in the sediments – "anchors" for the leaf-like straps that would have been exposed in the water column (Seilacher, 1992). On some horizons, the holdfasts are absent, even though fronds are aligned and apparently tethered, further suggesting that holdfasts were buried below the preserved surface and not subject to "death mask" casting by volcanic ash.

Fronds vary in length, width and shape. One small specimen shows identical arrangement of its branches to that observed in *Charniodiscus arboreus*, well known in Australia. Like *Charniodiscus* from Australia, these forms seem to have had a three-dimensional nature, which will require more detailed analysis before reconstruction and comparison with one suggested modern analogue, the pennatulid soft corals, can be made. These fronds certainly do not seem to be simply flat straps. There are at least three types of *Charniodiscus* known from the Mistaken Point and Trepassey formations, two of which are unique to the area, *Charniodiscus procerus* and *Charniodiscus spinosus* (Laflamme, Narbonne & Anderson, 2004), both of which have fewer branches than *Charniodiscus arboreus,* common elsewhere in the world.

Charnia masoni is present in the Mistaken Point Formation, preserved as external molds with low relief (Laflamme *et al.*, in press). Several specimens are associated with what appear to be discoidal holdfasts. The asymmetric or elongated appearance of specimens seems to be the result of deformation. Otherwise, the arrangement of branches and secondary divisions are indistinguishable from that seen in *Charnia masoni*, long known from Leicestershire, UK. *Charnia* is now known from England, Newfoundland, the White Sea and Siberia in Russia (Fedonkin, 1992) and Australia (Nedin & Jenkins, 1998). *Charnia* specimens from the White Sea show distinct morphological differences of dorsal and ventral sides of the frond. In the future, when the various preservational shapes are compared in detail, it seems likely that this fossil will be useful as a reliable index fossil for the Ediacara biota – hopefully for long-range correlation of rock sequences.

The recent discovery of a new elongate species up to 2 m long, *Charnia wardi* (Narbonne & Gehling, 2003), far down in the Drook Formation greatly extends the range of *Charnia* worldwide. A manuscript describing these and other species of *Charnia* from Mistaken Point is in press (Laflamme, Narbonne, Greentree, & Anderson, in press). A large and often-figured rangeomorph frond from Mistaken Point, sometimes informally referred to as the "fractal frond," is presently under study.

A number of tree-shaped forms have been found in the Mistaken Point Formation. These have a central "trunk" and rudimentary attachment discs, but differ from *Charniodiscus*-like fronds in that the main "trunk" branches itself, such that the resulting outline is spade-like or fan-shaped. The finer-scale structures are similar to those of the spindle-shaped forms. Trunk length and shape of the fronds are variable and may reflect the presence of two or more distinct species.

Bush-shaped forms given the name *Bradgatia linfordensis* (Boynton & Ford, 1995) from Leicestershire, UK, are confined to the Mistaken Point Formation on the south coast of Newfoundland, and on the SW side of Conception Bay, northern Avalon Peninsula. Preserved in positive relief, *Bradgatia* varies from oval shaped (4-12 cm long) with the confluence of branching at one end, to circular (10-16 cm in diameter) with branching originating from within the outline.

The bush-like forms share a similar fine structure with the tree-like and spindle forms. On some horizons they show orientations in two different directions, but in other cases their shape and orientation is quite variable, revealing no preferred direction. The bush-shaped forms lack any evidence of attachment discs so may have been free living.

Organisms with a pectinate shape are present but rare in both the Mistaken Point and Trepassey formations. Most specimens show a bow-shaped stolon, or central structure, preserved in positive relief with 6-12 branches along one side. In the largest specimens (20-50 cm long) each lateral branch can be seen to be frond-like, with secondary branches emerging at a low angle. Smaller specimens (8-25 cm) lack any detail beyond the pectinate arrangement of branches. There is no convincing evidence for an attachment stalk or preferred orientation of specimens, in either the pectinate or the network forms, which gives some support to Jenkins' reconstruction of these forms as rooted, bottom-dwelling organisms.

Organisms showing a network body form range from 30 to 60 cm in diameter. They are composed of a bipolar network of at least three orders of fractal branching from an original zigzag axis. Individual branches of these networks resemble those of the spindle-like forms, except that alternation of branching is more consistent; they lack the fine fractal bundles of elements. The variable preservation of the branches suggests that the pattern of growth was not confined to two dimensions. These network rangeomorphs are quite rare and confined to only a few horizons in the Mistaken Point Formation.

Irregularly ornamented discoidal impressions (5-60 cm in diameter) occur sporadically from the top of the Drook Formation to the top of the Trepassey Formation. They are common and quite varied in shape, especially within the Mistaken Point Formation. They have been referred to the Charnian fossil, *Ivesheadia lobata* (Boynton & Ford, 1995), known from Leicestershire, UK. *Ivesheadia* is preserved in both positive and negative relief. The smaller "lobate disc" morphs of *Ivesheadia* (10-14 cm in diameter) feature V-shaped ridges, pointing inward, and outlining three to five dumbbell–shaped and circular depressions. In some cases, the ridges merge with a smooth raised and relatively entire margin. The relief of this morph of *Ivesheadia* exceeds that of the thickness of the crystal tuff, and is greater than any other fossil forms in the assemblage.

The larger "bubble disc" forms of *Ivesheadia* (to 60 cm in diameter) are ornamented by rounded to oval craters that are largest near the outside margin. The lobe-shaped hollows are lacking in specimens more than 25 cm in diameter, and the arrangement of circular hollows becomes more regular with increasing size. Where these forms overlap spindle-shaped fossils, it seems that *Ivesheadia* rested over spindles at the moment of burial. Thus, it seems likely that these "morphs" of *Ivesheadia* were the result of the collapse of a chambered, three-dimensional body.

One form of *Ivesheadia* is called a "pizza-disc," and it is preserved in positive relief beneath thick volcanic ash at the top of the Drook Formation, ranging to the top of the Trepassey Formation. The surface of each pizza-disc is covered with closely spaced "pustules" 2-4 cm in diameter. The disc is bordered by a low-relief, sharp ridge. Despite its resemblance to a frisbee-like flat pizza, however, the discovery of a stalk extending away from some of the discs suggests

Figure 89. Pizza disc, Mistaken Point, Newfoundland. Scale in centimeters (P. Trusler). Total length about 35 cm.

Figure 90. Hiemalora, a star-shaped form that occurs sporadically through the Mistaken Point Formation (G. Narbonne). Width of specimen about 5 cm.

that *Ivesheadia* may well have been a very large holdfast onto which an even larger frond-like structure attached. There is no evidence at all, however, to support the idea that these were jellyfish, contradicting the idea of Boynton and Ford (1995).

Some discoidal forms, referred to as "annulate" forms, are made up of concentric, raised ridges. These are rare in the Mistaken Point and Trepassey formations (Anderson, 1978). The largest reaches about 28 cm in diameter. They resemble various unnamed, quilted, annulate discs from Ediacara assemblages in South Australia, NW Canada and Russia. Just what they were is at this point anyone's guess until better-preserved, more complete material is uncovered.

Other rarely occurring forms include simple, narrow triangular or arrowhead-shaped forms recently named *Thectardis avalonensis* (Clapham *et al.*, 2004). These are preserved in positive relief and are 10-12 cm long and 3-5 cm at the base. *Thectardis* is interpreted as a collapsed cone. *Thectardis* is one of the few taxa at Mistaken Point that is not obviously a rangeomorph, but its affinities are otherwise uncertain.

A radial form consists of a number of spokes extending from a small ring or crater, less than one-eighth the diameter (12-18 cm) of the array. The 12-16 spokes are usually simple, but in a few specimens they bifurcate. Like the pectinate forms, these radial

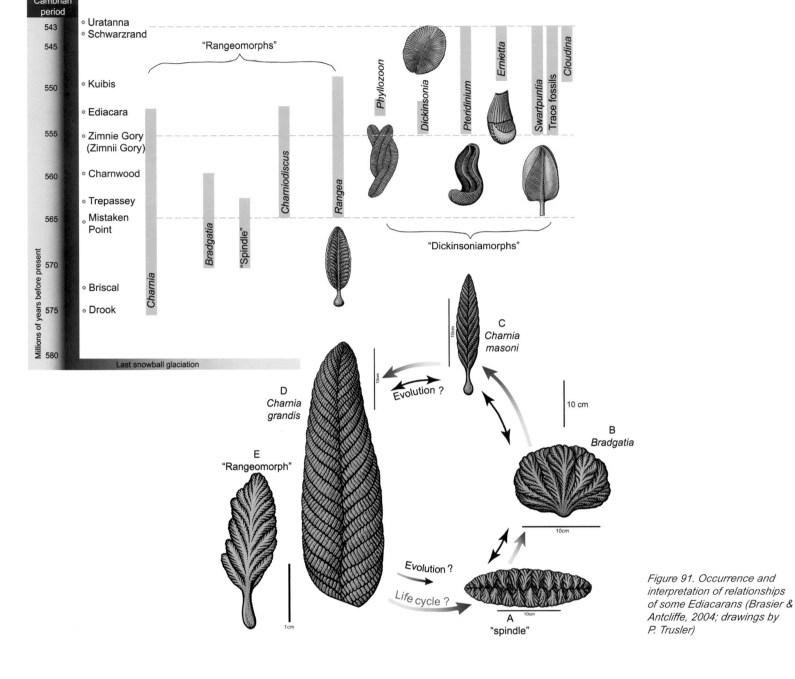

Figure 91. Occurrence and interpretation of relationships of some Ediacarans (Brasier & Antcliffe, 2004; drawings by P. Trusler)

impressions may represent poorly preserved radiating sets of frond-like forms. As indicated by Anderson (1978), there is a superficial resemblance to radiating burrow systems made by organisms digging into the sea floor – but the resemblance is only superficial. So far, no clearly identifiable trace fossils have been found in the Mistaken Point Assemblage. There is also some similarity with the "tentaculate" disc, *Hiemalora,* which is known from the Mistaken Point and Fermeuse assemblages as well as other Ediacaran sites around the world. *Hiemalora* is especially common and well preserved in Ediacaran strata on nearby Bonavista Peninsula, where there is evidence that it was the tentaculate base of a frond (O'Brien & King, 2004, 2005; Hofmann, O'Brien & King, 2005; O'Brien, King & Hofmann, 2006).

Fermeuse Assemblage

Aspidella was the first found of the Fermeuse fossils, one of the discoid types. *Aspidella* was originally interpreted as a fossil mollusk or "medusoid," a jellyfish. These interpretations were increasingly discounted by late 19th- and 20th-century paleontologists, who regarded this form as inorganic in origin. Some suggested it was a gas-escape structure, others a concretion and still others noted that it might be only a mechanical suction mark (see the excellent review in Hofmann, 1971). *Aspidella* was listed as a "pseudofossil" in authoritative compilations of the time, such as the *"Treatise on Invertebrate Paleontology"* (Hantzschel, 1962). Unfortunately, few of the published opinions seem to have been based on extensive field observations. For example, although most workers regarded *Aspidella* as a sedimentary structure, only Hsü in his unpublished M. Sc. thesis of 1972, carried out at the Memorial University of Newfoundland, attempted to consider *Aspidella* in its sedimentary context by actually observing it in the field.

Later discovery of unequivocal Ediacara-type fossils in the Mistaken Point Formation by Misra (Anderson & Misra, 1968), more than a kilometer stratigraphically below the Fermeuse Formation in which *Aspidella* was first found, led

Hofmann (1992) to reconsider *Aspidella* worthy of renewed study. He suggested that comparisons should be made with similar Ediacara body-fossils, a view reinforced by Bland (1984) and Richard Jenkins (1989, 1992). Based on their examination of *Aspidella* from localities across the Avalon Peninsula, Jim Gehling and colleagues (2000) concluded that *Aspidella* was most certainly organic and probably represented an attachment disc, an "anchor," of a frond-like Ediacara organism. In addition to *Aspidella*, Gehling's group found numerous specimens of other enigmatic, but characteristic, Neoproterozoic fossils such as *Palaeopascichnus*, *Intrites* and *Yelovichnus*, as well as a new fossil *Triforillonia costellae*. *Triforillonia* was the first Proterozoic fossil formally described from the Avalon Peninsula since Billings named *Aspidella* more than a hundred years earlier. Together, these fossils make up the Fermeuse Assemblage of the St John's Group, where fossils are preserved in a sequence of alternating shales and thin, sharp-based sandstones – essentially the same sedimentary context in which many other Ediacara assemblages are found around the globe, but differing in that the environment was instead deep-water, upper-slope or delta-front..

Aspidella* is known from the older Briscal Formation thought to be the younger upper part of the Fermeuse Formation, where it reaches its zenith. *Aspidella* is preserved in a variety of ways – resulting in three main types of morphology, depending on size of the specimen and the relative thickness of clay and sand preserving it (Gehling *et al.*, 2000). Specimens in the older Conception Group are generally flat and segmented structures. In other cases they appear to be the rounded anchors or holdfasts of "leaf-like" organisms, such as *Charnia*. On "bed E," in the Mistaken Point Formation, although segmented types are absent, discoidal fossils are present, attached as holdfasts to *Charniodiscus*-like fronds. They are preserved in positive relief with a smooth, concave outer rim and a central disc, suggesting that when their outer zone collapsed, the structure cast from below the attachment zone. The absence of other types of *Aspidella* shapes on beds preserving the complete bodies of *Charniodiscus* gives some support to the idea that *Aspidella* really is an isolated holdfast of perhaps more than one type of frond (Gehling *et al.*, 2000).

Several Avalon taxa are known only from the Fermeuse Assemblage. The tri-lobed *Triforillonia* occurs alongside *Aspidella* in Ferryland. Serial fossils of curved, sausage-shaped and bead-like forms, known variously as *Palaeopascichnus*, *Intrites*, *Yelovichnus* and *Neonerites renarius*, are common on some surfaces in the Fermeuse Formation. These were first described from the Vendian of the East European Platform in Russia and the Ukraine (Palij, 1976; Fedonkin, 1980, 1985) and have subsequently been found in late Neoproterozoic rocks of South Australia, Wales, Namibia and other locales around the world (Gehling *et al.*, 2000). The organic nature of these serial structures has never seriously been doubted, but just what they represent remains enigmatic. The original authors interpreted them as meandering and serially pelleted burrows, a view supported by most workers in the 1970's and 1980's. But this interpretation is not compatible with the branching system observed in some specimens (Haines, 2000) or with the ring-like collapse of elements commonly observed in these fossils (Gehling *et al.*, 2000). They appear instead to be impressions of some unknown soft-bodied form, perhaps bacterial colonies, algae,

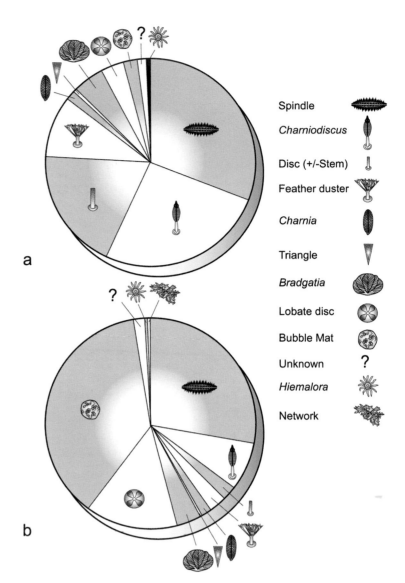

Figure 92. Relative abundance of organisms on E-surface, Mistaken Point, Newfoundland. A. census of the number of specimens per taxon; B. percentage area covered by each taxon (after Narbonne et al., 2001) (D. Gelt).

egg masses or even xenophyophores (Gehling *et al.*, 2000; Legouta & Seilacher, 2001) that grew on top of the sea bottom. It is of some interest to note that no true burrows of any kind have been observed in the Fermeuse Assemblage.

Neoproterozoic Environments Hosting the Mistaken Point Biota

Ediacara-type fossils occur through more than 3 km of sediments on the Avalon Peninsula of Newfoundland, from the top of the Drook Formation to the top of the Fermeuse Formation. Some existed for only a short time in the geologic record, others had longer time-spans. The distinctive frondose genus, *Charnia*, is restricted to terminal Proterozoic rocks in all known localities around the globe. As such it represents a potential index fossil for the terminal Proterozoic. The best index fossil is one that has a very short time range, has a wide global distribution and is very distinctive in its morphology so that it can be easily recognized. *Charnia* has all of these characteristics, though for some forms like *Charnia masoni*, the time range could be upwards of 20 million years (M. Fedonkin, *pers. com.*).

Figure 93. A cobble from the Gaskiers Formation dated at 580 million years, dropped by a retreating glacier. Striations on the rock reflect the multidirectional movement of the Gaskiers glaciers (G. Narbonne). Width of cobble on left about 10 cm.

Figure 94. Palaeopascichnus *was formerly regarded as a trace fossil, but now interpreted as a serial body fossil of uncertain origin. It is abundant in the Fermeuse Formation of Newfoundland (G. Narbonne).*

Figure 95. The Fermeuse Formation is the uppermost unit containing Ediacara fossils in Newfoundland. It consists of dark mudstones deposited in the upper part of a submarine slope environment and contains abundant fossils of the Ediacaran disc Aspidella. Jim Gehling for scale (G. Narbonne).

In the Avalonian succession the known stratigraphic range of *Charnia* is from the top of the Drook Formation in the Conception Group to the top of the Trepassey Formation on the south coast of the Avalon Peninsula. Although the pustulate discs referable to *Ivesheadia* range from the top of the Drook Formation to the top of the Trepassey Formation, they are more poorly understood than *Charnia*. Like *Aspidella*, they represent a range of shapes and may be holdfasts of larger organisms, so are not so distinctive and useful to the geologist attempting to date and map the rocks.

The distinctive spindle organisms are apparently nearly unique to the Avalon region, but are now known from the Bonavista area (O'Brien & King, 2004, 2005). Fronds that can be referred to *Charniodiscus* are confined to the Mistaken Point Formation with only rare occurrences in the Trepassey Formation. The common occurrence of *Charnia*, *Charniodiscus*, *Aspidella*, *Bradgatia* and *Ivesheadia* in both the Charnwood Assemblage of Leicestershire and the Mistaken Point Assemblage of the Avalon Peninsula gives geologists a basis for assigning them a similar age. Additional links can be made to rocks containing the Dyfed Assemblage of South Wales (Cope & Bevins, 1993) with those yielding the Fermeuse Assemblage – both share *Aspidella*, *Hiemalora*, and *Palaeopascichnus*.

The Future

One would think that, on an island where the first Ediacaran taxon was named more than 130 years ago (Billings, 1872), everything that could be done on these fossils already has been done. Far from it! The year 2004 rather dramatically showed that, despite the appearance of new papers on Ediacaran fossils in Newfoundland almost every year, significant new sites remain to be discovered. On the Bonavista Peninsula, Sean O'Brien and Art King (2004, 2005) discovered new localities hundreds of kilometers to the northwest of the classic localities on SE Avalon. Fossils are diverse and occur through many hundreds of meters of strata. Also in 2004, Narbonne reported the fossil Lagerstätten near Spaniard's Bay that elucidated the finest-scale architecture of Ediacara fossils for the first time. This demonstrated that most Mistaken Point fossils have a similar architecture consisting of fractally branching modules defining the "rangeomorphs." The major challenge for the future is to work out how these modules are shaped and combined to produce the myriad of rangeomorph constructions in the Mistaken Point biota.

At a basic level, most of the taxa remain to be described scientifically, to have their stratigraphic ranges documented and to be compared with Mistaken Point "survivors" in younger Ediacara assemblages. The Queen's University group is actively working on this in conjunction with Michael Anderson and Jim Gehling, and expects to complete the task within the next few years. Duncan McIlroy at Memorial University of Newfoundland is working independently on disc-shaped fossils in the Mistaken Point biota. High precision U-Pb dating by Greg Dunning at Memorial University of Newfoundland and by Sam Bowring at Massachusetts Institute of Technology will allow calibration of the Mistaken Point biostratigraphy radiometrically to estimate evolutionary rates and to correlate evolutionary stages during this critical phase of animal evolution with major Earth events and processes. There also needs to be a better understanding of conditions under which the Mistaken Point organisms lived, and sedimentological and geochemical studies are currently under way. Don Canfield, Simon Poulton and Guy Narbonne recently reported exciting geochemical evidence that the Mistaken Point biota evolved in response to a marked increase in atmospheric oxygen at the end of the Gaskiers glaciation, 580 million years ago (Canfield *et al.*, 2007). This discovery may help to provide an explanation for the sudden appeaarance of large and complex animals, a problem that confounded Charles Darwin, noted in Chapter 10 of the *Origin of Species* and known as "Darwin's Dilemma" ever since.

And, with further exploration of the windswept coasts of Newfoundland undoubtedly, in the future, understanding of such "forgotten" architectures will throw light on the early experiments that led to the modern biodiversity of which humanity is but a part.

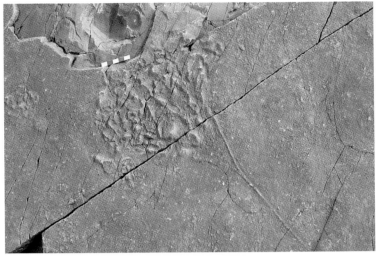

Figure 96. *Pizza-disc with stem, Portugal Cove South, Newfoundland (G. Narbonne).*

Figure 97. *Western Cove surface with Charniodiscus spinosus and rangeomorphs, Newfoundland (G. Narbonne). Scale in cm.*

Web Addresses of Interest

http://geol.queensu.ca/people/narbonne/recent_pubs.html
http://geolo.queensu.ca/people/narbonne/cur_research.html
http://geol.queensu.ca/museum/exhibitis/ediac/mistaken_point/mistaken_pt.html
http://geol.queensu.ca/museum/exhibits/dawnet.html
http://geol.queensu.ca/museum/exhibits/drook./
http://geol.queensu.ca/museum/exhibits/oldanim/oldanim.html
http://geol.queensu.ca/museum/exhibits/archean/archean.html
http://geol.queensu.ca/museum/exhibits/euk/euk.html
http://geol.queensu.ca/museum/exhibits/ediac/ediac.html
http://gsc.nrcan.gc.ca/paleochron/06_e.php
http://gsc.nrcan.gc.ca/paleochron/18_e.php
http://www.gov.nf.ca/parks/wer/r_mpe/index.html
http://www.gov.nf.ca/tourism/welcome/body.html
http://www.env.gov.nl.ca/p[arks/wer/r_mpe/directions.html

http://www.ucmp.berkeley.edu/vendian/mistaken.html
http://members.rediff.com/mistakenpoint/papers.html
http://members.rediff.com/mistakenpoint/paper1.html
http://members.rediff.com/mistakenpoint/paper2.html
http://members.rediff.com/mistakenpoint/paper3.html
http://members.rediff.com/mistakenpoint/gsi1.html
http://members.rediff.com/mistakenpoint/about.html

CHAPTER 4

The Nama Fauna of Southern Africa

P. Vickers-Rich

Namibia is a magic land of stark beauty and, for the geologist, a paradise. Because of the profound aridity of this country along the southwestern coast of Africa, rocks are exposed, not obscured, by the sparse cover of vegetation, plants toughened by an environment of extremes. Leaves, when present, are small, their cuticle or outer skin thick, their entire being protective of water loss. Many bear dangerous, elongate spines, warding off the hungry herbivores of today but developed for a similar past protection. It is this extreme environment that is so much a contrast to the world of 500-600 million years ago in this region when a sea of varying depths lay across a shrinking rift valley – water was everywhere and not in short supply. But still, the environment may have had its own extremes, at times piercingly cold; glaciers left their chaotic sedimentary signatures in the form of diamictites that punctuate the rock sequences of southern Namibia and northern South Africa. Cold waters held stores of oxygen in far greater concentrations than warmer seas, and it may be those cold conditions, a "cold cradle," that spurred later development of complex life, eventuating in the biodiverse assemblage of metazoans so well represented in the sandstones and claystones of the Nama Group.

A Cold and Violent Past

The Nama Group is made up of continental shelf sandstones, claystones and siltstones with some coarser types – conglomerates and diamictites, which may reflect the activity of glaciers. Sediments in the lower part of the Nama Group (the Kuibis Subgroup in particular) alternate between siliciclastics and carbonates, while higher up in the section (the Schwarzrand Subgroup), carbonates dominate. Thus, there seems to have been times of warmth and cold-alternating with one another through time, with perhaps warmer conditions more prevalent in younger times. These sediments range in age from Late Precambrian (Neoproterozoic) into the Cambrian. Some of the Neoproterozoic rocks reflect a period of violent volcanic eruptions that spread ashes widely across land and sea – ashes that give dates ranging from around 548 to 538 million years (Saylor *et al.*, 1995). The topmost of these provides the youngest date for any of the large, soft-bodied Neoproterozoic (Ediacara) metazoans, a date determined on an ash bed

Opposite page: Artist's recreation of Pteridinium *fossils in situ against a backdrop of the Nama Group outcrop on Farm Aar, Aus region, southern Namibia (P. Trusler)*

Figure 98 Outcrop of Nama Group rocks, Swartpunt, southern Namibia (P. Vickers-Rich).

Figure 99. Outcrop of Nama Group where Swartpuntia was discovered, southern Namibia (P. Vickers-Rich).

confined within a deep channel cut into the Spitzkop Member of the Urusis Formation, Schwarzrand Subgroup of the Nama Group. The metazoans that survived this late in time carry the names *Swartpuntia* and *Pteridinium*, contemporaries of the small carbonate-shelled *Cloudina*, which built great reefs in the limy waters of the Nama Sea. Above the Schwarzrand lies the Fish River Group, mainly of Cambrian age.

Figure 100. Swartpuntia *from southern Namibia (M. Fedonkin and P. Vickers-Rich). Length of main shaft of specimen about 11 cm.*

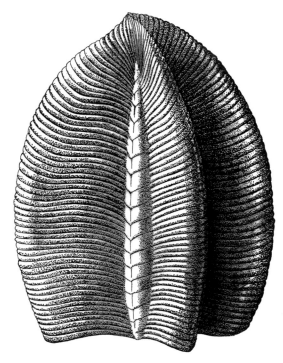

Figure 101. *Artist reconstruction of* Swartpuntia *(A Besedina and I. Tokareva).*

Figure 102. Outcrop of Nama Group rocks on Farm Aar, near Aus, southern Namibia (P. Vickers-Rich).

The Nama Group spreads across much of central and southern Namibia, thinned by a ridge in the middle, the east-west oriented Osis Ridge. To the south of this ridge occur the majority of complex metazoan fossils of Precambrian age, abundant in the Kuibis Subgroup, particularly in the Kliphoek Member of the Dabis Formation. The Nudaus and Urusis formations, subdivisions of the overlying Schwarzrand Subgroup, also contain complex metazoans. The Nama itself lies atop crystalline basement rocks, some of which surely stood above water, forming islands at the time the "Nama Sea" covered this part of southern Africa.

The Kuibis and Schwarzrand sediments are cyclic – cycles that begin with sediments likely laid down in a terrestrial environment or shallow intertidal to subtidal setting. These shallow-water settings were then slowly flooded, becoming near-shore marine environs – thus as a geologist climbs up the rock sequence, conglomerates, sandstones (Kanies Member, Dabis Formation) and carbonates (Mara Member, Dabis Formation) give way to silica-rich siltstones and sandstones of the Kliphoek Member, Dabis Formation, then grade back again to carbonates (Mooifontein Member of the Zaris Formation).

These sedimentary shifts mirror a rise and fall of sea level. It is the Kliphoek Member, reflective of a shallow-water marine environment – tidal channels and delta fans (Jenkins, 1992) – that at first nurtured, then preserved, the greatest variety of metazoans in Namibia – *Ernietta, Namalia, Orthogonium, Pteridinium, Rangea* and a number of new forms found by a joint Namibian Geological Survey-Monash University-Russian Academy of Science Expedition in 2003. A dated ash in the overlying Mooifontein Member of 548 +/- 1 million years gives a minimum age for this assemblage. A diamictite in the Kliphoek may reflect cool to cold glacial conditions, but may also simply represent massive downslope wasting of sediments off a delta fan and nothing more.

Figure 103. Charlie Hoffmann (Karl Heinz Hoffmann), chief geologist with the Namibian Geological Survey, Windhoek, who has mapped the Nama Group in detail, being interviewed for a television documentary in southern Namibia (P. Vickers-Rich).

Figure 105. Edward and his cousins, Farm Aar, southern Namibia. Edward, who worked on Farm Aar, has an excellent ability to find Ediacarans and he and his cousins are most helpful in the Nama language, to be used in naming of the new fossil material found on Farm Aar (P. Vickers-Rich).

Figure 106. Barbara Boehm-Erni (left) and Bruno Boehm, owners of Farm Aar, who have fundamentally facilitated the field-work in southern Namibia (M. Fedonkin).

Figure 104. Kombada and Aihey, two young mapping geologists from the Namibian Geological Survey, Windhoek, at work in southern Namibia (P. Vickers-Rich).

Diamictites can form in many ways and geologists are yet debating the interpretations of the Kliphoek diamictites.

The architecture of the Kliphoek sandstones, that is, the cross-bedding patterns, clearly reflect the strong currents that deposited these sediments – very likely offshore currents, which may have been related to a river entering the sea at some nearby point or to storm activity. This very architecture, because of its massive nature, requires water depths of several meters or more. And there is no evidence at all for subaerial exposure, such as salt deposits or desiccation cracks, so the conclusion is that the sediments formed under several meters of water that lay on a shallow, continental shelf.

The overlying Schwarzrand Subgroup also records a cycle of events – beginning with the near-shore to offshore shales and sandstones of the Nudaus Formation and basal

Figure 107. Ash beds in the Nama Group (M. Fedonkin).

Figure 108. Pteridinium, Nama Group, Farm Aar, southern Namibia (S. Morton). Basal width of block 25 cm.

Figure 109. Nemiana, Nama Group, Namibia (in collection of the Geological Survey of Namibia, M. Fedonkin and P. Vickers-Rich). Diameter of spheres averages 1.5 cm.

Urusis Formation with a number of "event beds" giving evidence of massive downslope avalanches. *Pteridinium*, *Rangea*, *Nasepia* and *Paramedusium* are characteristic of these sediments. These sediments grade upwards into the

carbonate-rich Huns and Feldschuhorn members of the Urusis Formation with their stromatolite reefs containing millions of small shelled *Cloudina*, surely representing a shallow marine environment on a carbonate platform at the continental edge that was influenced by tides and nearby river inlets. Characteristic of these rocks are the metazoan *Cyclomedusa* and the algal *Vendotaenia*.

The overlying Spitzkop Member of the Schwarzrand is made up of sediments reflecting the building delta as a substantial river flowed into the shrinking sea. Several ashes have been dated in these sediments, the oldest 545.1 +/- 1 million years and the youngest 543.3 +/-1 million years. *Pteridinium* and the unusual form *Swartpuntia* are characteristic fossils. The Urusis Formation is deeply incised and overtopped by the uppermost of the Schwarzrand sediments of the Nomtsas Formation – coarse conglomerates, diamictites, which filled the sometimes kilometer-deep canyons cut into the underlying sediments (Saylor, Grotzinger & Germs, 1995). Some researchers have suggested that such dramatic downcutting, clearly indicative of an equally dramatic drop in sea level, could have been caused by global glaciation. Ash beds in the Nomtsas Formation have been dated at 538 +/-1 million years. Nowhere else on Earth are there indicators of a glacial event at this time, so the jury is still out on just what this downcutting event represents. Could there have been a local glacial event in the narrowing rift valley? Was there tectonic uplift that brought about such incision? Whatever happened at this time, it brought an end to the Ediacarans that had been so abundant and biodiverse in this part of the world for more than 10 million years.

Nomadic Namibia

As the seas waxed and waned and left their silts, sands and limy sediments, which now stand as stark sentinels to their history, the world was not as it is today. The climate was of glacial conditions and quite dynamic – shifting from cold to warm, from glacial settings to warmer seas and then back again. Around 750 million years ago there may have been a supercontinent called Rodinia, which straddled the Equator. Namibia was a part of this possible supercontinent – and the word "possible" is used here because there is great discussion at this moment about just what reconstruction of past continental arrangements are the most likely. Older (Truswell, 1977; Hunter, 1984) as well as more recent reconstructions (see Chap. 2) place the Congo and Kalahari regions (called cratons, referring to the continental kernels represented by these regions) close to the Equator of the time. The northern part of Namibia today has as its framework the Congo Craton in the southeast, underlain by the Kalahari Craton. Sandwiched between those two stable continental blocks are the sediments that were laid down in the shallow sea, forming the Nama Group.

At the time the Nama sediments were being deposited in a shallow sea, given the name of the Khomas Ocean, the Congo and Kalahari cratons evidently were moving towards one another. A part of South America was moving towards both of them, narrowing the Adamastor Sea that lay between the African and South American continental cratons as well as closing the Khomas Ocean. Between 750 and 500 million years ago these seas disappeared as the three continental blocks converged, with the Kalahari diving beneath the Congo and the South American cratons, and formed a lengthy

Figure 110. Cap carbonate at top of glacial diamictite in Late Neoproterozoic sequence of southern Namibia (M. Fedonkin).

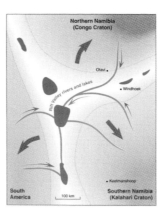
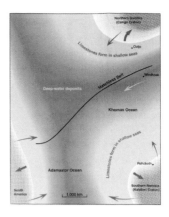
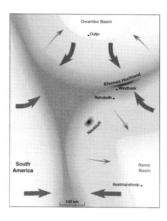
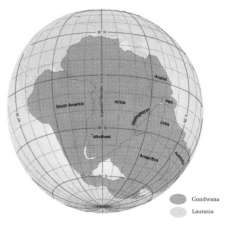

Figure 111 Geography of Namibia from around 850 to 180 million years ago. Upper left, 840 million; upper right, 750-55 million years; lower left, 550 million years; lower right, Gondwana, 550-180 million years (modified from Mendelsohn et al., 2002 with permission of New Africa Books).

mountain range, the Transgondwanan Supermountains (Squire *et. al.*, 2006) culminating in the consolidation of the supercontinent Gondwana.

Thus, Namibia today is a patchwork of past continental fragments that have merged through time and have moved about on the Earth's surface. In late Neoproterozoic times, what is now Namibia was first covered by seas that were influenced by rivers flowing off the continental fragments. The Adamastor and Khomas oceans ranged from shallow shelf seas to deeper waters. As these seas narrowed due to the movement of the Earth's rigid plates, mountains formed and sediments deposited during this time reflect the change from marine to terrestrial conditions. At the end of Neoproterozoic times the great supercontinent of Gondwana had formed, of which Namibia was a part, and the peregrinations of that continental mass led Namibia southward to occupy its current latitudinal position straddling the 24 degree south latitude parallel – ranging from close to 17 degrees south along the Angolan border to the north to close to 29 degrees south along the Orange River bordering South Africa. For more than 700 million years Namibia has been a nomad.

Ghosts in the Desert

Some of the first Precambrian metazoans were found in Namibia. Gürich put a formal name on the impressions, *Rangea schneiderhoehni* in 1930, recovered from the quartzites of the Kliphoek Member, Dabis Formation, Nama Group. But fossils had been found as early as 1908-1914 in the ruins of the small shelters built by German soldiers on the farms Plateau and Aar, soldiers who had been transported to this desolate outpost of the German Empire as the world went to war (Pickford, 1995).

Some soldiers, evidently bored with duties or just plain curious, found the first of the Nama fossils – vague outlines and impressions in the sandstone slabs. Just what these "ghosts of the desert" represented, and where they should be placed in the organic scheme of things, has been a point of debate ever since the young soldiers discovered them.

Figure 112 Walls of shelters built by German soldiers during World War I in southern Namibia. Many of the slabs of Nama Group rocks used to build these shelters contained fossils (P. Vickers-Rich).

Figure 113. Field conference, a joint trip of the UNESCO International Correlation Projects 478 and 493 to the Neoproterozoic sections in South Africa and Namibia in 2002 led by Claudio Gauchier (second from right, kneeling) from Uruguay, which included Precambrian and Cambrian researchers from around the world (M. Fedonkin).

And the debate still rages at scientific meetings, in the halls of academia and in the pages of scientific journals – good, healthy scientific activity.

Some paleontologists (Seilacher, 1989, 1992) consider the Nama fossils to belong to a homogeneous phylum (Vendozoa) or even a kingdom (Vendobionta), which died out by the beginning of the Cambrian – a lost biological design. Bruce Runnegar and Mikhail Fedonkin (1992) have summarized the variety of ideas. The Nama organisms have been called gigantic single-celled organisms that absorbed energy through body walls or housed photosynthetic symbionts, which produced energy for the "Namas" to feed on. Other suggestions are that the Nama fossils actually are related to more familiar groups known in the Phanerozoic – possibly precursors of the Coelenterata (soft corals, jellyfish, hydrozoans), the Conulata (conularids), Annelida (worms), Trilobitomorpha (trilobites and relatives) or even Chordata (chordates) (Richter, 1955; Glaessner, 1959,1979; Germs, 1973; Hahn & Pflug, 1985).

Another suggestion made earlier by Pflug was to place the group in its own subdivision recognizing a characteristic of many of these fossils, their leaf or frond-like morphology – the Petalonamae or "Nama petals" for "large, sessile, colonial and gregarious soft-bodied organisms" (Pickford, 1995), a group with a more complex physiology than that proposed by Seilacher. Pflug went further and recognized a number of subcategories of the Petalonamae, examples being the Rangeomorpha (Pflug, 1972) and the Psammocorallia (Seilacher, 1992).

When Martin Pickford (1995) summarized the knowledge to date of the Nama fossils he rightly commented, "In this report, I do not use supra-generic names, on account of the fact that no two authors appear to agree on supra-generic placement of any of the Nama Group fossils, a fact that not only highlights the complexities of interpreting such fossils, but also indicates how ignorant we are of the life forms that inhabited Earth in the late Proterozoic period. In question are the origins of at least two kingdoms, Plantae and Animalia. One cannot overemphasize the scientific value of the Nama fossils for settling, or more correctly, for stimulating the debates that revolve around the origins of organisms that dominate the Earth today."

A Plethora of Petalonamae

The Nama Group has produced a varied cast of fossil "characters." When Gürich first named *Rangea schneiderhoehni*, he noted there were two different forms – he called them *plana* and *turgida*. He also set up another species, which he called *Rangea brevior*, but this appears just to be a variant of *schneiderhoehni*. *Rangea* has a beautiful, leaf-like form with what is called fractal branching – primary, secondary and even tertiary splitting. Rangeomorphs, of course, are common forms in the older Newfoundland sequences (discussed in Chap. 3). *Charnia* from Charnwood Forest in England is also considered a rangeomorph. All of these ancient organisms seem to have been bottom dwellers, and they could have been filter feeders, often occurring in concentrations (Germs, 1973; Pickford, 1995). They appear to have been best preserved, along with *Pteridinium*, in storm-deposited sands, buried by sand avalanches loosened by the raging seas that moved masses of sediments down delta fronts, episodically burying all in their path.

Rangea has been interpreted in many ways. Richard Jenkins (1985, 1992) suggested that individual fronds were populated by "myriads of tiny polyps that derived their food from suspended organic matter." He viewed them as rather stiff organisms with a sort of hydrostatic inflation of the colony and suggested that they had a cnidarian level of organization, forms close to the ancestry of the living corals. Paleontologists continue to debate their true relationships. *Rangea* is known from the older parts of the Nama Group, from the Kliphoek Member, Dabis Formation, Kuibis Subgroup, upwards into the base of the Schwarzrand Subgroup in the Neiderhagen Member of the Nudaus Formation, so has a somewhat restricted distribution.

Jerzy Dzik (2002) has presented another, detailed analysis of *Rangea* and quite rightly noted that the "process of fossilization did not reproduce the original external morphology of the organism but rather the inner surface of collapsed organs," which he suggested represented a system of sacs connected by a medial canal. He pointed out that *Rangea* had four-fold symmetry and noted that this was also the case of the White Sea fossil *Bomakellia* and probably a number of other frond-like forms from the Neoproterozoic. The complex anatomy of *Rangea* was not reflected in the smooth outer surface, which is certainly well preserved in some specimens from Namibia. Dzik noted similarities of *Rangea* with Cambrian forms like *Thaumaptilon* and *Fasciculus* of the Middle Cambrian Burgess Shale and *Maotianoascus* from the Early Cambrian Chengjian fauna of South China, all of which he suggested may have affinities with the living combjellies, the ctenophores. Dzik proposed that the main reason Cambrian fossils are so different from those of the Neoproterozoic, besides the obvious presence of hard parts in Cambrian forms and the lack of them in the Neoproterozoic, is the manner in which each was preserved. He suggests that taphonomic styles (that is, how things are preserved after they die) clearly dictated what aspect of the organism was reflected by the fossils (Dzik, 1999; Gehling, 1999). Often, in Cambrian and younger organisms it is the external morphology that is represented, whereas the soft-bodied Neoproterozoic forms more often leave impressions of their internal anatomy.

Dzik (2002) does a great service to understanding the true nature of *Rangea* by suggesting paleontologists should not describe morphology using "loaded" terms, which

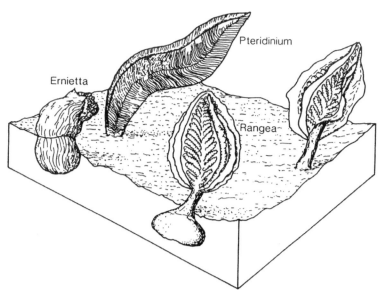

Figure 114. Reconstruction of several Petalonamae from Namibia (R. Jenkins).

Figure 115. Rangea schneiderhoehni, *Farm Kuibis, Aus region, Namibia (M. Fedonkin and P. Vickers-Rich). Width of specimen about 7.5 cm.*

Figure 116. Rangea *cf.* schneiderhoehni, *Farm Aar, Aus region, Namibia (S. Morton). Width of block about 16 cm.*

Figure 117. Ernietta *concentration, Nama Group, Farm Aar, Aus region, Namibia (S. Morton). Width of individual specimens about 2 cm.*

Figure 119. Pteridinium *cf.* schneiderhoehni, *Farm Aar, Aus region, Namibia (P. Vickers-Rich). See Figure 120 for scale.*

5cm

Figure 118. Pteridinium, *Namibia (specimen studied by Dolf Seilacher and Dima Grazhdankin, in collections of the Namibian Geological Survey) (M. Fedonkin and P. Vickers-Rich).*

automatically imply a particular affinity with another organism. So, he suggested that terms such as "petaloids" be used for the branching fractal subunits, that "frond" be used for the entire leaf-shaped collection of petaloids, and "rachis" be used for center of the frond. Halves of the individual fronds, he calls "vanes." These terms do not imply any particular relationship to any living or extinct invertebrate group. Dzik even uses the general term "bulb" for the end of the structure that appears to be a holdfast that anchored *Rangea* into the substrate. He notes that in all specimens of *Rangea*, especially those recovered from Namibia, there has been "no convincing evidence for the presence of any separate zooids," which would be expected if *Rangea* were a sea-pen or some sort of soft coral, a pennatulacean. He disagrees with Dolf Seilacher (1989), who suggested that *Rangea* and many other Neoproterozoic organisms be placed in the Vendozoa or Vendobionta based on the serial arrangement of transverse

units. Dzik favors classifying together the "frond-like forms, with pinnately arranged internal organs, well exemplified by *Rangea*" and using the name Pflug proposed, Petalonamae. This classification is also that used by other researchers, such as Fedonkin (1985) and includes *Charniodiscus*, *Bomakellia*, *Mialsemia* as well as a number of forms from the Avalon sequence in Newfoundland. And Dzik's suggestion that these all may be related to the living ctenophorans, the combjellies, deserves serious consideration. Today ctenophorans are not bottom dwellers, but in the Neoproterozoic, their ancient relatives could well have preferred a sedentary life, rather than the pelagic lifestyle they have today.

Pteridinium is not so time-restricted as *Rangea*, occurring in rocks from near the base of the Nama to the uppermost parts of the Schwarzrand Subgroup, in the Spitzkop Member of the Urusis Formation – in the uppermost sediments with Ediacaran fauna. *Pteridinium simplex* was described by Gürich in 1930, and it is one of the most common, large metazoan fossils in the Nama. It is preserved as a boat-shaped fossil that can reach up to more than 30 cm in length. Gürich placed *Pteridinium* in his Petalonamae, and it is indeed a weird form.

Individual fossil *Pteridinium*s often occur as single, bilateral units, called fronds, but consisting of three "veins," two of them usually preserved on the bedding plane, while the third is oriented oblique or perpendicular to the bedding plane (M. Fedonkin, *pers. obs.*). These veins in turn appear to be made up of tubes that run at right angles to the length of the vein, and these are filled with sand – in fact, the whole structure appears to be filled with sand. No specimen of *Pteridinium* actually displays terminations of the fronds, but the individual veins do narrow toward either end of the frond. This makes it impossible to determine if it was attached to the sea floor, or if it simply lay on the sea floor or even more bizarrely may have completely or partially lived in the sediments as suggested by Grazhdankin and Seilacher (2002). Many specimens of *Pteridinium* are bent and curved, but not broken, and their

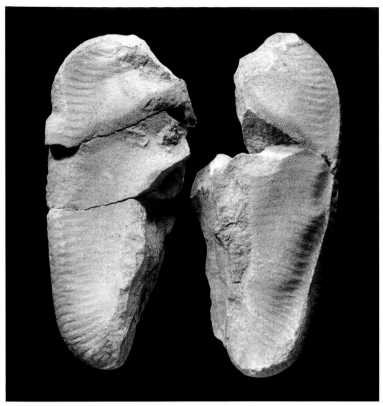

Figure 120. Pteridinium *cf.* schneiderhoehni, *Namibia, individual from block studied by Seilacher and Grazhdankin (M. Fedonkin and P. Vickers-Rich). Maximum width of specimens about 5.5 cm.*

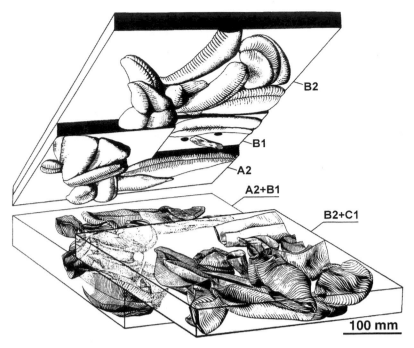

Figure 121. Reconstruction of the mode of preservation of Pteridinium *by Dolf Seilacher and Dimitri Grazhdankin (2002).*

preservational style indicates that in life they had a fairly tough, not jellylike, construction, which was indeed quite flexible, yet structured. They seem to be preserved as internal molds (Jenkins, 1992), with the living organism apparently being filled by sand near or after death, a bit like a plastic bag filled with sand being preserved and buried in beach sands today – if the bag were removed somehow and the insides preserved, perhaps this would be a good model of the preservation style of *Pteridinium*.

Pteridinium is truly an enigma at present. Was it a form that lay buried in the sediments as Seilacher *et al.*, (2003) suggested, or a giant protist that simply absorbed its food from its surrounds? Was it a colony of more complex organisms with polyps inhabiting the tubular structure that could have been filter feeders? For that matter did *Pteridinium* live within, on or part in and out of the sediments? These questions simply have not yet been answered. Fedonkin (2000) reconstructed *Pteridinium* as a tube with three longitudinal septae inside, which harbored a soft-bodied organism. Whatever it was, *Pteridinium* was abundant and had a long time range, occurring from near the base of the Nama Group to the very top, living alongside a changing physical and biotic landscape and surviving from the time *Rangea* (in rocks older than 548 million years) appeared until both it and the late-occurring petalonam *Swartpuntia* (Narbonne *et al.*, 1997) disappeared sometime after 543 million years ago.

And then there is a form called *Ernietta* – and *Erniobaris, Erniaster, Erniobeta, Erniocarpus, Erniocentris, Erniocoris, Erniodiscus, Erniofossa, Erniograndis, Ernionorma, Erniopelta, Erniotaxis,* etc., etc., etc., almost *ad infinitum*. Most, if not all, of these names appear to be what paleontologists call tapho-names. They are different forms of preservation, or different stages of growth, of one particular or a few species.

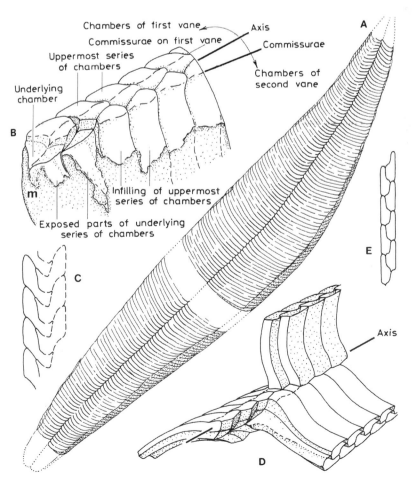

Figure 122. Reconstruction of Pteridinium simplex *by Richard Jenkins (from Lipps & Signor, 1992).*

Figure 123. Ernietta *in the collections of the Namibian Geological Survey in Windhoek. Many different species names were given to this material, now sorted in these drawers according to those names. They all may be just different preservational styles of the same species or just a few (M. Fedonkin and P. Vickers-Rich).*

Figure 124. Ernietta plateauensis, *Nama Group, Namibia (in collections of the Namibian Geological Survey, Windhoek; M. Fedonkin and P. Vickers-Rich). Maximum width of specimen about 9 cm.*

Almost all have been combined, or synonymized, with *Ernietta plateauensis* by Martin Pickford (1995), and this combining of names into one has been recognized by many (Runnegar, 1992, among others). *Ernietta* is an intriguing organism or a series of different organisms that apparently lived partly buried in the sediments, for they are found in place in sediments oriented perpendicular to the sediment-water interface and often occur in concentrated groups. They resemble sand-filled socks that apparently were composed of a series of parallel tubes, often constricted midway along the direction of growth by a prominent waist, but exactly what these structures reflect is still uncertain. Sometimes *Ernietta* specimens are found in place, and at other times they seem to have been transported. Some of the smaller forms, only a centimeter or two in diameter, occur as bulbous nodules that at a distance give the impression of a rhynchonellid brachiopod, likely only a superficial guise. They can be very abundant locally and thus far have only been found in the Kliphoek Member of the Dabis Formation, Nama Group.

And just what was *Ernietta*? There is wide disagreement among those who have studied this petalonam. Some think it was a coelenterate-like form (Glaessner & Walter, 1975), but Jenkins (in Lipps & Signor, 1992) noted that it shows no close resemblance to any known member of this group. Jenkins further noted that *Ernietta* seemed to have two phases of growth, separated by a distinct seam, indicating that "the basal part of the sack was complete or entire prior to upward growth of the side walls." He suggested that perhaps the juvenile form of *Ernietta* was rather discoidal in shape with up to 30 or so ribs or tubes on either side and may have been either pelagic or a bottom dweller. Once it reached maturity, perhaps it took up or continued its benthic or even partially infaunal lifestyle and grew upwards forming an elongate tube. He, like Martin Pickford, preferred to relate *Ernietta* to *Pteridinium*, placing both in the

Phylum Petalonamae, originally suggested by Pflug (1972).

In addition to these well-known and oft-abundant forms, many others are known, highlighting the biodiversity of this assemblage. *Orthogonium parallelum* was first named by Gürich in 1930, and at the time he suggested that it might be allied with crinoids. Later Bruce Runnegar and Mikhail Fedonkin suggested that it was one of the "quilted" petalonams, but with its likely loss during World War II, its true identity will probably never be known. *Namalia* is yet another name set up by Gerard Germs (1968) based on a conical fossil that he found on Farm Buchholzbrunn and later more on Farm Vrede in the Kuibis Formation. *Namalia* often occurs in large numbers, and individuals are round to oval in cross-section, with numerous longitudinal ridges. Sometimes *Namalia* is found within the sediments, narrow end down. Some have also been found in lag deposits on bedding planes, suggesting that they had been reworked and transported (Pickford, 1995). Bruce Runnegar (1992) suggested that *Namalia* was actually *Ernietta plateauensis*, simply one of its many growth or tapho-forms.

Other Petalonamae include *Paramedusium africanum*, again named by Gürich, *Nasepia altae*, set up by Gerard Germs (1973), and *Velancorina martina* of Pflug (1966), all of which apparently were forms preserved flat, perhaps in life reclining on the sea floor (Pickford, 1995), certainly like many of the rangeomorphs from Newfoundland.

Preferred Preservation

Jerzy Dzik (1999) has noted that the three most common fossils in the Nama sequence, *Rangea, Pteridinium* and *Ernietta*, are generally not found together. Each has a preferred style of preservation, exclusive to each form. *Rangea*, often with sand-filled basal discs and collapsed fronds, is found within sandstone that seems to represent mass flows (Jenkins, 1985). *Pteridinium* seems nearly restricted to storm- related sediments, massive sand layers in

Figure 125. Ernietta, *Nama Group, Namibia (Geological Survey of Namibia, M. Fedonkin and P. Vickers-Rich). Width of specimen about 4 cm.*

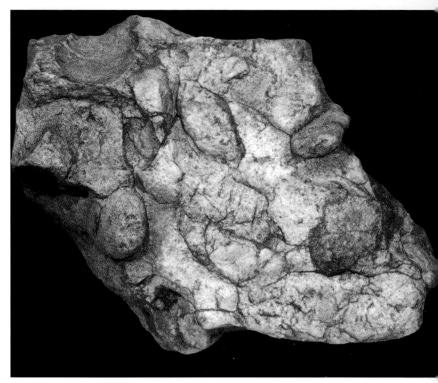

Figure 126. Ernietta *en masse, Nama Group, Namibia (in collections of the Namibian Geological Survey, Windhoek; M. Fedonkin and P. Vickers-Rich). Height of block about 12 cm.*

Figure 127. Namalia villersiensis *(holotype), Dabis Formation, Nama Group, Southern Namibia (J. Gehling). Length of specimen on left 5 cm.*

which fossils lack much of any preferred orientation (Jenkins, 1992; Seilacher, 1984). Dzik suggests that they were likely deposited together with the suspended sediment and "gradually loaded with sand while sinking." *Ernietta* is entirely different, often occurring in dark reddish mudstone/siltstone/ coarser sand, often with preferred orientations. Dzik notes that he observed only one block with *Ernietta* specimens in place, where they likely lived, but in 2003 a joint expedition of researchers from the Namibian Geological Survey accompanied by others from Australia and Russia found numerous blocks with *in situ Ernietta*, so this preservation style is not so uncommon as previously thought.

Dzik (1999) suggests that some of the different shapes taken on by a variety of *Ernietta* specimens could simply be due to a sand-filled organism sinking into a water-saturated bottom sediment that had just been emplaced by downslope avalanches during a storm-initiated event. How far the dead, sand-filled *Ernietta* sank would determine its final shape as a fossilized form. Dzik noted that the body wall of *Ernietta* would likely have been composed of a most flexible material, perhaps something like collagen, which would explain why it was able to "shape-shift," as it seems to have done. He notes that such a tissue as collagen is very durable, much more resistant to decay than many other soft tissues, and this may explain why *Ernietta* fossils so often maintain detailed morphology.

Following on with this reasoning he goes a step further, suggesting that the tubular structure so characteristic of *Ernietta* might have been three-dimensional "tubular boxes" surrounded by collagen, within which there were muscles – drawing an analogue with the myosepta present in the protochordate *Branchiostoma*. He suggests that these sorts of theoretical structures could have been preserved during Neoproterozoic times because collagen and polysaccharide decomposers were not present, and as they developed in the Cambrian, the type of preservation seen in forms such as

Ernietta and *Dickinsonia* became impossible.

The Odd Ones

Two rather interesting forms, *Kuibisia glabra* and *Ausia fenestrata* described by Hahn and Pflug in 1985, have been lumped together, but they may indeed be quite different organisms. Runnegar (1992) suggested that *Kuibisia* actually represents just another preservational form of *Ernietta plateauensis* and also noted in the same paper that another of Pflug's forms, *Petalostroma kuibis*, was so poorly preserved that one could not tell just what it was.

Ausia is another matter, and though represented by only one specimen with several individuals preserved on one rock surface, may indeed represent a group related to the chordates (Fedonkin, Vickers-Rich & Swalla, *in prep.*). *Ausia* has some similarity to material from the Vendian deposits of the White Sea of northern Russia.

And then there is *Protechiurus edmondsi*, first described by Martin Glaessner in 1979. He noted a set of ridges along the length of the "body," preserved in what he called a "cast,"

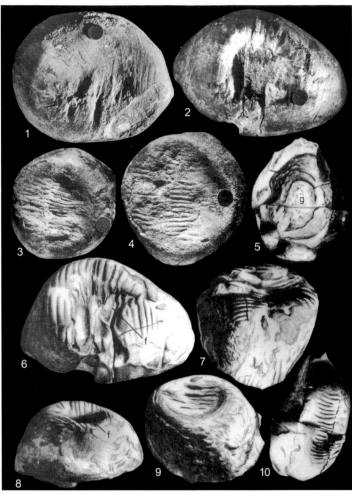

Figure 128. A plethora of Ernietta's from the Pflug monographs (1972-1974) in which many names were proposed for what is likely related to the manner in which one, or a very few, living species produced the fossils. All of the different taxonomic names are represented in these labelled reproductions of the original Pflug plates (courtesy of Palaeontographica).

Pflug Plate 27. Erniodiscus clypeus (holotype No 278) Fig 1 Erniofossa prognatha (holotype No 409) Figs 2, 6-7; Erniodiscus rutilus (holotype No 275) Fig 3; Erniofossa aff. prognatha (No 412) Figs 5, 10. **Pflug Plate 28.** Erniaster apertus (holotype No 238) Figs 1-3, 5-7; Erniocentris centriformis (holotype No 279) Figs 4, 8-9. **Pflug Plate 29.** Erniaster patellus (holotype No 239) Figs 1, 4, 8; Ernionorma peltis (holotype No 380) Figs 2, 5; Ernionorma abyssoides (holotype No 280). **Pflug Plate 30.** Ernionorma peltis (holotype No 380) Figs 1-7 and paratype Figs 2-6, 8-10. **Pflug Plate 31.** Ernionorma aff. clausula (No 370) Fig 1; Ernionorma clausula (holotype No 371) Figs 2-3; Ernionorma tribunalis (holotype No 191) Figs 4-8; Erniobaris epistula (holotype No 268) Figs 9-10; Erniobaris baroides (paratype No 423) Figs 11-12. **Pflug Plate 32.** Ernionorma corrector (holotype No 488) Figs 1-2 (paratype No 198) Figs 3, 5; Ernionorma rector (holotype No 257) Figs 7-9 (paratype No 498) Figs 4, 6; Erniobaris baroides (holotype No 282) Fig 11; Erniobaris aff. baroides (No 264) Fig 10. **Pflug Plate 33.** Erniobaris gula (holotype No 410) Figs 1, 2, 4; Erniobaris parietalis (holotype No 267) Fig. 5 (paratype No 266) Fig 3.6; Erniopelta scrupula (holotype No 236) Figs 7,10; Erniopelta aff. scrupula (No 235) Figs 8-9. **Pflug Plate 34.** Ernietta plateauensis (holotype No 277) Fig 4 (paratype No P.I.16) Figs 1-3, 6; Ernietta aarensis (holotype No 36) Figs 5, 7-8; Ernietta tsachanabis (holotype No 12) Fig 12 (paratype No 20) Figs 10-11. **Pflug Plate 35.** Erniocarpus sermo (holotype No 289) Figs 1-4. Erniocarpus carpoides (holotype No 290) Figs 5-9; Erniotaxis segmentrix (paratype No 393) Fig 10. **Pflug Plate 36.** Erniocoris orbiformis (holotype No 388) Figs 1-4; Erniocarpus sermo (holotype No 289) Figs 5-6. **Pflug Plate 37.** Erniotaxis segmentrix (paratype No 393) Figs 1-4, 7-8; Erniotaxis segmentrix (holotype No 396) Figs 5-6. **Pflug Plate 38.** Erniograndis sandalix (holotype No 192) Figs 1-2, 4; Erniograndis paraglossa (paratype No 385) Fig. 3. **Pflug Plate 39.** Erniograndis sandalix (paratype No 308) Fig 1; Erniobeta forensis (holotype No 287) Figs 2-4, 10; Erniobeta aff. forensis (No 355) Fig 5; Erniobeta scapulosa (holotype No 286) Fig 6; Erniograndis paraglossa (holotype No 384) Figs 7-8; (aff. paraglossa No 382) Figs 9-11.

Pflug, Plate 27

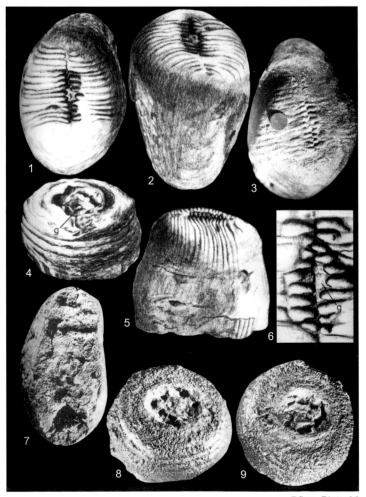

Pflug, Plate 28

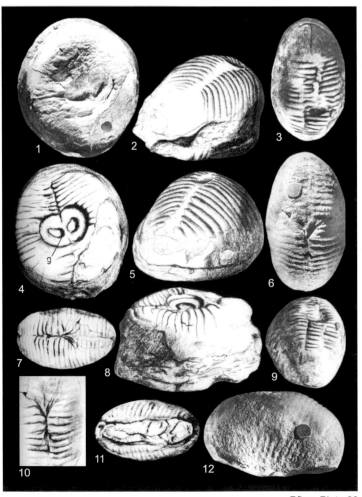

Pflug, Plate 29

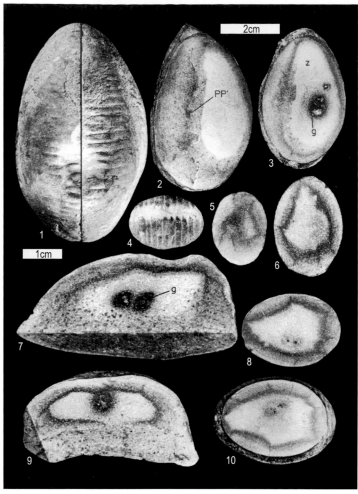

Pflug, Plate 34

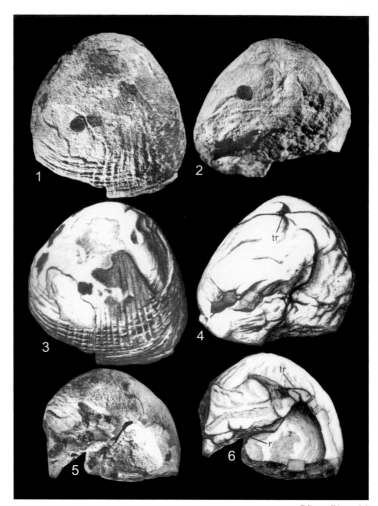

Pflug, Plate 36

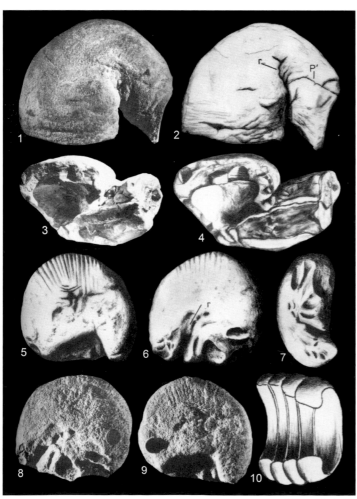

Pflug, Plate 35

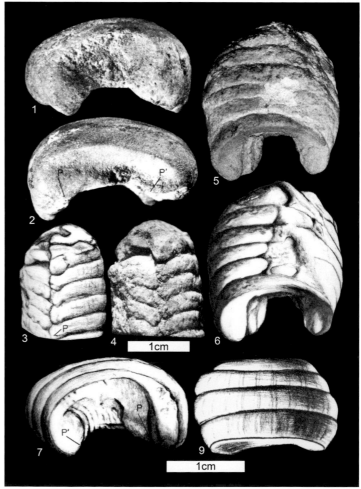

Pflug, Plate 37

Pflug, Plate 38

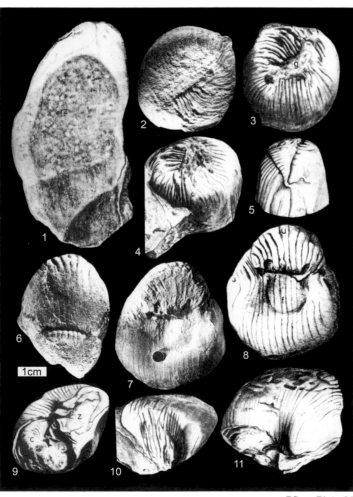

1cm

Pflug, Plate 39

and when compared to the living echiurid worms, he noted similarities. Living echiurids are marine or brackish water forms that can manage quite well in environments with low oxygen levels, feeding on the organic content of sediments. Glaessner noted that their feeding activities "comprise scraping of the sediment with the retractable ventral setae, peristaltic movements of the trunk, and pushing with the alternatingly thinned and thickened anterior end of the body." Glaessner, however, does point out one troubling fact – there are no trace fossils in the Nama sediments that match the sort of sediment processing traces left by modern echiurids – but then behavior could have changed in 570 million years! Runnegar (1992) dismissed Glaessner's suggestion and considered this form a "dubiofossil." *Protoechiurus* is certainly a form with a distinct morphology, but it will require further discoveries to clarify just what it is.

Just as Runnegar dismissed Glaessner's interpretation of *Protoechiurus*, Glaessner (1977) dismissed the suggestion that there were archaeocyathids in the Nama, a form called *Archaeichnium*. Archaeocyathids were important shelled forms during the early part of the Cambrian, their double-walled, coral-like skeletons forming great reefs the world around. Glaessner noted the tubular nature of *Archaeichnium* preserved in the quartzites from near the Hamm River at Grundoorn – again in the Kuibis Subgroup – and thought that they were cemented sand grains built by a worm-shaped animal. Because of their fragmentary nature and their restriction to this one locality, as well as their lack of morphological detail, he decided that it was impossible to determine their true affinity. The way the tubes were preserved is reflective of a thin-walled, compressible, yet flexible structure. The outer surface of these tubes is punctuated by fine, longitudinal ribs, only visible near the ends of the tube (Pickford, 1995). But, so far the true identity of *Archaeichnium* remains elusive.

Clearly, much is yet to be learned of the anatomy and biodiversity of the soft-bodied organisms of the Nama rocks – what is known is just a taste of what is to come as work continues in this area carried out by a number of International groups working with the Namibian Geological Survey in Windhoek.

And Then There Were Shells

At the beginning of the Cambrian, some 542 million years ago, many organisms developed hard parts. But these were not the first hard parts – certain groups of organisms, mostly small, already had shells. *Cloudina*, a tubular fossil in the Nama Group, in some places forms great reefs along with stromatolites. *Cloudina* is the oldest record of an organism with a truly mineralized skeleton – and this is older than 548 million years. *Cloudina* is tiny but still visible to the naked eye – you don't have to use a microscope to see it. Its skeleton consists of organic material that is weakly impregnated with calcite (Grant, 1990; Bengtson & Conway Morris, 1992), made up of a series of multilayered, tubular structures that close off at their bases (Hahn & Pflug, 1985; Grant, 1990). The tubes have periodic additions and did not grow continuously. The actual surface of the tubes can be quite complicated, having a number of wedge-like protuberances (Rozanov & Zhuravlev in Lipps & Signor, 1992).

What sort of animals deposited these first skeletons is a point of contention – some have labeled them annelid worms,

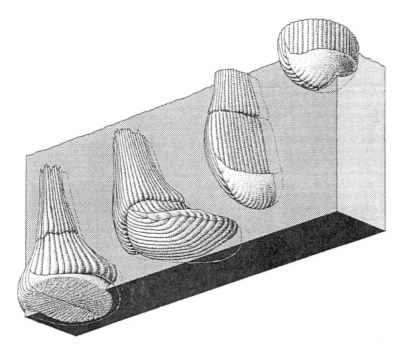

Figure 129. *Reconstructions of the many forms of* Ernietta *(courtesy of J. Dzik, 1999).*

Figure 131. Ausia fenestrata, *Nama Group, Farm Plateau, Aus region, Namibia (in the collections of the Geological Survey of Namibia, Windhoek; M. Fedonkin and P. Vickers-Rich).*

Figure 130. Kuibisia glabra, *Nama Group, Farm Plateau, Aus region, Namibia (in the collections of the Geological Survey of Namibia, Windhoek; M. Fedonkin and P. Vickers-Rich). Basal width of rock specimen 6 cm.*

In addition to *Cloudina*, another small form, *Namacalathus hermanastes*, populated the ancient reefs of Namibia. *Namacalathus* was certainly a weird and unique metazoan. It consisted of a stem topped by an outwardly-flaring cup – which was at maximum 2.5 cm across. It had a wide circular opening on the top, with an upward-curving lip. The cup itself was not solid but perforated by six or seven largish openings of a similar size and shape. *Namacalathus* was a bottom dweller, attached to and abundant on a rugged topography of microbial/algal biomats. It shows no particular relationship to any Cambrian or living group (Grotzinger *et al.*, 2000).

A much larger form that Wood and his colleagues (2002) named *Namapoikia rietoogensis* is known from the Omkyk Member of the Nama Group of the Friedoornvlagte pinnacle

Figure 132. Protechiurus edmondsi, *Nama Group, Farm Plateau, Namibia (in the collections of the Geological Survey of Namibia, Windhoek; M. Fedonkin and P. Vickers-Rich). Total length of specimen about 7.5 cm.*

others suggested they were algae or possibly large, single-celled Foraminifera (Germs, 1972; Glaessner, 1984) or even filter feeders of at least a cnidarian level of organization (Grant, 1990; Fedonkin in Lipps & Signor, 1992). At the moment, there is not sufficient evidence to confidently find a branch in the tree of life on which this group can "perch"!

reef complex near Rietoog in southern Namibia. A number of specimens have been collected, many measuring up to 1 meter in diameter. *Namapoikia* had a fully mineralized skeleton and seems to have been a cnidarian or poriferan, something related to corals or sponges. The skeleton itself consists of modules of joined tubules that remain about the same width throughout. *Namapoikia* shows some similarities to a number of the small Early Cambrian coralomorphs, especially a form called *Yaworipora* from Siberia and the calcified sponges, the chaetetids. As more and more is understood of forms making up the late Proterozoic reefs of southern Namibia, the biodiversity is increasing and the complexity of ecospace division is becoming more and more similar to that observed in the younger Phanerozoic reefs that have existed for more than 500 million years (Wood *et al.*, 2002).

Figure 133. Treptichnus (Phycodes) pedum, *Nama Group, Namibia (Geological Survey of Namibia, Windhoek) (M. Fedonkin and P. Vickers-Rich). Total length of trace on left about 2 cm.*

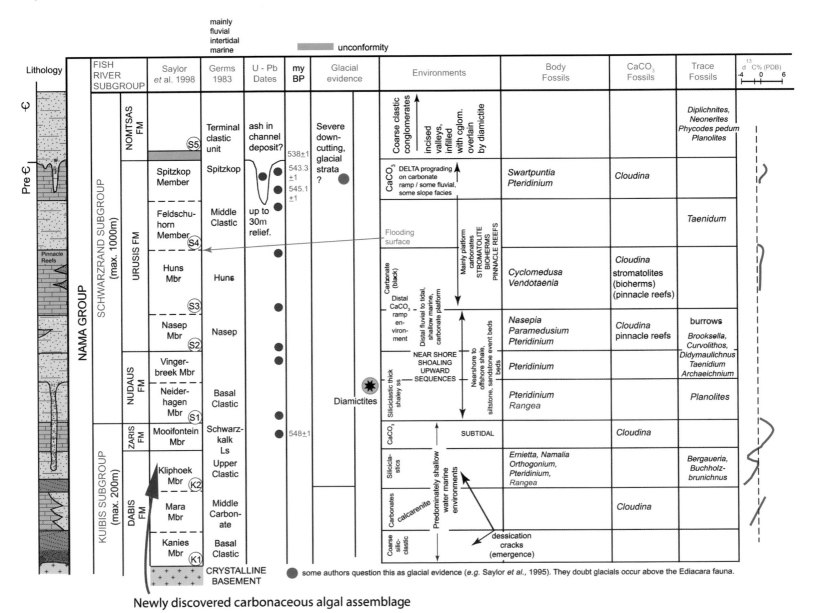

Newly discovered carbonaceous algal assemblage

Figure 134. Overview of the stratigraphy and fossil content of the Nama Group, Namibia (P. Vickers-Rich and K. H. Hoffmann, based on Saylor et al., 1995). (D.Gelt)

Figure 135. Cloudina *reef, southwest of Windhoek, Namibia (M. Fedonkin).*

Figure 136. Bag-like features, common in Ediacaran-aged rocks around the world. Nama Group, Farm Aar, Aus region, Namibia (M. Fedonkin and P. Vickers-Rich).

Figure 137. Tubular organism, Nama Group, Farm Aar, Aus region, Namibia (P. Vickers-Rich). Length of specimen about 5 cm.

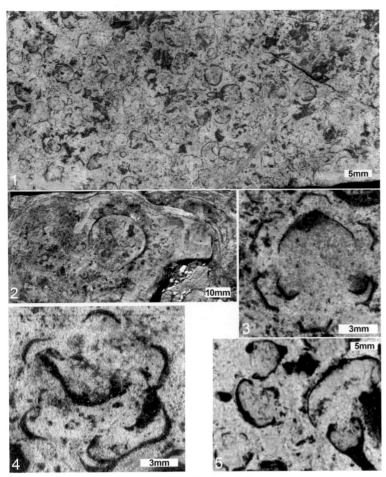

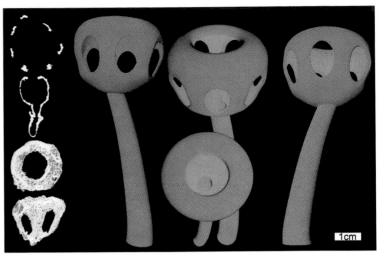

Figure 139. Reconstruction of Namacalathus from the Nama Group of Namibia. Diameter of bulbous "head" averages approximately 2 cm (from Grotzinger et al., 2000).

Figure 138. Namacalathus, *maximum diameter 2.5 cm, Namibia (from Grotzinger et al., 2000).*

Figure 140. Disc-like structure, possibly a holdfast, Nama Group, Farm Aar, Aus region, Namibia (P. Vickers-Rich). Diameter of specimen about 3 cm.

Other Parts of Africa

Although Namibia is one of the jewels in the crown of Ediacaran metazoan locales, a few other spots in Africa have produced a limited number of these fossils. It is likely that further prospecting will increase the number of these sites. One of the known spots is in West Africa – in Algeria (under study by Sarfati and colleagues) in the Taoudenni Basin. Sediments of the Cheikhia Group exposed southeast of Chenachane have yielded casts and impressions on the top of marine sandstones of forms such as *Medusinites* and *Nimbia* as well as a form possibly related to either *Beltanelliformis*. These sandstones are overlain by carbonates that contain Early Cambrian Small Shelly Fossils.

Thus far the sample is small, and definitive identification will need to await further collection. The entire continent of Africa, indeed, is a place ripe for future discoveries.

Figure 141. Modern-day northern Namibia at night. A mother and baby rhino cautiously approach a waterhole with the call of lions in the background. Etosha Pan, northern Namibia, where fossils of some of the earliest calcified metazoans have been recovered (P. Vickers-Rich).

CHAPTER 5

The Ediacara Hills

J. G. Gehling and P. Vickers-Rich

With the exception of reports informally describing structures that now seem likely to have been parts of Ediacara fossils (e.g. Hill & Bonney, 1877), studies of Neoproterozoic animals have generally been regarded as beginning in the mid-20th century with Reg Sprigg's reports in 1947 and 1949 of fossil soft-bodied "jellyfish" from the Ediacara Hills region of South Australia, or even somewhat before with Gürich's less publicized descriptions of soft bodied, frondose fossils from southern Africa in the late 1920's and 1930's. Both Sprigg and Gürich originally assigned their distinctive fossil assemblages a basal Cambrian age, presumably because at the time Proterozoic strata were generally assumed to be devoid of complex fossils. The Proterozoic age of the Ediacara biota was not established until 1958 when Trevor Ford's report of the soft-bodied frondose fossils Charnia and Charniodiscus collected from unequivocal Proterozoic strata at Charnwood Forest in central England led Martin Glaessner (1959) to propose a global "Ediacaran fauna" of Late Proterozoic age.

Opposite page: Rawnsley Quartzite with ribbons marking the occurrence of certain Ediacaran species, south of Ediacara Hills, Flinders Ranges, South Australia (artist Peter Trusler).

Figure 142. The blue-violet ridges of the Flinders Ranges, South Australia (P. Vickers-Rich).

North of Adelaide, only a few hours' drive, lie the blue-violet ridges of the Flinders Ranges. Long appreciated for their scenic beauty and wealth of indigenous and early European history, their real claim to fame is the role they play in throwing significant light on the early history of multicellular life – especially the Earth's first true animals. The exquisite impressions that cover the sandstone surfaces of the deep red rocks of the Flinders have provided a solution to Darwin's dilemma – what came before the first animals that have well-developed shells and skeletons, which appear quite suddenly in the rock record at the base of the Cambrian, some 542 million years ago?

The Ediacarans

In late March of 1946, Reg Sprigg, then a young government survey geologist, discovered the first fossils of the Ediacara biota in the Flinders Ranges. It was in these ranges that Ediacara fossils were clearly demonstrated to have a Precambrian age and quite distinctive character. Ediacara fossils had been found in England, Canada and Namibia long before. But it was not until these discoveries that an undisputed Precambrian age was established as well as the true organic nature of such fossils. These strange fossils represent a wide variety of multicellular animals, algae and a number of problematic forms that were preserved as mineralized "death masks" in shallow marine sediments rich in silica. Among the metazoans are possible cnidarians, forms allied with jellyfish and corals, as well as sponges and primitive mollusks, one form perhaps more closely related to snails. The joint-legged arthropods and spiny-skinned echinoderms appear to also be represented in the Ediacara biota. Certain trace fossils can be attributed to some of the Ediacara body fossils, clearly demonstrating that some of these ancient forms could move under their own power.

The excellent exposures of Neoproterozoic to mid-Cambrian rocks in the Flinders Ranges offer one of the most accessible and best-exposed records of this critical interval in the history of the evolution of the atmosphere,

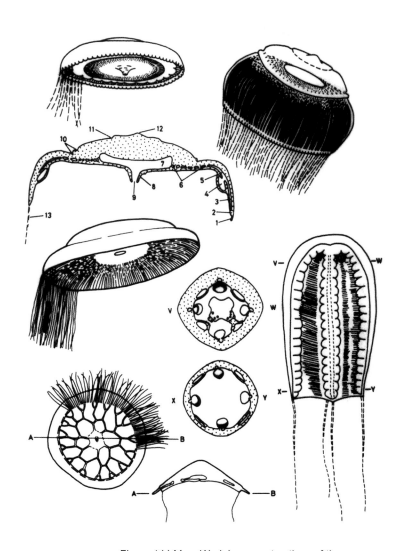

Figure 144 Mary Wade's reconstructions of the Ediacaran organisms (courtesy of Mary Wade's family and the Queensland Museum).

Figure 143. Reg Sprigg, discoverer of the Ediacaran fossils in the Flinders Ranges, South Australia (courtesy of Marg and Doug Sprigg).

the oceans and life. It was a time of successive ice ages and great fluctuations in climate, which characterized the late Neoproterozoic Era from around 800 to 542 million years ago. It is in these rocks that the key evidence for the "Snowball Earth" event is preserved, with Australia being affected by global ice ages even when it was positioned over the paleo-Equator. In the Adelaidean Sea of the time, sea-level ice sheets deposited thick diamictites, and dropstones were unceremoniously dropped by melting icebergs into the fine near-shore sediments, anomalies that allow geologists to interpret past climates from the rock record.

At the end of each of the glacial events at this time, thick cap carbonate sequences formed. One of these distinctive rock units, which records the end of an ice age named the Marinoan (also called the Elatina), was chosen for the Global Stratotype Section and Point (GSSP) defining the base of the Ediacaran Period (circa 635-542 million years ago). The rocks of the Ediacaran System in the Flinders Ranges are a 4 km thick succession of sediments, which record environmental changes and the rise and fall of global sea levels. On record in these rocks is evidence for a large meteorite impact event at around 570 million years, the Acraman impact horizon, that ejected debris for more than 300 kilometers, and may have had

Figure 145. Rhythmites representing daily couplets and half monthly cycles in tidal deposition in the periglacial facies of the Elatina Formation, Pichi Richi Pass, Flinders Ranges, South Australia (P. Vickers-Rich).

Figure 146. GSSP spike marking the base of the Ediacaran Period, Enorama Creek, Flinders Ranges, South Australia. The spike is emplaced above the glacial Elatina Formation (a diamictite) and at the base of the overlying Nuccaleena Formation cap carbonate, which signals the meltdown of the glaciers and return of global warmth (J. Gehling).

Figure 147. South Australian Premier Mike Rann dedicating the GSSP Golden Spike at the base of the Ediacaran Period, Enorama Creek, Flinders Ranges, South Australia (J. Gehling).

a profound effect on the composition of microorganisms living in the cool waters of an ancient Adelaidean Sea (see Chap. 12) – driving some to extinction while favoring others.

The oldest fossils representing the Ediacara biota occur in a thin band of rocks near the top of the Wonoka Formation, 1200 meters below the Cambrian strata. The richest and most varied Ediacara fossil horizons are in the lower to mid part of the ridge-forming Rawnsley Quartzite, some 400 meters below the earliest Cambrian strata and a considerable distance above the impact event, even farther above the main glaciation. The base of the Cambrian succession is marked by the first occurrence of complex trace fossils – burrows that penetrate the ancient sea floor along with a biodiverse fauna of animals with shells. Present, too, in the Early Cambrian limestone rocks are some of the first reefs built by metazoans, the archaeocyathids and the rarer tabulate corals, which constructed their edifices in a carbonate-rich environment, in a world that was likely warmer than that inhabited by the Ediacarans.

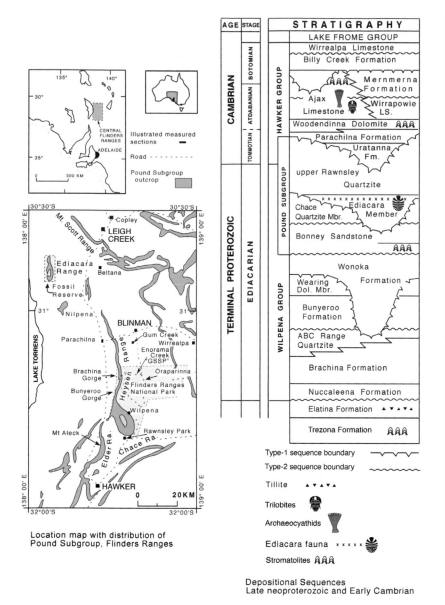

Figure 148. Stratigraphic chart and locality map of the rocks bearing the Ediacara biota, Flinders Ranges, South Australia (J. Gehling).

Figure 149. Wall of a building at the Ediacara Mine site. One slab of this wall has Ediacaran fossils in place and the miners who built this structure surely must have been aware of these strange patterns in the rocks. Jo Bain for scale (P. Vickers-Rich).

Mineral Wealth, Politicians and Challenging Ideas

The origin of the name "Ediacara" is still a mystery. Some say it was thought up by miners as a wrongly pronounced local Aboriginal name, "Idiyakra," with meaning interpreted either as a place of no water (Mincham, 1958) or the presence of hidden water – perhaps a soak long since lost beneath the shifting sand west of the plateau. "Ediacara" may be the anglicizing of two aboriginal names (*pers. com.*, John McEntee). Rather like the flattened hull of an ancient sunken ship, the Ediacara area is underlain by a shallow oval of rock formations some 8 kilometers long and 3 kilometers across. From atop the plateau, mine shafts cut down through the Ajax Limestone with its Cambrian sponge-like fossils, with the great domes of the Woodendinna Dolomite deposited by algae in the way the stromatolites today form in Shark Bay, Western Australia. Farther down, the mines slice through the manganese-stained, rusty red and yellow clay-rich sandstones of the Parachilna Formation, honeycombed with fossil worm burrows. The shafts bottom out atop the hard, silica-cemented sandstones

of the Rawnsley Quartzite, which preserves the soft-bodied impressions of the Ediacara animals. The top of these sandstones marks a major boundary in Earth history – the junction between the Proterozoic and the Phanerozoic eons, or what is more often referred to as the Ediacaran-Cambrian boundary.

The Ediacara region is bounded on the west by red sand dunes. This western edge is a fault, a vertical fracture in the Earth's crust, defining the sinking eastern edge of the Lake Torrens Basin, leaving Ediacara as a stranded "ship" on the margins of the Flinders Ranges. To the north, the Mt James Range is a higher plateau formed by even older rocks.

More than 120 years ago, intrepid or perhaps desperate prospectors journeyed to the Ediacara Hills, to the edge of their universe, in search of mineral wealth. They attempted to gouge lead, silver and copper ore from the rocky rim of Ediacara, up until about 1972. They built mills, smelters and houses of the barren local sandstone that underlay the mineral-bearing rocks. Today the remnant stone walls lie abandoned along with the health and fortunes of the long forgotten miners.

Figure 150. Ediacara Mine on the southern end of the
Flinders Ranges, South Australia (P. Vickers-Rich).

Were they aware of the fossil treasure trove that lay in the flagstones all around? It is hard to believe otherwise. A perusal of a crumbling sandstone wall of one of the mine engine rooms reveals a hand-sized specimen of *Dickinsonia costata* on a flat lain sandstone slab set in rough lime mortar. How could the mason, who collected, carried, hand-trimmed and cemented it into the wall, not have seen this distinctive fossil that is now an index for the Ediacaran Period around the globe? Perhaps in 1870, this fossil so recognizable today was considered a freak of nature, a mystery shape, and certainly of no value compared to the promise of unrealized mineral wealth. Minerals, however, were never destined to be the wealth of Ediacara.

Many of the first Australian geologists, be they professional or amateur, realized the possibility of discovering ancient fossils in the Proterozoic rocks of the Mt Lofty and Flinders Ranges in South Australia. But there were some classic blunders.

In 1926 Edgeworth David, the august professor of geology at Sydney University, in the company of a number of South Australian geologists, whose day-jobs were church clergy, discovered, named and published descriptions of what they thought were truly primeval fossils from the Adelaide Hills. They were impressed by the fact that South Australia preserved early Cambrian fossils in strata that lay atop even older rocks. Surely this was the place to discover ancestors of those animals that first appeared in the Cambrian?

Soon David and his colleagues discovered fossil gold – imprints of what they were convinced had been made by a eurypterid, an extinct arthropod previously known only from much younger rocks, 380 million years or younger. The South Australian specimens are still preserved in the South Australian Museum and are regarded as pseudofossils. These angular impressions are likely to be mudstone flakes ripped up

Figure 151. Cross-section of the Flinders Ranges
constructed by the famous Antarctic explorer
and geologist Sir Douglas Mawson in 1942, with
additions by J. Gehling (Elatina Formation, Nuccaleena
Formation) (courtesy of J. Gehling and the Royal
Society of South Australia).

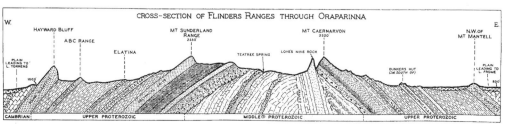

Mawson, D. 1942. The structural character of the Flinders Ranges. *Transactions of the Royal Society of South Australia* 66: 262-272.

93

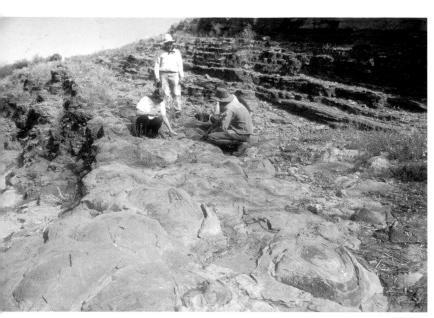

Figure 152. Hemispherical stromatolites, Trezona Formation, Enorama Creek, Flinders Ranges, South Australia, indicators of nearshore marine conditions rich in carbonate (P. Vickers-Rich).

Off early with Pat to mine workings to do some mapping starting about the south workings.
We had a wonderful view of the ~~Ediacara~~ Sand dunes to w of Ediacara. The red in the early morn light is wonderful. Later in day they became the "orange" of Albert Namat.
I started following limestone beds & underlying fossile quartz near South workings. Basin structure very obvious. ? mine is in kaolinised S.S. immediately under massive limestones. Faults readily visible on E & W. marked by massive quartz & manganese c little Cu.
In underlying fossile S.S. to S of main drive I found queer markings very suggestive of jellyfish also tracks & etc suggesting other creatures. Intend to collect more later.

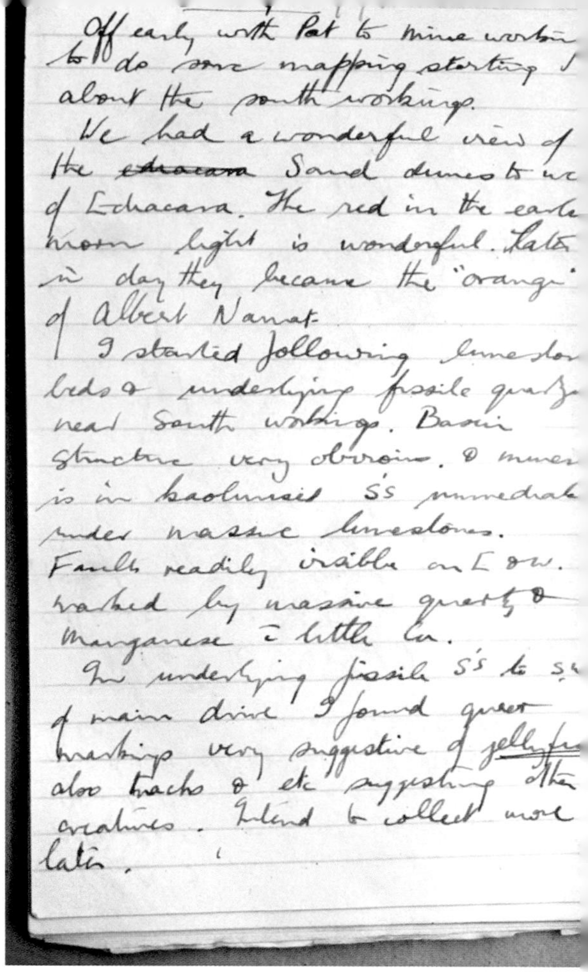

Figure 153. Field notes of Reg Sprigg, 31 March 1946, which record his finding of "jellyfish" near the Ediacara Mine, the first recognition of Ediacarans. It took time to convince others that these patterns in the rocks represented organisms that were of Precambrian vintage (courtesy of Marg and Doug Sprigg).

from a sea floor and subsequently buried in sand. Precambrian fossil animals remained just a dream prior to World War II.

In 1946, Reg Sprigg, a young government geologist from the newly formed Geological Survey of South Australia, visited Ediacara, ostensibly to re-evaluate the mineral field, but certainly spurred on by his thirst for discovery. Reg was not only curious, he was driven to ask profound questions about the origins of his native Australia. So advanced were his geological ideas that, in most cases, it took the mainstream geological community 20 years to accept their veracity. As a student of Sir Douglas Mawson, he knew the potential of South Australian Proterozoic rock record for understanding the environments of early animal life.

Mawson was one of Australia's great explorers and geologists. His expeditions to Antarctica are legendary and responsible for Australia's "tenure" over at least one-third of that ice-covered continent. He carried out pioneering geological descriptions of the enormous succession of Proterozoic rocks in the Flinders and Mt Lofty ranges of South Australia, which he referred to as the "Adelaidean System." His recorded traverse through the central Flinders Ranges from Brachina Gorge east to the Bunkers Ranges formed the basis for generations of more detailed geological mapping and description of rock formations in the Flinders Ranges. His cross-sectional diagram provided the first clear interpretation of the broad pattern of folding that gives such easy access to the record of rock formations spanning this important interval in Earth history.

Mawson described the domed, distinctly layered limestone reefs built by cyanobacteria – leaving behind structures called stromatolites, which still form in Western Australia's Shark Bay today. They form by the trapping of sediments by cyanobacteria. Mawson recognized evidence in the sedimentary record signaling at least two ice ages in this ancient Adelaidean sequence. However, while a strong advocate for the likelihood of the Proterozoic rocks harboring fossils, Mawson (who himself was a student of Edgeworth David) was a strong critic of speculative fossil finds. He was scathing of the claimed "arthropod fossils" discovered by a keen young student, Reg Sprigg, at Sellick's Hill, south of Adelaide.

Undeterred by such a discouraging opinion, Sprigg continued to search for Proterozoic fossils. It is clear from his published reminiscences that his review of mineral deposits at Ediacara was partly an excuse to search for more Proterozoic fossils. He was rewarded almost immediately with his finds of disc-shaped impressions in late March of 1946, which seemed to him to be the impressions of the most primitive animal life (metazoans) on Earth. His initial opinion on these Ediacara fossils was that they were coelenterates – fossil jellyfish. Mawson would not accept the evidence, preferring to think that these imprints were "pseudo" fossils – the marks left by weathering or chemical alteration during the compaction and solidification of sediments into rock. After showing the material to Mawson and receiving an unfavorable response, he took material to a meeting of scientists, the ANZAAS (Australia New Zealand Association for the Advancement of Science) conference in Adelaide in August 1946. Here, as his personal notes for the 19th of August 1946 detail, he "Exhibited fossil jelly fish from Ediacara. Nothing but discouragement – fortuitous markings according to Mawson. Teichert dubious. Glaessner against fossil origin. Ditto Fred Whitehouse."

Despite this, Sprigg held his views and dug into the literature, further testing his own observations and ideas. Shortly after the 1946 ANZAAS conference his own notes highlight this: "Following up references on fossil Scyphomedusae – pleased to find many similarities of my Ediacara find with those of Jurassic, but American Cambrian forms quite unrelated" (13/9/1946 Sprigg notes). He also returned to Ediacara, ostensibly as a member of the South Australian Geological Survey, to supervise the diamond drilling going on there, but took the opportunity to search for

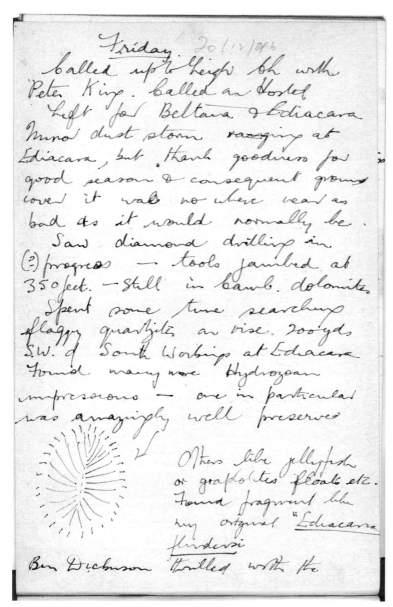

Figure 154. Field notes of Reg Sprigg, 20 December 1946, noting the naming of *Dickinsonia* (courtesy of Marg and Doug Sprigg).

MITCHELL ON THE BUDGET

"We seem to be fresh out of money—would you mind paying?"

Figure 155. Cartoon in Reg Sprigg's notes signalling the poor funding provided by governmental agencies at the time for geological reconnaissance, not an uncommon practice (courtesy of Marg and Doug Sprigg).

more fossils. His notes of 20 December 1946 highlight an important discovery: "Saw diamond drillings in (?) progress – tools jumbled at 350 feet – still in Cambrian dolomites. Spent some time searching flaggy quartzites on rise. 200 yards S. W. of South Workings at Ediacara found many more Hydrozoan impressions – one in particular was amazingly well preserved. Others like jellyfish or graptolites [*sic*] floats etc. Found fragmented like my original '*Ediacara flindersi*.' Ben Dickinson [head of the South Australian Geological Survey and after whom this new form to be called *Dickinsonia* was named] thrilled with the specimen (roughly figured) above – certainly no doubt whatsoever of its organic origin." Interestingly, in his notes of the same day, Sprigg suggested with regard to this new find and its bilateral nature that it could have been a "creeper" – thus a form that moved on its "own power." It was not until years later that specimens found in Russia and Australia of such forms as *Yorgia* and *Dickinsonia* showed this to be the case.

On this same trip in December 1946, only the second of Sprigg's visits to Ediacara, Reg was accompanied by the then premier of South Australia, Sir Thomas Playford,

investigating the possibility of re-opening the old lead-zinc mines at Ediacara. These mineral deposits lie within in the earliest Cambrian rocks, just up-section from the oldest evidence of deeply penetrating burrows in the sediments, one sort of trace fossils. Playford was completely unimpressed by Sprigg's fossils as his notes of 20 December belie: "Premier completely uninterested with jellyfish find – no money or practical value he said. Turned his back on them." Ironically, almost 50 years later, the fossils have become an economic target of illegal collectors who venture out to the isolated Ediacara plateau. Old and rare fossils now achieve values on the black market that far exceed any money ever made by the miners of old at Ediacara. But more important, the Ediacara fossils and their geological context now provide a new, and sustainable, resource for the state of South Australia and the country of Australia for that matter. The Flinders Ranges and the Ediacara fossils are prime resources for eco-tourism, and the current premier of South Australia, Mike Rann, is keenly aware of this. His visits to Ediacaran fossil sites in 2003 and to the GSSP site for its launch to the world in April 2005 are clear indications that he, unlike Premier Playford, understands the heritage, economic and scientific value of these first animals on Earth and sees them as part of the state's heritage to be treasured and nurtured. It is a pity that Reg Sprigg is not still around to take part in this celebration of his original find.

By the end of 1946 Sprigg's notes (31/12/46) indicate that even Mawson was beginning to think the Ediacarans were organic, in light of the variety of fossils that were now

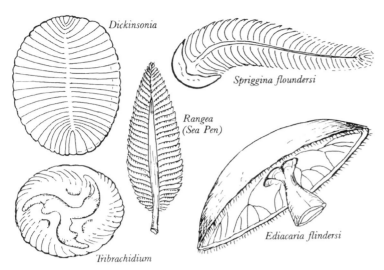

Figure 156. Reconstruction of Ediacarans by Reg Sprigg, some of the first attempts to understand these fossils as living organisms (courtesy of Marg and Doug Sprigg).

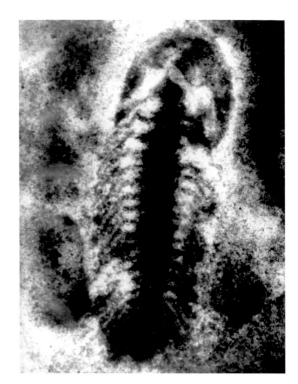

Figure 158.
X-ray of Spriggina
(M. Wade).

turning up at Ediacara: "Called on Prof. Mawson & yarned. Afternoon on jellyfish paper. Described 3 more of these fossils. Have decided on the following names. *Ediacara flindersi, Dickinsonia costata, Solenella davidi, Beltanella gilesi, Randelia eyeri.*" Other experts, too, began to take seriously Sprigg's interpretations of the Ediacara specimens as organic remains. Noteworthy was that they lay beneath rocks rich in what everyone clearly acknowledged as Early Cambrian fossils, the enigmatic ice cream cone like archaeocyathids, now thought to be primitive sponges. Still the question remained – were Sprigg's Ediacara fossils earliest Cambrian or latest Precambrian? This was yet to be clarified. And, in fact, the first two papers published on this material by Sprigg in the *Transactions of the Royal Society of South Australia* (1947, 1949) listed them as Early Cambrian in age.

Figure 157. Casts of Parvancorina's made by Mary Wade as she attempted to understand variation in this Ediacaran (courtesy of Mary Wade's family and the Queensland Museum; photo by S. Morton).

In these early days, a few people began to take an interest in the strange fossils from the outback of South Australia. Amateur collectors stepped in. There are rumors that amateur collectors sold large numbers of fossils overseas. But among this lot there were many that made major contributions to science. In 1956 and 1957, new and important finds were made by Hans Mincham, a school teacher based at Beltana in the Flinders Ranges, and Ben Flounders, a carpenter resident in the steel town of Whyalla, 200 km south of Ediacara.

Today, in his mid-nineties, Hans Mincham still has an encyclopedic memory of natural history in general and the history of South Australia in particular. He has authored several books on local history, which provide a framework for understanding the Flinders Ranges. Hans can recall every detail of his collecting trips to Ediacara. On one occasion, in mid-December, it was so hot that Hans and his colleagues could not even touch the rocks between 10 a.m. and 5 p.m.! Their solution was to go out after dark and scan the rock slabs using flashlights, looking for the "embossed" evidence of ancient animals. The result was the discovery of what may yet prove to be the most important fossils of the Ediacara biota: those small, sharp impressions that represent the earliest members of the arthropod and annelid phyla. Preserved in opposite relief to the disc-like discoveries of Sprigg, these bilateral and radial forms are the external molds of the original animals. These "amateur" finders are today immortalized in the species-name endings of two of the most important discoveries: *Parvancorina minchami* and *Spriggina floundersi*. These same amateurs prompted the South Australian Museum to mount an expedition to Ediacara under the leadership of Brian Daily (then the paleontologist at the South Australian Museum), and the results of that expedition confirmed that Ediacara fossils were biodiverse and plentiful.

This material inspired the professor of paleontology at the University of Adelaide, Martin Glaessner, to alter the direction of his research. Born in Bohemia and trained in Vienna, Glaessner had a distinguished career in micropaleontology. It was he who set up the geological exploration division in Stalin's USSR during the 1930's. He wrote the first comprehensive textbook on microfossils on his return to Austria from Russia. Then fleeing Hitler's Reich with his

Figure 159.
Martin Glaessner
(M. Wade).

Figure 160.
Brian Daily
(R. Tedford).

beautiful Russian ballerina wife, he worked for a time for Shell in New Guinea until once again evacuating his workplace to Melbourne following the Japanese invasion during World War II. It took five years after his appointment to the University of Adelaide for Glaessner to turn his attention from fossils of the Cenozoic basins and his reviews of the geological history of South Australia to the strange Ediacara fossils. At first he considered them to be pseudofossils, or shapes formed by some inorganic process, as Reg Sprigg was all too aware in 1946 when he exhibited his "jellyfish" at the ANZAAS conference. But the new discoveries made by Mincham and Flounders, as well as Sprigg himself, and reports of similar fossils from Charnwood in England changed all that. Glaessner (1958) rapidly published brief accounts of the new material in international journals, thereby establishing the primal importance of the Ediacara biota. The Ediacarans came to the notice of the general public when Glaessner published his cover article in *Scientific American* in 1961 – where he established the Precambrian age of these enigmatic fossils (Runnegar, 1992). Here he also noted their global occurrence and pursued the idea that they were related to modern groups of animals. Many papers followed, culminating in a seminal paper in *Science* in 1982 (Cloud & Glaessner, 1982) that suggested the name of the Ediacarian Period (not the spelling Ediacaran as officially sanctioned in 2004) and System as the basal division of the Phanerozoic Era lasting from 670 to 550 million years ago – at a "time during which metazoan life diversified into nearly all the major phyla and most of the invertebrate classes and orders subsequently known." Later many, including Dolf Seilacher in 1989, challenged the interpretations of the fossils, and interpretations of these early metazoans are still changing from one new find to the next (McMenamin, 1986; Valentine, 1986; Dzik, 2003; Fedonkin & Waggoner, 1997; and so on).

Late in 1958, Trevor Ford of Leicester University described the first Ediacara fossils from England in a brief paper in *Nature*. A paper describing the Ediacara fossils submitted by Sprigg to *Nature* in 1946 had been rejected, so knowledge of these fossils, even though thought at that time to be Early Cambrian, was not widely recognized. With Ford's paper it suddenly became obvious that these fossils might occur in Precambrian rocks worldwide. It jolted Glaessner into action. In the same year he quickly described and interpreted key fossils from the Ediacara biota of South Australia, and together with Brian Daily (soon to take up a lectureship in stratigraphy at the University of Adelaide), described the geological setting of the fossils at Ediacara (Glaessner & Daily, 1959).

Daily, who had completed his Ph.D. on the Cambrian stratigraphy of South Australia, initiated some of the first attempts to establish a global standard for the base of the Cambrian by visiting key geological sections around the globe. He and Glaessner understood the importance of the Ediacara biota for establishing the age of latest Proterozoic rocks. Students from the University of Adelaide were active in the search for new Ediacara fossil localities. Glaessner and Daily then published the first comprehensive paper on Ediacaran paleontology and its geological setting (Glaessner & Daily, 1959). They demonstrated that the Pound Sandstone (now called the Rawnsley Quartzite), which preserves the Ediacara fossils, was overlain by rock strata bearing Cambrian skeletal fossils and deep burrows. Despite this, the Ediacara fossil-bearing formation was still formally regarded as earliest Cambrian in age.

Somewhat later, Daily was able to show that a surface separating the Ediacaran and Cambrian rocks represented a break in sedimentation. The Uratanna Formation, bearing Cambrian trace fossils and microfossils, lies above the Rawnsley Quartzite in the northern Flinders Ranges, but is

Figure 161. Reconstruction of the Ediacaran Assemblage based on Mary Wade's and Martin Glaessner's interpretations (courtesy of the Queensland Museum).

Figure 162. Mary Wade (2nd from left) her parents and a friend at her graduation from the University of Adelaide, 4 April 1951 (courtesy of Mary Wade's family).

entirely absent farther south. This indicated to Daily that there had been a period of erosion between the Ediacara sediments and the Early Cambrian rocks – what geologists call an *unconformity*. Everywhere Daily found the earliest Cambrian rocks that contained complex burrows. Animals were definitely actively digging into the sea bed sediments. These vertical burrows and complex feeding burrows were not to be found in any of the Ediacaran sediments. This change in animal behavior, that is, from not churning up the sea bottom sediments to that of actively burrowing into them, was a behavioral revolution in marine ecosystems that set the stage for the next 542 million years of animal life. Glaessner authored his first paper in the journal *Nature* on Precambrian jellyfish from Australia, Africa and England in 1959 and then his popular review of the Ediacara biota in *Scientific American* in 1961, in both places firmly reiterating the Precambrian age of these Ediacara fossils. This publication set the pattern for interpretation of some of the Ediacarans as ancestors for most of modern marine animal phyla for the next 20 years.

Early in the 1960's, Bob Major discovered the first Ediacara fossils outside the Ediacara Reserve in the Red Range, 30 kilometers east, near Beltana. Undergraduate students such as Malcolm Walter and Richard Jenkins led collecting trips to the Flinders Ranges with their families. Mary Wade began a series of collecting expeditions in her collaboration with Martin Glaessner, her Ph.D. supervisor. At that time, it was rare for women to do fieldwork in geology. However, Mary, who had grown up on tiny Thistle Island on the west coast of South Australia, was accustomed to solitude and remote localities.

In 1966, Glaessner and Wade published a seminal paper in *Palaeontology* that demonstrated Glaessner's determined use of comparative morphology to interpret the Ediacara biota. It was not that he thought all Ediacara fossils could be accommodated within existing animal groups, but rather that he considered such an approach was important to establish the place of Ediacara organisms in the tree of life. It is common for paleontologists to describe unusual fossils and leave them unaffiliated with modern animal phyla, as "Problematica." A phylum is the main category into which scientists place an animal or plant. Mollusca, for example is the phylum that includes clams, snails and squid. These apparently different-looking animals share similar soft parts and show marked similarities, especially in their larval stages. However, if the various classes, such as the snails and squid in the Phylum Mollusca, were all extinct today and known only from fossil remains, it is quite possible that we would not recognize their close relationships and thus their common ancestry. The process of working out the phylum where a new group of fossils belongs is often quite difficult.

Meanwhile, almost 20 years after Reg Sprigg's first discovery of fossils at Ediacara, the main Flinders Ranges were thought to be barren of these much sought after fossils. Then, in 1967, a schoolboy came into the South Australian Museum to inquire about strange zigzag marks on a white slab of sandstone that he had collected in Brachina Gorge, one of the feature localities in what was to become the Flinders Ranges National Park. The zigzag cracks were caused by shrinkage of algal mats or clay – interesting but not particularly important. Hans Mincham, by then the museum information officer, saw what the boy had not, a specimen with a design like the radiator grill of a 1940's car. It was *Dickinsonia costata,* the most distinctive of all Ediacara fossils.

This chance discovery prompted Mary Wade, accompanied by Jim Gehling on his second fossil-collecting trip as an undergraduate volunteer, to explore Brachina Gorge in a rugged part of the central Flinders Ranges. This stunning canyon cuts through a range formed along the western edge of an eroded fold of the Proterozoic and Cambrian strata, exposing the layers like the edges of a tilted shelf of library books. From their camp in the center of the gorge they watched the sheer sandstone cliffs turn red as the sun rose on their first day in the field. They spent the entire day searching thousands of barren sandstone surfaces in the walls of the gorge, with no success. Disillusioned, they began to walk back to camp. It seemed hopeless, like finding a single sheet of paper in a library. Making their way back up the gorge, past progressively older layers, they were sure that the rocks were too old to host Ediacara fossils. But in the fading glow of sunset, they chanced to look in a gully where a band of red mudstone interrupted the normal white sandstones. The setting sun lit up a band of rocks on one side, backlighting the opposing bluff with a soft light that revealed every detail on the sandstone layers. Surfaces that were shrouded shadows in the middle of the day now produced the runes of ancient life.

This was one of the richest Ediacara fossil sites ever discovered. The sharp imprints of segmented mats and anchor shapes, disc- and globe-shaped forms were abundant on the under-surfaces of many layers of sand. But the fossil-bearing strata observed by Wade and Gehling were overlain by almost 500 meters of barren layered rocks before reaching the base of the Cambrian. Rock strata are measured perpendicular to their layering, so that even when tilted on their side by folding or faulting, their true thickness or "distance below" a certain key level can be measured. Driving through Brachina Gorge one can see the rock strata displayed like a pile of books tipped on their sides. So, was this Brachina Gorge discovery an older fossil record than recorded in the Ediacara Hills? The kind and variety of fossils were essentially the same as that at Ediacara, so there was surely not a significant difference in age. It turns out that at Ediacara, where the fossils are only a few meters below the Cambrian strata, the entire sequence of rocks that hosted the Ediacaran fossils had thinned to a wafer of its true thickness in the high ranges of the Flinders Ranges. Ediacara is near the western shoreline of the ancient Adelaidean Sea, while the strata in Brachina Gorge and all the high ridges of the central Flinders Ranges were deposited where the sea floor was sinking more rapidly. Brachina Gorge confirmed that the Ediacara fossils were indeed Proterozoic in age, at least several million years older than the earliest Cambrian fossils.

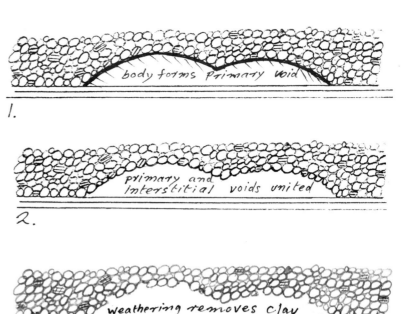

Figure 164. Interpretation of the mode of formation of Ediacaran fossils by Mary Wade (courtesy of Mary Wade's family).

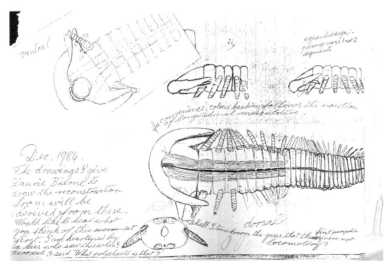

Figure 163. Reconstruction notes on Spriggina by Mary Wade (courtesy of Mary Wade's family).

In the next two years, Mary Wade traced the fossil-bearing layers over the greater part of the length of the western side of the Flinders Ranges. In 1971, Gehling located these same beds on the thinned eastern edge of the ranges, and soon after an M.Sc. student, Colin Ford, traced fossils south to Devil's Peak near Port Augusta. Simultaneously,

various exploration and survey geologists recognized possible Ediacara fossils much farther afield in the Punkeri Hills of the Officer Basin, in the northwest of South Australia, the Amadeus Basin near Alice Springs and also in the southern Georgina Basin of central Australia.

From the first, Sprigg was convinced that Ediacara fossils were the long predicted primitive precursors to modern animals, in particular, soft corals, jellyfish and sponges. He and many after him viewed the rippled sandstone slabs bearing impressions of discoidal fossils as evidence of stranding of jellyfish on beaches. This view was aired in Glaessner's Scientific American article of 1961. It was not until 1972 that this explanation was challenged. At the end of an almost fruitless expedition during a heat wave in December 1972, Gehling and Ford traced the fossil beds along the Heysen Range north of Wilpena Pound. They chanced upon some spectacular surfaces in the cliff face, covered with fossils. Embossed on the underside of one of these sandstone surfaces of a 30 cm thick bed they found a large leaf-like frond attached to a disc-shaped anchoring holdfast. Usually holdfasts are preserved on the under-surface of beds and the fronds on the top surface. But in this case, the sand had been winnowed away before both frond and holdfast were buried at the same level. This discovery challenged the long-held concept of jellyfish stranded and buried on tidal flats. Clearly most fossil surfaces were deeper sea floors where bottom-living animals were preserved in or near where they lived or transported only short distances.

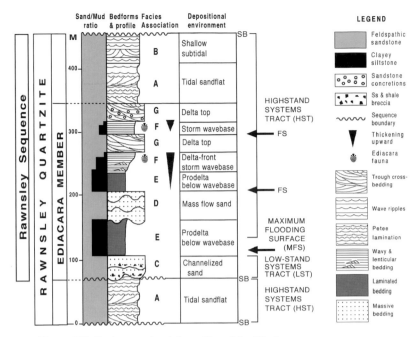

Figure 165. Environments of deposition of the Flinders Ranges Neoproterozoic rocks (J. Gehling).

This new interpretation was and still is not universally accepted. Even a study of Ediacaran sediments by a respected English sedimentary geologist, Roland Goldring, in 1969, interpreting the environment as shallow marine, had not been sufficient to change local prejudices. In 1968, Mary Wade proposed the first comprehensive explanation for how soft-bodied animals could be preserved in sand – a phenomenon unknown from all younger fossil deposits. How can an organism be buried by sand and leave a mold in the layer of sand that buried it, she asked. Surely once the animal decayed the sand would collapse into the space occupied by the buried body? But, based on observation, somehow the external form was preserved. Wade suggested that mud could be squeezed up into the space left by the rotting animal. If so, why did this not happen in younger rocks? The solution to this dilemma had to wait until later.

Mary Wade continued to explore for new localities until 1970, when Martin Glaessner retired. Mary then moved north and took on the position of curator of fossils at the Queensland Museum.

Richard Jenkins carried on with the Ediacara work. After completion of his Ph.D. on fossil crustaceans, Richard Jenkins was appointed as lecturer in paleontology at the University of Adelaide. He immediately began working on Ediacara fossils by comparing the South Australian forms with newly discovered collections from Namibia (then South West Africa), Newfoundland and England.

Jenkins recounts a collecting trip in Brachina Gorge when he was trying to extract an overhanging specimen of *Dickinsonia elongata* from the cliff face. The fossil-bearing slab suddenly broke away from the outcrop directly above Jenkins' head and most fortunately split so that the chunks each bearing half the giant specimen fell either side of him! This near-death experience was sublimated into some detailed reconstructions of *Dickinsonia* as a flattened, segmented worm. His artistic reconstructions have become the standards for interpreting Ediacara communities. Furthermore, Jenkins demonstrated the three-dimensional nature of many Ediacara

forms at a time when the prevailing view was that they represented mainly large, flattened ribbons, mats and fronds.

It was not until the late 1990's that any attempt was made to assess Ediacara community structure. For 50 years paleontologists had made collecting forays that were little more than stamp-collecting exercises. Fossil-bearing slabs were trimmed in the field for economy of transport, often resulting in fractures through the key specimen. In many cases re-cementing the fragments resulted in protrusion of glue across the specimens themselves, obscuring detail. This approach changed in 1972, when Jim Gehling took on a teaching position in earth sciences at the Murray Park College of Advanced Education (later to become part of the University of South Australia). Gehling then set out to learn as much as he could about modern marine communities, as one does by teaching about them in his courses at Murray Park. In the late 1970's, he launched a part-time M. Sc. project on the sedimentology of the Pound Subgroup in the Flinders Ranges. He measured section after section of rocks throughout the central Flinders Ranges and was able to clearly demonstrate that the Ediacara organisms were preserved where they lived, in offshore deltaic environments, buried in place by storm-induced avalanches into deeper water on the continental shelf. Not agreeing with Gehling's interpretation, Richard Jenkins still preferred a tidal lagoon setting as the environment of preservation and suggested that many of the organisms in the fossil assemblage had floated into their final resting

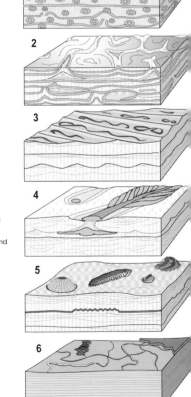

SUPRATIDAL
Petee lamination formed by desiccation and convolution of microbial mat bound sand laminae, with diagenetic growth of sulfate crystal rosettes.

INTERTIDAL
Petee lamination with disrupted and overturned microbial mat bound sand.

LOWER SHOREFACE
Flat bedded, wave and current rippled sand, with sand-fill of syneresis cracks in microbial mats within ripple troughs.

FAIR WEATHER WAVEBASE
Wave and current rippled sand with *Charniodiscus* preserved as convex, disk-shaped, holdfasts; fallen fronds rarely preserved in epirelief, usually torn away; current scours around frond bases. Thin microbial mat partings.

STORM WAVEBASE
Wavy bedded, medium-grained sand deposited by storm events. Mats developed during fair-weather, and colonized by benthic communities of Ediacaran organisms.

BELOW STORM WAVEBASE
Mat-coated, thinly bedded, fine-grained sand. Trails formed by small invertebrates, grazing mats below a thin veneer of sediment.

Figure 166. Different environmental settings of the microbial mat communities represented in the Ediacaran-aged rocks of the Flinders Ranges, South Australia. 1-2, Chace Range Member; 3-6, Ediacara Member of the Rawnsley Quartzite with increasing environmental depth from 1-6 (J. Gehling).

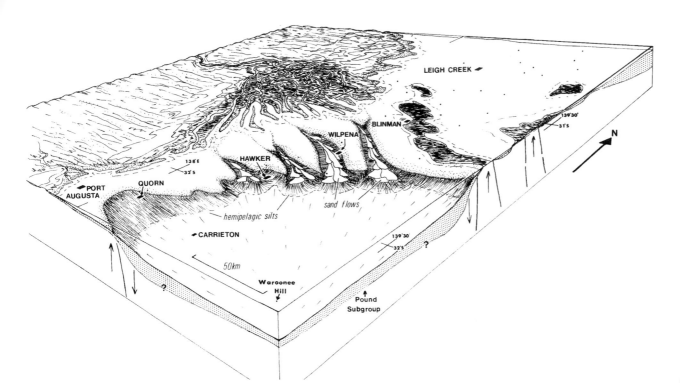

Figure 167. Reconstruction of the environments of deposition of the Ediacara Member of the Upper Rawnsley Quartzite, Pound Supergroup in the Flinders Ranges, South Australia (J. Gehling).

places. However, there is now quite compelling evidence from Russia, Namibia, Newfoundland and a number of other locales that Ediacara fossils represent bottom-dwelling communities rather than free-swimming and floating forms. Furthermore, they seem to have for the most part been buried in place and not transported far from where they once lived. With rare exceptions, most Ediacara organisms appear to have been sedentary or slow-moving bottom grazers or filter feeders.

The Pound Subgroup (previously referred to as the Pound Sandstone, after that great iconic landscape feature – Wilpena Pound, a remnant of a syncline, a downward buckle in the folded rocks formations) is made of two formations, the red-colored Bonney Sandstone and the lighter-colored Rawnsley Quartzite. These rocks are the principal ridge formers over the central and northern Flinders Ranges, a region some 300 km north to south and 50-80 km, east to west. In the many hundreds of kilometers of exposure of the fossil-bearing Ediacara Member new localities continue to be discovered. Gehling managed to survey many new localities with the help of enthusiastic students during mapping field camps. But detailed measurement and description of cross-sections through the 800-1500 m thick sandstone formations required solo work. As teaching duties allowed little time for fieldwork, Gehling traveled to the Flinders Ranges during periods of annual leave and on holiday weekends, camping near field sites with his wife Inara and two young sons, working his way around the many ridges of Rawnsley Quartzite. On Easter weekend in 1981, he made a chance discovery that re-directed his research efforts forever. This particular Easter he headed for the Chace Range. This area proved valuable in helping him reconstruct the nature of the fossil-bearing rocks in three dimensions, because the sedimentary layers in this sequence had been buckled into a vertical position so that their edges could be studied by looking at aerial photos.

Contrary to previous ideas, these sandstones were not simple monotonous sets of parallel layers. There were deep valleys cut into some layers during Ediacara times that had been later filled with finer-grained sediments. The fossil-bearing beds lay in the fill of these ancient valleys. Gehling needed to see a section on the eastern end of the Chace Range, but much of the outcrop had been cut up by faults. So, he set out walking until a clear section could be found. Time was short, so he began measuring without first doing the normal visual examination of the whole section. On coming to the level of the Ediacara Member he climbed a gap in a massive wall that was formed by a single bed, projecting vertically out of the hillside. The layers immediately above were thin and peeling away. He had put down his field notebook and staff to inscribe the accumulated layer thickness in chalk on a rock surface and begin a lateral search for fossils. Almost immediately he spied what looked like a squashed garlic clove, and then another, and then a whole bed surface covered with them. It was a first – something absolutely never seen before by anyone searching the sandstones of the Flinders.

A few meters along the outcrop he caught sight of a distinctive shape, a tiny, rimmed disc surrounding a five-pointed star. Echinoderms, which include sea stars and sand dollars, are the only group of marine animals with this pattern. Then there was another and another. While caching the labeled specimens for later collection he saw yet another unexpected fossil; it was a disc with a spiny rim and a surface that looked as though it bore the impression of insect mesh. Was this a sponge? In the space of a few minutes Gehling picked up fossils representing three previously unknown groups of modern marine animals in the Ediacara biota. Such occasions are gold for a paleontologist – to be the first to realize the oldest evidence of an entire group of animals is not something one does every day! They are the moments that cancel out the years of unrequited exploration. By comparison with certain other sites, this Chace Range locality was not rich, but besides the many new forms, it provided the key to understanding how soft-bodied animals could leave their sharp impressions in coarse sediment.

Some of the forms, which were preserved as flattened wheels in the Ediacara Hills, left high-relief, 3-D molds in the Chace Range sediments. Somehow these animals had

Figure 168. Dolf Seilacher (J. Gehling).

Figure 170. Slab of Rawnsley Quartzite with Dickinsonia, Kimberella *and* Radulichnus *(feeding traces of* Kimberella*), Flinders Ranges, South Australia (S. Morton).* Dickinsonia *specimen about 7 cm in diameter.*

suffered less flattening than elsewhere. The Chace Range sediments represented a slightly different environment than that preserved at other Ediacara localities in the Flinders Ranges. This was a clear message that the understanding of Ediacara paleobiology would require a solid knowledge of the sedimentary environments and mode of preservation of these strange fossils. Comparisons between sites, regions and countries were the only way to advance.

In the 1980's, visits by paleontologists from other countries helped to redirect Ediacaran research. Dolf

Figure 169
Bruce Runnegar
(J. Gehling).

Seilacher, a world-renowned paleontologist from Tübingen in Germany, came to study the Ediacara fossils. In 1984 he published a radical review of Ediacara paleobiology in which he emphasized the similarities between Ediacaran forms, rather than the differences. He claimed that most Ediacarans were members of a now extinct kingdom of giant, single-celled organisms, which he called "Vendobionts." Following his discoveries and descriptions of Ediacara (Vendian) fossils from the White Sea region of NW Russia, Mikhail Fedonkin visited South Australia in 1987 to compare his fossils with the Flinders Ranges forms. He has returned a number of times since.

In 1982, Bruce Runnegar, then professor of paleontology at the University of New England in New South Wales, surprised his fellow Australians by presenting the first biomechanical analysis of the most common and distinctive Ediacara fossil, *Dickinsonia*. Ten years later he teamed up with Jim Gehling (with Bruce as Jim's Ph.D. supervisor at UCLA), and they began a search of the more inaccessible localities in the high parts of the Flinders Ranges with the aid of a helicopter. Although this mode of transport was limited by a lack of possible landing sites, it enabled the collection of fossil-bearing slabs that might otherwise have been in danger from looting.

Runnegar and Gehling began the process of studying all of the fossils on one surface, as samples of what may have been sea floor communities. However, their attempts were always limited by the amount of surface that could be seen in the exposed layers on a cliff face; this was rarely more than a narrow band, no more than a few meters long and less than half a meter wide. Any wider and the rock had already fallen out of the cliff and shattered on the slope. It was impossible to try and expose more of each fossil layer without the cliff collapsing. Furthermore, there was no place to lay out the fossil-bearing surface, and most sites were a 1 to 3 kilometer walk from any place one could bring a vehicle. However, they had one lucky discovery. Fragments from a bed bearing pristine specimens of the segmented disc-like fossil, *Dickinsonia*, were discovered in a remote gorge. Clambering up the slope, they discovered the source, a fossil bed in the soil where the slope of the hill exactly coincided with that of the bed. Each slab they dug out carried more specimens of *Dickinsonia*. During the first season, they used a tiny Huges 300 helicopter to remove a few slabs. This belt-driven helicopter was so underpowered that it had to launch forward off the slope in order to even climb out of the craggy gorge. Imagine this noisy machine, with both doors removed, bearing an enormous slab of rock, perched precariously across the seat occupied by the pilot and sticking out both doors – launch, drop, gain altitude and then move out of the canyon to a waiting Toyota. This feat was repeated time and time again by a skilled pilot, Chris Collins, who removed at least 30 large slabs that were reconstructed like a giant jigsaw puzzle back in Adelaide. The result is that 15 years later this ripple-sided wall of sandstone is on display in the Simpson Room of the Origin Energy Fossil gallery of the South Australian Museum. With the slabs mounted vertically in a steel matrix, you can study the ripples on the top surface and on the opposite side are some 150 small specimens of *Dickinsonia costata*, along with a stunning array of other fossils. This surface, now proudly on public display, provided the first material available for a paleoecological study of Australia's Ediacara biota.

The year 1991 marked an unfortunate development in the history of Ediacara fossils. The discovery by Gehling and Ford, in 1972, of spectacular fossils high up in the Heysen Range of

Figure 172. Photograph of Richard Jenkins (right) and David McKirdy (courtesy of R. Jenkins).

Figure 171. Jo Bain, technician and expert Ediacaran model-maker from the South Australian Museum, Adelaide (P. Vickers-Rich).

the Flinders Ranges National Park, prompted calls to collect the best material for safekeeping. However, this was resisted by the National Parks and Wildlife Service, because there is a clear case for preserving such natural exhibits in place for posterity. So began the field use of silicone rubber molding material, which at least enabled impressions to be made of the field specimens so they could be described without having to collect them. In 1976, a BBC Wildlife Television crew, led by Sir David Attenborough, had visited many of these sites and

brought them to public notice in his "Life On Earth" series. At that time it seemed that such fossils were "protected" by being embossed on the bases of 30 centimeter thick quartzite beds, firmly fixed in cliffs hundreds of meters above and kilometers away from the nearest bush roads.

For almost 20 years after their discovery all seemed well. Then, in 1991, Ben McHenry, Jo Bain and Neville Pledge of the South Australian Museum made a surveillance trip to check Ediacara sites in the Flinders Ranges. Venturing up onto the Heysen ridge, high up on the western edge of the Flinders Ranges National Park, they discovered, to their horror, that the key slab bearing frond-like fossils of *Charniodiscus* was missing – quarried and removed from the rock face. The 1 square meter fossil-surface had been split off, leaving the rest of the slab lying precariously on the steep slope. The valued portion, still weighing about 160 kg had somehow been hauled down a 1 km scree slope to the gorge below and then carried some 2 to 4 kilometers to the nearest access point by a vehicle. The missing slab prompted an Australian Federal Police search for fossil looters. After police raids in South Australia and Western Australia, the case resulted in the apprehension of a group of semi-professional looters who had targeted key fossil localities throughout Australia and exported the fossils to overseas buyers without permits.

Ediacara and Cambrian material from South Australia was traced to merchants in Japan. Some of these fossils, including the famous frond-bearing slab, were thankfully repatriated to South Australia by the Japanese government. Still it was not until 2001 that the perpetrators could be tried and convicted for their thievery. This problem remains today. Ediacara fossils are by definition both old and rare, and that makes them collector's items. While harsher penalties may help to dissuade some looters, the best means of conservation of such fossils is to convince local custodians in the Flinders Ranges that this material is part of their heritage that once lost cannot be replaced. The interest in ecotourism promises real reward to local entrepreneurs who are able to manage remote fossil sites for specialist visitors, a strategy that is working in other parts of the world with rare fossil deposits such as the Cambrian Burgess Shale fossils of British Columbia, the La Brea tar pits in Los Angeles, dinosaur sites in Utah and Colorado and even with the wild game of Africa.

One excellent example of how this approach works well is on a property on the western side of the Flinders Ranges. Late in the 1980's, Pam Hasenohr, an amateur geologist, visited friends on a sheep station, where the owner, Ross Fargher, took her out to the quarry where he commonly collected ripple-marked slabs for use as paving on his property. She noticed some Ediacara fossils on slabs. Ross, who had crossed paths with Richard Jenkins in the past and discussed Ediacaran fossils, had previously noted the curious shapes when driving across the rocky slopes near his homestead on his motorbike and had made the initial discovery of this important new site. These fossils included discs and fronds, some of which were preserved within the slab, not just as surface impressions. This was an important discovery for South Australia, as the only previously known cases were from Namibia. Pam Hasenohr reported this new discovery to Richard Jenkins. For the next 16 years Richard Jenkins, Pam and his Ph.D. student, Chris Nedin, along with numerous other students, had surveyed this new fossil field, which is still being carefully guarded by Jane and Ross Fargher and their station employees. Since 2002, the area has been studied intensely by Jim Gehling (South Australian Museum), Mary Droser (University of California, Riverside), Sören Jensen from Spain and their field associates. Their studies focus on understanding detailed ecology and preservation *in situ* fossil assemblages.

Figure 173. Jim Gehling (far left), Patricia Vickers-Rich and Ross Fargher, Nilpena Station, Flinders Ranges, South Australia (F. Coffa).

Early Interpretations of the Ediacara Biota

From the late 1950's onwards, discoveries of Ediacara fossils were made in late Proterozoic rocks around the globe. Fossils discovered in the early 1900's in southwestern Africa, particularly in Namibia with discoveries as early as 1908, were recognized as Proterozoic in age. In 1957, a schoolboy, Roger Mason (now a professor at China University of Geosciences), had discovered Ediacara fossils in Charnwood Forest near Leicester in the English midlands. Ediacara fossils were found in profusion on the Eastern European Platform in the Ukraine and various parts of Russia from 1973 onwards. Discoveries spread to SE Newfoundland, then in NW Canada as well as on the east and west coasts of the USA. It was, however, the discoveries in the Ediacara Hills that gave impetus in the beginning to the recognition and acceptance of the ancient age of these fossils and the work of such people as Sprigg, Glaessner and Wade that spurred on research.

The approach to understanding the Ediacara biota adopted by Glaessner and Wade (1966) was to compare the body plans of the most prominent forms with existing groups of soft-bodied animals in a quest for the origins of modern phyla. This was later described by Stephen Jay Gould and others as an attempt to "shoehorn" Ediacara taxa into modern groups. In practice, both Martin Glaessner and Mary Wade were fully aware that many Ediacara forms were not related to any of the living animal phyla. Rare specimens and those strange forms with poor preservation were left unnamed, and many remain so today. Glaessner, who was a qualified lawyer, saw his role as an advocate for the case that the Ediacara biota included some forms related to familiar phyla. In turn, this encouraged paleontologists who specialized in certain fossil and living groups to at least consider some Ediacarans as potential ancestral roots of their animal subjects that could have occurred before hard parts developed. A similar approach has been followed by Mary Wade, Richard Jenkins and Jim Gehling from the Adelaide school of thought. But to many casual scientific observers there was something wrong with the Ediacara model that Glaessner had pushed so determinedly.

Trevor Ford, who described the sparse assemblage of Ediacara fossils in England, thought the frond-like forms were actually not animal, but instead brown algae. His idea was dismissed with little ceremony by Glaessner. In 1972, the German paleontologist, Hans Pflug, became convinced that the strange ribbed Ediacara fossils of Namibia were representatives of an extinct phylum of unknown affinities, that he called the "Petalonamae." A decade later Seilacher argued that all but a few sponge and coral-like forms were members of the 'Vendobionta,' a lost kingdom of single-celled animals constructed of tubular units, arranged in various fractal patterns to give the various forms. The reaction of Martin Glaessner to this was far from indignation. He saw this challenge in legal terms as the case for the "prosecution." In 1988, the year of his death, Glaessner confided that he considered that the deciding evidence would have to come from new observational techniques and newly discovered Ediacara fossil assemblages, such as those of Russia and China.

Geological History of the Flinders Ranges

The Flinders Ranges are part of the library of Earth history. The rock ridges that are admired by tourists for their scenic beauty are revered by geologists because they are the exposed pages of a most important chapter in the history of Earth. Each layer of sandstone, shale and limestone provides clues that allow the reconstruction of past climate change and the evolutionary patterns of animals, just as the "pages" of the Dead Sea scrolls record the recent history of humanity. Just as with the Dead Sea Scrolls, the scholars attempting to understand the events they record must work with fragments – not the whole "book."

Much of the story originally recorded has been lost, or in the case of the sedimentary rocks of the Flinders Ranges lies hidden deep beneath the surface. But of the record exposed in the rocky outcrops of the Flinders, much can be learned, especially when studied by those who know how to read the "words" recorded in the rocky pages – these "readers" being the geologists, paleontologists, geochemists and others that have tramped over the Flinders Rocks for more than 50 years, carefully searching, carefully recording what they see there.

Figure 174. Living microbial mats, creek bed, Flinders Ranges, South Australia (P. Vickers-Rich).

There are few other places where this particular time between around 650 and 540 million years is so well studied and documented. The modern scenery of the Flinders actually masks a much deeper beauty – the evidence of a time when the first complex animals on Earth were getting their start. The strata of the Flinders Ranges provide insights into a time when, after 3 billion years of Earth history, the buildup of oxygen in the atmosphere, due to the millions of years of photosynthetic activity of microbial life together with the particular arrangement of continents at the time resulted in extreme cooling, setting the stage for the development of complex animals. Volcanic activity went on in spite of the climatic extremes, so that even during the depths of the global ice ages, carbon dioxide and other Greenhouse gases eventually restored a Greenhouse atmosphere at the end of this long period of repeated glaciations, leading to a climate that would dominate most of the time from around 580 million years ago to near present.

The rocks of the Flinders Ranges were deposited for the most part in a shallow sea from around 850 to 500 million years ago. On the edge of an ancient and massive continent (incorporating India, Africa and Antarctica and the western two-thirds of Australia), the Adelaidean Ocean stretched from what is now Kangaroo Island into central Australia, a sea about the size of the present-day Mediterranean Sea. Rivers dumped billions of tons of sediment stripped off even more ancient rocks that lay to the west onto a continuously sinking ocean floor. From time to time sea level rose and then fell, thereby interrupting the sequence of layers, sometimes exposing the sea bed sediments to aerial erosion that cut deep canyons into sediments laid down on the bottom of an ocean. Each change in depositional environment and change in climate is clearly recorded by differences in thickness and coarseness, mineral composition and color of the sedimentary layers. Like tree rings, the rocks themselves record the events taking place in these ancient marine environments, and for that matter events of a global nature.

As it happens, this chapter in Earth history recorded in the Flinders Ranges rocks was the time when marine bacteria and microscopic algae were generating oxygen that was accumulating in greater and greater quantities in the atmosphere. Stromatolites, sedimentary structures characterized by the rhythmic domed layers formed by microbes, built great reefs at this time. Increasing oxygen in the atmosphere is reflected in the red color of the rocks resulting from the oxidizing of iron minerals. Eventually these

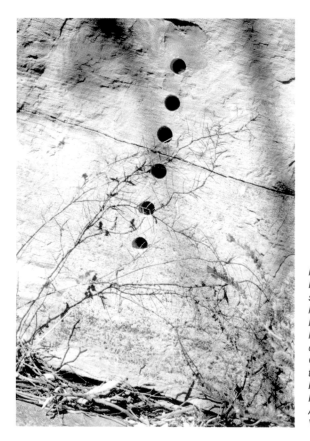

Figure 175. Palaeomagnetic sample drill holes in the Nuccaleena Formation (cap carbonate), GSSP site for the Ediacaran Period, Flinders Ranges, South Australia (P. Vickers-Rich).

microbes were so successful that they lowered the carbon dioxide content of both ocean waters and atmosphere and interfered with the prevailing Greenhouse atmosphere, resulting in the growth of icecaps. Episodes of such global cooling leading to glacial times are clearly recorded in the sedimentary debris, called tillite (a kind of diamictite), resulting from melting continental glaciers – the dropstones from melting ice and erratics emplaced on the sea floor of the Adelaidean Ocean clearly seen today in the outcrops of the Flinders Ranges. The waxing and waning of ice ages is recorded in the distinctive tillites and other sedimentary features – permafrost wedges, fine sands characteristic of seasonal freezing and thawing. Although the end of one of the more intense ice ages marks the beginning of the Ediacaran Period, at least one smaller ice age is recorded in early Ediacaran sediments, before the first diverse fossils of the Ediacara biota appear. During what is called the "Snowball Earth" ice ages, activities of life were probably all but shut down except in refuges near the Equator and around undersea-volcanic vents.

The magnetic field of the Earth also left a record in the Flinders Rocks – it oriented the magnetic minerals in the sediments being deposited in the cold Adelaidean Sea, like needles in a compass. Geophysicists who measure this magnetic signature of the Neoproterozoic rocks of the Flinders Ranges, by measuring the angle of tilt of the magnetic field recorded in ice age rocks of the time, found that Australia lay across the Equator. Limestone deposits that lay on top of the glacially derived sediments suggest that thawing of the ice was rapid, the climate moving from an Icehouse world to Greenhouse – a most dynamic time in climatic history. These very times of instability are often the times of origin of novelty – and this certainly seems to be the case with the rise of multicelled animals, the metazoans, as recorded in the Flinders Ranges.

Figure 176. Debris of Gawler Range Volcanics in the Acraman Impact Event layer, Flinders Ranges, South Australia – evidence for a major asteroid strike in the Gawler Range, 300 km west of the Flinders Ranges, during Ediacaran times (P. Vickers-Rich).

To add to the dynamism, during this period there were rare events that literally impacted on the biota. One of these was the collision of a 4 kilometer wide meteorite with the Earth around 580 million years ago. This impact is clearly recorded in the layered rocks of the Flinders Ranges as well as by a massive scar, the circular crater of Lake Acraman, north of the Gawler Ranges on Eyre Peninsula and 300 km west of the Flinders Range. So great was the energy released that the unique rocks at the impact site were blasted

Figure 177. Mawsonites spriggi (holotype SAM 24595), Flinders Ranges, South Australia (J. Gehling). Diameter about 12 cm.

into the Precambrian atmosphere, perhaps causing another global environmental crisis producing a distinct layer in the Flinders Ranges made up of the impact debris thrown out of the Acraman crater. Not only did this extraterrestrial visitor leave its scar on the countryside, but it dramatically affected the marine microbiota, causing a complete change in the composition of the acritarch fauna (see Chap. 13).

The Precambrian-Cambrian Boundary. The Verdun Syndrome

At a very well-defined level in the Flinders Ranges record, the first evidence of deep burrowing occurs. This record of animals churning the sea floor marks the most important time-line in Earth history: the Ediacaran-Cambrian boundary, a level above which modern ecological systems begin. These lowly traces of burrows into the sea floor are a clear sign that this was a turning point in the history of life. Soon after, rocks become crowded with the fossils of animals with shells, scales and platy armor. The rapid appearance of burrowing traces and animal armor represents the invention of predators. The "arms race" had begun. Large, unprotected Ediacaran animals, lying on the sea floor, could not survive the onslaught of animals looking for a short cut to obtaining food. Some animals evolved armor for added protection and also to allow more efficient use of muscles. Others turned down into the sea floor for protection and burrowed – both strategies of the Verdun Syndrome discussed later in Chapter 14. In so doing, burrowing animals were able to exploit a new source of food buried beneath the sediment. Many Ediacaran organisms were doomed – in two ways. They had no armor, and their buried bodies were either exhumed or consumed before they could join the fossil record.

The Cambrian rocks of the Flinders Ranges, Kangaroo Island and western New South Wales record a spectacular array of armored organisms, including the familiar trilobites, lamp shells (brachiopods), mollusks, ancient ornate sponges (archaeocyathids) and the oldest stony corals on Earth. On the fringes of the Flinders Ranges the oldest animal-framework reefs occur. The complexities of marine ecology that are typical of tropical reefs today had their origins in the Cambrian of southern Australia.

This chaotic chapter of Earth history, covering the emergence of modern animals, ended around 510 million years ago, in the Middle Cambrian. The finale came when drifting continents collided with ancient Australia, closing the Adelaidean Ocean and buckling the sea floor. Some 12 kilometers of layered sedimentary rocks were thrust up into a mighty mountain range. To this day the wearing-away of this mountain range continues. The current Flinders and Mt Lofty ranges are a mere ghost of the ancient Adelaidean Ocean and mountain range formed in its wake.

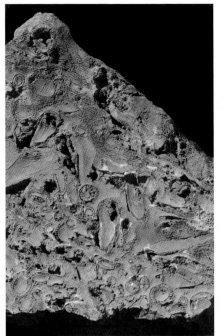

Figure 179. Left. Archaeocyathids, Ajax Limestone, Ajax Mine, Flinders Ranges, South Australia of Early Cambrian age (S. Morton). Width of block about 15 cm. Right. Reconstruction of an archaeocyathid (courtesy of the Paleontological Institute, Moscow).

Australia was then a part of a giant continent, Gondwana, which also included Africa, Antarctica, South America and India. In the Early Cambrian (542-510 million years) as this continent drifted across the Equator, new skeleton-bearing marine animals had evolved in the warming ocean, and some, the archaeocyathids, built great limestone barrier reefs. By the Middle Cambrian, the seas retreated from the Flinders region, giving way to river-plains and dune fields. Finally sedimentation ended with a continental collision that buckled and thrust the Flinders sedimentary strata into a mountain range. Today the eroded edges of the folded strata are exposed. This record would have remained hidden if it were not for the folding and consequent attrition by rain and wind. Each band of strata can be traced laterally across the landscape. The hard rocks composed of limestone and sandstone resisted weathering and made the high ranges. The soft rocks, such as shales and siltstones, formed the valleys and hollows.

Ediacaran Biology

Ediacara fossils are still a source of much controversy. Many questions remain unanswered. How could animals without shells or skeletons be preserved as impressions in sandstone? Why is this type of preservation almost unknown in younger rocks? Were Ediacara animals the oldest members of modern phyla – or were they failed, early experiments in the history of evolution? In other words, are Ediacarans our most distant ancestors or not?

As mentioned before in this chapter, there are two schools of thought on this topic. Dolf Seilacher, Jan Bergström, Jerzy Dzik and the late Stephen Jay Gould see them as a "lost construction." They regard fossil animals of Early Cambrian age (from 540 million years), as the earliest evidence of modern phyla. Following in the footsteps of the late Martin Glaessner others, like Richard Jenkins, Jim Gehling and Simon Conway Morris, interpret most Ediacarans as unusual

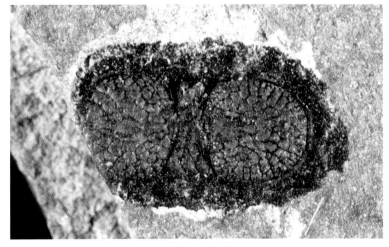

Figure 178 Agnostid trilobite from Beetle Creek, Queensland, Cambrian (S. Morton). Length of specimen about 1.5 cm.

members of modern phyla, if not our direct ancestors. Clearly, some Ediacara forms did not survive into the Cambrian. Others have strong similarities with living groups of animals. Some left trace fossils, such as those imprinted by *Kimberella* and *Dickinsonia*, abundant in the Ediacara fauna, and are evidence that moving animals had evolved by 555 million years ago.

Paleontologists and evolutionary biologists alike would like to know why Ediacara animals evolved when they did, some 3000 million years after the earliest life appeared on Earth. Did it take that long for oxygen to build up in the atmosphere so that large animals could live without having to manufacture their own food? Was there some environmental change that stimulated the evolution of multicelled (metazoan) organisms? Whatever the answers to these profound questions, there is a biodiverse assemblage of Ediacarans known from the Flinders Ranges in late Neoproterozoic rocks. They include body impressions, called body fossils, as well as some trace fossils.

Ediacara fossils from the Flinders Ranges were formed by the preservation of imprints of soft-bodied organisms that varied in size from meter-long mat and frond-shape forms to quite tiny bilateral and disc-shaped forms. It is possible that some of these organisms could have had the soft bodies reinforced with cellulose, as in living (and likely extinct) tunicates, acquiring

this ability through lateral gene transfer, as mentioned later in the text (Vickers-Rich, 2006), certainly facilitated by their close association with microbial mats that would have been built by organisms with the ability to produce this substance.

Some of the Ediacarans are preserved as molds on the bases of sandstone beds, and others in opposite relief, as casts. Newly discovered fossils, some from Namibia and Newfoundland, are preserved in three dimensions in massive sandstones.

Figure 180.
Eoporpita,
Rawnsley Quartzite,
Flinders Ranges,
South Australia (P.
Vickers-Rich).

Figure 181.
Charniodiscus arboreus
(left) from the Rawnsley
Quartzite, Flinders
Ranges,South Australia,
compared with a living
soft coral, Sarcoptilus
grandis (J. Gehling).

The Ediacara fossils may represent many animal groups.

1. Soft corals (cnidarians) such as polypoid and anemone-like animals that lived on the sea floor [*Mawsonites, Inaria, Conomedusites, Eoporpita*]; frond-shaped pennatulids (sea pens) with branches bearing small polyps [*Charniodiscus*] and discoidal holdfasts [*Aspidella*, including various preservational forms such as *Cyclomedusa* and *Ediacaria*]. Another interpretation is that these forms [*Charnia, Charniodiscus*] belong in the phylum Petalonamae – with no clear relationships to later or living phyla. Many of these forms now seem to be holdfasts for frondose forms, rather than separate taxa.

2. Sponges (poriferans) with a meshwork of spicules [*Palaeophragmodictya; Rugoconites*].

3. Worm-like forms (annelids) with segments and large, flattened bodies [*Dickinsonia*], but relationships to modern worms are certainly not definite.

Figure 182. Dickinsonia costata *and tubular fossils, in field block, Rawnsley Quartzite, south of Ediacara Hills, Flinders Ranges, South Australia (F. Coffa).*

4. Segmented animals (arthropods) with heads (*Spriggina, Marywadea*) and shield-shaped bodies (*Parvancorina*).

5. Mollusks with limpet-like, soft shells and feeding traces (*Kimberella*) seem to have grazed the leathery algal mats on the sea floor, leaving fan-shaped sets of paired radular traces.

6. Sea stars (echinoderms) shaped like small discs with five feeding grooves (*Arkarua*) preserved as impressions, showing no evidence of the mineral plates common in later echinoderms.

Figure 183. Arkarua adami, *reconstruction, Flinders Ranges, South Australia (J. Gehling). Diameter of specimen about 1.5 cm.*

Many Ediacarans seem to have no known living representatives. Their constructions appear very bizarre:

1. A number of hollow forms, built from walls of tubular construction and showing no evidence of muscular contraction or internal organs, may have lived partly buried in the sea floor sediments. They seem to represent body plans and constructions unknown in the modern world, but recent studies (Droser, Gehling & Jensen, 2005) note that these tubular forms share "some pattern elements with more complex Ediacaran body fossils, and may represent a simpler grade of organisms constructed in a similar manner." The most common are canoe-shaped forms with three walls of tubular construction [*Pteridinium*]. Bag-shaped forms with tubular walls [*Ernietta*], like *Pteridinium,* lived partly buried in the sand. Dimitri Grazhdankin and Dolf Seilacher (2002) interpreted *Pteridinium* as living within the sediment but the lack of sediment disturbance around these forms requires explanation. Perhaps they were primitive internal skeletons bearing more delicate sponge or soft coral tissue. Though present in Australia, Russia and elsewhere, they are particularly common in Namibia.

2. Discs with three arms [*Tribrachidium, Albumares, Anafesta*] have little in common with any modern phyla. They have been compared with jellyfish, echinoderms and sponges. They share very well-defined, but delicate, ribbed outer coverings, masking resilient internal organs. The very regular construction of these forms with three-fold body division is more suggestive of complex animals such as brachiopods. However, they were more likely representatives of now extinct groups unrelated to living phyla.

Ediacaran animals did not dig deep burrows (see also Chap. 12). Rather, they only shallowly furrowed the sea floor or tunneled through microbial mats. These irregular scribbling

Figure 184. Tribrachidium heraldicum *and shallow, horizontal trace fossils, natural epirelief cast of external mold, Rawnsley Quartzite, Bathtub Gorge, Flinders Ranges, South Australia; this form ranges from 0.3 to 5 cm in diameter (S. Morton). Slab width about 25 cm.*

traces were apparently made by animals too small to leave body fossil impressions and demonstrate that some Ediacara animals were mobile and had well-developed sensory organs for feeding on the sea floor.

Environment, Ecology and Preservation Styles of the Ediacarans

There is no evidence that Ediacara animals ate each other. Rather, they seemed to have been passive feeders that filtered microbes or absorbed nutrients from the microbial mats from the water, some perhaps using symbiotic algae in their tissues to generate energy. Those leaving traces like *Kimberella* probably grazed on the sea floor mats and organic detritus or like *Dickinsonia* simply absorbed nutrients through their outer wall from the organic-rich mats on the sea floor.

Ediacara fossils are found mainly on the bases of ripple bedded sandstone slabs. Most Ediacarans were clearly unable to plow and churn the sea floor – a process geologists and modern ecologists call bioturbation – although some did, leaving groove traces and sets of scratch marks in the mat-grounds. As a result, microbial communities living in or on the sea floor developed as thick organic mats trapping the sediment in the shallow Ediacaran seas, at depths too great for regular disturbance by waves, but within the zone illuminated by sunlight. Burying and preservation of some of the living Ediacara animals occurred when storms sent sediment avalanching down the slopes of deltas and marine shelves, thus smothering the mat-dwelling communities. Bacterial decay of both the sand-blanketed mats and the animals lying on the sea floor allowed a mineral crust or "death mask" to preserve the body forms, often in exquisite detail.

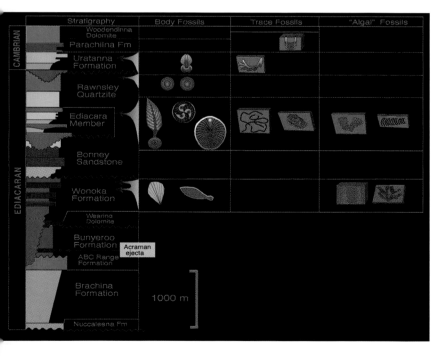

Figure 185. Distribution of body fossils, trace fossils and algal remains in the Ediacaran sequence, Flinders Ranges, South Australia (J. Gehling).

One of the dilemmas faced by scientists trying to understand the fossils of the Ediacara biota is the fact that soft-bodied organisms, lacking mineral skeletons or armor, were preserved as cast and molds in coarse sediments. This

was one of the main reasons why Dolf Seilacher proposed his Vendobionta theory (Seilacher, 1984, 1992). He was convinced that such organisms had a unique construction using biological materials that were particularly resilient. However, numerous observations have demonstrated that the Ediacarans were not particularly resilient. They were preserved with folds, creases and tears. They drape other structures on the sea floor. Many of these pliable organisms were preserved as external molds, for the most part. Most are negative impressions on the bases of sandstone slabs.

Mary Wade suggested that this type of preservation happened when mud on the sea floor replaced the decaying organism after it had been buried under a layer of sand. However, excavation of specimens has shown that in some

Figure 186. Ripple marks indicating the presence of water and direction of current flow – quite likely stabilized by microbial mats, Rawnsley Quartzite, Flinders Ranges, South Australia (P. Vickers-Rich).

cases, molds were filled by clean sand and not clay or silt. Most Ediacara fossils are now known from specimens where there is a "part" (an impression on the base of a bed) and a "counterpart" (the corresponding cast or mold) on the bed top below. So, how could such biological forms be preserved as external molds, with sand casting the molds from below?

Figure 187. "Elephant-skin" structure indicating the presence of microbial mats in sands, Rawnsley Quartzite, Flinders Ranges, South Australia (P. Vickers-Rich).

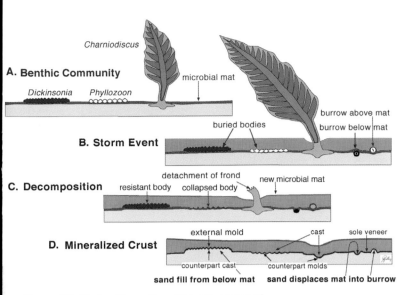

Figure 188. Mode of formation of "death masks" of Ediacaran organisms (J. Gehling).

One of the characteristics of Ediacara fossil surfaces is their texture. These surfaces display a wide array of patterns that resemble fine leather; they are creased and grooved on a very fine scale. Others are characterized by dimples and fibrous patterns. These types of textures are termed "old elephant skin" or "rhino hide." They provide draw cards for the paleontologist in search of sandstone beds likely to produce fossils. What produced such surface textures?

The answer came to Jim Gehling when he was studying some of the unfossiliferous sandstones of the Rawnsley Quartzite. These were characterized by weirdly deformed layers within each slab. In these odd sediments, ripples were incomplete and layering curiously disrupted. Sometimes the internal layers were folded over or rolled up. This sort of structure is characteristic of certain tidal flats on dry and hot coastlines. The waters on these flats are too salty for snails and crustaceans to enjoy life in – places where sand is tightly bound by microbial mats making them tough and flexible when wet and brittle and cracking when dry. So, when there are no animals to graze the seashores and sea floors, sand can be trapped as in a carpet. The textures that were seen as dimples and wrinkles in Proterozoic sandstone surfaces were most certainly reflecting slime-coated sea floors. These were sea floors that were undisturbed by waves or strong currents, or for that matter grazing animals. The sands were "infiltrated" by slime and microbial filaments. Then, when rapidly buried by storm-induced sedimentary avalanches, the buried sand layers were separated from the incoming disaster by a kind of "organic food wrap." But what has this to do with preserving fossils?

The best clue to understanding fossil preservation in these circumstances came from a study of matching surfaces. The bases of Ediacara fossil beds in South Australia, and in Namibia as well, are red – sediments cemented by iron oxides, quite smooth and resilient. In many cases, the interior of the sandstone layer is light in color, even white, indicating that the iron is concentrated on the base or sole of the bed. The opposing surface, or counterpart, is often even lighter in color and has a "sugary" texture. As a result, the counterpart is easily eroded and quickly lost to weathering. Something has selectively cemented the sole-veneer or basal surface of

each fossil bed. Perhaps this was because a mineral deposit formed over the organic-rich surface after it was buried. Jim Gehling (1999) suggested that the microbial mats contained surface bacteria that could alter iron in sand to iron sulfide and thereby produce a "death mask" of pyrite or "Fools Gold" that encrusted the top of the buried animals – and for that matter the entire organic-rich surface.

In 2002, Mikhail Fedonkin noted that the sedimentary surface in the White Sea outcrops of Ediacaran age were coated with filaments that had been replaced by pyrite, which rapidly converted to "rust" when exposed to the atmosphere during excavation of fossils. He further noted that fossils like *Kimberella* were often found with pyrite filling their external molds. Evidently, the reason that this occurs in Russia and not in Australia is one of geologic history. The Australian landscape has been long subjected to enormous change in climate, as Australia moved from the wet southern latitudes into the tropics during the last 100 million years – and so the continent is deeply weathered, as far down as [300 or more meters] below the present land surface. In addition, the Australian Ediacaran sequence is mainly porous sandstones, not the impervious clays of the White Sea of northern Russia, which has protected the enclosed fossils from the leaching effects of acidic ground waters in permeable sediments. So, the rocks of the Flinders Ranges were long ago exposed to oxidation – thus "rusting" – whereas those of the White Sea and northwestern Canada have remained gray-green because these rocks have only been exposed at the surface within the last few thousand years.

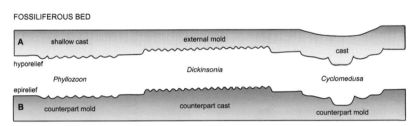

Figure 189. Different styles of preservation of Ediacaran organisms (J. Gehling).

Back in the Ediacaran Period, slow-moving and sedentary metazoans thrived on the sea floor below reach of fair-weather waves. Seasonal storms or cyclones stirred up sand in the shallows and sent avalanches of sand-laden water down the slope, instantly smothering the communities below. Once buried, the Ediacara metazoans were unable to dig their way out – they were trapped on top of the mat and below the layer of murderous sand. Oxygen was quickly depleted and sulfur-loving bacteria began their decay of the buried organic matter. The bacteria extracted sulfate from water trapped in the pore spaces of the sand, combined it with the rich supply of iron in the same sand and deposited pyrite. Organisms that were non-muscular or hydrostatic or gel filled quickly collapsed and were preserved only as casts. More resilient organisms, such as *Dickinsonia*, resisted decay long enough to allow a pyrite crust to form in the sand overlaying their bodies – thus a "death mask" formed as an external mold.

Eventually, both the mat and the buried body of *Dickinsonia* rotted away. But instead of the sand collapsing into the space from above, as the sandy sediments were compressed when more were piled atop the sequence, sand was pushed upwards from below the mat where the pyrite had not formed. And this underlying sand formed a cast of the external mold of *Dickinsonia,* from below. The same fate awaited organisms that had been carried into deeper water by the storm surges and buried within sand layers. Pyrite formed on the surface of such buried organisms and created a parting along which the rock would later break when weathering or a geologist hit it!

Once burrowing organisms evolved in the Early Cambrian, this form of preservation of soft-bodied organisms gradually almost came to an end. Only special circumstances in the Phanerozoic sequences on Earth have produced rare homologues. Burrowing animals are essentially "grave robbers." Being scavengers, many live on buried bodies alone. However, a majority of burrowing animals prefer live, infaunal organisms for their diet. Only animals with mineral armor or internal skeletons could be easily preserved in post-Ediacaran times. Still the Cambrian Period revealed another form of preservation of unmineralized bodies – in black mudstones where oxygen was excluded and scavengers could not survive – two outstanding examples being the Burgess Shale biota in the Rocky Mountains of Canada and the Chengjiang biota of China.

Events Fueling the Rise of Animalia

• The record of Earth history in the Flinders Ranges begins soon after the first complex organisms evolved. They are preserved in silica within algal reefs built of structures called stromatolites. A number of "forces" fueled the development of these first metazoans, well exemplified in the Flinders Ranges.

• The most catastrophic ice ages in Earth history froze the ocean at least twice, and probably several more times, even though the Adelaidean Ocean straddled the Equator. These were the so-called Snowball Earth events, when the Earth almost froze over.

• A distinctive cap-layer of limestone, the Nuccaleena Formation, records the reaction of global oceans to melting of the ice and precipitation of lime as the ocean warmed. The Global Stratotype Section and Point (GSSP) for the geologic period called the Ediacaran Period lies at the base of this cap limestone in the Flinders Ranges.

• The Acraman Event, which left a crater 300 kilometers west of the Flinders, produced by the impact of a meteorite 4-5 kilometers in diameter, sprayed rock debris into the deep waters of the Adelaidean Ocean. This distinctive layer of red and green rock chips can be traced all around the Flinders Ranges and beyond.

• Single-celled organisms reacted to extremes of climate at this time 600-570 million years ago, by evolving multicelled bodies. These oldest animals making up the Ediacara biota represent the *invention of death*. Instead of reproducing by simple cell division, the Ediacarans produced offspring and then died. The advantage of being an animal is an extended life cycle for the individual. The disadvantage is that after reproducing by sexual exchange of genetic information, animals get old and die. Single cells just continue dividing.

• Ediacarans came in many sizes – some large and quite flat *e.g. Dickinsonia.* This may have been an adaptation for absorbing light and nutrients to farm microbes in their tissues, or because they lacked respiratory and excretory organs.

• The base of the Cambrian Period is marked, in the Flinders Ranges and worldwide, by the sudden advent of burrowing into the sea floor. At this time, there was a dramatic increase in the variety of traces left behind by animals modifying the sea bottom. Animals had evolved new lifestyles. At about the same time, Small Shelly Fossils became common in limestones and shales. These fossils replaced impressions of the larger soft-bodied animals, the Ediacarans.

• The base of the Cambrian represents the evolution of modern ecosystems. It was the beginning of an "arms race." Animals first appeared which ate each other! The responses of the prey animals were many and varied. Soft, muscular animals burrowed for protection and feeding. Mollusks and arthropods developed mineralized scales and spine-like armor to ward off predators. Annelids hid from the same predators in tubes. Echinoderms evolved skin impregnated with mineral plates. As a result of predation, which both posed danger and fueled deep burrowing, large, unarmored and rather passive Ediacarans quickly became extinct. The small bilateral animals, with concentrated sense organs and the ability to crawl or swim, were the survivors.

• The invention of burrowing activity destroyed the "death masks" of soft-bodied organisms that had so characterized Ediacara times. Bacterial decay and the accompanying mineral precipitation that formed these masks over time could not take place in an environment where burrowing not only disrupted the sediments but allowed oxygen to penetrate to buried organic matter.

• Just above the base of the Cambrian, fossils record a massive expansion of new organisms, mainly small marine animals. In particular, arthropods developed armored appendages that allowed them to exploit many lifestyles in the world's oceans. These amazingly varied animals are best preserved in the Chinese Chengjiang and Canadian Burgess Shale fossil deposits of Cambrian age.

Figure 190. Dickinsonia costata, *Flinders Ranges, South Australia (S. Morton). Specimen on left about 5 cm wide.*

Figure 191. Aspidella terranovica, *Flinders Ranges, South Australia (S. Morton). Central disk about 5 cm in diameter.*

Figure 194. Rangea *sp., Flinders Ranges, South Australia (S. Morton). Block about 18 cm wide.*

Figure 192. Dickinsonia *showing corrosion surface, Flinders Ranges, South Australia (S. Morton). Width of specimen about 8 cm.*

Figure 195. Possible chordate – fossil and cast pulled from fossil impression; may be a taphomorph of Kimberella *(Australian one-dollar coin for size = 2.5 cm), south of Ediacara Hills, Flinders Ranges, South Australia (J. Gehling). Coin about 2.5 cm.*

Figure 193. Inaria karli, *Rawnsley Quartzite, Flinders Ranges, South Australia (S. Morton). Diameter of specimen about 6 cm.*

Figure 196. Holdfast and attached branching stem, *Rawnsley Quartzite, south of Ediacara Hills, Flinders Ranges, South Australia (P. Vickers-Rich).*

CHAPTER 6

The White Sea's Windswept Coasts

M. A. Fedonkin and P. Vickers-Rich

Today, the Winter and Summer coasts of the White Sea in northern Russia are two of the best places on the planet to find the oldest multicelled animals. These finds came late in the search for Precambrian life, life older than 542 million years. The very first specimens collected in this region did not immediately herald that much of interest would be recovered from rocks on the cold, north shores of Russia. These enigmatic fossils were thought to be much younger than they actually were.

Detailed study of what are called the Vendian deposits in the southeastern White Sea region began in the 1930's as part of a regional geological mapping program. Interestingly, among the pioneering geologists exploring this harsh region were a few women, who left a bright track of discovery. Imagine endless swamps and taiga, not amenable to motor vehicles. The only way in and out of these areas was in small boats – and everything had to be carried on one's back after leaving the water. Add to that clouds of mosquitoes and black flies and at times freezing conditions. Life was lived in the tents with long separation from family and friends. Geology was not easy work even for the strong and physically fit.

The Winter and Summer Coasts

One of the intrepid women geologists was Ekatarina A. Kalberg, who in 1940 published a small, seemingly insignificant book titled *Geological Description of the Onega Peninsula*. In 1936-1937, she discovered imprints of plants and what appeared to be worms in rocks of the Onega Peninsula west of Arkhangel'sk. Similar fossils had been found previously in cores pulled from shallow boreholes drilled on the Onega Peninsula, along the tributaries of the Nenoksa River.

At the time of Kalberg's mapping survey, the age of the clays and sandstones in this region was thought to be Late Devonian – around 350-370 million years. But the rocks had such a "fresh" appearance that it was difficult to believe they were even this old, even harder to contemplate an older Precambrian age. They looked as if they could have been deposited only a few years ago!

A Devonian age for the rocks exposed along the White Sea was originally suggested by the British geologist R. I. Murchison when he visited Russia in 1840. Murchison together with Adam Sedgwick had defined the Devonian System two years earlier. There is no evidence that Murchison actually visited the White Sea region during his 1840 visit, but after publication of his monograph *Geological Description of*

Figure 197. Taiga covers the ground to the west of Arkhangel'sk, northern Russia (P. Vickers-Rich).

Figure 198. Boris Sokolov, founder of the Laboratory of Precambrian Organisms, Russian Academy of Sciences, Moscow (F. Coffa).

Opposite page: Imaginary Yorgia rising over Cliffs of Zimnii Bereg, White Sea coast, northern Russia where rich collections of Ediacara-aged metazoans have been recovered over the past decades (artist Peter Trusler).

115

the European Russia and Ural Range in 1849 (in Russian), a Devonian age was assigned to the White Sea sediments, and this assignment persisted for more than a hundred years. In his historical and autobiographic study, *Essay on the Advent of the Vendian System,* Boris Sokolov mentions that until 1956 most geological maps continued to cite a Devonian age for these rocks (Sokolov, 1997).

In the late 1940's, the first deep boreholes were drilled through sediments down to the Archean crystalline basement near Nenoksa, Arkhangel'sk and Ust'-Pinega. Near Nenoksa, 10 km to the east of the Suz'ma River, one borehole bottomed out on ancient, biotite-rich granite at a depth of 615 m. Atop this granite lay a reddish to yellowish sandstone, later named the Nenoksa Formation, more than 300 meters thick. It had no surface exposures. The top of this group of rocks had been cut into by erosion and deposited on top of it were bluish-green to gray sandstones and clays. These, unlike the Nenoksa sediments, were exposed on the surface, along much of the White Sea coast and in the valleys of the rivers to the east and west of Arkhangel'sk.

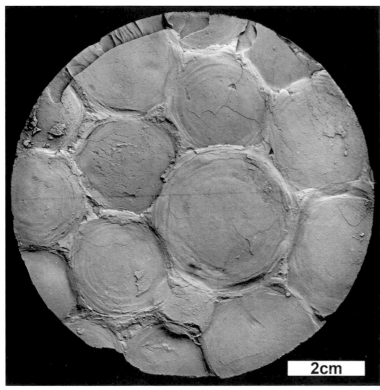

Figure 199. Beltanelloides sorichevae *from a core into the Vendian sequence of northern Russia, Yarnema borehole, 118.4 meter depth, Ust'Pinega Formation, Lyamtsa, Zimnii Bereg, White Sea (M. Fedonkin).*

Kalberg was the first to work out the sequence of these rocks – which were deposited first and which last. She gave them names: from oldest to youngest, the Lyamtsa, Arkhangel'sk, Verkhovka, Zimnii Gory and Yagry formations. In 1939-1940, she and her colleague G. I. Ershova found organic tubes in these rocks in one of the boreholes, fossils common in these rocks – *Saarina* and *Calyptrina*, named by Boris Sokolov. Also discovered was the first fragment of a form originally thought to be closely related to soft corals, the sea pens (Sokolov, 1997, plate XII), but now being reconsidered.

In 1948, A. I. Zoricheva, another well-known lady geologist, discovered the disc-shaped organism now known as *Beltanelloides* in one of the Nenoksa boreholes. It is quite noteworthy that these first Precambrian fossils were all found

in borehole cores even though the chances of encountering them with such small "bullets" into the ground were very low. How to interpret these strange finds? Their true significance was not realized immediately, because the rocks were still thought to be of Devonian age, more than 200 million years younger than they actually were – and at a time that had a wealth of animals, complex animals, most with hard skeletons that were far more interesting to study than these vague, enigmatic forms. No one really could be bothered.

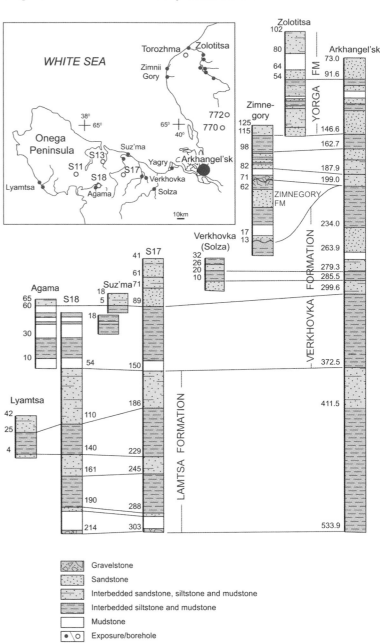

Figure 200. Cross-section of Vendian sequence of the Onega Peninsula and Zimnii Gory coast, northern Russia (modified from papers by Dima Grazhdankin). (D. Gelt)

The true age of these rocks was not worked out until after the war, and this was directly related to the deep drilling program carried out in the European part of Russia. The main aim of this program was to find oil to bolster the Russian economy, just recovering from the devastating effects of World War II. These deep drill holes encountered a new sequence of sediments, which later were named the Vendian Complex and found to be of latest Precambrian age. Directly overlying these were the younger Baltic Group of Early Cambrian age, already known in both the western and northwestern parts of the Russian Platform.

Geologist Boris Sokolov announced the antiquity of these sediments in an article published in the Geological Series of the academic journal *Izvestiya Akademii Nauk SSSR* (Sokolov, 1952). At the time, he was head of the Paleontological Laboratory at the All-Union Oil Science Research Geological Prospecting Institute (VNIGRI) in Leningrad and a respected researcher in the Paleontology Department of Leningrad State University. He was a geologist with broad experience, having already led a number of expeditions in Russia and remote Central Asia. He had spent the years 1941 to 1943 in Xinjiang Province of western China, under rather unsettled conditions. His Ph.D. research had dealt with Carboniferous faunas of the northeast Ukraine and adjacent areas. So, he had broad experience in many parts of the world and many parts of the Geological Time Scale.

The Vendian

The name Vendian was derived from the ancient name of Slavonic people who lived south of the Baltic Sea more than 2000 years ago (as mentioned by Tacitus). The artifacts of their culture are found as far south as the Volyn' Region. The Baltic Sea itself was called the Vendian Sea by Ptolemy. This name is still used occasionally by some ethnic groups – for example, the Luzhitsa Serbians (Sokolov, 1997). The Vendian assemblages from the Eastern European Platform lived during a time now named the Ediacaran Period, just prior to the Cambrian Period.

Understanding of the antiquity of the rocks given the name Vendian brought about re-assessment of the stratigraphy and paleogeography of the area from the Black Sea to the White Sea and from the Baltic to the Ural Mountains. Natalia Igolkina (1956) was probably the first to suggest that the sands and clays exposed in the cliffs along the Winter Coast of the White Sea of northern Russia to the east of Arkhangel'sk, were about the same age (latest Precambrian and earliest Cambrian) as the sediments exposed in the area around Leningrad. It is now clear that the White Sea deposits are older, but in 1956 it was a brave step to surmise these rocks were so ancient – much older than the Devonian – even though older sediments had already been reported from boreholes in the area.

In the core of the boreholes, Igolkina had discovered some organic-walled, tubular fossils resembling the Early Cambrian, *Sabellidites cambriensis,* first described by M. E. Yanichevsky in 1924. These fossils had been collected and previously described from the "Cambrian blue clay" known near Leningrad (St Petersburg). Later studies of new borehole cores by Mikhail Fedonkin, Anatoly Stankovsky and Marina Gnilovskaya revealed that organic tubular fossils are very common in Vendian and older deposits (Stankovsky *et al.*, 1990; Gnilovskaya, 1996, 1998; Gnilovskaya *et al.*, 2000).

After these initial studies, a more in-depth understanding of the Vendian succession in the White Sea area resulted from a systematic geological survey carried out by Stankovsky and his colleagues from 1965 till 1985 (Stankovsky *et al.*, 1990). Stankovsky was born in 1934 in Leningrad, and during World War II (Russians call it *Velikaya Otechestvennaya Voina* – the "Great Fatherland War") when Leningrad was blockaded by the Germans, this seven year old boy was evacuated along with hundreds of thousands of other children to Perm, a city on the Kama River, west of the central Ural Mountains. Many towns

and villages east of the front lines became shelters for millions of people. Anatoly was one of these and learned much from his stay in these very mountains that later would yield up an impressive collection of Precambrian fossils. He was unaware of the geological significance of the surrounds at the time. Once the war ended, Anatoly returned to a ravaged Leningrad, which was being rebuilt. He was educated at Leningrad State University, gaining a diploma in geology. Rocks of the Russian north were his first choice of research.

Stankovsky loved world literature and poetry, to such a degree that, after a long day in the field, he would read through the night by the light of the campfire. He went to the north and discovered a spectacular diamond field. His field was economic geology, but as he mapped the rocks of the far north, figuring out their stratigraphy, which laid the groundwork for later research on Vendian paleontology and stratigraphy. For years he was the principal advisor for paleontological research in the region, significant in that he was head of the Northern Territorial Fund for Geological Information (the name of which changed a number of times in the transition from USSR times to those of the CIS – the Commonwealth of Independent States). Stankovsky consistently shared his immense knowledge of the regional geology with and opened extensive archives to younger colleagues from the Precambrian

Figure 201. Andrey Ivantsov, Zimnii Bereg, White Sea. Ivantsov has been a major field leader in the expeditions to the White Sea and Siberia, beginning his career in the Ordovician as a trilobite researcher. His tireless and determined search of the Vendian sequence has led to an in-depth understanding of behavior and relationships of the Ediacarans (P. Vickers-Rich).

Laboratory of the Paleontological Institute in Moscow – Andrey Ivantsov, Dimitri Grazhdankin and many others. He was a major contributor to advances in Vendian research and understanding of evolutionary changes of the Vendian biota, as well as the environmental and climatic changes underway in the region more than 550 million years ago.

Stankovsky and the Paleontological Institute of the Russian Academy of Science (PIN RAS) crew found sand- and clay-rich deposits of Vendian age widely exposed over the southeastern margin of the Baltic Shield. The underlying crystalline basement had probably formed emergent islands in the Vendian ocean more than 560-570 million years ago, around which the clays and sands were deposited. The PIN crew found exposures of these sediments all across the southeastern White Sea region from the Onega River to the Kuloi Plateau. The thickness of the sands and clay varies significantly, from a few meters at one place to hundreds at another, and the exposure also varies – from restricted river cliffs to the magnificent outcrops along the Zimnii Berg of the Winter Coast of the White Sea to the east of Arkhangel'sk. Correlation of the rocks from one isolated outcrop to another is not easy because of lateral facies changes reflecting the changing environments at the time of deposition. After the deposits were laid down, vertical displacement due to faulting of the buried sediments made correlation even more difficult.

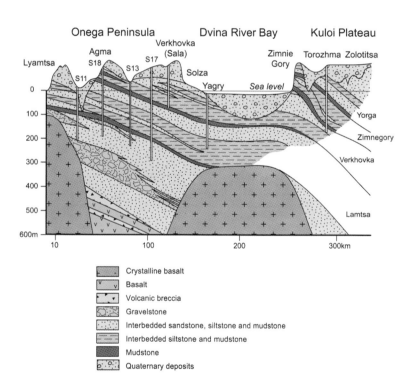

Figure 202. Location and correlation of boreholes drilled into the Vendian sequence on the Onega Peninsula and Zimnii Gory coast, White Sea, Russia (modified from papers of Dima Grazhdankin). (D.Gelt)

What makes the correlation possible are the rocks recovered from those very boreholes that led to the discovery of the first Vendian fossils. And, in addition, volcanic ash deposited in the shallow seas at the time Vendian animals were alive serve as marker beds, which can be recognized both in the cores and in outcrop over great distance – hundreds of miles. Some of these ashes can also be dated radiometrically, yielding a 555.3 +/- 0.3 my on the volcanic

event that produced this ash (Martin *et al.*, 2000). Besides these two methods of correlation, sometimes the actual sequence of the sediments is quite unique and the techniques of sequence stratigraphy (that is, the order in which different sediments occur) can be applied. Fossils preserved in the sands and clays, including those of algae and tiny microfossils (acritarchs, for example) and metazoan imprints along with the tracks and trails left by ancient animals, all help tie a sequence of rocks from one place to that of another. Because animals and plants have a distinct time of existence, a time range, the matching up of rocks from many different parts of the world becomes possible. If the fossils are similar, the time during which the rocks were deposited is likewise similar.

Figure 203. The White Sea, whose erosive force into the Zimnii Gory coast exposes the fossiliferous Vendian sequence of late Neoproterozoic age in early Summer (artist Peter Trusler).

The total thickness of Vendian deposits in the southeastern White Sea region reaches around 550 meters. These deposits rest almost horizontally on an ancient, eroded crystalline basement in most places, but occasionally they overlie those of younger Riphean age: the Solozero and Nenoksa formations, which themselves are mostly sandstones, gravels and conglomerates, interspersed with basalts and dolerites, part of the Solozero Formation. During the Riphean (the time between basement formation and Vendian deposition), volcanoes and subsurface magmatic activity was underway, and both sediments and volcanic rocks fill deep

Figure 204. Lyamtsa River mouth, White Sea region, northern Russia (V. Kushlina).

Figure 205. Suz'ma locality, White Sea region, northern Russia (P. Vickers-Rich).

channels cut into the crystalline basement (Stankovsky *et al.*, 1972, 1990; Yakobson *et al.*, 1991), indicating a period of subaerial exposure and erosion prior to their being emplaced.

The very best exposures of the Vendian rocks occur in the sea cliffs of the Onega Peninsula and along the Winter Coast (Zimnii Bereg) of the White Sea. Smaller exposures crop out in the valleys of many rivers intersecting the coast – the Lyamtsa and Purnema rivers that incise the Onega Ridge; the Nizhma, Agma, Suz'ma, Verkhovka, Solza and Kinzhuga rivers that flow across the western part of the Dvina Bench; and the Torozhma and Zolotitsa rivers that flow in deep valleys on the Kuloi Plateau (Grazhdankin, 2003). The thickness of the Vendian sequence increases towards the northeast, east and southeast, where it is in turn overlain by Paleozoic sediments, younger than 540 million years. In these younger rocks are the first animals with hard skeletons.

As noted above, the first Precambrian metazoans in Russia's north were recovered from boreholes drilled in 1930. The first Vendian metazoans recognized in outcrop came much later, only in 1972, discovered by V. A. Stepanov, a graduate student from Moscow State University, during a summer mapping exercise. Stepanov prospected outcrops exposed on the eastern bank of Suz'ma River about 5 km upriver (south) from the old village of Suz'ma, which lay at the river's mouth. This is a remarkable locality. Here the river makes a complex loop and exposes a steep cliff of clay, about 15 meters high, on the eastern bank. Sandstone lenses could clearly be seen in this clay-rich bank, probably the remains of old sand channels. A few rock slabs lying scattered along the river's edge contained strange imprints that resembled a series of regular ribs. Stepanov wondered what in the world had made these shapes.

Stepanov collected a few of these "fossils," stuffed them into his field pack and brought them back to Moscow – first with a 5 km hike, then by boat to Arkhangel'sk and finally by overnight train to Moscow – no small journey. He took the specimens to Professor Boris Keller, head of the Laboratory of Geology and Geochronology of the Precambrian based at the Geological Institute (GIN), Russian Academy of Science in Moscow. The GIN was the best place to start, for this is where paleontologic and biostratigraphic research on the Proterozoic had been underway since the early 1960's. Keller,

Figure 206. Arkhangel'sk, northern Russian port on the White Sea and a staging area for Russian expeditions to the Vendian sequence (P. Vickers-Rich).

who headed this lab, was a world authority on stratigraphy of the Proterozoic. He had gathered around him a dynamic research group, which included a number of bright, young and enthusiastic students whose projects dealt with many aspects of Precambrian paleobiology and its paleoenviroment.

When Stepanov discovered his Precambrian metazoans, Keller knew about the fossils recovered from cores. One had been collected from a depth of 1552 meters in the Yarensk borehole. This oval-shaped bilaterally segmented fossil had been identified by Vladimir Menner as a trilobite-like organism, but one without any remains of an organic carapace. Menner was a well-known paleontologist and stratigrapher in Russia. In 1969 when it was restudied, Boris Keller formally named it *Vendia sokolovi*, after the Vendian Period and its author B. S. Sokolov, but it remained an enigma in terms of its relationships with other known animals.

119

Figure 207. Boris Maksimovich Keller (M. Fedonkin).

Figure 208. Tatiana and Timofei Burych and Paleontological Institute field assistant, Suz'ma, Russia (M. Fedonkin).

Tatiana and Timofei
Tatiana Andreevna Burych (*née* Ermolina) and Timofei Antonovich Burych, like so many people in the region around Suz'ma, lived difficult lives over years of turbulent Russian history. Timofei's first wife died young and left him with a number of children to rear on his own. He married Tatiana, who only a few months after her marriage had been widowed when her husband was killed in action at the very beginning of World War I. Tatiana was left with a small daughter. Life in a northern Russian village exposed to the open sea was not easy. Much time must be spent simply gathering supplies for winter – hay must be cut and dried just to keep the cattle alive. Piles of wood must be chopped to size just to heat the house. The garden must be cultivated during a very short growing season, fish must be caught and salted, mushrooms and berries gathered from the forest – in the winter there was no food there for the picking.

Years of revolution, collectivization, another war – all that only added to the hardship. Widespread repression also touched this family, as it did so many others in the early Soviet years. In 1938, Timofei was sentenced to 10 years of hard labor in one of the hundreds of camps around the USSR. The formal charge was the "murder" of four raccoons, locally farmed for fur in the kolkhoz (the collective farm). Just before the "crime" Timofei had been sent to work at the local lumber mill. There he became seriously ill and was brought back to the village where he was immediately arrested, based on the "trumped up" charges of raccoon slaughter – a rather stiff sentence for the loss of four pelts!

Regardless, Timofei was taken to a prison camp in Puksa, in the Plesetsk District of Arkhangel'sk, where he was assigned the task of woodcutting. Tatiana, alone, raised the children. Timofei finally returned in 1943. Russia was in the midst of a war, and every able-bodied male had been sent to the front. Perhaps it was just as well that he had wood to cut, for even in this isolated part of Russia, there were enemy incursions from this long and bloody war – one of his sons, Mikhail, was injured during aerial attack on the way to school. There were many problems, but for Tatiana and Timofei, they almost went unnoticed. They were again together after all those years of painful separation.

Timofei died peacefully one winter night, as did Tatiana a few years later at 88. In Russia such a death is thought a gift for good people. Both Tatiana and Timofei lie buried in the village cemetery on the grassy hill overlooking the River Suz'ma, surrounded by birch and rowan trees. Tatiana's son, Mikhail Nekrasov, a local ranger now retired, spends the summer months with his wife in Suz'ma village, where both were born. His family and a new generation of paleontologists inherited both the friendship and memories of their parents.

of an old couple, Tatiana and Timofei Burych. Keller and his crew were graciously received in a beautiful, ancient, but solid wooden house typical of the region. Then taking advantage of the high tide each day, when the stream flow literally stops and the tide pushes water upstream, Keller and crew borrowed a boat and made their way upstream. There, on the eastern bank of the river, they revisited the outcrop where Stepanov had collected his specimens, and they spent the rest of their time excavating. What surprised Keller when they began to examine the Suz'ma cliffs was that the rocks were extremely soft, almost non-lithified. Sandstones could easily be broken with the light touch of a hammer – truly unusual for sediments of this antiquity. The first few days of their expedition were not auspicious; the weather was inclement – snow, and if not that, rain. This was June, thought Keller, summer! But such is the nature of Russia's northern coasts. Despite this, fossils were found in abundance and particularly interesting was the feather-like form *Pteridinium* (probably *P. simplex*) as well as some discoidal forms.

The first preliminary scientific paper describing this small collection was published quickly, in 1974, an announcement of global significance. Before this discovery, *Pteridinium* was known only from the rocks in Namibia, likely of comparable

With all this in mind, when Keller saw Stepanov's specimens from Suz'ma, he immediately realized their significance. They were the very first Vendian metazoans collected in outcrop! So, in 1973 Keller returned with his own expedition to Suz'ma. On this crew, in addition to Keller and Stepanov, was Nikolai Chumakov, world expert on glacial deposits, paleoclimates and stratigraphy. It was too cold for camping; Keller inquired about accommodation in the home

Discovery of Summer Coast Metazoan Biodiversity

In 1975, Nikolai Chumakov returned to Suz'ma with Mikhail Fedonkin, who was a young paleontologist based in Keller's laboratory at the GIN at that time. He studied tracks and trails left by metazoans, known as trace fossils, and was intrigued at the possibility of identifying the track-makers. He asked Keller if he could work on the White Sea Coast to follow up what he had seen with Chumakov. Keller said "yes!"

It was a warm, dry summer in 1976. As a result, the water level of the river dropped so much that it was possible to walk right across in many places. So, the lowermost parts of the outcrops were exposed, not previously investigated because generally they were not accessible. In addition, a tectonic contact between the clay-rich and sandy members was noticed. Both rock units of the Suz'ma sequence turned out to be fossiliferous. Fedonkin found new fossil assemblages, one exposed on top of another – "in superposition," as geologists say.

What interested him as he prospected the outcrops in 1976 was that the sandy member, known for some time, contained fragments of the clays. These fragments had probably not traveled very far, for they were not well rounded as happens with longer transport. Some of them preserved their original laminations. These had accumulated in quiet waters, then for some reason had been ripped up and transported as part of downslope movement, a sand "avalanche" along some ancient coastline. What caused the movement was likely a storm. When the sand finally came to rest on the shelf of the shallow sea, it buried within it the clay fragments and any animals that were swept away or were living on the ocean bottom. These sandy sediments contained body fossils but very few trace fossils. There was little order in these sandy sediments; everything was mixed up and reflected the chaotic nature of formation.

The purer clays and silts were characteristically laminated with layers from 0.5 to 15 cm thick. The laminae had very distinct lower surfaces and less distinct upper ones showing graded bedding – that is, the bottom of the silt layer contains the largest particles and the top the finer. This results from silt gradually settling out of the water, so the first grains to settle are the biggest and heaviest, the last to settle the smallest and lightest. Repeated episodes of such graded bedding made it quite clear that submarine avalanches had occurred again and again along the shores of the ancient Vendian sea, and they were happening even in the deeper waters, well below the storm base. Silt was being transported down to the deeper, muddy sea floor, remote from the shallow waters, and in these less disturbed environments, trace fossils such as *Planolites*, *Neonereites* and *Nenoxites* were preserved along with rare body fossils like *Tribrachidium* and *Cyclomedusa delicata*.

After the first excavations of 1974, followed up by Chumakov's and Fedonkin's more extensive work in 1975, annual excavations along the Suz'ma River revealed a diverse assemblage of soft-bodied metazoan forms, despite the limited outcrop. Along with the endemic three-fold, discoidal *Albumares* and the bilateral *Onega* and *Vendomia*, the fossil assemblages had much in common with those found in Namibia (*Pteridinium*, an *Ausia*-like form, *Ernietta*) and South Australia (*Dickinsonia*, *Tribrachidium*). Martin Glaessner, who visited Moscow in the early autumn of 1974, confirmed the identity of these known Ediacara species. The same autumn Boris Sokolov and Yuri Tesakov, inspired by these new discoveries, visited the Suz'ma fossil locale. Every year of excavation yielded new species of both body and trace fossils (Keller & Fedonkin, 1977; Fedonkin, 1977, 1981). The villagers of Suz'ma assisted the geologists in many ways; the family that had helped Keller initially, the Burychs, formed lasting friendships with the crews from Moscow. The Burychs were always ready to help. Their traditional Russian longhouse became the staging area for these expeditions for more than a decade, and the enduring friendship with this family greatly facilitated scientific exploration.

Figure 209. Dickinsonia *sp., Zimnii Bereg, White Sea, Russia (F. Coffa). Total length of specimen about 8 cm.*

Figure 210. Eoporpita *medusa, Zimnii Bereg, White Sea, Russia (F. Coffa). Diameter of central protuberance about 1 cm.*

age. These fossils provided the possibility of using fossils to correlate rocks over long distances – a tool called biostratigraphy. And this was a fundamental discovery for fossils in rocks of Precambrian age, for it was a technique that was outstandingly useful in the Phanerozoic, the time from 542 million years to present. It was not a technique thought to be useful for the most part in the Precambrian.

In 1977, Fedonkin, Natalia Bochkareva and Alexander (Sasha) Kochedykov worked the Suz'ma locale and discovered abundant trace fossils, unrecognized before. Significantly, on this trip, body fossils of more metazoans were located – imprints of completely soft-bodied animals – none had shells or skeletons. As the blue clays were washed from the gray-green or rusted surfaces of the resistant sandstones with the icy waters of the Suz'ma, the metazoans appeared as if by magic.

Each ensuing season, new species were collected at Suz'ma, for with spring melting ice floating down the river cut deeply into the banks, literally bulldozing the flat-lying river outcrops. Since the sediments were soft, water-soaked clays, new exposures appeared each year. It was crucial to inspect and remove the newly exposed specimens and wrap them in plastic or impermeable paper so that they dried slowly and did not split with desiccation. If they dried too quickly, fossils would be destroyed. The bulldozing river was truly an ally,

Figure 211. Pteridinium nenoxa, *from the Vendian sequence at Suz'ma, Arkhangel'sk region, northern Russia (F. Coffa). Width of specimen about 12 cm.*

Figure 212. Natasha Bochkareva at work in the Vendian sequence, Zimnii Bereg, White Sea, northern Russia (M. Fedonkin).

Figure 213. Mikhail Fedonkin (right) and A. Kochedykov during the early exploration of the Vendian sequence along the Suz'ma River, Arkhangel'sk region, White Sea, Russia in 1977 (Natasha Bochkareva).

but paleontologists had to move quickly behind the onslaught of the spring thaw or the weathering process and erosion would destroy the precious treasures, which the river had just exposed.

But even though each year's excavations at Suz'ma uncovered new metazoans, exquisitely preserved, there was a need to search for more extensive outcrop, farther afield. The Suz'ma locality represented only a small part of the total Vendian sequence. Rocks of slightly different ages or deposited in different environments, so forming different facies, would surely give new opportunities for finding additional forms.

A publication by Boris Vasilievich Timofeev was the spark that initiated a wider exploration program. Timofeev was a pioneer of Precambrian micropaleontology (see Schopf, 1999). In 1966, he published a paper including detailed descriptions

and photographs of the cliffs along the Winter Coast (Zimnii Bereg), northern Russia, which lay to the east of Suz'ma and Arkhangel'sk. The age of the sediments forming these cliffs was thought to be Early Cambrian (Baltic Stage) by Nikolai S. Igolkina (1959) and A. I. Zoricheva (1963), based on micro-algae preserved in the coastal clay. Both noted, however, that the lower part of these outcrops might possibly belong to the Vendian sequence, based on their correlation of sediments exposed along the cliffs with those pulled from the deep boreholes of this region in cores.

Figure 214. Alexander ("Sasha") Kochedykov, longtime field assistant to the Paleontological Laboratory of Precambrian Organisms (M. Fedonkin).

Figure 215. Vendia rachiata (holotype), Solza, Arkhangel'sk region, White Sea, northern Russia (M. Fedonkin).

Figure 216. Micropaleontologist Tamara Herman (left) and Boris Timofeev (M. Fedonkin).

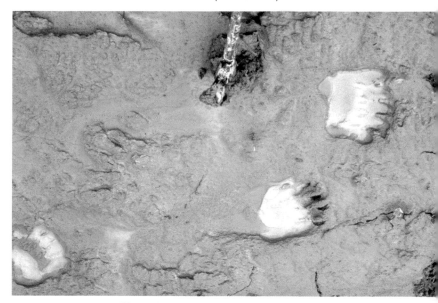

Figure 217. Large bear prints in the eroding Vendian muds on landslide, Zimnii Bereg Coast, White Sea, northern Russia (P. Vickers-Rich).

The cliffs of Zimnii Bereg had been prospected by very experienced geologists, including Professor Keller. Shells of bivalves (clams) and snails and of many-legged trilobites are normally abundant in the Paleozoic marine deposits. In addition to the lack of typical early Paleozoic shelled fossils, despite the in-depth geological work carried out in this area, Fedonkin mentally noted that the microfossils recovered from these sediments did not look just right for the Cambrian and was determined to prospect the Winter Coast for macro-metazoans and traces.

Fedonkin discussed this plan in the spring of 1977. Keller had visited those sections years before and, in spite of a determined search for fossils, had found nothing. He had thought that the age of these rocks was Late Vendian (Keller, 1969), an age supported by the radiometric dating of

the mineral glauconite abundant in some of these sediments (a potassium-argon date of 600 million years had been determined). But, because he had found no macrofossils, despite a concerted effort, Keller was not very enthusiastic about finding a Vendian fauna on Zimnii Bereg (the Winter Coast). He had noted this in his field notebook, beside the

First Discoveries at Zimnii Bereg

The field season of 1977 began with excavations on the Suz'ma River and concerted prospecting along the 20 km upstream. Nothing of any great interest was found.

From Suz'ma the team from the Paleontological Institute returned to Arkhangel'sk, and after some negotiation, hired passage on an old ship called the "Yushar," which regularly traveled along the Winter Coast delivering supplies to the lighthouses and isolated villages. Their destination was Nizhnaya Zolotitsa, a small village at the mouth of the River Zolotitsa. The trip was to take a day. Not long after leaving Arkhangel'sk, the ship was pounded by an intense storm, not so unusual for this part of the world. It took three days to reach Zolotitsa! Instead of being able to "jump ship" at Zolotitsa, the Yushar had to continue along the coast to a more remote port with a safe harbor and wait out the storm. Along the way, everyone was profoundly seasick – there was no relief from the continuous roll of the boat. Not an auspicious beginning. The storm passed and the Yushar hove off Zolotitsa. A smaller boat was sent out from the village to collect crew and cargo. The expedition had finally reached its starting point.

Nothing but sand dunes – that was all that Fedonkin and his two teammates saw on arriving at Zolotitsa. Those who lived in Zolotitsa were fishermen part of the year, but during the spring, they harvested thousands of fur seals on the sea ice. And during the long "White Nights" of the endless sun of summer, people never seemed to sleep – the light completely obliterated night – and with the nagging knowledge that the coming winter was long and, for the most part, dark, doing as much as possible during the light of summer "days" obsessed everyone. Amidst this frantic activity of the summer, Mikhail Fedonkin, Natasha Bochkareva and "Sasha" Kochedykov set up camp in the long "white" night, gained their land legs and the following morning began prospecting.

For nearly a week, Fedonkin and his crew of two meticulously inspected outcrops around Zolotitsa. Absolutely nothing was found. Disappointing, but not daunting. But Fedonkin and Natasha Bochkareva, the last remaining crew, decided to travel south from the village to inspect the highest outcrops atop the 100 meter cliffs of Zimnii Gory near the Zimnii Lighthouse. They hitched a ride on a small boat (a dora), slow-going but very reliable, even in heavy seas. They could see that all along the coast, usually by the mouths of the rivers, there were a variety of wooden huts used by fishermen who came for the salmon during summer and early autumn. Small boats from the village regularly collected fish from these seasonal fishermen and returned to Zolotitsa with the iceboxes overflowing. After being dropped off, they had another 12 km walk along the sloping, rocky beach of the White Sea from the Veprevsky Lighthouse northward – not a Sunday afternoon stroll by any accounts, especially with the constant tilt of the beach. Not only that, the "beach" was made of rounded stones that constantly moved as they walked on for kilometers and kilometers. And, then, of course, the tide came in and at times there was no room between cliff and "beach" and so they simply had to tramp on with water lapping at their knees.

As they walked, the cliffs of the Winter Coast climbed higher and higher. They were struck by the lack of alteration of these horizontally stratified sediments, more than half a billion years old, one layer stacked neatly atop another. They looked like they had been deposited only yesterday.

The only place to camp was the pebbly beach itself – really too narrow. But, tents were set up at the base of the cliff wall. Not a good idea! It was clear that with even a slight storm, waves would swamp the camp for sure. But there was an even more present danger - landslides. The clays themselves, saturated with water, were known to flow like lava with little warning. Mud avalanches went on day after day, but the flows were small, rather insignificant – mostly. However, the great danger came when giant slices of the cliff, sometimes more than 100 meters long, still clothed in forest on top, would suddenly break away and rapidly glide down the clay-lubricated slope. And if your camp happened to be below, it could be fatal.

So, the next day camp was moved to the mouth of Medvezhii Ruchei (Bear Creek) as far from the cliffs as one could get, thanks to the river's erosion of the landscape. Medvezhii Ruchei cut a valley that had cliffs of 100 meters or more. And conveniently these cliffs were climbable. Atop these was a flat plateau clothed with a dense, dark and very damp magical forest – a mixture of evergreens and birch, covered with moss and floored by a variety of primitive plants, including the stunning Club Moss *Equisetum*.

With the Bear Creek camp established, Mikhail and Natasha began to prospect the cliffs along this watercourse. On the fourth day, Bochkareva found the first fossil, identifiable as *Ediacaria*, a palm-sized specimen on a sandstone cobble, which had been rounded by water transport. Clearly the fossil had fallen from higher up the cliff, and so that same evening, in the low-angled light of the polar white night, Fedonkin climbed to the top of the 120 meter cliff in search of the source of Bochkareva's first find. In the uppermost beds, he almost immediately found *Ediacaria*, *Dickinsonia* and *Cyclomedusa*, familiar late Precambrian body fossils on one single slab. This slab was made of sandstone that occurred as lens-shaped bodies of rock surrounded by the ever-present clays, deposited in nearshore marine waters with a significant bottom current. They are not the result of quiet waters where clays were laid down. Now that their eyes were "tuned" to the local situation, fossils were found not only in the sandstones, but also in the clays. Large blocks were split, and dozens of *Charnia* and *Dickinsonia* were recovered, more than 200 excellent specimens, as well as many other forms collected from rocks of a similar age from the Flinders Ranges of South Australia by Reg Sprigg, Martin Glaessner, Mary Wade, Richard Jenkins and their colleagues.

Figure 219. "White Nights." Sunset at 2:30 a.m. on the Zimnii Gory coast (N. Hunt).

Figure 218. Zolotitsa River, White Sea region, northern Russia (Y. Shuvalova).

Figure 220. Loading boats with fossils and field gear, Ershika Creek, Zimnii Gory coast, White Sea region, northern Russia (E. Serezhnikova).

Figure 221. Ediacaria, *the first metazoan form found along the coast of the White Sea, northern Russia, and also a form common in Australia. Preserved diameter of specimen about 17 cm.*

Figure 222. Dickinsonia *sp., Zimnii Gory, White Sea, Russia (F. Coffa).*

Figure 223. *Quarry excavation in the cliffs at Yelovy Creek, Zimnii Gory coast, White Sea, northern Russia (E. Serezhnikova).*

sketches of the cliff section of sediments and notes, as well as numerous squashed mosquito "mummies." He noted as well: "Galina is afraid of bears." Bears were evidently often sighted walking through the forest, close to the tent where Keller and his wife Galina, also a geologist, were camping near the mouth of a narrow canyon named Medvezhy Ruchei (Bear Creek). Interestingly, the very place where Keller had spent time became the major campsite for Fedonkin's field team for years. Fossils were everywhere. But in 1977, not a metazoan had been found.

Figure 224. *Helicopter base near Arkhangel'sk and Mi-5 helicopter preparing to depart with field crew to Suz'ma locale to the west (P. Vickers-Rich).*

The eyes of the geologists, were not yet tuned to picking the soft-bodied imprints that had left their marks in the clays along with the footprints of the yet living bears!

Excavation into the cliffs of the White Sea had revealed four important aspects of the newly discovered fossil field. First, faunal remains occurred from the bottom to the top of the outcrops. Second, different kinds of sediments contained different kinds of fossils. Third, there was potential for collecting over a fairly expansive area – at least 20 km along the coastline.

Figure 225. *Andrey Ivantsov sawing fossiliferous blocks along the cliffs of the Zimnii Gory coast, White Sea, northern Russia (M. Fedonkin).*

Figure 226. View out cockpit window of Mi-8 helicopter preparing to land on massive landslide along Zimnii Bereg coast, White Sea, northern Russia (N. Hunt).

And fourth, the lowermost and uppermost sediments were very different, reflecting progressive environmental change in the region over a significant period of time. After the modest collections made at the Suz'ma locale, Zimnii Bereg provided almost unlimited opportunities. It truly opened the door on the Vendian Period in the far north of Russia.

The next year, 1978, Fedonkin returned with a larger crew, putting in by boat and remaining for three months. His team included Bochkareva and two technicians who helped in excavations. Literally hundreds of fossils were collected from different levels of the section and from a number of locales along the coast. A month into this expedition, a second larger field team (12) led by Boris Sokolov arrived by helicopter and included specialists from several geological institutions in Moscow, Ekaterinburg and Arkhangel'sk

Figure 228. Precambrian researchers on the Zimnii Gory, White Sea coast: left to right, Mikhail Fedonkin, Jere Lipps and Jim Valentine.

including A. F. Stankovsky, who was keen to study the rocks themselves in detail, his specialty being lithostratigraphy. Alla L. Ragozina and A. F. Veis collected clay samples for microfossils. A. Krasnobaev collected samples to analyze for radiometric dating of clay minerals by using the rubidium-strontium isotope technique. Certainly this trip demonstrated that Ediacara-type fossils were not rare but were abundant, occurring in many different facies; thus apparently the Vendian organisms had lived (or at least were preserved) in many different kinds of environments – from dynamic, near shore shelf seas to the deeper recesses of the slope and perhaps even deeper. Fossils occurred in massive sandstones and in finely laminated clays. They occurred on the soles (undersides) of the sandstone lenses and on the tops.

Together, the Fedonkin and Sokolov crews boarded a helicopter and flew nearly a 100 km inland to check out the topmost part of the Vendian section. This had previously been mapped by Stankovsky in the upper part of Zolotitsa River Basin. The variegated color and peculiar sedimentary structures proved, unfortunately for the paleontologist, that these deposits were probably non-marine. At this time in Earth history, some 560-580 million years ago, the land had not yet been colonized by anything but bacteria, fungi and probably algae. Even those life forms would have lived in very restricted environments. So, none of the large metazoan fossils recovered from the marine sediments along the White Sea were

Figure 227. Members of the Laboratory of Precambrian Organisms, Paleontological Institute, Russian Academy of Sciences, Moscow. From left to right, Mikhail Burzin, Tatiana Suvorova, Andrei Ivanovsky, Alla Ragozina, Igor Babkin, Mikhail Fedonkin (current head), Larisa Voronova, Nadezhda Kireeva, Svetlana Solovieva, Natalia Bochkareva and Boris Sokolov, founder of the Laboratory, about 1982 (M. Fedonkin).

September Creek falls into Autumn River
Nights are darker now, full of stars
And in mornings, hoar frost on the grass is blue
It's time to move my canvas house
Which smells of smoke and nostalgia
Back to the fumes and race of the city
Going back, I feel like a stranger
And as I step off the train
My mind turns back to a camp fire
Which may already be covered by the first snow

Fedonkin, White Sea, Russia, 1990's

Fossil Prospecting: Strategy and Techniques

The strategy that paleontologists have when they go in the field is to explore the various sedimentary facies (that is, the many different kinds of rocks of a region that they are interested in). This is just what Fedonkin and his crew did with the Vendian rocks over a geographically vast area, including places in the upper reaches of the Onega River, as well as a number of other localities on the Onega Peninsula and extensive sea cliffs on the Winter Coast of the White Sea. Once fossils began to show up, the prospectors tried to determine the distribution of the fossils up and down the rock section as well as over a geographic area. This gave insight into when and how long in geologic time they lived as well as how widespread they were geographically. Paleontologists attempt to determine how much the occurrence of certain fossils, such as *Dickinsonia*, is controlled by the environment (thus the rock types they were found in) and how much by the time (where in the stack of rocks did they occur) or by the geographic place where they lived. In order to be sure it was not just chance that a particular fossil form was found, the crews had to carry out extensive prospecting and excavations at each fossil locality in order to be sure that they had a large and representative sample.

In addition to simply collecting on the surface, it is important to have every bit of information available on the rocks from Russia's far north. Anatoly Stankovsky gave Fedonkin and his colleagues access to the cores from deep boreholes drilled in the region including very recent ones, and this added to the information wrenched from the cliff outcrops. It also gave unweathered, fresh samples and much longer successions of the sediments than were exposed in outcrop. These cores contained a few species of tubular and frond-like fossils as well as algae and microfossils, adding to the fauna and flora collected on the surface. Deep boreholes drilled in this region revealed that below the rocks exposed in the cliffs along the Winter Coast there were a few hundred meters of the older sediments of the Vendian, and perhaps even older rocks. Using the borehole data to tie sections of rocks from different areas together, to correlate them, Fedonkin and his team were able to suggest that about 100-200 meters of the younger Vendian sediments were mapped by geological surveys on surface rocks of the Kuloi Plateau. These, thus, lay above the rocks exposed on the Winter Coast. In the future these sediments give possibility for further discoveries and, importantly, quite recently Cambrian deposits have been recognized.

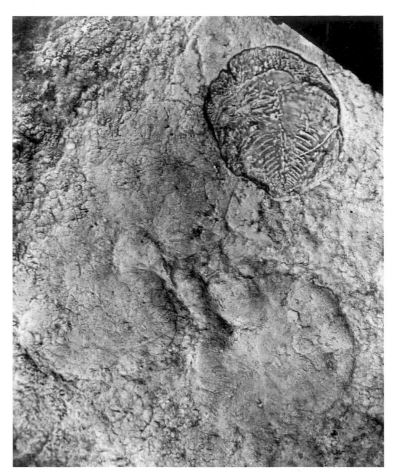

Figure 230. Yorgia waggonerni and "footprint," Zimnii Bereg, White Sea, northern Russia (A. Ivantsov and M. Fedonkin). Length of body fossil about 5 cm.

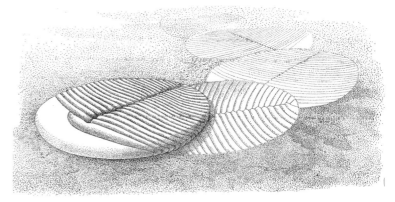

Figure 231. Yorgia waggoneri, reconstruction of paleobehavior which produced the body fossil impression and the trace fossils (artists A. Besedina and I. Tokareva, courtesy of A. Ivantsov and M. Fedonkin).

to be found. Mikhail and his crews had already plied the waters of the Zolotitsa River in their rubber boats, inspecting every outcrop of the Vendian section. Results were negative. And so absence of metazoan fossils in the non-marine, ancient river and lake deposits was further proof that megascopic animal life during the Vendian Period was likely confined to the oceans.

Figure 229. Andrey Ivantsov and Yana Malakhovskaya preparing the rock surface for collection of Vendian material, Zimnii Bereg, White Sea, northern Russia (P. Vickers-Rich).

Since 1978, the cliffs of Zimnii Bereg along Russia's White Sea have been visited annually by the paleontologists from the Paleontological Institute (PIN) based in Moscow. A number of major excavations have been carried out at many places along this coast and at several stratigraphic levels. One of the spectacular discoveries along this coast was made by Andrey Ivantsov on a joint field trip with Dimitri Grazhdankin in 1994 when two concentrations of *Kimberella* were located, clearly associating body fossils with feeding traces reported by Fedonkin and Waggoner (1997). In 1995 the first large-scale quarry operations were coordinated by Ivantsov, leading to the discovery of *Yorgia* and other large metazoan trace fossils.

This methodology continues to provide insights into the mobility and feeding strategies of many Ediacarans.

In addition to the Russian crews, people from around the globe have visited and taken part in the work along this coast. It has became the focus of a wide international collaboration – currently exemplified as a major part of a UNESCO-sponsored project, IGCP493 (International Geological Correlation Project of the International Geoscience Program), on the Vendian (Ediacaran age) biota.

For each fossil field there are unique problems. The problems encountered from the very beginning of work on the White Sea were those of how to collect and get the material back to the laboratory in Moscow with a minimum of damage.

Figure 232. Charnia *sp., White Sea, northern Russia (M. Fedonkin). Width of specimen about 3 cm.*

First, one needed to carefully search debris at the base of the cliffs and identify just which layer contained the fossils. Once found, natural bedding planes need to be exposed. In this manner the "footprints" of *Yorgia* were unearthed and without this careful exposure of large natural surfaces, such traces of motion likely would never have been discovered.

The Winter Coast provided almost unlimited possibilities for effective collecting. The cliffs were constantly being eroded by the waves and wind of the White Sea itself. Records kept by those who manned the meteorological station at Zimnegorsky Lighthouse over decades note that the rate of retreat of the coastline, due to this erosion, had been on

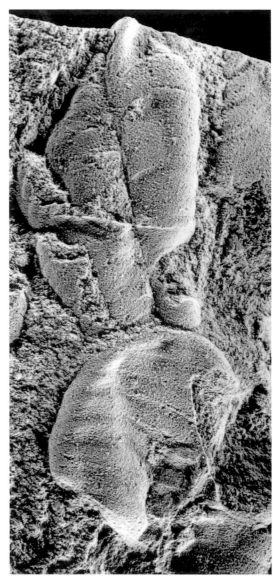

Figure 233. Inkrylovia lata, *Suz'ma, Arkhangel'sk region, White Sea, Russia (M. Fedonkin).*

average around 60 cm per year. So, it was critical to monitor this area every year, first collecting from the surface of new landslides and avalanches to find new locales, then excavating at the most productive locales. Even surface collecting is valuable, for new species appear in the landslide sediments every year.

It was also clear from the beginning of work along the Zimnii Gory coast that it was critically important to collect material as quickly as possible after it is exposed. Unless this is done there is a significant danger that the fossils will be destroyed by the very erosion and weathering that has exposed them. Sediments are so soft that large slabs of siltstone and clay can turn into a pile of fine debris and crumble in the hand if left for a year or two. Even sandstone, the most durable of rocks, breaks down by the action of water and temperature fluctuations. Fine fossil imprints are also particularly at risk because they contain pyrite. This mineral is easily oxidized, so if a specimen is left on the surface, the fossil simply fades away as the pyrite oxidizes, unless it is removed from the field and stabilized.

The most effective manner of collecting these fossils is to excavate, exposing new bedrock. This produces the best-preserved fossils, and most important gives the chance to record the exact position of the fossils in the rock succession in relation to one another on a single bedding plane. This is the technique used by Andrey Ivantsov systematically since 1995 along the White Sea coast and recently by Jim Gehling and

Daily Life on the White Sea Coast in the 1970's. A Vendian Quarryman's View

Life along the White Sea coast is quite basic and simple. An example is one man, who collected rocks for a living. When Fedonkin and his first Precambrian crew alighted from their rubber boat after weathering the storms at sea, off in the distance they saw a lone figure. He was walking along the cliffs, approaching them – but then stopped and began to raise a terrible din. He did not approach any farther – so the crew set about putting up tents and went about the business of setting up a permanent camp. Still the man did not approach. Odd, they thought, but they had other concerns than this oddly behaving individual in the distance – perhaps a crazy man.

It was not until they lit their "evening" campfire to cook that he walked towards them and made contact – he was in his 60's, a short, muscular, balding man, clean-shaven. He was nearly camouflaged in his green/gray clothing. From his distance, and with poor eyesight, he had mistaken them for bears! It was Saturday and he had been on his way back from the local lighthouse at Zimnii Gory after his weekly steam bath (banya) – where he not only had the pleasure of warm water and steam, but beating himself with thin birch branches, a custom in these northern areas! He actually lived some distance from the lighthouse in an abandoned fisherman's hut at the mouth of Bear Creek. One window, a bed made of planks, and a table were his only embellishments. His bed had been "decorated" by pieces of a sail to keep the mosquitoes at bay, the sail suspended by ropes taken from a boat that had washed ashore, abandoned – which served as a source of materials for a decade while the fossil crew prospected the White Sea cliffs. But just as quickly as the boat had arrived on these shores, one night, in a ferocious storm, it disappeared, leaving not a trace.

This man led a simple life – each year he spent the long summer days mining the Vendian sandstone – a stone quite valuable for its volcanic ash content and the manner in which it weathered. He carried out his work with two axes and an iron bar – the bar for rolling slabs down the slope and the axes for "releasing" the rock from its layered beds. In the autumn, a boat would gather the sandstone and then transport it away to be used as a polishing material for metal cylinders and lenses. After the boat left, the quarryman would himself leave, to winter in Arkhangel'sk. He made sufficient funds quarrying the ancient Vendian sandstones during the summer to survive the long, dark winter in warmth and with more frequent steam baths! Meanwhile, a few miles inland, more valuable treasures had been discovered – the first diamonds found in this region, lodged in kimberlite pipes, tubes that intersected hundreds of feet of Vendian deposits.

Figure 235. Vendian sequence cliffs along Zimnii Bereg, White Sea, Russia (M. Fedonkin).

Figure 236. Ventogyrus chistyakovi, Vendian sequence, Yarnema, Onega River, White Sea, Russia (M. Leonov).

Figure 234. The keeper of the Zimnegorsky Lighthouse, Valery Stanislovovich Lisetsky, on the White Sea (left) with Terufumi Ohno from Kyoto University, Japan, in a field camp "kitchen," 2003 (M. Fedonkin).

Mary Drosser in the Flinders Ranges of South Australia. From such quarry mapping paleontologists gain valuable insights into the mode of life, the interrelationships of different species in the biota, as well as the way in which they were preserved (a whole area of study called taphonomy). Each kind of sediment requires its own excavation and preservation technique.

There was real value in this approach to collecting, even after the first field season. When comparing the collections of fossils from sediments vertically separated by only 120 meters, they found at least two distinct fossil assemblages, which suggested that such differences reflected an evolutionary change through time (Fedonkin, 1977). The sediments were much the same in each of these quarries, probably ruling out environment being the cause of the differences. The quarry in the lowest sediments contained the leaf-like form *Charnia*, known at that time from Great Britain and Newfoundland. The topmost clays contained a few species, known only from South Australia.

Figure 237. Fisherman's hut along Zimnii Gory coast, often used by paleontological field crews as base camp (N. Hunt).

Other Fossil Localities of the White Sea Region

Sometimes mistakes can lead to wonderful discoveries. One of the most remarkable fossil localities yielding Vendian fauna was discovered far from the White Sea cliffs and rivers – in the valley of Onega River – by a young geologist, V. G. Chistyakov from Leningrad University. During the period from 1974 to 1977 Chistyakov made several field trips to study the rocks and fossils of Devonian-aged deposits along the Onega River. The outcrops are very limited and primarily silica-rich carbonates of Middle Carboniferous age (approximately 320 million years old). Below the carbonates are a few meters of clay and siltstone of various colors. At the time, these were considered Devonian in age. The siltstones contained some fossil fragments, tentatively identified as remains of fish.

Figure 238. IGCP493 field crew preparing to depart from Arkhangel'sk helicopter base for work at Suz'ma (2003): left to right, Andrey Ivantsov, Yulia Shuvalova, Yana Malakhovskaya, Alexander ("Sasha") Kochedykov and Mikhail Fedonkin (P. Vickers-Rich).

In 1976, Chistyakov invited Boris Sokolov to examine his collections. From the first glance it was clear to Sokolov that the fossils resembled some known from the Vendian of the White Sea. This collection was subsequently donated to the Laboratory of the Precambrian Organisms at the Paleontological Institute in Moscow. Description of the material was published, years later by colleagues, after Chistyakov's tragic death (Chistyakov et al., 1984). In this paper they described many Vendian forms from the Onega that were already known from elsewhere – Inkrylovia, Cyclomedusa, "Arborea", Neonereites, as well as a new taxon Yarnemia ascidiformis, thought by St Petersburg paleontologist Lev Nessov to be the oldest record of the sea squirts, or tunicates. Tunicates belong in the group Chordata, which includes the backboned animals, the vertebrates – cats, dogs and us humans.

Upon learning of these finds, Fedonkin brought them to the attention of Andrey Ivantsov and Dimitri (Dima) Grazhdankin. Their subsequent excavations yielded a unique series of fossils, preserved in three dimensions, the most remarkable of which was Ventogyrus, published on by Ivantsov and Grazhdankin in 1997 and reinterpreted by Ivantsov (2001, 2003) as a triradial coelenterate - ctenophoran, that is a combjelly. Since then the Onega has been visited and collected many times, and the rock sequence studied in detail by Dimitri Grazhdankin, who has carefully logged several drill strings in his attempts to correlate these rocks with those along the Zimnii coast.

To reach these fossil localities requires some ingenuity. One must rent a heavy truck from the railway station at Plesetskaya in order to reach the village of Yarnema. After that, a boat, preferably with motor, is needed to access the outcrops about 8 km upstream from the village. Outcrops on the Onega, as at Suz'ma, are about 10-15 meters high and exposed over a series of 100 meter-long cliffs along the river. The upper part of these Vendian deposits crop out along the upper reaches of the Onega River also near Yarnema Village itself, between Somba and Teksa creeks. Again, as along the Winter Coast, the almost horizontal beds of soft, semilithified clays are altered constantly by landslides, some up to 50 meters long, so prospectors must be constantly on the alert not so much for dangerous earth movements, but for making sure where the fossils actually originated in the rock sequence – were they in place, or were they part of a landslide that had slipped away for the outcrop?.

As on the Zimnii Coast, the Vendian sequence contains laminated siltstone and clay with occasional sandstone lenses. The exposed part of the section lies in the lower part of what is called the Ust'-Pinega Formation, an equivalent of rocks included in the Redkino Regional Stage of the Vendian System (Stankovski et al., 1990). But, just where in the more than 600-meter-thick Vendian section the rocks of the Onega fit has not yet been determined.

This Onega area has produced an abundance of the unique and endemic form, Ventogyrus, described at first as a boat-shaped, sea pen-like metazoan by Ivantsov and Grazhdankin (1997; Dzik & Ivantsov, 2001). Recently it has been reinterpreted as a complex, egg-shaped pneumatophore made up of three identical parts. This reinterpretation was prompted by specimens recovered from the Onega region, critical because specimens are frequently preserved in three dimensions, rare at other Vendian locales.

The Onega fauna contains the oldest known conulariid-like organism, whose conical form has six faces and three-fold symmetry (Ivantsov & Fedonkin, 2002), tubular forms with radial septa resembling the chitinous tubes of the recent scyphozoan polyps, a giant bilaterally symmetric, soft-bodied organism about 0.5 meters long, which has serially arranged, comb-like elements, and enigmatic bag-like fossils and other new species yet to be studied. In addition to some new taxa, numerous tubular forms similar to *Calyptrina* and the complex, serially arranged *Swartpuntia*, previously only known from the youngest Neoproterozoic rocks in southern Namibia (Narbonne *et al.*, 1997) were collected as well. Perhaps this site represents one of the youngest parts of the Vendian sequence, but for now, further study is needed to fully clarify just where these fossils occur in the sequence of events that took place more than 550 million years ago in the north of Russia.

Recently the geologist Chernov discovered Vendian fossils along the Solza River. In 2000 Andrey Ivantsov and Yana Malakhovskaya discovered and excavated an enormous concentration of imprints, with not only feeding trails of *Kimberella* but movement trails as well. Detailed study of this locality by Grazhdankin and Bronnilov (1997) had previously noted that the fossil assemblage from this area has much in common with that described from the lowest part of the sections exposed along the Winter Coast, including *Charnia*. So, here is yet another locale that offers more promise in unraveling the sequence of events in the White Sea region.

From 1974 through 2006, Fedonkin's team worked the White Sea cliffs every year –for 30 years. He and his crews searched a line of outcrop swept by the north winds stretched over more than 300 kilometers. These beds lay one upon the other, almost horizontal, with a slight tilt to the south, stacked layers offering a unique opportunity. One can follow along the length of the beds (along strike) and carefully plot the changing environments over a broad area. And with the cliffs standing over 100 meters high with their northern exposure, one could look at the changes through time as you climbed and mapped from base to top. Time had been gentle to these rocks, the organic microfossils in them were light brown in color – indicating that they had not been much altered as they lay buried, the sediments had not sunk deep into the Earth, no more than 1 kilometer – often much less. So, deformation and alteration of the organic content was little – the algae and bacteria, as well as the macrofossils like *Dickinsonia*, remained much as they had been in life, for their imprints had not been contorted.

Ancient Environments and Their Time

The tall cliffs of the White Sea's Winter Coast and the river cuts of tributaries draining into this area are unique for their Neoproterozoic fossil content. Nowhere else on Earth in rocks of this age is there such a diverse assemblage of early metazoans boasting high quality of preservation in such a variety of different environmental situations. Exposure of the fossil-bearing siliciclastic sediments (that is, sands, silts and muds) is outstanding in itself, for in some places several hundred meters of nearly horizontal clay, silt and sand layers are constantly being eroded and exposed as are the more limited exposures along the rivers that empty into the White Sea itself, both east and west of the city of Arkhangel'sk in Russia's far north (N. Butterfield & D. Grazhdankin, *pers. com.*, 2005).

Figure 239. Field camp of Dima Grazhdankin and his crew at Ivovik Creek, White Sea, in 2004. Dimitri has been a tireless collector and mapper of the Vendian sequence in Russia and is now working with Nick Butterfield in the Urals and the Ukraine, following up on earlier work) (Y. Shuvalova).

Figure 240. Use of satellite phone on the White Sea Coast, Russia, which has improved coordination and logistics in this isolated area in the 21st century (P. Vickers-Rich).

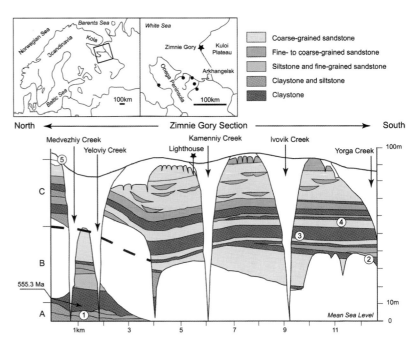

Figure 241. Stratigraphy of the Vendian sequence along the Zimnii Gory coast, White Sea, Russia (modified from Grazhdankin, 2004) (D.Gelt)

The oldest sediments in this region were deposited in a sea that lay alongside a stable continental shelf. But that changed through time, and the youngest sands and clays were laid down in a much more dynamic setting atop developing deltas that splayed forth from rivers bearing fresh water into the more saline seas – rivers that brought loads of sediments generated by the uplift of a mountain range to the northeast from the Timan region of today. During this time, too, there

SEQUENCE A	SEQUENCE B	SEQUENCE C
	Body fossils	
Anfesta	*Charnia*	*Anfesta*
Beltanelloides-like		*Brachina*
structures	*Ovatoscutum*	
Bonata	*Staurinidia*	*Cyclomedusa*
	Molds of soft	
	tubular structures;	*Rangea*
Charnia	simple circular	*Charnia*
	impressions	*Dickinsonia*
Cyclomedusa		*Ediacaria*
Dickinsonia		*Eoporpita*
Ediacaria		*Inaria*
Eoporpita		*Irridinitus*
Hiemalora		*Kimberella*
Inaria		*Mawsonites*
Irridinitus		*Nemiana*
Kaisalia		*Parvancorina*
Kimberella		*Ovatoscutum*
Nimbia		*Tribachidium*
Protodipleurosoma		*Vendia*
Tribachidium		*Yorgia*
Three-dimensional		Others, including numerous
molds of tubular		new forms
structures and others		
	Trace fossils	
Simple radial feeding		Diverse radial feeding
burrows; backfilled		burrows; traces of
burrows with circular		crawling mollusk-like
path behavior; traces	Vertical burrows	organism; fan-shaped
of crawling mollusk-like		sets of scratch marks;
organism; fan-shaped		diverse vertical
sets of scratch marks		burrows; simple tunnels
	Enigmatic biological	
	structures	
Yelovichnus		*Yelovichnus*
Palaeopascichnus		*Palaeopascichnus*
Intrites		*Intrites*

Figure 242. Subdivisions of the Vendian sediments that
crop out along the Zimnii Gory coast of the White Sea
and their fossil content; see extent of subdivisions A-C in
Figure 237 (modified from Grazhdankin, 2004) (D.Gelt).

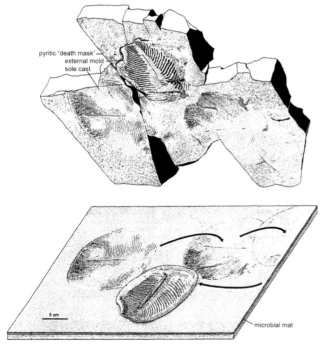

pyritic "death mask"
external mold
sole cast

5 cm

microbial mat

Figure 243. Reconstruction of the process of
fossilization of Yorgia waggoneri as proposed by Dzik
(2003). Dzik interprets them as "death tracks" made
by a dying animal. The tracks may, instead, be prints
left by a "healthy" organism absorbing the underlying
microbial mat down to a level where no further useful
nutrients were available and then moved on to
"absorb" another part of the mat. Both the absorption
"tracks" and the organism itself were later buried by a
submarine landslide and both left their marks on the
overlying, suffocating sand, which then hardened into
sandstone (from Dzik, 2003).

were significant global glaciations. Perhaps as a result of this
and the rise of a gigantic mountain range, sea level fell and
then rose again, producing a variety of environmental settings
that waxed and waned through geologic time (Nikishin *et al.*,
1996; Grazhdankin, 2004; Grazhdankin *et al.*, 2005; Squire *et
al.*, 2006).

The most recent work on the sediments of the White
Sea region has been carried out by Dimitri Grazhdankin
(2003, 2004; Grazhdankin *et al.*, 2005). He has spent
years mapping the sediments, which reach up to 550
meters in thickness exposed along the Onega, Letnii and
Zimnii coasts, but also along major rivers of the area – the
Nizhma, Agma, Suz'ma, Verkhovka, Solza, Kinzhuga,
Torozhma and Zolotitsa – and in numerous boreholes
drilled in the region in search of kimberlite pipes bearing
diamonds. These sediments do not at first sight appear to
be old. They are not much deformed or even tilted – they
lie almost horizontally in the cliff outcrops and are often
quite soft to the touch, considering their ancient age of
more than 550 million years. Grazhdankin built on the work
of Kalberg, whose investigations in the late 1940's of both
outcrops and deep bore data had labeled these sediments
as Devonian. Further analysis, however, by Boris Sokolov
(1952) and others, made it clear that these deposits were
older, based on fossils of organic-walled tubular fossils
similar to *Sabellidites cambriensis* (Igolkina, 1956, 1959;
Zoricheva, 1963; Grazhdankin, 2003), and in fact they were
Neoproterozoic in age, not Cambrian or Devonian.

Geologists divide up their sections of rocks and give
them names – names that describe their color, texture and
structure and that reflect past environments and events. To
the non-geologist this can become a complex web of names.
Similar-aged rocks can vary laterally as environments change.
If exposures of rocks are not continuous, or if a geologist is
examining rocks taken from deep bores and tries to relate
these to rocks exposed in sea cliffs or river cuts, things can
get quite complicated. Such is the case with Neoproterozoic
rocks on the Eastern European Platform that contain metazoan
fossils critical to understanding evolutionary developments
at this time. Grazhdankin (2003) has presented a good case
for the connection of the different bits of the massive data
base compiled on this region since Kalberg's, Stankovsky's
(Stankovsky *et al.*, 1990) and a few others' work.

The White Sea region boasts many different sorts of rocks
– and each has a story to tell about the environment in which it
was deposited. Reflected, too, in the sequence of rock types are
trends in the kinds of environments dominant through time – and
indeed from bottom to top conditions did change in the White Sea
region. These changes reflect prevailing topography and climate
– basically the surrounds that affected those animals and plants
that were alive at the time, driving the development of complexity
during these late Neoproterozoic Vendian times.

Grazhdankin (2004) was able to recognize at least
five different sorts of rock sequences in the Vendian, what
geologists call "lithofacies:" 1) laminated shales, 2) shales
and siltstones alternating with one another, 3) sandstones
and shales interdigitating with each other, 4) interstratified
sandstones, 5) massive or "amalgamated" sandstones. Each
of these sequences has certain fossils associated with it, which
Grazhdankin nicely outlines.

The geologist can read these rock layers like the pages of a book and they translate as follows:

1. **Laminated Shales.** Deposited in quiet, calm waters. Main sediments are thin beds of shale or siltstone. Some fine-grained sand is present and always has sharp boundaries with ripple patterns, suggesting that it represents sediment moved downslope off the front of a delta into deeper waters. These sediments sometimes contain volcanic ash beds, which are useful in providing an absolute age in millions of years before present and are continuous over several kilometers, making good marker beds for defining a time plane.

Fossil Content: Only a single Ediacaran, *Inaria*, occurs and appears to have been entirely enclosed in the sediments with only a tube protruding above the surface, perhaps an attachment for a frond-like organism.

2. **Alternating Shale and Siltstone.** Deposited in quiet, calm waters that were periodically "invaded" by coarser sediments loosened from upslope by storms. The siltstone and shale occur in pairs and have sharp boundaries, the siltstones showing erosional bases and graded bedding – that is, larger particles at the base grading up into finer particles or clay – the result of slow settling of particles out of water, heaviest first, lightest last. These beds do not continue far laterally.

Fossil Content. Ediacaran forms include *Charnia*.

3. **Interstratified Sandstone and Shale.** Deposited in a more energetic environment, perhaps on a delta front slope. The sandstone "packages" may have been deposited as pulses of sand moved off the building delta, first eroding the surface over which it slumped off into deeper water, finally coming to rest downslope from its origin. Such slumping could have been caused by storms or even increased sediment supply affected by either seasonal variation or climatic cycles over time.

Fossil Content. Ediacaran forms are quite varied, showing the greatest biodiversity in the White Sea region: palaeopascichnids, *Eoporpita*, dickinsoniids, yorgiids (*Yorgia, Andiva, Archaeaspis*) and vendomiids (*Vendomia, Vendia*), trilobozoans (*Anfesta, Albumares, Tribrachidium*), *Kimberella, Parvancorina, Brachina, Charniodiscus*, discoidal holdfasts (*Cyclomedusa, Ediacaria, Paliella, Protodipleurosoma, Irridinitus, Medusinites*) and many other forms yet to be named. Seldom are any of these prodeltal assemblages dominated by one species (Grazhdankin, 2005). The underside of many bedding surfaces (soles) shows a pattern termed "elephant skin," indicating the presence on the top of the sea bed of organic mats or biofilms. These were probably the main food source of forms like *Yorgia, Dickinsonia* and *Kimberella*, each of which left trace fossils behind giving clues to their feeding styles (Seilacher, 1999; Ivantsov & Malakhovskaya, 2002).

4. **Interstratified Sandstone.** Deposited in an energetic environment often as single flood events or saturated sand overflow from channels, laid down near the mouth of a river as it flowed into the sea. The sandstone represents channel fills and shows current-induced cross-bedding and lens-like forms, reflecting the channels in which they were deposited.

Fossil Content. Ediacaran fossils tend to be preserved in three-dimensions. *Ventogyrus* very common. Present are *Rangea* and *Swartpuntia*. All are found in thin, lens-shaped sandstone beds. Other forms are confined to thick channel sandstones: *Onegia, Bomakellia, Palaeoplatoda* and a form related to the Namibian *Ausia*.

Figure 244. Archaeaspinus fedonkini, *Vendian sequence, White Sea region, Russia (M. Fedonkin).*

Figure 245. Paravendia janae, *Zimnii Bereg, White Sea region, northern Russia (M. Fedonkin).*

5. **Amalgamated Sandstone.** Deposited in more energetic conditions, such as the sand filling of channels distributing across a delta issuing from a landward river. The presence of such nodules of a mineral called siderite and evaporitic crystal impressions (pseudomorphs) have suggested to Grazhdankin (2004) fluctuating salinity of the waters. The bottom of the beds shows significant erosive surfaces, and often these sandstone beds contain "small mudstone pebbles," which seem to be the result of sand movement across a mudstone surface. The current carrying the sand eroded the underlying mud, picked up bits of the mud and rolled this along the sea bottom until the fragments became rounded. Then, when the current slowed and dropped its sediment load,

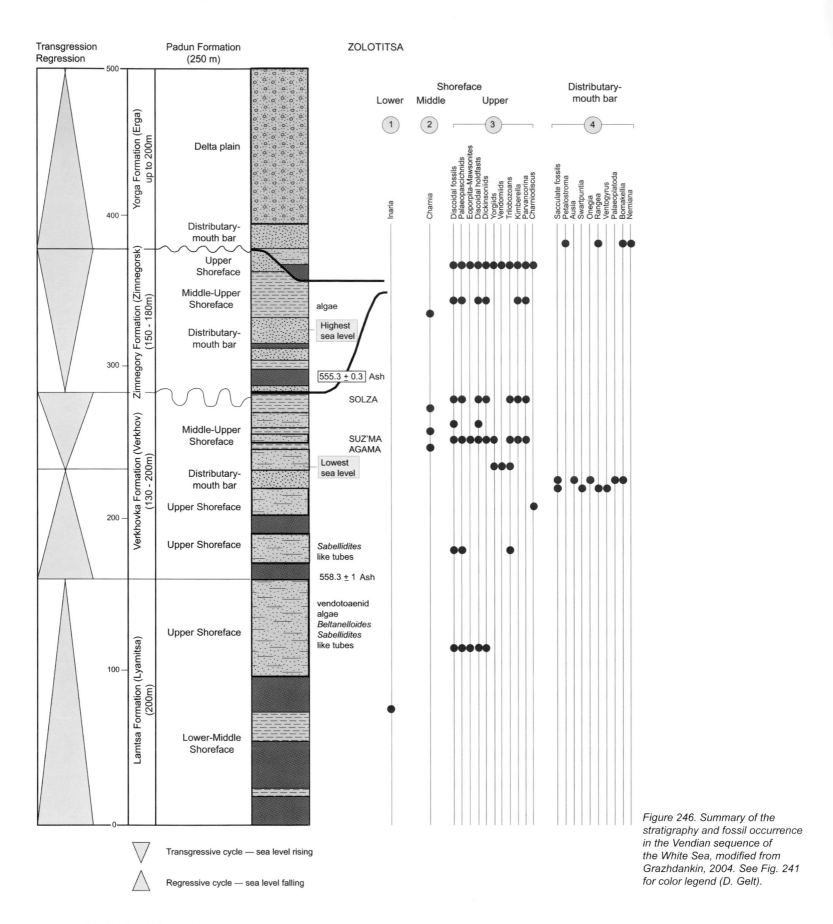

Figure 246. Summary of the stratigraphy and fossil occurrence in the Vendian sequence of the White Sea, modified from Grazhdankin, 2004. See Fig. 241 for color legend (D. Gelt).

the "mud balls," which later become pebbles, were all mixed up with the sand. Upwards these sand lenses show cross-bedding patterns, which indicate significant current flow.

Fossil Content. Only *Arumberia*, which is likely a pseudofossil interpreted as a sedimentary structure produced by swift-flowing bottom currents acting on microbial-mat bound sediments, producing sharp ribs and smooth hollows.

Geologists give names to the distinctive rock units that they map – called formations. This allows them to lump together a number of different sorts of rocks into a "package," which represents events that have occurred in the past that can actually be traced out on a map. Beginning with Kalberg in the 1940's this has been done over and over again, with each new set of names a refinement on the understanding of past events that took place and environments and biotas that have characterized the White Sea and its surrounds from around 560 to 550 million years ago. Radiometric ages determined on the volcanic ash beds preserved along the Zimnii Gory coast have yielded the dates of 558+/- 1 to 555.3 =/-0.3 (Martin *et al.*, 2000). Not only do these ash beds provide dates, but they also reflect significant volcanic activity in the region, associated with the rising of a gigantic mountain range off to the northeast of the White Sea region. And the ash dates coincide nicely with dates on granite intrusions which built the mountains that in turn provided the source for the White Sea sediments – those dates ranging from 550 to 560 million years in age (Gee *et al.*, 2000; Squire *et al.*, 2006).

Grazhdankin's work on the Vendian sequence is the latest in a number of geological studies and he divided the rock sequence into four major units: the lowest being the Lyamtsa Formation, the next being the Verkhovka Formation, followed by the Zimnii Gory Formation and finally by the Yorga Formation. He has further noted (Grazhdankin & Ivantsov, 1996; Grazhdankin & Seilacher, 2002; Grazhdankin, 2000; and Grazhdankin & N. Butterfield, *pers. com.,* 2005) that each of these formations contains sediments that represent at least three to four different kinds of environments, each hosting a different fauna, certainly the case in the world today and clearly the case during Phanerozoic times (the last 542 million years). Three to four principal faunal associations can be linked to different environmental conditions: continental shelf-shallow water coastal plain (with both an inner shelf and an outer shelf fauna), delta (or prodelta) and channel mouth-bar (distributary-mouth bar). Geologists like to use many names, often for much the same thing (Grazhdankin, 2004).

The continental shelf-shallow water coastal plain environment (called strand-plain facies by Grazhdankin & Butterfield, *pers. com.*, 2005, modified from Grazhdankin, 2004) is mirrored by the clays that are rich in organic-walled fossils like the vendotaenids, *Beltanelloides*, sabelliditids, probably holdfasts like *Inaria* and swing marks (*Kullingia*) that could have been made by erect forms anchored to the sediments due to submarine current variations brought about by occasional storm surges mirrored in the rock record by sandstone interlayers within mudstones. This sort of environmental setting is only known on the East European Platform along the White Sea Coast, best represented by the Lyamtsa (or Lyamitsa) Formation, but certainly may have counterparts in Namibia and perhaps other places. Due to the presence of algal fossils, the Lyamtsa sediments were likely laid down in a zone where sunlight penetrated. The middle shore face sediments of siltstone and shale couplets contain attached forms like *Charnia* and *Thaumaptilon*, which are often preserved within the sandstone beds – likely overwhelmed by avalanches that buried these erect forms. Perhaps some of the discoidal forms preserved here are the remains of holdfasts or "anchors" that secured these organisms to the sea floor, themselves buried in the underlying sand, while the frond-like structure stood atop the sediments.

The ***deltaic environment*** (termed the "prodelta facies" by Grazhdankin and Butterfield, *pers. com.*, 2005), characterized by a mixture of sandstone and siltstone, was the residence of a wide range of metazoans. The sea floor was covered by microbial mats as reflected in the "elephant skin" textures of the tops of sedimentary beds. Immobile forms, such as *Tribrachidium*, and the mobile "grazers" *Kimberella*, *Dickinsonia* and *Yorgia* are preserved as Jim Gehling's (1999)

Figure 247. Dickinsonia costata, *Zimnii Bereg, White Sea, northern Russia (M. Leonov).*

"death masks." This assemblage is very reminiscent of the Ediacaran communities of South Australia's Flinders Ranges. It not only flourished in the present White Sea region (best seen in the upper parts of the Verkhovka [Verkhov] Formation and the basal part of the Yorga [Erga] Formation, but also in the Urals and Ukraine (Grazhdankin & Ivantsov, 1996; Grazhdankin and Butterfield, *pers. com.*, 2005). In the White Sea region, it was a time of marine transgression – the sea level rose, reaching an all-time high when the basal part of the Yorga Formation was deposited.

Finally, the *channel **mouth-bar environment*** ("distributary mouth bar facies" of Grazhdankin, 2004) is represented by the coarser sands that occur in lenses, reflecting downcutting by currents, followed by rapid deposition, best represented by the lower parts of the Zimnii Gory [Zimnegorsk] Formation (Grazhdankin, 2003). These were probably deposited in estuaries, places where rivers meet the sea, providing a unique environment with salinity greater than freshwater but less than the open ocean and often quite variable. Metazoans that occur in these sandstones are *Onegia*, *Ernietta*, *Pteridinium* and *Rangea*-like organisms, usually preserved in three-dimensions. Some researchers (Grazhdankin & Seilacher, 2002) have interpreted these forms as having lived entirely within the sediments. Not all specialists would agree, some suggesting that these forms were "rooted" in the sediments but fed above the sediment-water interface. Certainly forms recently collected in Namibia (Vickers-Rich, Fedonkin & Ivantsov, *pers. obs.*, 2004) clearly demonstrate this lifestyle for some *Ernietta*. Their preservation within the sediments appear to be the result of organisms being overwhelmed by masses of incoming sand. In Russia, this assemblage is only known along the White Sea, where sediments and fossils compare favorably with those in Namibia.

Figure 248. Tribrachidium heraldicum, *field specimen, Vendian sequence, Zimnii Bereg, White Sea, Russia (P. Vickers-Rich). Specimen about 2.5 cm in diameter.*

Figure 249. Vendomia menneri, Suz'ma River, *Vendian sequence, Arkhangel'sk region, White Sea, Russia; latex peel (left), original (right) (M. Fedonkin). Specimen about 3 mm wide.*

In summary, the Vendian sediments of the White Sea region paint a picture of a coastal region with a rising mountain range off to the northeast punctuated by volcanic activity that periodically showered the area with fine ash. Just how high these mountains rose is unknown but they were the source of the sand, silt and clay deposited in the shallow waters of a continental margin sea, first very immature with high feldspar content, maturing with time to produce well-sorted and clean sands at the end of Vendian times when rivers flowed into that sunlit sea in several places, building deltas, and their fresh waters mixed with the sea in a number of coastal estuaries. During the time that the Lyamtsa and Verkhovka formations were accumulating, sea level was in general falling ("regressing"), perhaps because of the building of the deltas as the rivers added more and more sediment shed off the nearby growing mountains, but also perhaps because during the building of mountains,

seas can be forced to regress as the lands bordering the seas, along with the mountains, rise. Add to that the possibility of a global glacial event that could have tied up seawater in ice caps at the North and South poles and over the seas, which they "fed on" for water. The lowest sea level for the White Sea region recorded in the Vendian rocks occurred when sediments near the middle of the Verkhovka Formation were being deposited, sometime after 558 and before 555 million years ago. Then the sea "transgressed," reaching its deepest during deposition of the lower part of Yorga Formation, sometime after (younger than) 555.3 million years ago (Martin *et al.*, 2000).

During the time that sediments in the Lyamtsa, Verkhovka, Zimnii Gory and Yorga formations were being deposited, with the overall trend of regressing seas followed by a major transgression, there were minor sea level rises and falls. The main switch from advancing to retreating sea occurred at the boundary between the Zimnii Gory and the Yorga formations – a switch from deltaic and coastal marine shelf to estuarine and river mouth environments. Sea level reached its highest during the early part of Yorga times, since the beginning of deposition of the Lyamtsa Formation. It was in these offshore sediments affected by occasional sand avalanches from the growing delta above that the most biodiverse Vendian assemblages are preserved, as well as some of their associated traces. In the varied environments represented by the Vendian rock successions metazoans developed and prospered – their abundance in the rocks, however, perhaps more controlled by the styles of preservation and their environmental preferences than by their true distribution in time and space.

Grazhdankin (2004) has notably pointed out that there are compelling reasons to suspect that Ediacarans could be more controlled by environment than by time. Forms like *Ernietta* and *Pteridinium* seem closely tied to distributary channel deposits at river mouths, dickinsonids, charniids and *Kimberella* to middle and upper shoreface, *Inaria* to outer shore depths and the rangeomorph forms that dominate the Newfoundland Avalon biota living in even deeper waters on the continental slope. His idea is worthy of serious consideration and only further detailed field investigations will support or bring into question these ideas. It may well be, that as Dimitri suggests, innovation during this 20 million year period, just prior to the Cambrian explosion, may not have been spectacular – and metazoans at this time may have been quite conservative.

The White Sea Metazoan Fauna

Of all those places in Europe and Asia, the White Sea sediments have produced the most biodiverse assemblages by far. Fedonkin (1985) noted no less than 50 different sorts of metazoans were known from either the Zimnii Gory coast or the Letny coast of the White Sea. This compared more than favorably with the rich fauna from the Flinders Ranges of South Australia, certainly far more diverse than those of Newfoundland and Namibia. To some extent this is likely due to prospecting and excavation and philosophies of classification. Nonetheless, the White Sea biodiversity is perhaps the highest of any place on Earth. And because of the fine-grained sediments which preserve this fauna, the anatomical detail available in the White Sea fossils is outstanding, by far the best on the globe.

A wide range of body plans are represented in the White Sea metazoan faunas. Some radial forms once thought to be

Figure 250. Parvancorina minchami, *Vendian sequence, Zimnii Bereg, White Sea, Russia (M. Leonov).*

Figure 251. Mialsemia semichatovi, *Vendian sequence, Zimnii Bereg, White Sea, Russia (M. Fedonkin, but recent research suggests that this is a form of* Rangea.) *Specimen about 5 cm wide.*

jellyfish now appear to be holdfasts for frond-like forms that are classed by some in the Petalonamae. Originally these radial forms were placed in a group called the Radialia that was further subdivided on the basis of such features as lack of radial elements, presence of radial elements, presence or absence of concentric differentiation of the body and symmetry. Some forms were trilobate, such as *Albumares*, *Anfesta* and even *Tribrachidium*, while others had four-fold symmetry (*Staurinidia*, *Conomedusites*). *Albumares* has been allied to many groups, the coelenterates, echinoderms as well as the sponges – it is a truly enigmatic form.

Besides the forms with radial symmetry, there are an abundance of forms with bilateral symmetry – that is, one side is a mirror image of the other. These were appropriately called Bilateria. This group is dominated by a variety of segmented forms. *Platypholinia pholiata* is one of the simplest of these, being a flat, rather leaf-shaped form with a midline ridge and what could be a mouth. Little more is known of this form, however, and it is rare. Another similar form with more anatomical detail is *Vladimissa*, and it together with *Platypholinia* have some similarities to some kinds of flat worms, in particular the rhabdocoelian tubellarians.

The majority of the White Sea bilaterians are distinctly segmented, but to whom they are related is still not resolved. *Dickinsonia* and *Palaeoplatoda* are just two examples. These are generally flat and thin, bilaterally symmetric forms that have large numbers of segments. The segments do not line up perfectly along the midline, however, but are offset, are alternate, a kind of architecture called "gliding symmetry" by the Russian researchers. These forms seem to grow by adding segments at the rear or posterior end of the body.

Another, rather peculiar group with segmentation which possesses the symmetry of "gliding reflection" are the members of the Vendomiidae. The group includes *Vendia*, *Vendomia*, *Pseudovendia*, *Onega* and *Praecambridium*. *Praecambridium* is a typical member of this group, with its oval-shaped, disc-like body with segments meeting along a midline. In well-preserved specimens there seems to be a headshield – which has led many workers to ally this group with the arthropods, perhaps even more specifically to the trilobites, which appear in the Cambrian, perhaps even a form intermediate in some ways between polychete worms and arthropods (Fedonkin, 1985).

Figure 252. Rangea *from the Vendian sequence, Zimnii Bereg, White Sea, Russia (M. Fedonkin).* Specimen is 63 mm wide.

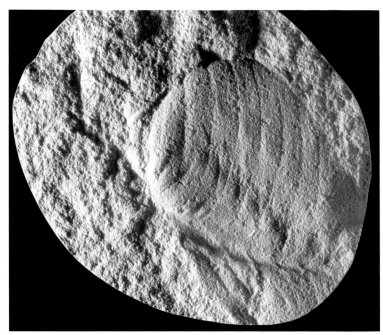

Figure 253. Inkrylovia lata, *Vendian sequence, Suz'ma, Arkhangel'sk region, White Sea, Russia (F. Coffa).*

Figure 254. Kimberella quadrata, *Vendian sequence, White Sea, Russia (M. Fedonkin)*

Likely related to members of the Vendomiidae are forms such as *Spriggina* and *Parvancorina*, which also have affinities with the joint-legged arthropods, again more specifically with the trilobites – Fedonkin (1985) noted some similarity between *Parvancorina* and the Middle Cambrian *Marella* from the Burgess Shale fauna. *Spriggina* does take on the guise of a trilobite-like animal – with its regular, angular segments that meet along a midline groove. Its U-shaped "head shield" with tines that sweep backwards resembles the cephalon of a trilobite and the prominent central bulge is even reminiscent of the glabella of that group. For the moment, it is classified as a member of the Arthropoda but placed in its own family, the Sprigginidae.

Bomakellia and *Mialsemia* are two other forms in the White Sea fauna. Though originally thought to be related to trilobites (Fedonkin, 1985), recently Dzik (2002) has suggested they may be rangeomorphs and might better be classed in the Petalonamae.

Other undoubted frond-like forms are quite abundant in the White Sea fauna, many of which are now known to be connected to holdfasts, radial forms that in the past had been classified as jellyfish – for example, the frond-like form *Charnia*, with a discoidal attachment. Others like *Pteridinium*, known so well from Namibia, and the strictly White Sea *Archangelia* and *Inkrylovia* (probably a synonym of *Pteridinium*), are also a significant part of the White Sea fauna – but just what they are related to, as mentioned many times before, is still not clear. Some (Fedonkin, 1985) have classified them as coelenterates – forms related to soft corals – while others have remained less certain and placed them in their own group, the Petalonamae, awaiting further study and discoveries to resolve relationships.

The White Sea sediments also contain a variety of colonial forms that occur in closely spaced concentrations. *Nemiana* is an example. These fossils are preserved on the underside of beds as a series of convex bumps, which appear to represent the filling of cups that perhaps housed polyps, living much as colonial corals do today. *Beltanelliformis brunsae* also occurs in some concentrations, but individuals

Figure 255. Kimberella quadrata, *Vendian sequence, White Sea, Russia (M. Fedonkin). Specimen about 8 cm in length.*

are not so closely spaced and are rather randomly distributed. They may be the remains of planktonic algae that were preserved on the sea floor leaving radial and convex imprints also on the underside of beds (Leonov, 2004).

Some forms, previously thought to be the remains of once living metazoans, seem to be the result of escaping gas bubbles or holdfasts being ripped from ocean floor sediments (*Pseudorhizostomites*) or sedimentary structures produced by high-velocity bottom currents affecting a microbial mat.

Despite some forms being found to be identical, and thus their names synonymized with one another, and some turning out to be inorganic features, the White Sea metazoan fauna is still strikingly diverse. Such diversity will undoubtedly increase as more collection and research proceeds. Some forms are clearly related to younger and even living phyla. These provide insights into the long fuse that led to the Cambrian "explosion" of life, when skeletons made them almost instantaneously much more visible and understandable. *Kimberella*, *Yorgia* and *Ventogyrus* are some of these forms, which, because of the detail, and in the case of *Kimberella*, the significant number of specimens, are keys to understanding the early evolution and behavior of the metazoans.

Personae in the Vendian Play

Kimberella was originally described from the late Precambrian Pound Quartzite of Ediacara Hills, South Australia, as a problematic fossil, possibly belonging to a group called the Siphonophora (Glaessner & Daily, 1959). At this time, it was reconstructed as a medusa of uncertain affinities (Glaessner & Wade, 1966). Known from only a few specimens, it was later interpreted as a pelagic medusa, most closely related to living boxjellies (Wade, 1972), some said most similar to the chirodropid cubozoans or sea wasps (Glaessner, 1984; Jenkins 1984, 1992; Gehling *et al.,* 1991). With this interpretation, *Kimberella* was then used as one of the best examples of a metazoan lineage crossing the Precambrian-Cambrian boundary with little morphological change from then to the present day. As such it was further employed to

Figure 257. Marine snail from below (ventral view) showing pattern of muscular foot, Seto Marine Labs, Shirahama, Japan (E. Savazzi)

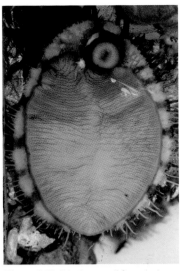

Figure 258. Marine snail from below (ventral view) showing pattern of muscular foot, Seto Marine Labs, Shirahama, Japan (E. Savazzi)

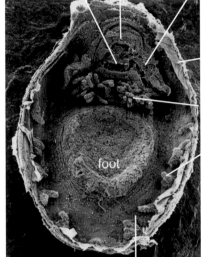

mouth region   anterior lip   velum

periostracum

postoral tentacles

circular respiratory tract with internally projecting gills

foot

pallial groove

Figure 259. A living monoplacophoran, Laevipilina rolani, *from 1000 m depth on the Galicia Bank, NW Spain (courtesy of A. Warén) showing structures analogous, perhaps even homologous, with* Kimberella.

Figure 256. Helix, *a living gastropod and perhaps a distant relative of* Kimberella, *likely a stem group.* mollusk (U. Shuvalova).

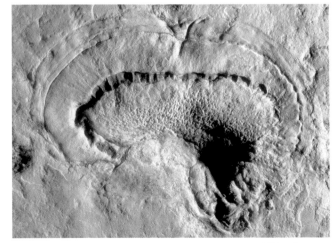

Figure 260. Perhaps Kimberella quadrata, *Vendian sequence, White Sea, Russia (F. Coffa). Specimen about 15 cm in length (right to left).*

Figure 261. Solza margarita, *latex peel of sole of bed, Solza, White Sea, northern Russia, perhaps an isolated "shell" of a form related to Kimberella (M. Leonov).*

Figure 264. Kimberella quadrata, *latex peel of sole of bed, White Sea, northern Russia (M. Fedonkin). Length of specimen about 2 cm.*

Figure 262. Kimberella quadrata, *shell impint on sole of bed, White Sea, northern Russia (M. Fedonkin). Specimen is 25 mm long.*

Figure 263. Kimberella quadrata, *several specimens juxtaposed on sole of bed, Vendian sequence, White Sea, northern Russia (M. Fedonkin). Center specimen about 4 cm in length (trace and body fossil).*

counter suggestions that Ediacaran metazoans had little, if anything, to do with living groups, but were instead lost body plans (Seilacher, 1992). If *Kimberella*, indeed, was anything like modern cubozoans, this had major ecological implications. Living chirodropids are fast-swimming predators with powerful venom. *Kimberella* would then have been one of the first larger predators on record (Jenkins 1992).

Additional specimens of *Kimberella* were discovered in the 1980's in both outcrops and borehole cores of the Vendian rocks of Russia. At first these fossils were identified as enigmatic, bilaterally symmetric animals, but were given no name. Only after years of exploration and extensive excavations along the Winter Coast of the White Sea was the true nature of *Kimberella* determined – more than 800 specimens from this area are now in hand, and with this large sample, *Kimberella* was found to be something quite different than originally thought. It certainly was not a box jelly.

Kimberella has been found in a variety of sediments, but is most commonly preserved on the undersurfaces (called the "soles") of ancient channels (called "gutter casts") cut into a muddy bottom of the Vendian sea. These channels were filled with sand that was moving downslope off the edge of a delta, probably initiated by storms that destabilized the edge of a large sediment wedge issuing forth from a massive river system. Such fossiliferous sand lenses have been named "*Kimberella* lenses" because of the concentration of this particular fossil in the channels. The fossils represent imprints of the dorsal or top side of the organism and show that at least parts of *Kimberella* were quite soft and deformable. Lying atop the soft parts, larger specimens, thought to be adults, have a high, shield-like dorsal shell, which grew up to 15 cm long, 5-7 cm wide and 3-4 cm deep. The shell was stiff, but not mineralized to any great extent.

It appears that *Kimberella* had a foot, rather like that of a living snail, sometimes leaving its imprint in the fine sands and muds in which it was preserved. Peripheral to the impression of what has been called the "shell," which sometimes could even be the imprint impressed from below of the foot, is a circular tubular, or perhaps folded, system, which may have been an analogue in the respiratory system of monoplacophorans . And beyond that there is an impression of soft tissue that extended even farther peripherally – perhaps an analogue to the periostracum of mollusks. Much more is to be learned about this species as the large sample is further studied.

With the finding of new fossils *Kimberella* was interpreted as a mollusk, certainly not a box jelly – showing many analogous structures to monoplacophorans, chitons and gastropods – still not placed in any particular group. *Kimberella* was depicted as a mollusk-like, triploblastic animal with a high dorsal, non-mineralized shell (Fedonkin & Waggoner, 1996, 1997). There was further speculation that it had a feeding groove around the outside of the foot (circumpedal) lined with ciliated cells on the dorsal side. With such an arrangement, according to the theory, when the foot was stretched out beyond the shell, the broad dorsal surface could have served as a food collector, trapping particles with its covering of mucus. Trapped food particles could then have been transported down to the feeding groove and there, in a current created by cilia, "dinner" could be directed toward a mouth at the anterior part of the body, which must have lain under the shell. There was a suggestion that numerous miniature ctenidia (respiratory organs) were situated on the upper side of the feeding grooves, organs that could have had a double function – respiration in addition to filtering particles selectively caught and removed by the intensively secreted mucus. Further studies and more specimens did not support this interpretation.

Kimberella is often associated with such fossils as *Cyclomedusa, Dickinsonia, Tribrachidium, Parvancorina* and other well-known forms from Ediacaran localities in South Australia and the White Sea of Russia. Absence of any size sorting of these body fossils, most apparently preserved in life position, along with the presence of abundant bottom-dwelling filamentous algae (*Zinkovioides, Striatella*) and the net-like *Orbisiana* (Ivantsov & Grazhdankin, 1997) all indicate that submarine channels had much to do with the preservation of these diverse metazoan communities – sediment that was transported down these channels, probably loosened by storms, provided the entombing material that preserved communities in place with very little transport. Volcanic ash beds interfingering with the sand along the Winter Coast of the White Sea clearly establish an approximate date of when *Kimberella* lived – around 555 million years ago (Martin *et al.*, 2000).

With more than 800 body fossil specimens of *Kimberella* collected in the Vendian siliciclastic rocks exposed along Zimnii Bereg (Winter Coast) of the White Sea, this is by far the largest collection of any one Ediacaran, and it is known over a wide geographic area from the Onega Peninsula (the Suz'ma, Karakhta and Solza rivers) to the west of the Zimnii Bereg occurrences as well as in the Flinders Ranges of South Australia. The White Sea material revealed intriguing new information about how *Kimberella* moved and fed and even some new details about its internal and external construction. Several specimens preserved imprints of the soft body parts, and individual shells disassociated from soft body impressions have also been found. Fan-shaped feeding tracks, crawling trails and movement structures were found directly associated with *Kimberella*. Such detailed information is unknown for other Neoproterozoic metazoans (Fedonkin, Simonetta & Ivantsov, 2004).

In summary, *Kimberella* had a tall dorsal "shell" covered with numerous semihemispherical knobs, a hood-like structure at the front end of the animal and a peripheral zone of soft

Figure 265. Kimberella quadrata *associated with feeding traces, Vendian sequence, White Sea region, Russia (M. Fedonkin). Body fossil about 2 cm in length.*

tissue, perhaps the mantle or the periostracum. The knobs may represent supportive sclerites, which in later animals may have fused to form shells. Beyond the elongate, bag-like body and the "mantle," *Kimberella* seems to have had an extensive "foot." This was likely quite strong, for as discussed below, the mode of feeding would have required *Kimberalla* to have been firmly anchored at the time it was taking in nutrients.

Andrey Ivantsov has collected a large number (800+) of *Kimberella* specimens, and this collection has been the base for understanding this form in detail. For some time, in both Australia and the White Sea, scratch marks arranged in a fan-like pattern had been observed on the soles of sedimentary beds. What organism made them or how they were made was unknown. Ivantsov found these scratches directly associated with *Kimberella* – they were feeding marks evidently produced by paired hooks that were housed in a retractable proboscis, impressions of which are preserved in some specimens (Fedonkin *et al.*, 2002). Evidence of the paired hooks, which must have functioned like the radula of snails, lies in the "grazing" patterns left in the sea floor sediments, fan-shaped patterns defined by a series of fine, always paired grooves.

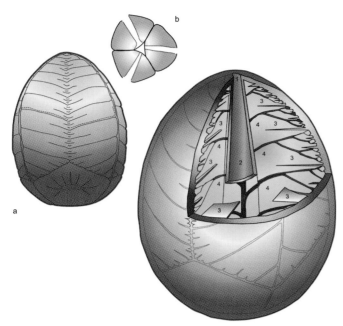

marks are overrun with movement trails, and thus it seems that *Kimberella* backed up when it finished feeding on one plot of microbial mat.

The large sample of *Kimberella* specimens also provided some other insights. Animals with body plans that most closely resemble *Kimberella* are primitive mollusks, the monoplacophorans and to a lesser extent the chitons. Both of these groups, unlike other members of the Mollusca, such as snails, clams and squids, possess a number of segmented muscles. Monoplacophorans, indeed, with this arrangement and other aspects of their structure have often been used to link mollusks to other major groups of animals, such as

Figure 266. Ventogyrus chistyakovi, *Vendian sequence, Yarnema, Onega Peninsula, White Sea, Russia, a reconstruction (modified from Ivantsov, 2001 by D. Gelt).*

One interpretation of how *Kimberella* fed is that it rested in one place, extended its proboscis, pushed it down into the sea floor algal and bacterial mat and then drew it back towards the body and the mouth. As the proboscis with its tough hooks engaged the mat, it left behind its "farming furrows," taking with it bits of the mat to the mouth. Once done, the process was repeated over and over again, producing the fan-shaped patterns. Apparently, after the mat was thoroughly farmed and most of the nutrients removed, *Kimberella* put itself in reverse, backed away from the "feeding field" until it reached "greener pastures" – a mat that was fresh and untouched – and then began the process all over again. This would have been a most efficient way of feeding, for *Kimberella* would have wasted no time and energy moving across unnutritious sea floor. None of the feeding scratch

Figure 268. Ventogyrus chistyakovi, *internal structure, White Sea Region, Russia (M. Fedonkin). Width of specimen about 8 cm.*

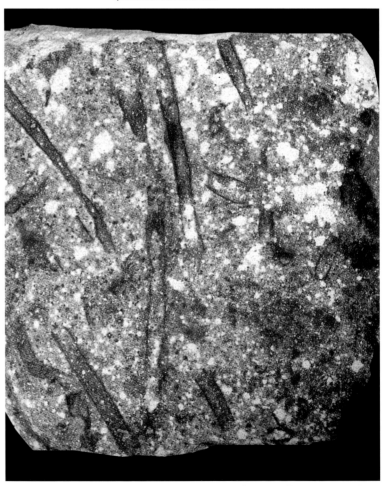

Figure 267. Ventogyrus chistyakovi, *Vendian sequence, Yarnema, Onega Peninsula, White Sea, Russia (F. Coffa). Individual specimens about 3 cm in width.*

Figure 269. Tubular fossils from the Onega Peninsula, associated with Ventogyrus *material, White Sea Region, Russia (M. Fedonkin). The height of the rock specimen is 15 cm.*

Figure 270. Nemiana simplex, *Zimnii Bereg, White Sea, Russia (M. Leonov). Each individual is about 1 cm in diameter.*

Figure 271. Reconstruction of Nemiana, *a common fossil in Neoproterozoic sequences globally (artists A. Besedina and I. Tokareva).*

worms and even arthropods – they are a group with a mosaic of characters that in some ways make them hard to place in a classification. This is exactly the sort of animal to be expected in ancient rocks, and monoplacophorans are known from the Cambrian, in a way the "missing links" that give some idea of how major groups of animals are related.

In addition to the mosaic nature of the soft part anatomy of *Kimberella,* its "shell" also serves as a good precursor to true mineralized shells, which appeared near the end of the Neoproterozoic and became standard issue in the Phanerozoic. Some of the first examples of shelled animals are the tiny Small Shelly Fossils (SSF's), so abundant during the early part of the Cambrian. As mentioned before, the small knobs or protuberances on the *Kimberella* shell could have been

organic precursors to these later mineralized microsclerites. This development of shells occurred nearly simultaneously in a number of different animal phyla in Early Cambrian times.

The presence of *Kimberella* in the Late Precambrian has a number of significant implications that have been highlighted by Fedonkin, Simonetta and Ivantsov (2004).

1. If *Kimberella* is indeed a mollusk, this means that other related, primitive mollusks must have appeared prior to Ediacaran times, and precursors to forms that "suddenly" appear in the Cambrian explosion must have already been around – but lacked skeletons to announce their presence. The Cambrian explosion must have had a long fuse, an idea that is consistent with molecular clock predictions, which place the origin of many metazoan groups farther in the past than the current fossil record suggests.

2. The wide geographic distribution of *Kimberella* (along with the other Ediacara-type metazoans) requires that these organisms had planktonic larvae, specialized eggs or some other mechanism allowing their broad dispersal. They certainly could not have "crawled" to all of those geographically dispersed areas!

Ventogyrus chistyakovi is another of the White Sea metazoans that has been found in large numbers and is preserved in three dimensions. Ivantsov and Grazhdankin first described this unique form in 1997, based on a collection from the Onega River between the mouth of Somba River and Yarnema. Fossils of this new form were entirely confined to sandstone lenses within an otherwise monotonous sequence of laminated siltstones of the Teksa Member in the Ust'-Pinega Formation.

Figure 272. Vendoconularia triradiata, *Vendian sequence, Yarnema, White Sea, Russia (F. Coffa). Length of specimen about 7 cm.*

Figure 275. Hiemalora stellaris, *Vendian sequence, Zimnii Bereg, White Sea, Russia (F. Coffa). Width of specimen about 4 cm.*

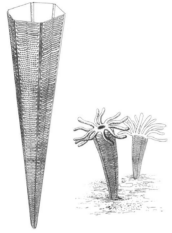

Figure 273. Vendoconularia triradiata, *reconstruction, Vendian sequence, Yarnema, White Sea, Russia (reconstruction by Ivantsov and Fedonkin, artists A. Besedina and I. Tokareva, courtesy of A. Ivantsov).*

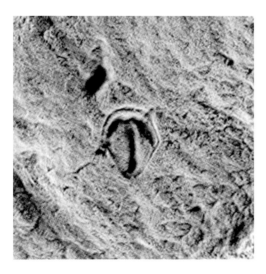

Figure 274. Parvancorina minchami (PIN 399316191), Zimnii Bereg, White Sea, Russia (M. Leonov). Length of specimen about 1 cm.

Most of the few hundred specimens were recovered from one of the largest sand lenses ever found in the Teksa sediments, a concentration originally thought to be an erosional pothole cut into the sea floor. This lens was about 60 cm wide and could be traced for more than 2 meters in length, grading into the surrounding siltstones. The lens itself is a white, fine-grained sandstone containing rare, but large, claystone pebbles, and rather than a pot hole accumulation, the lens appears to be an erosional feature incised by a storm-

loosened sand avalanche coursing down the side of a delta, eroding a channel or "gutter" and then within that channel, as the current slowed, sand and the accumulated load of organisms included within it were unceremoniously dumped as a fossiliferous sand lens discovered and collected by Ivantsov hundreds of millions years later.

The fossils are so numerous and concentrated in the sand lens that individuals often overlap, sometimes deforming each other. Sometimes it is difficult to identify where one body stops and another begins! Fossils often break into component parts, which appear to be sediment casts of the chambers that formed the internal part of *Ventogyrus*. One longitudinal and two sets of transverse baffles or walls compartmentalize the internal space of *Ventogyrus*, making up three modules. The overall organism was egg-shaped. Each module, at its broader end, bares a single, unpaired chamber, whereas the rest of the module was constructed of pairs of chambers that lay either side of a midline. The chambers on one side are slightly offset relative to the chambers on the other side: a symmetry given the name of "gliding reflection." This sort of symmetry is characteristic of forms like *Pteridinium*, so common in the rocks of Namibia, but the same symmetry is documented in many Ediacaran bilateral organisms, examples being *Vendia* and *Yorgia*. When first reconstructed, Ivantsov and Grazhdankin suggested that *Ventogyrus* was constructed like a boat, but as more specimens were collected and studied, this model did not seem correct.

A shift in interpretation of *Ventogyrus* came when Fedonkin discovered a new locality along the Onega River – a small outcrop just across the river from the original locale. It was clear from the rock succession here, only about 5 meters thick, that the sediments from fine-laminated clay incised by poorly sorted, medium-grained sandstone in lenses with rare casts of *Ventogyrus* were preserved in the same manner as in the first locality. At the base of the lens that Fedonkin discovered there was what geologists call a clay-interclast conglomerate. This is a rock type that consists of sandy sediments containing abundant flattened clay pebbles of several different sizes – indicating that the waters which washed these sediments into a channel first incised a clay-rich surface, ripped up some of that clay, rolled it along until it formed somewhat rounded pebbles and then dumped it along

Figure 276. Charnia cf. masoni, *Vendian sequence, White Sea region, northern Russia (F. Coffa). Length of specimen about 25 cm.*

with the fossils and sand into the confining channel that had just been cut. As with the original locale the fossiliferous lenses were elongate with a V-shaped cross section, smooth top surface and wavy cross-stratification of the channel sands. In other words, they seem to have formed in exactly the same way as the sediments at the original locale.

In this new channel, there were a few specimens preserved in vertical position within a small channel. These specimens had broken apart, splitting along the midline septa, and thus exposed a far more complex internal structure of *Ventogyrus* than had been seen in the original specimens. In the new specimens, three major longitudinal septa connected the outer wall with a newly discovered axial stem. From these new specimens it was quite clear that *Ventogyrus* had three-fold symmetry, quite an unusual symmetry in living animals but seemingly common in many Ediacaran metazoans, which have traditionally been emplaced in the group Radialia, called trilobozoan coelenterates. The body plan of *Ventogyrus* was not boat-shaped at all, but rather more egg-shaped. Its internal space was clearly divided into three major spaces or modules arranged along the axis of this "egg." Each module was then subdivided by a longitudinal half-septum and two sets of transverse septa into many chambers, which decreased in volume towards the narrow end of the egg-shaped body.

Another structure noted in the new material has been interpreted by some as a "float" at the broad end of the organism, whose purpose may have been to keep this metazoan suspended in the water column. Also visible were impressions of three channels preserved along the axis that branched laterally in an alternating pattern. These branches split twice before intersecting the body wall, where a longitudinal and a transverse septum intersect.

Ventogyrus may have been a free-swimming, soft-bodied coelenterate, but its affinity is currently debated. Fedonkin has noted marked similarity of the overall body plan with the living siphonophores, such as the Portuguese Man of War, colonies of polyps that have undergone "job specialization" and division of labor as they float in the oceans. He notes that *Ventogyrus* could represent the pneumatophore of such a siphonophore colony, a single, large specialized polyp that forms a float which buoys up the rest of the colony. This presumed float is egg-shaped and covered

with a layer of soft tissue. Like living siphonophores this float could have been filled with gas, quite probably carbon dioxide.

Other relationships have been suggested for *Ventogyrus* – Ivantsov reconstructs *Ventogyrus* as an egg-shaped ctenophore, related to the living comb jellies, also coelenterates. Another idea has been put forward recently by Fedonkin, who originally thought it was related to siphonophores. Noting the three-fold symmetry and the hollow triangular axial structure of *Ventogyrus* he suggested that there are significant similarities with some Early Cambrian Small Shelly Fossils, such as as anabaritids or angustiochreids, many of which in reality were not the external shells but rather internal skeletal structures. The amazing three-dimensional preservation revealing fine details of both the internal anatomy and external morphology of *Ventogyrus* makes this organism a "model object," which helps in the understanding the diversity of preservation styles and the diverse body plans which existed during Vendian times. Many other features of this intriguing metazoan may be revealed when tomographic studies underway are completed and once and for all, perhaps, just perhaps, the true relationships of this well-preserved and abundant Vendian metazoan will be sorted out.

An International Reference Point

The sites along the White Sea have become one of the few areas on Earth to serve as reference points in the late Precambrian. Preservation in all of these areas is superb and biodiversity is high. Because of the Russian work along the Winter Coast, a number of international expeditions have either gone there or visited the collections in Moscow, including a list of "stars" well known for their long-term involvement in Precambrian studies – Ellis Yochelson and Doug Erwin from the Smithsonian Institution in Washington, groups from the University of California, Berkeley (Jere Lipps, Jim Valentine, Ben Waggoner, Allen Collins) and Los Angeles (students of Bruce Runnegar), the California Institute of Technology (Joe Kirschvink and Dave Evans), the Massachusetts Institute of Technology, Rob Hengeveld from Holland, Dolf Seilacher from Tübingen and from the University of Adelaide, home of the Ediacaran name , Martin Glaessner, Jim Gehling and Richard Jenkins. Al have viewed the Russian collections. The most recent of international visitors was the 2003 expedition, which included Australians and Japanese colleagues. In 2002 IGCP493 entitled "The Rise and Fall of the Vendian Biota" was

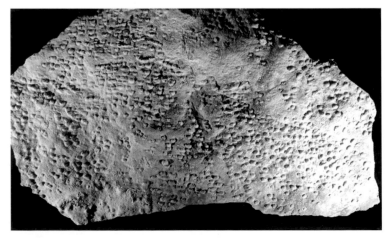

Figure 277. *Shallow burrows on the sole of a bed, Vendian sequence, White Sea region, northern Russia (M. Fedonkin). Width of block about 20 cm.*

Figure 278. Field crew including, from left to right: Yuri Gevorkyan, Natasha Bochkareva, Anatoly Stankovsky and Andrey Ivantsov, at Zolotitsa River, Vendian sequence, White Sea region, northern Russia in 1978 (M. Fedonkin).

Figure 279. International field (from left): Andrey Ivantsov, Yuri Nikolaev, Dimitri Grazhdankin, Allen Collins, Larisa Zakharchenkova (school teacher from Arkhangel'sk), Ben Waggoner, (both Allen and Ben from the University of California, Berkeley). Vendian sequence, Winter Coast of the White Sea, northern Russia (Y. Nikolaev).

funded for five years by UNESCO jointly organized by Mikhail Fedonkin, Patricia Vickers-Rich and James Gehling. The work continues.

Future Exploration

After a quarter of a century, the Winter Coast has given up some of her secrets. But, much remains to be found – many areas are still unexplored. The region is remote, moderately difficult to access and for that reason alone a wonderful challenge that will be met by few. Based on cores that have been recovered from bores in the region, the Vendian here ranges from 500 to 1000 meters in thickness. But outcrops rarely reach 100 meters. This succession, for the most part, is marine or estuarine in nature, but the top of the sequence was laid down in freshwater – and so far, no fossils have been recovered from there.

Tying the isolated outcrops to one another using the information from bores, working out how the environments changed through time, working out just where the volcanoes were erupting, just how stable the continental shelf was and why marine conditions gave way to freshwater during the time the precursors of modern animals and plants were living and evolving in these environments – all of this is the challenge of the future. Lifetimes will be spent seeking the answers, and these ancient rocks and their contents are very likely to give hints about how to deal with the future intelligently and ultimately sustainably.

Quite recently Cambrian deposits have been identified on the Kuloi Plateau, White Sea. Like the Vendian deposits below, the Cambrian sediments, made up of siltstones and clays, are rather soft and fresh in appearance. They contain some trace fossils and body fossils that are under study, with the aim to determine the precise age as well as analyze the fossil content. This discovery provides a new window on this dynamic time that may help detail the events taking place when the organic world underwent a revolution – the transition from mainly soft-bodied Ediacaran assemblages to the much more varied, and often "armored," Cambrian biota.

Figure 280. A possible tunicate fossil from the Suz'ma locale, White Sea Region, perhaps with affinities to the Namibian Ausia. Two specimens of this form have been found at the Suz'ma locale and are currently under study by M, Fedonkin, P. Vickers-Rich and Billie Swalla (M. Fedonkin). Measure of specimen from top to bottom is about 9 cm.

Opposite page: Precambrian Shadow *by Peter Trusler. The Vendian sequence forms the cliffs along the Zimni Bereg coast, White Sea, northern Russia.*

Podolia's Green Valleys

M. A. Fedonkin and P. Vickers-Rich

The White Sea has by far the richest record of Neoproterozoic metazoans for all of Europe, but finds in the Ukraine have added more pages to the book of life for this time period. Marine sediments with possible early metazoan remains have been known in the Ukraine since the early 1900's, at first thought to be raindrop prints. It was not until enthusiasm and determination brought a young paleontologist, Vladimir Palij, from Kiev to the green valley of the Dniester River, that the significance of Podolia was recognized. Palij carried out his Ph.D. research, noting a number of disc-shaped fossils. Then, with the construction of a major hydroelectric facility along the Dniester, the oldest Vendian sediments in the area were laid bare. In 1979 the biodiversity of this fossil assemblage was vastly increased and described – stimulating further research. A youthful geologist now plans to return and study not only the fossils but the environments that nurtured them.

Raindrops, Jellyfish or Anchors?

The Ukraine gave up the first hints of Precambrian animals when A. B. Krasovski found disc imprints there in 1916. At first they were thought to be raindrop prints, but later, in 1939, L. F. Lungersgausen suggested that they could be the patterns left by air bubbles formed when waves oscillated back and forth over a gently inclined beachfront. By 1928, O. K. Kaptarenko had already ruled out that possibility, simply because of the detail of the pattern that they left behind. In fact, he thought that they could be the impressions of jellyfish medusae – he even suggested that they were very similar to the genus *Medusina* named by Charles Walcott, the American geologist who carried out significant work on the Precambrian and Cambrian rocks of North America, while describing in detail the Cambrian Burgess Shale fauna (see Yochelson, 1998).

Figure 281. The green valleys of Podolia, underlain by the fossiliferous Mogilev Formation, Dniester River Basin, Podolia, the Ukraine (M. Fedonkin).

Opposite page: Castle along Dniester River, Podolia, the Ukraine, where Ediacaran fossils occur (Y. Shuvalova).

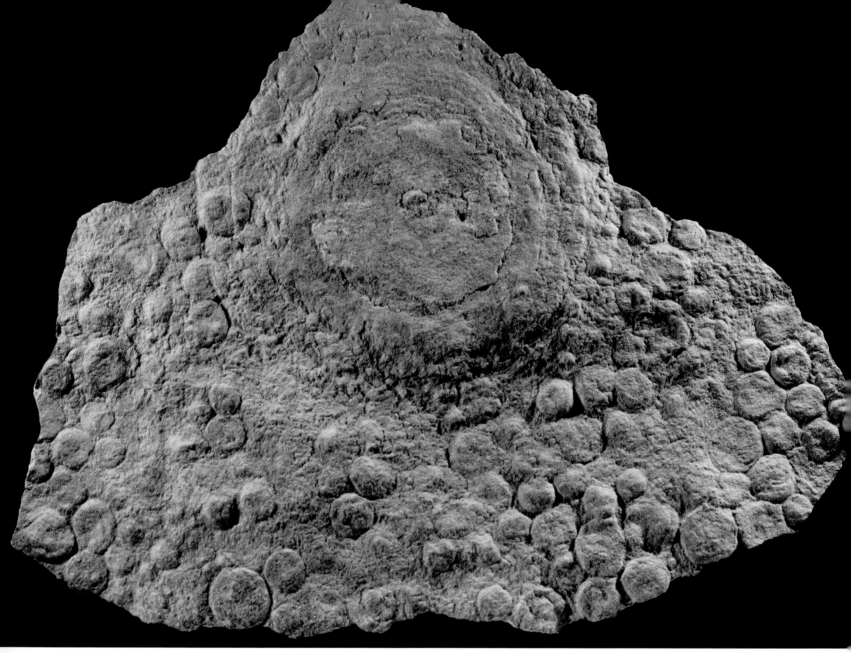

Figure 282. Ediacaria *and* Nemiana, *Yaryshev Formation, Dniester River Basin, Podolia, the Ukraine (F. Coffa). Rock specimen about 55 cm in width.*

Just as the rocks in the White Sea region had many ages assigned to them before absolute dating settled the issue, those of the Podolia region had originally been assigned a Silurian age (434-410 million years ago).

Little more thought was given these vague impressions until the mid 1950's, when O. N. Voznesenski took up an interest – and from the beginning of his investigations he considered the age to be Precambrian. His thought, however, was that these impressions represented trace fossils – bioturbations – the end result of organisms burrowing and "eating" their way through the sediments, leaving no internal stratification. He suggested that this sediment "mining" was carried out perhaps by the oldest chordates, ancestors of backboned animals, or maybe even crustaceans, relatives of crabs and their kin, who lived in ancient rivers, lakes or floodplains of the present-day Ukraine. Many other paleontologists considered these strange markings in the 1960's and 1970's – V. C. Zaika-Novatsky, V. A. Velikanov and V. M. Palij and a bit later on Boris Sokolov, Y. A. Gureev and Mikhail Fedonkin (Zaika-Novatsky *et al.*, 1968, Palij, 1969, 1974, 1976; Palij, Posti & Fedonkin, 1979) and due to their

contributions, the significance of these fossils became clear – they were pre-Cambrian metazoan remains. However, up to this point, the fossils recovered had been mainly discs of varying sizes and were given names like *Nemiana, Cyclomedusa, Medusinites, Tirasiana* and *Planomedusinites*.

In 1979, coincident with the work on the White Sea that was developing into a major program, Mikhail Fedonkin decided to visit some of the Ukraine sites with a small team of three – in spite of the fact that Vladimir Palij, who had worked on the Podolian fauna for his Ph.D., discouraged him from going there. Fedonkin was advised that nothing new had been found there for some time, and thus was asked what was the use of such a trip? With his normal skepticism about advice, Fedonkin was convinced he had to go. Such advice has standardly been a challenge.

The Banks of the Dniester River

And again, "Lady Luck" was on his side. First to be checked were the well-known locales, studied by Palij, who freely provided detailed information on precise localities, and more

Figure 283. Cyclomedusa cf. plana, *Yaryshev Formation, Dniester River Basin, Podolia, the Ukraine (F. Coffa).*

important, how to gain access. Pallij was correct – the known fossil licality did not yield any new species, but it so happened that a new hydroelectric power station, the Novodniestrovskaya Power Station, was about to be built. The draw card to this area, of course, was the massive quarry that was excavated for the major construction on the "right bank" of the river. This quarrying operation exposed new rock surfaces never available before – they had all been underground up to this point. All the sedimentary rocks had been removed from the quarry, and only the crystalline basement remained, glinting in the Sun on the quarry floor. Nothing organic was to be seen in that basement – but the next step, for the practiced prospector, was to check the sediments that had been removed and piled up in great mounds nearby. Unfortunately, after a thorough check of the area with absolutely no luck, the crew was about to leave empty-handed – the field trip was to end the next day, and all the crew and the driver were quite ready to return to Moscow. The driver was especially keen to leave,since he had not seen wife and family for more than four months. Unfortunately, for the driver, plans were to dramatically change.

As the sun was moving towards the horizon on this last day, Fedonkin was having a last prospect alongside the road leading from the quarry to spoil piles where the sedimentary rocks had been unceremoniously dumped. Suddenly, he could not believe his eyes as the dimming, low-angled light washed across one of the slabs that had clearly fallen "off the back of a truck." He spied on one slab the beautiful outline of *Hiemalora*, a "medusoid" disc with long "tentacles," now thought by many to be a complex hold-

fast or anchor for an organism that fed above the ocean floor. There was no doubt that this was an imprint of a once living organism– many of these had been found elsewhere in rocks of Precambrian age.

Fedonkin was caught between the proverbial "rock and a hard place" – on the one hand he had promised the driver his freedom to go home, on the other he had just made one of the major discoveries in his life. He simply had no choice – he immediately hid the fossil under his coat and asked the driver to return for the final night at camp. However, once Mikhail climbed into the vehicle, the smile on his face gave him away – the driver bleated out, "I am on strike."

That night seemed excruciatingly long – Mikhail could hardly sleep. He could only just contain his excitement – but nothing could be done in the October darkness. The next morning the small crew headed immediately to the quarry where the erosion surface developed during Ediacaran time on the underlying granite basement was clearly to be seen. To his frustration, Fedonkin saw that further collection of any large amounts of the overlying Vendian sediments was impossible. Material could also not be collected in place – it had all been completely stripped out by the quarry operations and transported. Momentarily, his heart sank – were these precious and unknown treasures, now recognized, all gone, dumped somewhere and no longer collectable? To his relief, one of the construction engineers pointed to a meadow in the distance, about 2 kilometers away, alongside the Dniester River, noting that is where all the trucks dumped the quarry spoil.

The crew set out at high speed across those 2 kilometers. With the low autumn Sun shining across the great slabs of silvery-gray rock that lay spread across the green, green carpet of the meadow – familiar Vendian animals almost jumped out to greet them. Over the next few days, despite the despondency of the driver yearning to rejoin his family, Fedonkin and his biology student apprentices (including Evgeny Rogaev, a well-known geneticist today) collected a substantial number of metazoans new to this region – the disc-like *Ediacaria* and *Eoporpita*, *Conomedusites* with its four-fold symmetry, the segmented *Dickinsonia*, the "three-armed" *Tribrachidium*s, Ediacarans that had never been found in the Ukraine before, along with a variety of new forms.

Finally, the crew left and en route to Moscow, a stop was necessary in Kiev, the home base of Palij. Fedonkin simply had to show him the new finds and convince him to continue the work on this now lucrative area. Unfortunately, the Ukrainian geologists were unable to carry on this work and most was carried on by crews from Moscow.

Fedonkin returned the next summer, 1980, with a crew of four and continued collecting from the hydro-spoil heaps. It was clear from the beginning that this material was weathering, and the fossils were not so crisp and distinctive on the rock surfaces as they had been when first recognized in 1979. With weathering, the rocks expanded and began to crumble – they were full of the minerals cerussite and mica. All the samples collected in 1980 had to be impregnated with glues to preserve them, and handled with care.

Prospecting, Local Flavor and International Conferences

1980 was a year of further prospecting. Soon it turned out, the spoil heaps were not the only source of fossils. The setting

Figure 284. Outcrops of the Lomosov Beds of the Mogilev Formation, Dniester River Basin, Podolia, the Ukraine, with Novodniestrovskaya Power Station in the background. It was excavations related to the construction of this facility that led to the discovery of Ediacaran fossils in the Ukraine (M. Fedonkin).

Figure 285 Unnamed fossil fragment from Dniester River Basin, Podolia, the Ukraine (M. Fedonkin). Rock specimen about 10 cm wide.

Figure 286. Lomosov Beds, Mogilev Formation, Dniester River Basin, Podolia, the Ukraine (M. Fedonkin).

Figure 287. Dickinsonia, Yaryshev Formation, Dniester River Basin, Podolia, the Ukraine (F. Coffa).

for this work was so very different from that of the White Sea. The Dniester River and its tributaries had cut deep valleys into the siliciclastic sediments of the Mogilev-Podolsk Group crystalline basement – mainly made up of granite. The present climate of this region is one of warm summers and cold winters, and the valley sides and hills are covered with broadleaf forests – not the stunted taiga and tundra of the Russian far north. All along the valley walls are outcrops of Precambrian rocks. The soils on the top of the rolling hills of Podolia that border the rivers are deep and black – so rich that the locals said "if you put a stick in it, the stick will sprout." Out on the grasslands, storks cavorted and called, while in the leafy forests, a myriad of birds provided a dawn and dusk orchestra. Forget about sleeping late or for that matter going to bed too early! It was a peaceful and biodiverse environment enjoyed by the fossil hunters who hiked the valleys to find ancestors of the life surrounding them in their venture.

Like the wildlife, the people living in these valleys were fascinating. They had neat, whitewashed houses and gardens full of fruit trees – but unfortunately many had been forced to move by the rising waters created by the hydroelectric scheme, the very source of the paleontological treasures that Fedonkin and his assistants sought.

The second field season lasted just over a month – until the middle of October, with prospectors taking advantage again of the low-angled autumn Sun. All outcrops around the area as well as the sediments exposed along the quarry walls were thoroughly checked. A third visit was the last to this area and was organized as part of an international conference. This trip brought international scholars including Jim Gehling and Richard Jenkins from Australia, the expert on the Ediacara fauna of South Australia, Guy Narbonne from Canada and Bruce Runnegar, now involved with NASA in the search for extraterrestrial life, but with a lifelong interest in early metazoans, to the Ukraine. One result of this trip was that Richard collected a few volcanic ash samples for dating that were taken back to Australia, which yielded an age of about 560 million years – a reasonable date to expect for the oldest metazoans known elsewhere.

Biomats, Food and Preservers of Early Metazoans

Back in Moscow, studies of the fossils collected over three seasons gave some rather unexpected insights into the fauna that lived near the end of the Precambrian in the Ukraine. The biomass as well as the biodiversity of the fauna was high – the fauna was living on a sandy bottom in rather crowded conditions. Many forms appeared to be attached in dense arrays onto sand that was "glued" together, most likely by microbial mats. These fossilized sands, sandstones, often preserved an "elephant-skin" texture typical of biomats today, and certainly a similar environment of deposition was clearly a significant preserver of some of the richest assemblages of the White Sea.

Such mats also provided nutrients that the Ediacarans such as Dickinsonia could graze upon. These animals lived in a changing environment, and the biodiversity reflected this. As one collected upsection, from bottom to top of the 30 meter sequence of the Lomosov beds, species diversity waxed and waned. The Lomosov sediments alternated from dark silvery gray, micaceous siltstones to coarse-grained arkosic sandstones, which were clearly deposited in shallower waters.

Figure 288. Egg-like structures and elephant-skin texture reflecting the past presence of microbial mat, Lomosov Beds, Mogilev Formation, Dniester River Basin, Podolia, the Ukraine (F. Coffa).

Figure 289. Prospecting for Ediacarans, Lomosov Beds, Mogilev Formation, Dniester River Basin, Podolia, the Ukraine (M. Fedonkin).

Before the Paleontological Institute crew began working along the Dniester, Zaika-Novatsky and Palij had done most of their collecting from the Yampol Beds, which lay above the Lomosov sediments. To Mikhail's surprise, the Lomosov Beds contained a rich array of metazoan fossils, the number of species were far greater than those in the overlying rocks. At least 30 different species were collected from the Lomosov Beds. The layers above had only five species of body fossils and had been bioturbated (Gureev, 1987). What was also interesting was that the species in the Lomosov Beds were the same or very similar to animals known from the Flinders Range of South Australia – the Ediacara fauna. As work progressed after this on the White Sea, similar metazoans were found – also associated with the sea floor algal and bacterial mats that occurred there too.

The Lomosov Beds contain abundant metazoans. They are part of the Mogilev Formation (as are the Yampol Beds), itself included in the Mogilev-Podolsky Series (Velikanov, 1990). This series is made up of mudstones, siltstones that are often finely laminated. Within this sequence are lenses of sandy siltstones and sands, and occasionally even conglomerates, especially near the base of this sedimentary stack. Velikanov suggested that these sediments were deposited under conditions occasionally affected by "slightly agitated water" resulting from input of nearby land and even islands in this sea that had "rugged bottom topography" probably developed on the underlying crystalline basement rocks. He further suggested that such islands could have also provided some protection from storm-induced downslope avalanches, which emplaced the sandy lenses, leading to depositions of "fine terrigenous material and stagnation" – in other words, the development of the sea floor microbial mats that require calm waters for some period of time.

The overlying Yampol Beds differ considerably from the Lomozov sediments, in having a rather monotonous makeup – mainly sandstones with a massive appearance. The

Figure 290. Lomosov Beds, Mogilev Formation, Dniester River Basin, Podolia, the Ukraine (M. Fedonkin).

sandstones have a sugary appearance, with a fine to medium grain size and are only occasionally varied by the inclusion of coarser sands, all suggesting a shallow marine shelf environment of deposition.

In addition to the metazoan body fossils, algal remains (Gnilovskaya, 1990; Gnilovskaya et al., 1988) and other microfossils, including acritarchs, were discovered (Aseeva, 1983; Volkova et al., 1990).

Future Exploration

As work continued on the White Sea, study of material from the Ukraine stalemated. Little progress was made for more than two decades. Dimitri Grazhdankin has now begun work in this area, a fresh face with experience not only as a paleontologist but also as a sedimentologist. Working with Nick Butterfield and other paleontologists from Kiev (A.

Figure 291
Elasenia aseevae,
Lomosov Beds,
Mogilev Formation,
Dniester River
Basin, Podolia,
the Ukraine (M.
Fedonkin).

Figure 292. Kaisalia reconstruction, Vinoz, Vinnitsa district, the Ukraine (artists A. Besedina and I. Tokareva).

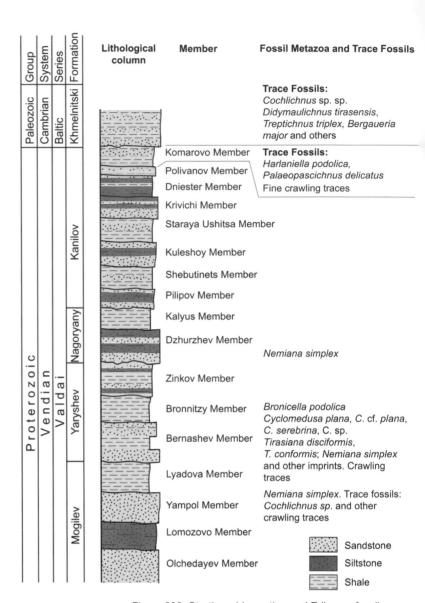

Group	System	Series	Formation	Lithological column	Member	Fossil Metazoa and Trace Fossils
Paleozoic	Cambrian	Baltic	Khmelnitski			**Trace Fossils:** *Cochlichnus* sp. sp. *Didymaulichnus tirasensis, Treptichnus triplex, Bergaueria major* and others
					Komarovo Member	**Trace Fossils:** *Harlaniella podolica, Palaeopascichnus delicatus* Fine crawling traces
					Polivanov Member	
					Dniester Member	
Proterozoic	Vendian Valdai	Kanilov			Krivichi Member	
					Staraya Ushitsa Member	
					Kuleshoy Member	
					Shebutinets Member	
					Pilipov Member	
					Kalyus Member	
		Nagoryany			Dzhurzhev Member	*Nemiana simplex*
					Zinkov Member	
		Yaryshev			Bronnitzy Member	*Bronicella podolica Cyclomedusa plana, C.* cf. *plana, C. serebrina, C.* sp. *Tirasiana disciformis, T. conformis; Nemiana simplex* and other imprints. Crawling traces
					Bernashev Member	
					Lyadova Member	
		Mogilev			Yampol Member	*Nemiana simplex.* Trace fossils: *Cochlichnus* sp. and other crawling traces
					Lomozovo Member	
					Olchedayev Member	

Sandstone
Siltstone
Shale

Figure 293. Stratigraphic section and Ediacara fossil occurrence of the Vendian sequence in the Ukraine (courtesy of M. Fedonkin) (modified by D. Gelt).

Menasova, and V. P. Gritsenko) and Moscow (M. Fedonkin, A. Yu. Ivantsov, M. Leonov), this group plans to approach this area examining both biological and geological aspects – trying to work out the age based on ash dating and paleontology. In addition to the metazoans and rocks, the microfossils and algal remains should be integrated into the picture of this area so a whole biota and their environmental setting will result. New radiometric dates combined with use of microfossil biostratigraphy should much more precisely tie down the dates of evolutionary events in this part of the Eastern European Platform and closely compare it with other areas, such as the White Sea, the Urals and Siberia.

Already Grazhdankin (*pers. com.,* 2005) has noted that there are at least three fossiliferous assemblages in the Ukraine. The most biodiverse and fossiliferous are the pro-delta rocks so well known in the White Sea with fossilized imprints of biomats, "deathmask preservation," and hosting forms such as *Tribrachidium, Dickinsonia* and a variety of discoidal fossils as well as frond-like forms and the trace fossil *Harlaniella.* A second assemblage populated the shallow sandy environments of the very shallow seas – such forms as *Nemiana.* And a third assemblage preferred the muds that may have been deposited in brackish waters, perhaps in estuaries, now preserved in shales and mudstones: comb-like, chitinous *Redkinia,* spiral *Cochleatina* and a variety of vendotaenids (Gnilovskaya, 1988; Grazhdankin & Butterfield, *pers. com.* 2005).

Figure 294. Podolimirus mirus, Lomosov Beds, Mogilev Formation, Dniester River Basin, Podolia, the Ukraine (F. Coffa). Rock slab 15 cm wide.

Figure 295. Nimbia occlusa, *Yanpol and Lomosov Beds, Mogilev Formation, Dniester River Basin, Podolia, the Ukraine (M. Fedonkin).*

Figure 296. Valdania plumosa, *Lomosov Beds, Mogilev Formation, Dniester River Basin, Podolia, the Ukraine (F. Coffa). Specimen about 7 cm wide.*

Figure 297. Palaeopascichnus delicatus, *Vendian sequence of Suz'ma, Arkhangel'sk region, White Sea, Russia. Similar forms occur in the Ukraine (S. Morton). Basal width of slab 10 cm.*

CHAPTER 8

The Siberian Tundra

M. A. Fedonkin and P. Vickers-Rich

Discoveries of early metazoans in Siberia came late in the history of investigation of these organisms. The first possible Vendian fossils were collected in the 1960's. It was not until the mid-1970's that anything more than just the odd specimen turned up. An intriguing feature about this area is that fossils are preserved in both carbonate- and silica-rich sediments. New taxa were slow to come because of the difficulty of access and harsh conditions field crews had to deal with in order to collect in this land of the summer midnight Sun. It was not until 2000 that the significant diversity of this biota was recognized.

The First Finds

Siberia can be a forbidding terrain. Those who work there must have real determination. Just getting there can be difficult, if not nearly impossible, and always has a real edge of risk. In spite of this, geologists began finding some Late Precambrian fossils as early as the 1960's, when the rather problematic discoidal fossils *Suvorovella* and *Majaella* were found in the lime-rich deposits in the Yudoma-Maya region (in the upper parts of the Udoma Formation). Vologdin and Maslov (1960) reported these as possible medusoids – jellyfish. Others disagreed and thought that the regular rhomboldal structure of these imprints was more like that of stromatoporids.

Then in 1975 in the Yenisei Kriazh and Batenev Kriazh regions (and a number of other places), mainly in the silica-rich clastic sediments, fossils more characteristic of late Precambrian animals began turning up. Boris Sokolov described what he called a jellyfish-like fossil, *Cyclomedusa*, in 1975 from the Ostrovnaya Formation, which was exposed along the Angara River near its confluence with the Irkineeva. And the list of spot finds went on – *Ediacaria* was discovered and reported on by N. M. Zadorozhnaya in 1985 in the central Altai-Sayan region of the Azartyl Range (Batenev Kriazh) in the upper part of the Matyukhin Formation – again in silica-rich sediments – and so on.

The most remarkable collection of Late Precambrian animals from this region, was described by Boris Sokolov in 1973 – these came from the bottom of the section of the Kurtun Formation. Sediments lay exposed along the Malyi Anai River and contained impressions of animals closely resembling, or identical to, such forms as *Pteridinium, Baikalina, Cylindrichnus* and many, many more. At the same time, Sokolov noted a rich assemblage of algae, associated with the tubular remains of the metazoan *Paleolina*, from rocks of a similar nature that were exposed along the Khindusa River near Lake Baikal.

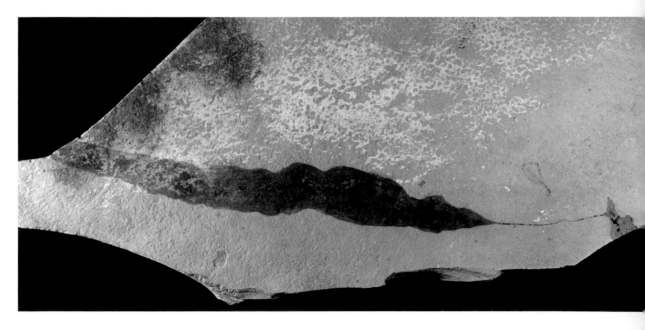

Figure 298. Possible algal frond, Khatyspyt Formation, Khorbusuonka Group, Yudomian, cropping out in cliffs along Anabyl Creek, Khorbu-suonka River Basin, Olenek Uplift of Arctic Siberia (F. Coffa). Length of fossil 55 cm.

Opposite page: Helicopter transport landing near the cliffs of the Khatyspyt Formation, Khorbusuonka Group, Yudomian, cropping out in cliffs along Olenek River in Olenek Uplift of Arctic Siberia (M. Fedonkin).

Although there was a lull in collecting after this first discovery, these sites are now being worked by Dimitri Grazhdankin and Nick Butterfield as well as Mikhail Fedonkin and Andrey Ivantsov from the Paleontological Institute – their work will likely yield a rich array of new forms in the future as well as a better understanding of this unusual mixture of siliciclastics and carbonates.

Between the Lena and the Olenek

Well before the Precambrian work had begun in the White Sea area, Boris Keller, along with Igor N. Krylov (GINRAS – Geological Institute, Russian Academy of Science) and Mikhail Fedonkin, then a graduate student, had been exploring Arctic Siberia as early as 1968. The field crew first went to Yakutsk, and then by boat downstream on the Lena River to a small village, Chekurovka, that lay far above the Arctic Circle. The aim of the trip was documentation of the stratigraphic succession and fossil record of the Proterozoic rocks of the Kharaulakh Mountains. Most of the fossils were represented by microbial sedimentary structures, such as stromatolites and microphytolites, which became the subject of Fedonkin's diploma thesis and subsequent joint publications. Fedonkin's future interests, however, lay to the west of the Lena River.

Figure 299.
Igor N. Krylov
(M. Fedonkin).

In north central Siberia, midway between the Lena and the Olenek rivers, lies the Olenek Highland. Here the same succession of deposits are laid bare along the upper part of the Lena River. Olenek was a place far less accessible, but much more interesting. In 1959, geologist T. N. Kopylova from the Institute of Arctic Geology in Leningrad (now Sankt Petersburg) had collected an Ediacara-like fossil from the fine-laminated dolomite of the Khatyspyt Formation. This enigmatic fossil, after some years of discussion among paleontologists in St Petersburg, was finally positively identified by Boris Sokolov (Sokolov, 1965, 1971, 1973) as a new species of *Rangea, R. siberica*. But, only in 1981, 22 years after the first discovery by Kopylova, a paleontological expedition, including experienced Vendian paleobiologists, was finally organized to revisit this

site, with the intention of finding a metazoan assemblage, not just single fossils.

Only from one Moscow airport, Domodedovo, can one catch a direct flight to Tiksi. For the crew that set out on this expedition, the six-hour flight was a welcome relief from the sweltering, humid heat of a Moscow summer. Even so, the contrast between Moscow and Tiksi was quite a shock for first-time visitors. No grass or trees were anywhere to be seen upon landing in Tiksi. The air had a sharpness to it – it was cold and the T-shirts that had served everyone well as they left Moscow were little comfort here – welcome to the Arctic summer!

Not only the cold, leaden air was a shock, but the Sun was up all day and night! There was no night, just one long day. Fortunately, the windows in the hotel where the crew collapsed afterwards with profound fatigue due to the days of preparation to exit Moscow, the flight itself and then the confrontation of the Arctic were graced with thick, black curtains. Only with this blackout assisting exhaustion was sleep possible. The worth of one night in a modest Tiksi hotel before going to the tundra for a month or two was not fully appreciated until the crew returned weeks later. You might not like the room service or your bed, but your attitude to this place certainly changes dramatically at the end of several months in the damp, always cold environs of the Arctic with attendant mosquitoes and black flies.

This first expedition to the Olenek Highlands began with a river trip, which set off from Tiksi. Later expeditions also launched from here, but helicopters became the workhorses in later times. Tiksi, a small Arctic town, is perched on the rocky shore of the Arctic Ocean – and is well known for its cold, cold climate, with temperatures sometimes reaching – 50° C. The 24-hour darkness of winter stretches across months.

Tiksi is a northern port in Russia, connecting the Arctic Ocean with the Lena River. Supplies can be sent south on the river from here and produce can alternatively come north to be shipped out. Ready access to points south makes Tiksi an ideal site for the military base situated in the region. Other than the military presence and attendant jobs, Tiksi is not an easy place to make a living – because of its truly severe environment. This was quite apparent to the geologists as they drove from the airport to their hotel – the sea, even in the summer, was covered with ice, the countryside shrouded in mist. Winds in this region are fierce

Figure 300. Field camp and radio antenna at Khorbosvonka River, Olenek Uplift, Arctic Siberia, for crew from Paleontological Institute (M. Fedonkin).

158

Figure 301. Field crew discussions in Siberia: S. Serebryakov (far left), Nikolai Chumakov (center) and Mikhail Semikhatov (far right). All worked in the laboratory headed by Boris Keller (M. Fedonkin).

– as evidenced by the telephone poles, some of which had clearly been bent nearly horizontal. The telephone poles were not standard ones either – their bases were made of steel railroad rails!

Jokes about Tiksi abound – "This place has 12 months of winter and the rest is summer" or "Fortunately this summer coincided with the weekend, so people had the opportunity to collect mushrooms in the tundra!" Despite the inhospitable environment, the chance of finding late Precambrian fossils in this area, so far the only place where Ediacara-type fossils were preserved in carbonate rocks rather than in the standard siliciclastic sandstones or claystones, was alluring. In 1981 Boris Sokolov and Mikhail Fedonkin, accompanied by their Siberian colleagues Boris Shishkin, Anatoly Val'kov and Peter Kolosov, set out to see what they could find.

Mi-8 Helicopters

If you are lucky and if the meteorology forecast gives the OK, pilots of the hardworking Mi-8 helicopters pull back on the stick and up one goes – with fingers crossed. The voluminous innards of the helicopter are generally filled with logistical items – tents, sleeping bags, rubber boats, boxes with food, iron stoves, backpacks – and of course the field team along with all the flammable fuels. All members of the team had to be in the best of health and prepared to be totally independent for a considerable period of time. If you forgot your toothpaste, you just did

Figure 302. Mi-8 helicopter lifting off the landslide drop point for field crew (N. Hunt).

without, for the helicopter, once it dropped the team, would not be back until the appointed time, months later – if you were lucky. These first expeditions were run at times when there were no satellite phones and thus essentially no outside communication. The big helicopters, capable of lifting and transporting a considerable load, had to carry in addition a large barrel of aviation fuel, just to get them

Figure 303. Slabs of fossiliferous claystones, Vendian sediments, Khatyspyt Formation, Khorbusuonka Group, Yudomian, Olenek River in Olenek Uplift of Arctic Siberia (M. Fedonkin). Slab in foreground about 15 cm high.

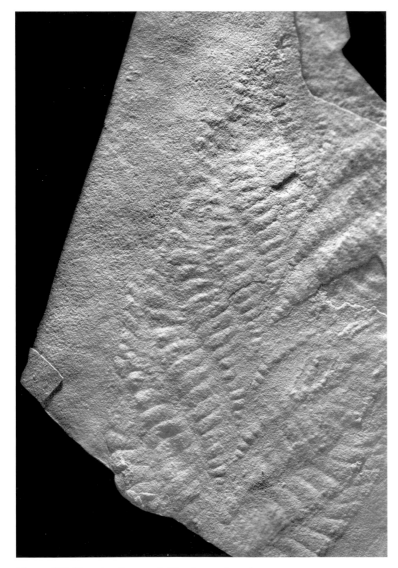

Figure 304. Charnia cf. masoni, Vendian sediments, Khatyspyt Formation, Khorbusuonka Group, Yudomian, cropping out in cliffs along Olenek River in Olenek Uplift of Arctic Siberia (M. Fedonkin). Width of frond about 15 cm.

back to base. And that was sandwiched in somewhere between the backpacks and the sleeping bags as well as the passengers.

In a few minutes after liftoff, the first crew to use helicopters crossed the Lena River near where it flowed into the Arctic Ocean. The river here, even from a few thousand feet, is intensely black, colored by the tannins of the birch tree cover. The flight then moved across endless, hardly changing polar tundra: moss, numerous lakes and rivers with patches of ice and a very smooth relief. After about an hour and half of flying, the flatness and monotony was broken by some low hills, which grew in height, and the valleys became deeper and deeper. This was the approach to the Olenek Uplift, the Olenek Highlands. And finally, after all this travel, the first paleontological group to visit the area sighted the pale yellowish cliffs in the valley cut by the Khorbusuonka River.

Challenge of the Permafrost

For four seasons, Paleontological Institute crews from Moscow worked in this area, the last being in 2002, absorbing as much local knowledge from seasoned geologists of the region as possible. The crews quickly learned some important secrets about "comfortably" camping on the permafrost – in tents! Insulation from cold and moisture from below was absolutely critical. Some of the crew used a plastic sheet, and then added a piece of plywood or wooden planks to form the floor of the tent. If they did not do this, the permafrost below the tent would melt in the tent, and they would end up living in a lake! Everything would be soaked with icy, cold water. If you were lucky enough to be near a pebbly bank of the river, conditions were better, but one had to be careful to watch the water level in the river, for it could rise quickly during a rainstorm or during "black water" days when mass melting of permafrost released quantities of water from the slopes of the valley and the tributaries. Like the water level in the river, the weather patterns, even at the height of summer in July, were erratic. Temperatures could range from 25°C to 0° C, or below, in a few hours. Snow was always possible. It was a luxury to have a stove in one's tent, but wood was not always available. One solution was to use the dark brown sub-fossil wood that washed out of the peat. Prior to the last glacial period, some 14,000 years ago, significant, yet stunted, forests had grown in this area, thus leaving their fossil "skeletons" behind.

Prospecting Arctic Cliffs

The Olenek region is spectacularly virgin territory – more or less untouched by the hands of humans, much less paleontologists. The wildlife and nature of the environment is almost pristine and deeply inspiring – albeit isolated and testing.

But to the seasoned geologist, the major interest of the region, of course, are the rocks. And the rocks in this region have a rather unique appearance. The Arctic climate and limited vegetation have a particular effect on the finely laminated carbonate rocks. The rocks have a very low degree of chemical and biological weathering but have been severely affected by the freezing and thawing that occurs each year. This, together with the initial mode of deposition, produces the style of preservation of fossils, and results in a kind of "natural" preparation, not quite preserving the fine detail of fossils along the White Sea.

The high cliffs produced by fluviatile erosion beautifully expose the Riphean, Vendian and younger Cambrian succession underlying the more recent deposits of the Siberian Platform. Unlike the terrestrial nature of this region today, for most of the Neoproterozoic, this area was covered by a shallow, carbonate-rich sea. Lime-rich sediments were precipitated from this sea, very often assisted by cyanobacteria, which produced the reef-forming stromatolite structure. Today the ancient environment is represented by limestones and dolomites that reflect the shallow marine conditions on the edge of a low-lying continent that produced them more than 600 million years ago. They also reflect how sea level rose and fell as the carbonate-rich deposits were repeatedly replaced by cross-bedded sandstone and siltstone. Fossil assemblages change from metazoan-dominated communities to massive accumulations of columnar stromatolites, then back again to metazoan-dominated, through the Cambrian.

Often the carbonate rocks are made up almost entirely of stromatolites, those structures accumulated by microbial mats initiating the precipitation of layer upon layer of carbonates that form convex upward columns, domes and sometimes extensive flat-lying layers, when formed in relatively undisturbed waters, in the shallows of the sea. Algae played only a minor role, if any, in Precambrian stromatolite construction – formation was mainly brought about by the action of cyanobacteria.

Between 1.5 and 0.5 billion years ago this area of Siberia functioned rather like a frying pan turned on low heat: like the pan, the bottom of the continental edge was almost flat. As a result, there was little erosion of the adjacent terrestrial environs of the continent and little input of sediments. With the heat, there was a high evaporation rate of these shallow marine waters, which encouraged carbonate precipitation. This was an ideal area, similar to Shark Bay in Western Australia today, for the growth of stromatolites. This continental region was situated at low latitude, with a subtropical or even tropical climate, during the Vendian (Ediacaran) and into the Cambrian Period.

Familiar and Unique Species – Low Biodiversity

In 1981, when Boris Sokolov, and his colleagues B. B. Shishkin, P. N. Kolosov, and A. K. Val'kov, arrived, no one expected to find much. Many geologists had worked in the area. But within days of Sokolov and his crew's arrival, a significant collection was gleaned from Khatyspyt Formation – including abundant *Nemiana*, *Ediacaria*, *Charnia* and *Paleopascichnus*. All fossils come from the valley of the Khorbusuonka River. Trace fossils from the Precambrian-Cambrian transition beds in the Kessusa Formation were collected in the Olenek River Valley, where this transition is well represented by siliciclastic, interrupted occasionally by carbonate lenses bearing the most ancient of the Small Shelly Fossil (SSF) fauna. At the end of the trip, a representative collection was returned to the Paleontological Institute, Russian Academy of Sciences, in Moscow for study and was soon formally described.

The insights gained from study of this material coincided with those inspired by fossils from other sites around the world, insights concerning the emergence of complex animal life. During a severe glaciation called the Varanger, which fundamentally affected this region, sea level dropped significantly and an immense part of the continental platform,

Figure 305. Ediacaria, Vendian sequence, Kooten-Boolgok, Zagoro, Altai region, Siberia (F. Coffa). Slab base width about 15 cm.

normally covered with shallow marine waters, was exposed to subaerial weathering – what used to be underwater became dry land – for millions of years. As the great ice sheets began to melt, sea level rose, flooding the broad platform with marine waters again, waters warm enough to encourage the deposition of carbonate-rich sediments that eventually solidified into limestone. Life began to repopulate this carbonate-rich sea, not any earlier than 560 million years ago. Abundant polyps of *Nemiana* formed colonies on the soft sea floors, probable attachment discs, such as *Ediacaria*, sea pen-like *Charnia* with its graceful frond along with a variety of other invertebrates spread rapidly across the northern part of this continental shelf and upper slope. Biodiversity of metazoans was greatest here and not in shallower or deeper waters of this ancient continental seaway.

Figure 306. Present-day stromatolites at Shark Bay, Western Australia (M. Fenton).

Figure 307. Neoproterozoic fossil stromatolites cropping out in the Northern Territory of Australia, like those so common in the Siberian sequences (J. W. Warren).

Although there were many familiar forms, nearly or exactly the same as those known from Australia and the White Sea region of Russia – *Charnia*, *Hiemalora*, *Ediacaria*, *Nemiana* – there were also some new forms found which were unique to the region – *an* example being *Khatyspytia* (?= *Charniodiscus*), a narrow, frond-like organism, which likely grew attached to the sediments. Most striking, however, was the total lack of animals with bilateral symmetry – animals with one half of their bodies a mirror image of the other, such as *Dickinsonia* - it was nowhere to be found here in Siberia.

And even though fossils were quite abundant, total biodiversity of the Siberian fauna was rather low – certainly there was not the variety of forms that were later found on the Winter Coast of northern Russia or the desert ranges of South Australia, or even Namibia. Perhaps this could be partly explained by the lack of collecting carried out in an area difficult to access. But by the end of the first field season, over

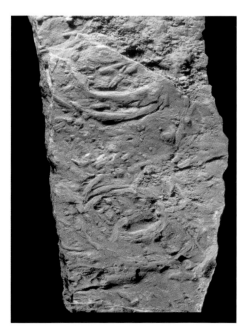

Figure 308. Circulichnus, a trace fossil, Vendian sediments, Khatyspyt Formation, Khorbusuonka Group, Yudomian, Olenek River in the Olenek Uplift of Arctic Siberia (F. Coffa). Base of slab width about 7 cm.

a hundred good specimens had been collected, and Sokolov and his colleagues began to question why the diversity was so low. Could it be explained by the nature of the local environment? Was the warmer temperature of the Siberian seas, the latitude, and the lack of sedimentary input part of the answer? This certainly seems the case for the Neoproterozoic carbonate deposits along the Yangtze Gorge of South China (Sun, 1986; Shen *et al.*, 2004). The first hint came from the way in which the fossils were distributed in the rock sequence (Khomentovsky in Sokolov & Iwanowski, 1985; Sokolov & Fedonkin, 1984; Fedonkin, 1990, 1992).

Stromatolites vs. Metazoans

The Vendian fossils of Siberia were recovered from the Khatyspyt Formation, a bitumen-rich limestone more than 150 m thick. This limestone lies under the Turkut Formation, which is rich in stromatolitic dolomites, some 200-230 meters thick. In this region, as in most others, the Ediacaran metazoan communities did not persist into the Cambrian, which is also well exposed. Stromatolites and metazoans are not often found together in the same rocks. "Stroms" are structures formed primarily by cyanobacterial communities and they seem to have "taken a kind of revenge" and "wiped out, "or rather replaced, the metazoan biota for at least 10-15 million years, as is recorded nicely in the Turkut Formation. To put it another way, environmental conditions changed from the time the chemical sediments were being deposited to form the Khatyspyt Formation into those times when the Turkut Formation was being laid down. In Turkut times conditions were perfect for stromatolite-forming communities to thrive – shallow intertidal sediments of this formation mirror this. Metazoans, on the contrary, may have been excluded or literally driven to local extinction.

Recolonization of the Olenek seas by the metazoans did take place at the very end of Vendian (Ediacaran) times, and they have dominated marine environments ever since. This event is clearly documented in the silica-rich sediments of the overlying Kessusa Formation, which contains limited carbonate lenses. In the Kessusa rocks, the explosive growth in trace fossil diversity and gradual increase in the number of species of the microscopic Small Shelly Fossils through the Precambrian-Cambrian transition beds highlight the increase in biodiversity from older to more recent times. This change in biotic content and increase in biodiversity of complex organisms also reflect changing ocean conditions, environments and climate.

After this, during the Cambrian, great reefs on the edge of the Siberian continental platform were formed by both calcareous algae and archeocyathans. Eukaryotic life was so abundant and biodiverse that there seems to have been no opportunity for stromatolite-building microorganisms to return to their former dominance. At this time animals with eyes and a penchant for grazing hungrily on algal and bacterial mats (as well as some hunting one another!) would have taken their toll on the stromatolite-builders, and today stromatolite constructors only survive in harsh environments which exclude their major destroyers, such as the radula-wielding gastropods who graze faster than the stromatolite-builders can lay down their fine layers of calcite!

Return to a Familiar Place

The Olenek Uplift of Siberia provides an opportunity for study of both environmental and biotic change during the late Precambrian and Early Cambrian. It allows detailed study of the pioneering colonization of a carbonate basin, where

Figure 309. Mikhail Fedonkin in front of field tent, Khorbusuonka River, Siberia (M. Fedonkin).

environmental change and evolutionary innovations led to the demise of a long-lasting environment and ecosystem that most favored the building of stromatolites. This ecosystem was then supplanted by something entirely new to the world – one of shelled metazoan supremacy, associated with deep burrowing and massive bioprocessing of sea floor sediments by 545 million years ago.

The fossil record of the Precambrian in Siberia is nearly unique in that most of the rocks are carbonates – elsewhere silica-rich sediments usually dominate. Because of this the stromatolite record is long, representing nearly a billion years of cyanobacterial activity, ranging from about 1.5 billion to 545 million years ago. The oldest soft-bodied Ediacaran invertebrates in carbonate sediments occur here with almost no analogues anywhere else in the world. Other Ediacaran fossil occurrences are in silica-rich clastic sediments, likely to have been deposited in cold or cool waters, unlike most carbonates, which are usually laid down in warmer waters. Carbonate rocks are also significant in that they can be dissolved with acids that delicately etch fossils, such as minute phosphatized embryos and eggs of metazoans, and preserve fine details that are not recorded by other types of fossilization. One should point out, however, that there is some evidence to support some carbonates forming in cooler waters. The Egan Formation in the Kimberley region of Australia has interbedded carbonate and diamictite. The stromatolites in these carbonates grow encrusted on diamictite clasts within a diamictite matrix. These peculiar carbonates appear to be laid down about the same time as carbonates that immediately underlie the Ediacara fauna-bearing horizons in Australia, some of warmer water origin. All contain the same stromatolite, *Tungussia julia*, with an estimated age of around 560 million years, considerably younger than the main Marinoan glaciation so well known in Australia.

In spite of the difficult access and high cost of expeditions, Siberia is certainly one of the important keys to understanding early metazoan evolution, particularly the initial colonization of warm carbonate seas by metazoans.

Fedonkin returned to the Olenek region three times. The last trip, in 2000, changed his attitude about just how biodiverse the fauna actually was. On this trip, work was carried out in the valley of the Anabyl River, a tributary of the Khorbusuonka. He returned to a locale he had visited more than 20 years before, where he had found absolutely nothing. Despite that, something intuitive drove him back there, even though nothing had been found before. Just getting there was not easy – it entailed a 12 kilometer walk from base camp. What was it that drew him here? It was impossible to know.

Upon arriving and finding nothing yet again, he approached a wall of rock, sat down, made a pot of tea and had lunch – and sat thinking, taking a break before returning empty-handed, 12 kilometers to camp. When he got up to begin the return trek and stepped slowly over the debris pile that lay at the base of the cliff, suddenly he saw fossils everywhere! Was it the angle of the light or the quality of the light? Was it experience he had gained on numerous trips to the White Sea or the spectacular finds in the Ukraine or even Spain? Who knows? But the fossils were there. There was even a new form that could be seen from a 100 meters distant – a worm-like, tubular fossil that stood out in dark color. Fossils here were actually three-dimensional in outcrop, and even some organic material was preserved. They appeared as black shadows on a light gray limestone – algae, tubular fossils, segmented animals – preserved in the same fashion as those in the famous Burgess Shale in the high mountains of North America, Middle Cambrian in age.

Everything worth collecting was gathered until very late in the night – at least it was summer with the long stretch of sunlight. With his best specimens – more than 25 kg – Fedonkin managed to tramp back into camp around midnight. His crew was still awake and rather excited by what the local Ded Moroz (a Russian name for Father Christmas) brought back in his knapsack!

With the new find, the whole crew moved camp the next morning – but this was no small task. The move had to be made across tundra – and walking across soaking wet tundra is not like taking a Sunday walk in the park. It is more akin to walking across a sponge, even on the gentle slopes. And in order to set up their camp on this seemingly endless expanse of soaking wet tundra, the intrepid "campers" had to lay down plastic, then plywood on top of that and then in sleeping areas one or two foam mattresses – to deal with both the wet and the cold of this place.

Camp was set up inside a big canvas tent – a stove set up to allow heating and drying of the ever-wet clothes and sleeping gear. The fire had to be kept alight most of the time for both cooking and drying, and food intake had to be high calorie to deal with the constant cold – just to maintain a reasonable body temperature. Wood was gathered from the south-facing, more protected slopes of the hills as well as the fossil "wood" preserved in the peat that is abundant in this far north region. The team was surprised to find trees up to 40 cm across preserved in the peat, and these were easy to split along the annual rings and burn as you would that from living trees – only the smell was different from that of fresh wood. The wood preserved in the peat was left over from the time of the last glaciation in the area when vast forests once grew here – now, if lucky, the deciduous gymnosperms that remain alive reach up to 5-6 meters in height, only shadows of their predecessors which thrived when great glaciers glided south from the North Pole, leaving corridors where forests maintained a foothold.

Isolation – Rewards and Challenges

Camp was quite isolated and to get into this region was moderately dangerous. The way out, as was the way in, was via Russian military helicopter – many of these aging machines sit on tarmacs around the country today and to get one to fly usually requires salvaging parts from others that themselves are not in commission. As alluded to above, it is better not to think about using them at the moment – especially when airborne! But once the crew did arrive alive and alight from the one reliable helicopter they had the fortune to ride in, the camp was supposed to have radio contact with the outside world. The first camp was about 400 kilometers from the nearest habitation. Radio transmissions between the field teams were not always reliable because of deep valleys that cut into the plateau. To get in touch with Tiksi or any other place one must use a reliable radio with sufficient power, employing a tall antenna and heavy-duty batteries, like a truck battery. Additionally someone had to be hired in the nearest town who had a radio receiver to receive incoming calls and who would be there when they arrived – not always possible! Recent satellite phones delightfully resolve most of these problems – but not so on all of the expeditions to Siberia up to and including the 2002 expedition.

Such isolation was not without its rewards. Humans had seldom, if ever, frequented these parts of the world, and wildlife was abundant. As an example, one day the countryside seemed to be covered at a distance with white moss and the shadows of the clouds moving lazily across the landscape. But, on closer inspection with binoculars, this moss and cloud shadows turned out to have legs. The "shadows" were a massive herd of reindeer – and this herd continued to cross the river for the entire, long summer day. The stream-banks were covered with the fur of the passing horned "moss eaters" painting the picture of a carpet merchant who had just made a fortune installing his "wall to wall stream-bank coverage."

But not only herbivores were about – there were wolves

Figure 310. Slabs of fossiliferous shales, Vendian sequence, Khatyspyt Formation, Khorbusuonka Group, Yudomian, Olenek River in Olenek Uplift of Arctic Siberia (M. Fedonkin).

as well. A few were seen each year, and for a time the crew carried guns – but as time went on, permits for firearms became difficult. One year the team had only a dog – a useful companion, not only as a friend but because she could also be sent out to bring down game and return it for the dinner pot. One night, however, a large wolf approached and the big dog's scruff went up. She quite sensibly remained entirely quiet as the wolf approached and then silently joined the crew in the tent until the wolf retreated. Her assessment of the situation was that silence would lead to less trouble than barking. And that was exactly the attitude the crew took as well.

The Olenek and the Maya River Basins

To date more than 1000 excellent specimens have been recovered from the Olenek region of Siberia, from along the Olenek River, from the Khatyspyt Formation, more than 200 meters thick, exposed in the cliffs of the Khorbusuonka River.

Figure 311. Aldanella crassa, NW Anabar region, Kotuy River, Medvezhja Formation, Early Cambrian of Siberia (P. Yu. Parkhaev and A. Rozanov).

Most of the fossils have been recovered from the upper layers. They include metazoans and macroscopic algae, which thrived in a small enclosed, shallow sea – about 30 kilometers wide – encircled by stromatolite-dominated reefs (Yakshin, 1987). This lagoonal sea existed for a few million years and its waters were saturated with carbonate. As mentioned above, the Olenek Sea was one of the few places on Earth where metazoans lived in such lime-rich, relatively warm seas and are now preserved in the laminated, bituminous limestones of the Khatyspyt Formation, laid down as the ocean transgressed onto a nearly flat continental edge. At least three distinct Ediacaran assemblages have been recognized in this succession, the most diverse coming from near the top and including such familiar forms as *Charnia, Ediacaria, Cyclomedusa, Hiemalora*-like forms, *Paliella*, palaeopascichnids and some trace fossils – *Nenoxites* – as well as such endemic forms as *Khatyspytia* and a number of yet unnamed segmented tube-like forms as well as large strap-like forms, likely alga, preserved as meter-long organic films with holdfasts. The lower part of the Khatyspyt Formation has primarily forms that resemble *Hiemalora* that appear to be rather bag-like in construction, preserved in many different, often three-dimensional modes (Vodanjuk, 1989). Perhaps this biota represents one of the first attempts of metazoans to invade such an environment. Perhaps, too, it is just the luck of the draw, and one of the only places so far that metazoans have been preserved in this sort of rock. Maybe in the future more will be found elsewhere; only time will tell.

The rock sequence in this restricted sea shows clearly that metazoans were around there for a while and then disappeared. The youngest and topmost sediments of the Vendian in the Olenek region lack these multicelled forms. The Turkut Formation, overlying and sometimes interfingering

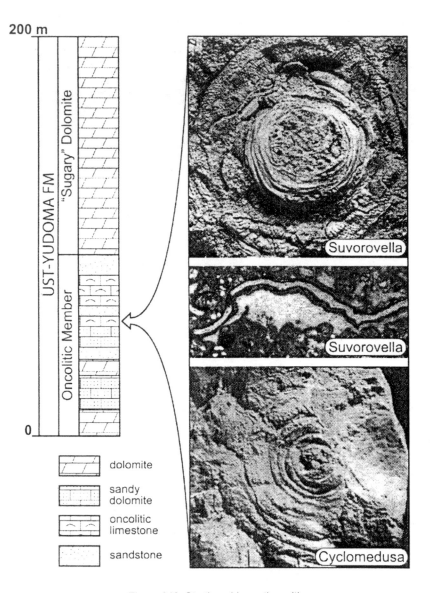

dolomite

sandy dolomite

oncolitic limestone

sandstone

Figure 313. Stratigraphic section with occurrence of Suvorovella *and* Cyclomedusa, *Yudoma River, Siberian Platform, Russia (courtesy of D. Grazhdankin).*

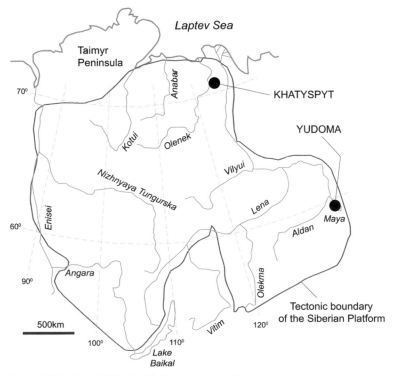

Figure 312. Map of Siberian Platform where major outcrops of late Neoproterozoic carbonates occur (modified from Grazhdankin). (D.Gelt)

Figure 314. Charnia *and* Khatyspytia *from the Khorbusuonka Group, Siberian Platform (courtesy of D. Grazhdankin).*

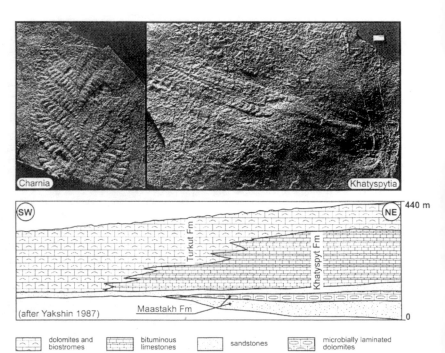

dolomites and biostromes

bituminous limestones

sandstones

microbially laminated dolomites

with the fossiliferous Khatyspyt Formation, is dominated by reef-forming stromatolites in more than 200 meters of carbonate, where they sometimes have accumulated in reef-like structures called biostromes. Cyanobacteria that built the reef structures were definitely once again in control. Perhaps over time the varying levels of oxygen available either favored the metazoans, which needed higher levels of oxygen, or did not when oxygen levels fell, and stromatolite builders once again took over. Along with the stromatolites, remains of Small Shelly Fossils have also been found and these belong to what is called the Nemakit-Daldyn SSF (Small Shelly Fossils) assemblage. But not a single soft-bodied Ediacaran species has been found at this level.

Above the Turkut Formation stromatolite reefs lies the Kessyusa Formation, a collection of siliciclastic rocks with interdigitated lenses of carbonate. These rocks contain no Ediacaran metazoan body fossils but show a distinct increase in their biodiversity by the presence of trace fossils. Biodiversity of the Small Shelly Fossils (deposited by tiny, biomineralizing metazoans) increases as well. This formation significantly contains many volcanic ash layers, one of which has yielded a lead-uranium age of 543 +/- 0.3 million years (Bowring & Schmitz, 2003), thus giving a minimum age for the Khatyspyt metazoans. Further dating of multiple ash layers in this sequence is planned by Dimitri Grazhdankin and Nick Butterfield, and this has good potential for timing events in a region with its unique carbonate setting hosting Ediacaran metazoans.

Another region in Siberia that has yielded Ediacara fossils is to the southeast of the Olenek region, and that is the Maya/ Yudoma River Basin. These occur primarily in a unit, called the "Oncolitic Member" (Semikhatov *et al.*, 1970; Butterfield &

Grazhdankin, *pers. com.,* 2005) – a unit with both siliciclastic and carbonate rocks, which could form a fundamental link between the two major types of environments in which Neoproterozoic metazoans are found – the carbonate of the Olenek region and the more characteristic siliciclastic rocks of Australia, the White Sea and Namibia. Such familiar forms as *Cyclomedusa* and *Medusinites* are known from the Oncolitic Member as are such enigmatic, endemic forms as *Suvorovella* and *Majaella* whose relationships are not currently understood – they have been allied in the past with archaeocyathids, receptaculitids and algae, among others (Vologdin & Maslov, 1960; Sokolov, 1976). Also represented are typical Nemakit-Daldyn Small Shelly Fossils (Khomentovsky & Karlova, 2002) and microscopic acanthomorphic acritarchs that resemble Doushantuo forms from China and Pertatataka Formation forms from Australia. Dating of this unit using lead-lead analysis has yielded 553 +/- 23 million years old (Semikhatov *et al.*, 2003), which needs further refinement, but it does not rule out an Ediacaran age for these fossils. This date, however, is surprisingly young since the acritarch assemblages noted above, in both China and Australia, are about 570 to 560 million years old. The large spiny acritarchs do not seem to be present this high in the succession anywhere else in the world. Usually by this time there has been a return to the simple leiospheres (see Chap. 13). Perhaps the Siberian forms belong to different taxa from those in Australia and China or the dating is incorrect.

The current work being carried out by Dimitri Grazhdankin and Nick Butterfield in this area holds great promise for understanding the breadth of tolerances and the early evolution of metazoans.

Figure 315 (opposite). Vendian sediments, Khatyspyt Formation, Khorbusuonka Group, Yudomian, cropping out in cliffs along Olenek River in Olenek Uplift of Arctic Siberia (M. Fedonkin).

CHAPTER 9

The Urals

M. A. Fedonkin and P. Vickers-Rich

Neatly dividing Europe from Asia, the Urals stretch from the Arctic Ocean in the north to the desert steppes of Kazakhstan in the south. It was in these mountains that the English geologist Sir Frederick Murchison found rocks like no others he had seen before bearing fossils that defined a new time period in the Geological Time Scale, the Permian. But not only Permian-aged rocks occur here – much older ones are to be found. Proterozoic rocks called Riphean (after the ancient name of the Ural Mountains) are widely exposed from the polar Urals to the southern Urals. They are overlain by younger rocks in which Ediacaran fossils were first discovered in the Middle Urals in 1972 by Yuri Becker. These are being investigated meticulously at present with renewed enthusiasm, which undoubtedly will yield new discoveries in the future.

Land between Europe and Asia

The much older rocks beneath Murchison's Permian were first studied in some detail by Yuri Rafailovich Becker. Becker was based in St Petersburg (VSEGEI - Vserossiskii Geologicheskii Institut, All-Russian Geological Research Institute) and began his fieldwork on the western slope of the Central Urals in 1972, mapping Riphean-aged rocks. Although concentrating on the geological succession there, he discovered a series of circular body fossils in the Sylvitsa Group, specifically the Cherny Kamen (Chenokamen of Grazhdankin *et al.*, 2005) Formation along the Kos'va River Valley (Becker, 1977). He also found more Ediacaran fossils about 100 km distant from his first finds, on the Koiva River in the Ust-Sylvitsa sediments – also in the Sylvitsa Group, and lying above the Cherny Kamen rocks. The fossils found on the Kos'va River differed from those he collected on the Koiva, but in both locales, what was most outstanding was the three-dimensional nature of the metazoan body fossils. Becker also found rather large fossils, but noted the rarity of tracks and trails left by organisms. This was very different from the situation later discovered in the southern Urals, where the metazoans were mostly flattened and trace fossils abundant.

After Becker's initial discoveries, the local geologists of the Uralian Mapping Expedition – V. G. Varganov, A. G. Grigor'ev, N. I. Tristan and V. I. Krivosheyev, a young geologist

Figure 316. Russian geologists examining the Vendian sequence in the Central Urals (M. Fedonkin).

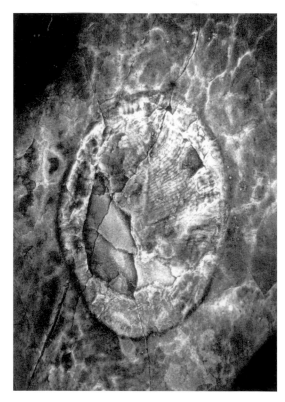

Figure 317.
Dickinsonia sp.,
Vendian sequence,
Ural Mountains,
Russia
(M. Fedonkin).

Opposite page: The wooded mountains of the Urals, the divide between Europe and Asia. Outcrops of Ediacaran-aged sediments occur in river canyons (M. Fedonkin).

169

Figure 318. Organic matter, Vendian sequence, Ural Mountains, Russia (M. Fedonkin).

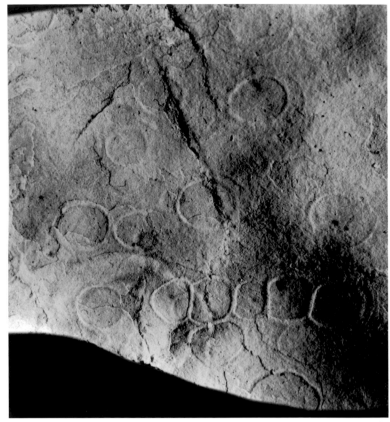

Figure 319. Similar to Beltanelloides sorichevae, Cherny Kamen Formation, Sylvitsa Series, Ural Mountains, Russia (M. Fedonkin).

from the Urals Geological Survey "Uralgeologia," followed up and managed to increase the diversity of metazoan body fossils to over 200 Ediacaran fossil imprints as well as make a large collection of trace fossils. Victor Krivosheyev, in particular, made a major discovery – he recovered biodiverse and abundant Ediacaran fossils from sediments that were exposed along the Sylvitsa River, a completely new series of locales. At this time, and for years, however, Krivosheyev, Vagranov and their colleagues considered Serebryanka and Sylvitsa Series to be early-middle Paleozoic – Silurian or Devonian (Maslov et al., 1997). They collected over a number of field seasons and brought material to the Paleontological Institute in Moscow to identify it. The collection was photographed and described. After these discoveries, Mikhail Fedonkin visited Sverdlovsk (now Ekaterinburg) and continued studying material there jointly with Victor Krivosheyev.

Preservation of the fossil material was exquisite, in three dimensions, and many forms were familiar: *Beltanelloides*, *Dickinsonia*, *Protodipleurosoma*, *Eoporpita*, *Cyclomedusa*, *Neonerites*, *Palaeopascichnus* – the list went on. A joint paper on this fossil assemblage was drafted, but one obstacle delayed the official description of this important material. Specimens must be housed in an institutional collection and given a unique number of some sort before a scientific journal will accept a submitted manuscript. The fossils at this time were part of a private collection, not housed in an official institution – and so publication was impossible then.

Another difficulty in the publication of the material was that Krivosheyev and his senior colleagues thought that the age of the fossiliferous rocks was, in fact, Devonian (410-354 million years old) and not Precambrian. But the fossils were so similar to the Ediacara-type fossils, which are time markers for the Vendian sequences then known from the White Sea region, that Fedonkin could not agree with this age. Such Ediacaran forms had never been found in younger rocks. The trace fossils found in the same area seemed, however, to be quite consistent with a Devonian age.

But still, dating of the metazoan body fossils as Paleozoic seemed to hinge on the siliciclastic nature of the rocks from which Victor Krivosheyev had collected both the body fossils and the trace fossils. The environment of deposition that had preserved both the traces and the metazoan body fossils was similar – they were preserved in the same kind of rocks. But were they of the same age? What did the *Dickinsonia* and

other body fossils have to do with the trace fossils? They were not found on the same slabs or in the exact same beds. When both the traces and the body fossils were first found, the regional geology of this area of the Urals had not been clearly worked out. There were numerous faults and folds in the stack of rocks, making it difficult to determine the position in that sequence where the fossils fit. Fossil occurrences always need to be placed in some sort of vertical order to determine which came first, which is younger and which is older – or if they are indeed the same age. Even today it is not clear if the trace fossils and metazoan body fossils actually occur at the same level, or if one is younger or older than the other. By 2002, Krivosheyev's collections from the Central Urals were deposited in a museum in Ekaterinburg, the specimens had collection numbers, and the time was right for further study of the material. Some of the material was even placed on display in the Urals Geological Museum in Ekaterinburg.

Fedonkin photographed the Urals metazoans and passed the information on to Dmitri Grazhdankin, a colleague in Fedonkin's Laboratory of Precambrian Organisms at the Paleontological Institute in Moscow. Picking up the torch, Grazhdankin worked in the Urals in collaboration with local geologists A. V. Maslov, M. T. Krupenin and S. V. Kolotov from the Institute of Geology and Geophysics, Urals Branch of the Russian Academy of Sciences, from 2002 to 2003 and continued this work in 2005 with a longer planned program and grant support in cooperation with Nick Butterfield. The field team has been exploring Vendian outcrops in the valley of the Sylvitsa River and currently is examining in detail a new locale on the Belaya River to the south of Ekaterinburg discovered recently by Yuri Polenov and Victor Krivosheyev, which has yielded a variety of discoids, all on the western side of the Central Urals.

Figure 320. Cyclomedusa *cf.* davidi, *Cherny Kamen Formation, Sylvitsa Series, Ural Mountains, Russia (M. Fedonkin).*

Figure 321. Eoporpita *sp., Cherny Kamen Formation, Sylvitsa Series, Ural Mountains, Russia (M. Fedonkin).*

The Central Urals Fauna and Its Environmental Setting

The finds of Ediacaran fossils in the Urals are of major scientific interest. The Urals are far from the White Sea, and the faunas there are confined to siliciclastic sediments. The Urals biota is preserved in the same sorts of sediments – but the distance between the productive locales gave the possibility of studying the impact of geography on faunal composition. Both areas boast layers of volcanic ash and so absolute dating is feasible – and thus because of the precise timing it was possible to compare events from the White Sea and the Urals.

The Cherny Kamen (Chernokamen) Formation (in Grazhdankin *et al.*, 2005) comprising shallow marine siltstone and fine-grained sandstone exposed along the Kos'va River produced a variety of impressions that were called medusoids at the time of discovery. A number of species of a form called *Tirasiana*, reported by Palij (1976) from the Vendian of the Ukraine, were described by Yuri Becker in his summary of the Urals metazoans in Boris Sokolov's *The Vendian System* (1985). Becker had shown the material to Martin Glaessner by 1985 and Glaessner had recognized a marked similarity between these forms and the "medusoids" from the Flinders Ranges in South Australia. With further collection and study most paleobiologists now assign these sorts of "body fossils" to one species, *Aspidella terranovica*, thought to be related to living soft corals. These rounded discs are likely to have been the base, or anchor, of the "feather-like" organisms with bases buried in the sea bottom and fronds gaining their nourishment from the water column above.

And, besides these forms, others are known. Grazhdankin and his colleagues (2005) have noted three distinct fossil assemblages in the Cherny Kamen (Chernokamen) Formation and just as in the White Sea sediments these assemblages seem tied to certain lithologies.

Grazhdankin's recent fieldwork has not only yielded more fossils but has located a number of volcanic ashes associated with fossil-bearing rocks, in particular near the base of the Cherny Kamen Formation. For the first time in this region, there is the possibility of absolute dating. For the first time, too, metazoans have been discovered in the Perevalok and Staropechny formations. This success is being pursued further by Grazhdankin, in particular, in the vicinity of the Shirokovskoye Reservoir and along the Serebryanka and Mezhevaya Utka rivers (Perm District), where the siliciclastic Vendian deposits are extremely well exposed.

Figure 323. Ediacaria *sp.*, *Cherny Kamen Formation, Sylvitsa Series, Ural Mountains, Russia (M. Fedonkin).*

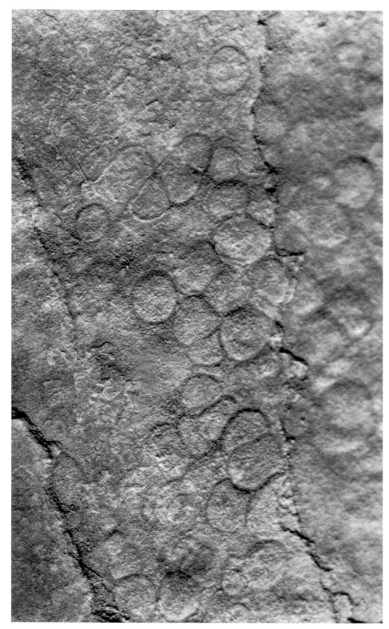

Figure 322. Beltanelloides sorichevae, *Cherny Kamen Formation, Sylvitsa Series, Ural Mountains, Russia (M. Fedonkin).*

One assemblage is dominated by the garlic-shaped *Inaria* with occasional concentrations of *Beltanelloides* – hosted by finely layered shale with lenses of sandstone with erosional bases. As in the White Sea this seems to represent deeper water sedimentation that is affected by occasional downslope sand avalanches, which form the channel sands.

A second assemblage is the most diverse and includes the discoidal forms *Cyclomedusa* and *Ediacaria*, palaeopascichnids, *Eoporpita*, *Charniodiscus*, *Vaizitsinia*, many with discoidal holdfasts attached, and the bilaterians *Dickinsonia* and *Yorgia*. *Palaeopascichnus* is an intriguing form – occurring widely across the globe. It has been interpreted in many ways – as a trace left behind by an organism moving across the sea floor, but some paleobiologists now suggest that it could be a series of collapsed tubes or beads that might be related to xenophyophores. Xenophyophores today are abundant on deep sea floors and are gigantic single-celled protozoans – reaching up to 25 centimeters across and are enclosed

by a branching tube system. Their tubes consist of particles derived from surrounding sediment and fine debris (fragments of skeletal material) glued by a polysaccharide mucus. However, this interpretation is yet problematic and just where the xenophyophores sit on the eukaryotic family tree is yet controversial. This second assemblage is preserved in a series of alternating shales, siltstones and sandstones with typical "elephant-skin" textures indicating the presence of algal or bacterial mats, so typical of some units in the White Sea and Australia. One more assemblage, similar to the first, hosts only one kind of fossil, *Nemiana*, found in channel sands and preserved in life position and in three dimensions.

The overlying Ust-Sylvitsa Formation, exposed along the Koiva River, produced another type of "body fossil." Yuri Becker noted this form in his early collections and called it *Arumberia banksii*, similar to a form known from Central Australia and southern Africa. This fossil, when first discovered, was assigned to the cnidarians – which today includes the living anemones and corals. Jim Gehling and others now think that these may be features produced by swift currents acting on a sea bottom stabilized by algal or bacterial mats. So, *Arumberia* may not be a metazoan at all but a sedimentary structure that is encouraged by the microbiological content of the sea bottom sediments. Sedimentary structures of a great variety are known in the Ust Sylvitsa – raindrop prints, desiccation cracks and a variety of ripple marks – many of these again probably nurtured in their preservation by the microbial mats that held the sediments and their impressions together more firmly than abiotic deposits.

The Southern Urals

In 1986-1987, Yuri Becker led an expedition to some of the tributaries along the Belaya River on the western side of the southern Urals. This group located more Ediacaran material, especially at the Ryauzyak locality on the Ryauzyak River (Becker & Kishka, 1989; Becker, 1992). Fossils were recovered from the Basa Formation, a part of the Asha Series,

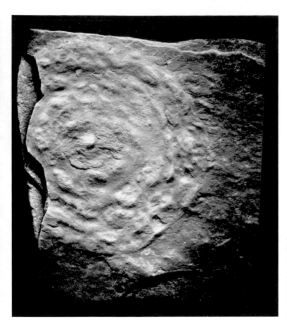

Figure 324.
Mawsonites sp.,
Cherny Kamen
Formation, Sylvitsa
Series, Ural
Mountains, Russia
(M. Fedonkin).

and in some ways these were similar to the rounded, flat forms recovered from the Central Urals – but they were much smaller in size. Unlike the three-dimensional fossil from the Central Urals, these were flattened and not so much detail was preserved.

Forms such as *Paliella patelliformis*, *Medusites* and *Protodipleurosoma* appear to belong in the category of *Aspidella terranovica* – holdfasts for some frond-like organisms. Others, like *Pseudorhizostomites howchini*, may be either holdfasts or gas escape structures, and *Kullingia* appears to be a "scratch-circle" (Jensen *et al.*, 2002) formed when an attached structure, which likely was an anchored tubular organism, possibly a sabelliditid, was twisted by the bottom currents, leaving a circular "trace" of its motion. Trace fossils are more abundant in the Basa Formation, such as *Neonereites uniserialis* (Becker, 1985), but some once thought to be traces, for example, *Palaeopascichnus delicatus*, may indeed belong to the protozoan xenophyophores.

Effects of Environment, Geography, Climate and Time

The Ediacaran-aged rocks of the Urals hold great promise. They have good outcrop over significant areas. They show a variety of different environments that host distinct assemblages of Ediacaran fossils, with many similarities to rocks and faunas of the White Sea though they are separated by more than 1200 kilometers. There are volcanic ashes that can be dated, just as in the White Sea, and both areas have deep bore data that can be used to tie the areas together not only with rocks, but in the future with geochemical, microfossil and biostratigraphic data and, most important, radiometric dates. In addition and unlike in the White Sea region so far, the base of the section of the Sylvitsa Group, the Staropechny Formation, consists of glacially deposited sediments, boulders and cobbles making up a diamictite that filled valleys in an glacially incised landscape. Although these sediments do not yet show cap dolomites that could specifically signal the end of glaciation, the sequence in the Urals may, with further intense investigation by a variety of specialists hold some keys to understanding how changing climate, biogeography and environment affected early metazoan evolution. This was just prior to the great biodiversity event at the Precambrian-Cambrian boundary, also preserved in the Urals.

Figure 325 Protodipleurosoma paulus, *Cherny Kamen*
Formation, Sylvitsa Series, Ural Mountains, Russia
(M. Fedonkin).

CHAPTER 10

The Canadian Cordillera

Guy M. Narbonne

Supercontinent Rodinia began to break apart 800 hundred million years ago, creating new shorelines, shelves, and even oceans. Nowhere is this more evident than in the mountains of western Canada, where the Windermere Supergroup records the breakup of Rodinia and the formation of the proto-Pacific Ocean. Ediacarans flourished in the shallow seas, deep slopes and stromatolite reefs that flanked the western margin of ancestral North America. Ediacaran fossils in the Mackenzie Mountains of the Northwest Territories, in the Wernecke Mountains of the Yukon and in the southern Rocky Mountains of British Columbia lived in very different environments. Collectively these fossils tell us about some of the factors that control Ediacaran fossil assemblages worldwide.

A thick succession of late Proterozoic sedimentary rocks, the Windermere Supergroup, crops out discontinuously along the length of western North America from Sonora in Mexico to the Canada-Alaska border. The Windermere sedimentary succession begins with Neoproterozoic glacial tillites and ends at the sub-Cambrian unconformity. Generations of researchers since the time of Walcott have regarded this succession as one of the best prospects for finding Precambrian animal remains anywhere in North America. However, in contrast with Newfoundland, where Ediacara fossils have been known since the late 19th century (Chap. 3), no one had been able to find these fossils anywhere in the Windermere succession until recently.

In the early 1980's, Hans Hofmann and his colleagues dramatically extended the known geographic range of the Ediacara biota by reporting new localities in three widely spaced localities of western Canada: the sub-Arctic Mackenzie Mountains in the western Northwest Territories (Hofmann, 1981); in the Wernecke Mountains 250 km to the west in the Yukon Territory (Hofmann, Fritz & Narbonne, 1983); and in the peaks of the Rocky Mountains in the southern Canadian Cordillera (Hofmann, Mountjoy & Teitz, 1985). The Ediacaran assemblages from these three areas overlap in age, but occur in markedly different lithologies and settings ranging from sediments laid

Opposite page: Mackenzie Mountains Camp of Guy Narbonne and Canadian colleagues called the "Valley of the Five Glaciers." Quite a contrast with the setting of other Ediacara sites, but the fossils recovered from the surrounding mountains are uncannily similar to those from the White Sea of Russia and the deserts of Australia and Namibia (G. Narbonne).

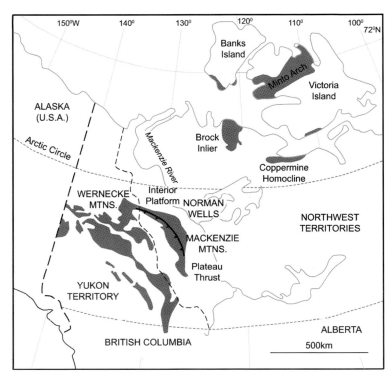

Figure 326. Distribution of Neoproterozoic strata in NW Canada. Ediacan strata and fossils occur in the Wernecke and Mackenzie mountains (from Narbonne & Aitken, 1990, 1995) (modified by D. Gelt).

down on shallow shelves to those deposited on deep slopes. This makes the Windermere Supergroup one of, if not the best, succession in the world to study firsthand environmental factors that control the distribution of Ediacara fossils.

The Mackenzie Mountains

One of the thickest, most geologically significant but unfortunately least accessible of any Ediacaran fossil locality lies in the Mackenzie Mountains. This sequence represents the northwards extension of the Rocky Mountains into the western Northwest Territories of Canada. The Neoproterozoic sediments are more than 10 kilometers thick, and exposure of these strata is nearly 100% in this mountainous, sub-Arctic region. A review paper published by Narbonne and Aitken in 1995 remains the best overview of this remarkable succession.

Neoproterozoic strata in the Mackenzie Mountains consist of two supergroups of differing style. The older, Mackenzie Mountains Supergroup, is a 4 to 6 kilometer-thick succession of carbonates and mature sandstones deposited in the shallow seas that formed on supercontinent Rodinia 1000

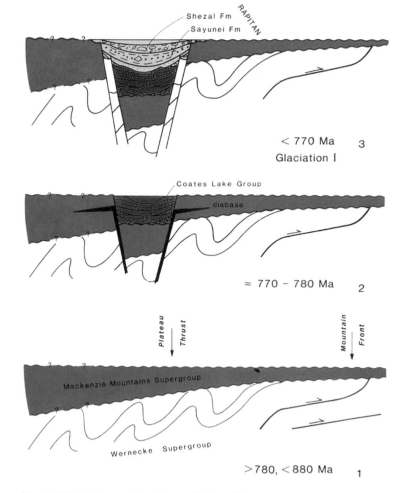

Figure 327. Tectono-sedimentary evolution of the Neoproterozoic of NW Canada. Part 1: Cratonic sediments and rifting. 1, deposition of the Mackenzie Mountains Supergroup. 2, rifting during the breakup of Rodinia and early stages of rift fill. 3, final fill of rift grabens with Raptian (Sturtian) glacial deposits (J. D. Aitken).

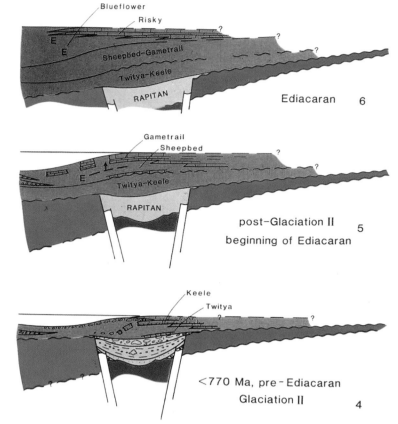

Figure 328. Tectono-sedimentary evolution of the Neoproterozoic of NW Canada. Part 2: Deep-water sedimentation on a passive margin. "E" marks the position of Ediacara fossils. 4, deposition of the Twitya-Keele Grand Cycle and Ice Brook (Marinoan) glacial deposits. 5, deposition of the Sheepbed-Gametrail Grand Cycle (Early Ediacaran). 6, deposition of the Blueflower-Risky Grand Cycle (Late Ediacaran) (J. D. Aitken).

to 780 million years ago. The Little Dal Group, the topmost unit in the Mackenzie Mountains Supergroup, is best known for two remarkable occurrences of pre-Ediacaran fossils. In 1979, Hans Hofmann and Jim Aitken reported carbonaceous megafossils, including the discoid fossil *Chuaria* and the newly named sausage-like fossil *Tawuia*, which probably represent compressed and carbonized algae. Subsequent work by Hofmann (1994) demonstrated a wide diversity of taxa among these carbonaceous compressions, an important addition to our knowledge of pre-Ediacaran eukaryotes.

Equally important, the Little Dal Group and its equivalents along the shores of the Arctic Ocean and on Victoria Island in the Canadian Arctic Archipelago contain a spectrum of early Neoproterozoic microbial reefs, ranging across an environmental spectrum from near shoreline, through carbonate ramp to basinal settings (Aitken, 1988; Batten *et al.*, 2004; Narbonne *et al.*, 2001; Turner *et al.*, 1993, 1997, 2000). The most spectacular deposits are basinal reefs up to 3 kilometers in diameter and 500 meters high, which represent the first reefs in Earth history composed of calcified microbes (calcimicrobes). Aprons of boulder-sized blocks, which clearly tumbled off the reefs during drops in sea level, attest to the fact that these microbial reefs were rigid pinnacles that stood several hundred meters above the surrounding sea floor. Calcimicrobes may be insignificant in size, but were major reef

builders throughout the Paleozoic and younger eras. It now appears likely that the calcimicrobial Little Dal reefs represent a template that, with the addition of skeletal animals in the Cambrian, evolved into a Phanerozoic reef ecosystem.

The breakup of Rodinia around 800 million years ago put an end to this Early Neoproterozoic pattern of wide, shallow, carbonate seas in this part of the world. The overlying Windermere Supergroup 5-7 kilometers thick and reflects rifting of Rodinia and subsequent continental drifting that produced the proto-Pacific Ocean (Ross, 1991). Initial rift sediments consist of fanglomerates, redbeds and evaporites, which are typical of rift deposits of any age from nearly any place in the world. These are overlain by glacial tillites of the Rapitan Group (Sturtian, probably dating at about 740 million years old) and the Ice Brook Formation (Marinoan, probably about 635 million years old). These were probably the two greatest glaciations Earth has ever known, with reliable evidence from several continents of ice sheets marching on the Equator at sea level. Both of the tillites contain cap carbonates, which are remarkably similar to their probable stratigraphic equivalents in Australia and Namibia (James *et al.*, 2001). The intervening Twitya and Keele formations are about 2 kilometers thick. The sporadic presence in these formations of relic crystals of the cold-water carbonate mineral ikaite (glendonite) implies that temperatures were near

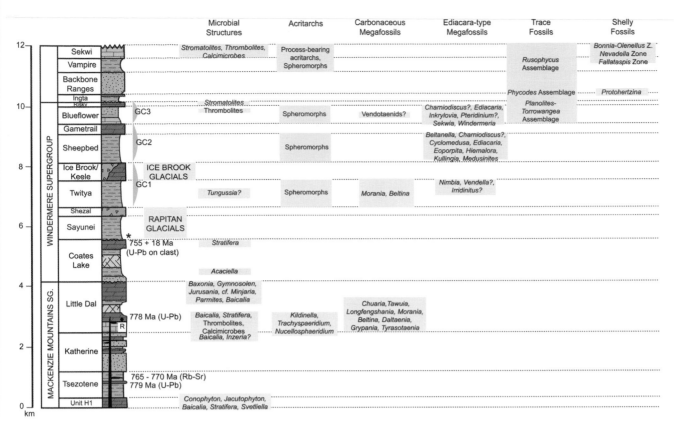

Figure 329. Stratigraphic occurrence of fossil taxa in the Neoproterozoic and Early Cambrian of NW Canada (modified from Narbonne & Aitken, 1995; redrafted by D. Gelt).

freezing much of the time from the beginning of the Sturtian until the end of the Marinoan glaciation (James *et al.*, 2005), a period of more than 100 million years. This cold-water period, aptly named the "Cryogenian," is important, since it was the immediate precursor to the Ediacaran radiation.

Throughout the Canadian Cordillera, sediments of the Rapitan Group fill rifts that reflect the initial breakup of Rodinia. The nature of the block that rifted away from the western margin of ancestral North America is uncertain, but various geologists have proposed that it was Australia, Siberia or China. Subsequent tectonic drifting moved these blocks progressively farther apart to produce the proto-Pacific Ocean. Post-Rapitan deposits in western North America were laid down on a tectonically inactive (passive) margin facing into this newly formed ocean, and the different parts of this margin are preserved in different areas of western North America (Ross, 1991). Ediacara fossils are common in the marine marginal sediments of the Windermere Group of western Canada, and each of the major fossil localities represents a different environment with a different preserved assemblage.

The first Ediacara body fossils and trace fossils discovered in western North America were found in the Mackenzie Mountains during a field excursion by the International Union of Geological Sciences (IUGS) Working Group on the Precambrian-Cambrian boundary (Fritz, 1980). The trip was led by Jim Aitken, a well-known Cambrian stratigrapher, who was mapping the area for the Geological Survey of Canada. Hans Hofmann (a professor at the University of Montreal and the best-known Precambrian paleontologist in Canada) was also on this trip, and he discovered and collected numerous specimens of Ediacara megafossils and trace fossils from a site near Sekwi Brook in strata that are now known as the Blueflower Formation. Other participants on this field trip (including a young Mikhail Fedonkin) also collected specimens and donated them to Hofmann. In 1981, Hofmann described these fossils in an article published in *Lethaia*. This was the first report of

Ediacara fossils from anywhere in western North America, and Hofmann's new animal, *Sekwia*, was the first Ediacara fossil described from North America since Billings had named *Aspidella* in Newfoundland more than a century earlier.

Things quieted down for a while, until Jim Aitken invited Guy Narbonne (a paleontologist, who had co-discovered the Ediacara biota in the Wernecke Mountains in 1982 while working as a post-doc with Hans Hofmann and who was now a relatively new professor at Queen's University) to work with him while Jim mapped some apparently unfossiliferous strata half a kilometer stratigraphically below Hofmann's Ediacaran discovery. The result was the discovery of an even more diverse assemblage of Ediacara body fossils, this time from the Sheepbed Formation (Narbonne & Aitken, 1990). A joint study by all three workers in 1990 extended the range of Ediacara-style fossils even farther into the Twitya Formation, a stratigraphic range of 2.5 kilometers for the Windermere Ediacara biota, second only to Newfoundland in thickness.

Beds of volcanic ash have not been yet found in the Neoproterozoic of NW Canada, so the section and its fossils cannot be dated radiometrically. However, geochemical analyses by Jay Kaufman and Andrew Knoll at Harvard University demonstrated that this succession has one of the most complete and well-preserved carbon- and strontium-isotope signature records of any Neoproterozoic succession worldwide. Using this geochemical information allows correlation of the Ediacara fossil assemblages in NW Canada with other assemblages worldwide.

Sedimentological studies demonstrate that Ediacara organisms from the Mackenzie Mountains lived on deep-water slopes (Aitken, 1989; Dalrymple & Narbonne, 1996; MacNaughton *et al.,* 2000). Evidence of a deep-water slope documented by these authors includes thick successions of stacked, deep-sea-type turbidites with paleocurrrents that are directed downslope (to the southwest); abundant slump structures, slide masses and debris flows of boulders

Figure 330. Little Dal Reefs. Two calcimicrobial reefs in the Little Dal Formation (about 850 million years old). The reefs contain a resistant core (shown by the castellated peaks) surrounded by an apron of house-sized boulders that tumbled off the reef as it grew. The reef is 200 meters high (G. Narbonne).

Figure 332. Mackenzie glacial features. Exposure of the Rapitan tillite (Sturtian, about 740 million years old), a typical Neoproterozoic glacial deposit, showing disorganized fabric and abundant glacial striations on the largest clast (G. Narbonne). The largest block width is about 18 cm.

Figure 331. Exposure of the middle part of the Windermere Supergroup in the central Mackenzie Mountains. The white band in the lower right is the Ravensthroat cap carbonate on top of the Ice Brook glacial tillite, which marks the base of the Ediacaran System (N. P. James).

Figure 333. Jim Aitken examining a debris-flow conglomerate at Sekwi Brook in the Mackenzie Mountains in June 1988. The light-colored clasts are boulders of shallow-water Ediacaran thrombolites that have tumbled downslope into this deep water deposit (G. Narbonne).

reflecting downhill transport on an unstable slope; contourite paleocurrents which are directed to the northwest along the slope contours and the complete absence of any indications of storms, waves or light affecting the sea bottom. A deliberate attempt was made to find Ediacarans that had been transported downslope on slide masses or within turbidites in order to determine the nature of the contemporaneous shallow-water biotas. None could be found. All Ediacaran fossils appear to be completely in their original life positions on the slope. Pustular textures ("old elephant skin") and carbonaceous or pyritic coatings on soles are similar to those found in the Flinders Ranges and the White Sea (Chaps. 5 and 6), and imply that the sea bottom was coated with a microbial mat. As with the deep-

water environments in Newfoundland (Chap. 3), it seems most likely that this mat was composed of sulfur-oxidizing or heterotrophic bacteria.

Three distinct assemblages of Ediacara fossils are present in the Mackenzie Mountains. The oldest consists of structures commonly called the "Twitya discs" that occur in beds of the upper part of the Twitya Formation below the Ice Brook Tillite and that were reported by Hofmann, Narbonne and Aitken (1990). Twitya discs are the only pre-glacial tillite Ediacara-type fossils known anywhere in the world, and their discovery created quite a stir. Twitya discs consist of centimeter-scale discs and rings preserved in positive relief on the bases of turbidite beds. Three genera were tentatively recognized: *Nimbia*, consisting of simple

Figure 334. William Fritz and Guy Narbonne enjoying lunch in the Mackenzie Mountains, July 1982 (G. Narbonne).

Figure 335. Hans Hofmann on section in the Wernecke Mountains, July 1984 (G. Narbonne).

rings 10-35 mm in diameter; *Vendella*, represented by low hemispheres, 4-14 mm in diameter; and a single specimen attributed to *Irridinitus?* about 25 mm in diameter with radial, concentric ridges and a central tubercle. Despite their age, their organic nature has been accepted by most workers, but it is difficult to determine the affinities of such simple structures. They are preserved in a way that is very like that of other Ediacarans and is quite dissimilar to almost anything else. Beyond that little more can be said.

As with other localities hosting the Ediacara biota worldwide, large and diverse fossils do not appear until after the meltdown of the last of the Neoproterozoic glaciers. Narbonne and Aitken (1990) discovered abundant fossils in the post-glacial Sheepbed Formation, which contains abundant discs and discs with partial stems, including the forms such as *Beltanella*, *Cyclomedusa* and *Medusinites*, all now regarded as different preservational morphs of *Aspidella* (Gehling *et al.*, 2000; see Chap. 3). The tentaculate discs *Eoporpita* and *Hiemalora* occur rarely, and like *Aspidella* may represent the bases of fronds. A single rangeomorph frond is known, but awaits formal description. Burrows are conspicuously absent, perhaps implying that bilaterians had not yet evolved, or had not yet become a part of deep-sea ecosystems or perhaps were simply not yet capable of burrowing.

Sedimentological analyses by Dalrymple and Narbonne (1996) demonstrated that the fossil-bearing strata were deposited on a continental slope more than a kilometer deep. Black shale and pyrite are common throughout most of the Sheepbed Formation, implying low oxygen levels at the time of deposition. Importantly, all of the Ediacaran fossils were found in lighter-colored beds close to sedimentary deposits called "contourites" (Stow & Lovell, 1979) – fine sandstones and siltstones that were formed by underwater currents flowing horizontally along the contours of the slope. These underwater currents would have carried food and oxygen into this otherwise poor-quality environment, thus facilitating colonization of the deep-sea floor by these Ediacarans. Prior to the Ediacaran discoveries at Sekwi Brook in the Mackenzie Mountains, many paleoecologists had thought that Ediacaran deep-water environments were anoxic and that significant

Figure 336. Abundant rings of Nimbia *on the sole of a turbidite in the Twitya Formation in the Mackenzie Mountains (Geological Survey of Canada, GSC 98293) (G. Narbonne). Diameter of spheres average about 2 cm.*

colonization of the deep-sea floor did not begin until the Ordovician. This Ediacaran discovery, and subsequent work on the Cambrian trace fossil record worldwide (Seilacher, Buatois & Mangano, 2005), has confirmed a much earlier origin for deep-sea animal communities.

The Blueflower Formation contains the youngest Ediacara fossils in NW Canada (Hofmann, 1981; Aitken, 1989; Narbonne & Aitken, 1990). Simple bilaterian burrows, such as *Helminthoidichnites,* are abundant, and include some meandering forms, such as *Helminthoida*. These may represent a precursor to the complexly meandering traces characterizing burrows of modern deep-sea floors. Discs include *Ediacaria* (a taphomorph of *Aspidella*) plus numerous specimens of *Sekwia*. A frond tentatively identified as *Inkrylovia* and another identified as *Pteridinium* (two taxa that Runnegar now regards as synonymous) are also present.

Figure 337. Irridinitus? *is the most complex of the Twitya discs, with an annular ring, radial striations and a central tubercle, Twitya Formation, Bluefish Creek, Mackenzie Mountains, probably about 630 million years old (Geological Survey of Canada, GSC98295) (G. Narbonne). Diameter of fossil about 2.5 cm.*

Figure 338. *The Sheepbed Formation contains abundant discs. These three small specimens of* Aspidella *have displaced each other laterally during growth (Geological Survey of Canada, GSC 95898-95900) (G. Narbonne). Width of fossil concentration about 3 cm.*

Figure 339. Sekwia excentrica, Mackenzie Mountains, *NW Canada (from Hofmann, 1981). Width of individual fossils about 1.5 cm.*

Figure 340. *The tentaculate disc* Hiemalora *is a cosmopolitan Ediacaran form that is known from numerous localities in Europe, Asia, Australia, Newfoundland and the Mackenzie Mountains, Sheepbed Formation (Geological Survey of Canada, GSC 102373) (G. Narbonne). Width of central sphere about 3 cm.*

Figure 341. *A tiny specimen of* Windermeria aitkeni *(holotype, Geological Survey of Canada, GSC 102374) from the top of the Blueflower Formation is the only dickinsoniid known from the New World, Sekwi Brook, Mackenzie Mountains (G. Narbonne). The length of body fossil about 1.8 cm.*

Sedimentological analyses confirm that most of the Blueflower Formation was deposited under deep slope conditions but also show that shoaling near the end of Blueflower deposition times brought some parts of the sea floor environment within the range of storm waves (MacNaughton *et al.*, 2000). A single specimen of *Windermeria* occurs in these shallow-water deposits at the top of the Blueflower Formation (Narbonne, 1994). *Windermeria* is a small (16 mm long), bilaterally symmetrical, segmented fossil that, in contrast with all other Windermere taxa, occurs in negative relief on the bed sole. It represents a dickinsoniid, and as such is the only dickinsoniid known from outside of Australia/Russia.

The Wernecke Mountains

The Wernecke Mountains are located 250 kilometers west of the Mackenzie Mountains along the eastern edge of the Yukon Territory. Ediacaran fossils were first reported

Figure 342. *Inkrylovia sp. (Pteridinium) Mackenzie Mountains, NW Canada (from Hofmann, 1981). Maximum width of body fossil about 7 cm.*

Figure 343. *Fronds, such as this specimen of* Pteridinium *from the Blueflower Formation, are extremely rare in NW Canada (Geological Survey of Canada, GSC68463) (G. Narbonne). Width of body fossil about 4 cm.*

Figure 344. *The simple, subhorizontal burrow* Torrowangea *shows annular constrictions suggesting that it was produced by a bilaterian worm moving through the sediment using peristalsis, Blueflower Formation (Geological Survey of Canada, GSC 95931) (G. Narbonne). Base of slab @ 8.5 cm.*

Figure 345. *The irregularly meandering burrow* Helminthoidichnites *is known from Ediacaran successions worldwide. Preservation in both positive and negative relief on surfaces implies that it was burrowing within the microbial mat, Blueflower Formation (Geological Survey of Canada, GSC 98924) (G. Narbonne). Base of slab width about 22 cm.*

from the region in a paper by Hans Hofmann, Bill Fritz and Guy Narbonne (1983). Narbonne had wanted to study Ediacara fossils for some time, and his readings suggested that Yukon contained rocks of the right type and age. His post-doc supervisor Hans Hofmann linked him up with Bill Fritz, a well-known Cambrian trilobite worker at the Geological Survey of Canada, trying to determine the position of the Precambrian-Cambrian boundary in NW Canada. They deliberately extended their studies far below the boundary, and discovered possible Ediacara fossils in three different sections in the Wernecke Mountains. Hofmann worked on these specimens, and the three of them published a short note of

their findings in *Science* in 1983. Narbonne and Hofmann returned to these sites the following year and collected literally thousands of specimens, which they published in *Palaeontology* in 1987. Australian paleontologists were especially excited by the new finds in the Yukon, which contained many of the same taxa as they had described from the Flinders Ranges. This developed into international cooperation, which continues to the present day, and has significantly benefited global understanding of the Ediacara biota.

Detailed sedimentological studies were carried out by Leanne Pyle and her colleagues (2004). The Windermere Supergroup in the Wernecke Mountains contains the same formations as in the Mackenzie Mountains. However, in contrast with its deep-water origin in the Mackenzie Mountains, the Blueflower Formation in the Wernecke Mountains contains abundant evidence of fluvial, tidal-flat and storm-dominated

Figure 346. Sekwi Brook from the air. Sekwi Brook was the discovery site of the first reported Ediacarans in western Canada, and remains one of the most important Ediacaran sites in NW Canada. The basal Cambrian boundary is located at the top of the prominent cliff-forming dolomite, and most of the underlying 1.3 kilometers of section visible in this photo contain Ediacaran fossils. Ediacaran strata exposed at Sekwi Brook are (from bottom to top) the upper Sheepbed Shale, the Gametrail Dolostone, the Blueflower Formation (sandstone and shale) and the Risky Dolostone (H. J. Hofmann).

Figure 347. Numerous specimens of Beltanelliformis from the Wernecke Mountains exhibiting both flat (Beltanelliformis-style) and hemispherical (Nemiana-style) preservation on the same surface (G. Narbonne). Diameter of larger spheres about 1.5 cm.

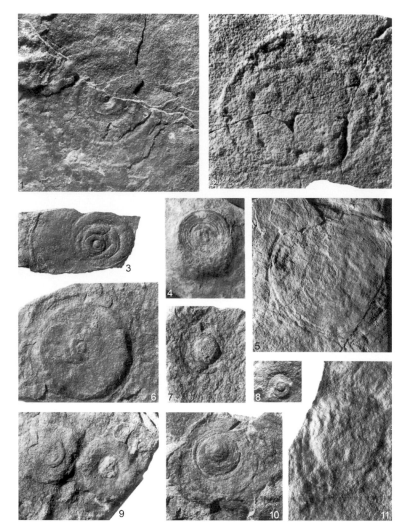

Figure 348. A variety of Ediacaran taxa from the Blueflower Formation of the Wernecke Mountains, NW Canada, as illustrated in Plate 73 of Narbonne & Hofmann, 1987 and described using their original caption. Nadalina yukonensis was a new genus and species which is not known from any other Edaicaran locality, but subsequent work has shown that many of the other forms illustrated on this plate are preservational variants of Aspidella. 1, Cyclomedusa sp. (Geological Survey of Canada 83021); 2, Nadalina yukonensis (GSC83022); 3, Cyclomedusa plana (GSC83023); 4, Spriggia wadea (GSC 83024); 5, Kullingia? sp. (GSC 83025); 6, Beltanella gilesi (GSC 83023); 7-9, Medusinites asteroides (GSC99095, 99042, 83028); 10, Spriggia annulata (GSC83030); 11, Rugoconites? sp. (GSC 83031)

shallow marine deposition (Pyle *et al.*, 2004). Ediacara fossils occur only in the shallow-marine part of this formation, commonly in association with storm turbidites and hummocky cross-stratification (HCS).

The most common fossils are discoid to hemispherical protrusions 1-3 cm in diameter that occur in profusion on the underside (the soles) of sandstone beds. Narbonne and Hofmann (1987) were able to demonstrate a complete gradation between hemispherical forms (elsewhere in the world generally referred to as *Nemiana*) and discoid impressions with concentric wrinkles (elsewhere in the world generally referred to *Beltanelliformis* or *Beltanelloides*) and

regarded the two as preservational variants of a single biological taxon, which they referred to *Beltanelliformis*, a view supported by many but not all subsequent workers. Morphological similarities with modern benthic green-algal balls are evident (Xiao *et al.*, 2002), whereas taphonomic features are perhaps more consistent with anemone-grade cnidarians (Narbonne & Hofmann, 1987). *Beltanelliformis* is probably the most common soft-bodied megafossil in shallow-water Ediacaran deposits worldwide, and occurs abundantly in sedimentary layers of the White Sea, the Ukraine, Flinders Ranges and Namibia, probably as an opportunistic species that took advantage of fluctuating environmental conditions in these near-shore environments. Larger discoid fossils are also abundant in the Wernecke Mountains and have been described under a variety of names previously defined in

Figure 349. As in the Mackenzie Mountains, fronds like this Charniodiscus are rare in the Ediacara biota of the Wernecke Mountains (G. Narbonne). Width of block base about 6 cm.

Figure 350. A large specimen of Aspidella exhibiting Ediacaria-type preservation in sandstone, Blueflower Formation, Wernecke Mountains, NW Canada (G. Narbonne). Width of fossil about 13 cm.

Australia and the White Sea (*e.g. Beltanella, Cyclomedusa, Ediacaria, Medusinites, Spriggia* and *Tirasiana*), most of which have now been synonymized with *Aspidella* (Gehling *et al.*, 2000). Pustular discs (*Rugoconites* and the new genus *Nadalina*), simple burrows (*Helminthoidichnites*), fronds (*Charniodiscus*) and carbonaceous algae (*Vendotaenia*) occur sporadically. Thrombolite reef mounds up to 15 meters high occur in the Blueflower Formation in the Wernecke Mountains (Pyle *et al.*, 2004) and as boulder-sized clasts in equivalent deep-water deposits in the Mackenzie Mountains (Aitken & Narbonne, 1989).

The Rocky Mountains

More than 1200 kilometers to the south, in the Rocky Mountains straddling the British Columbia-Alberta boundary west of Edmonton, the upper Windermere Supergroup consists of sandstones of the Miette Group underlying and flanking carbonate platforms and reefs. In 1985, Hans Hofmann, Eric Mountjoy and Martin Teitz reported the discovery of Ediacara fossils in the Miette sandstone. The fossils consisted of a single large slab containing a variety of discs including *Cyclomedusa, Irridinitus* and *Protodipleurosoma*, all of which would now be referred to *Aspidella* following the suggestion of Gehling *et al.* (2000). Subsequent reports also included simple trace fossils and a specimen of the rangeomorph *Bradgatia* that remains to be described (Hofmann *et al.*, 1985; Hofmann & Mountjoy, 2001).

The Ediacara fossil-bearing sandstone beds underlie and flank stromatolitic platforms and reefs, one of which contains calcified fossils (Hofmann & Mountjoy, 2001). The fossils occur as shell-rich pockets and lenses within stromatolite biostromes, and consist of the *Cloudina*, a tapering, calcite

tube up to 4 mm across and 5 cm long, and *Namacalathus*, a goblet-shaped calcite fossil up to nearly 2 cm in diameter. These fossils represent the oldest calcified metazoans in North America. They also provide a link to the Ediacaran succession in Namibia (Chap. 4), which also contains Ediacara fossils in sandstone facies and the calcified metazoans *Cloudina* and *Namacalathus* in associated microbial reefs.

Conclusions

The Windermere Supergroup contains a wide array of sedimentary environments, each of them characterized by subtly to markedly different assemblages of Ediacara fossils. Shallow-shelf environments of the Wernecke Mountains are characterized by an extreme abundance of the discoidal to hemispherical fossil *Beltanelliformis*, which is known from shallow-water Ediacaran settings worldwide and has become an unofficial index fossil for these deposits. Deep-water slope deposits in the Mackenzie Mountains contain discs similar to those of Australia and the White Sea, along with abundant trace fossils that give the first hints of the behavioral complexity that was to characterize the infauna of Phanerozoic deep-sea environments. Stromatolite reefs in the Rocky Mountains contain calcified metazoans (*Cloudina* and *Namacalathus*) that are indistinguishable from their famous counterparts in Namibia. All of these occur on the western margin of the Laurentian tectonic plate, implying that biogeographic separation may have been less important than environment and preservation in controlling the nature of Ediacaran fossil assemblages.

Beyond the Major Sites

P. Vickers-Rich and M. A. Fedonkin, with assistance of Xiao Shu-hai

Although the majority of specimens of early metazoans come basically from four areas on Earth, the White Sea of Russia, the Outback of Australia, the deserts of Namibia and the misty shores of Newfoundland. However, there is a scattering of locales in many parts of the world which have yielded material of importance to understanding evolutionary pathways at this early stage in the development of major animal body plans. Some of these locales are becoming increasingly important as they are further collected and studied by a growing number of Precambrian paleobiologists.

Areas where Precambrian metazoans are known besides the classic biodiverse sites include more than 25 locales: Sweden, Charnwood Forest in Leicestershire, Carmathen in Wales; the Carolina Slate Belt in the southeastern USA, the Great Basin of the western USA, China, Mexico, Argentina and Brazil and scattered possible occurrences in Norway, Poland, Iran, India and Mongolia (Runnegar & Fedonkin, 1992).

Odd Places, Odd Structures – Pseudo- or Dubiofossils

In the search for early animals, odd structures attract attention, but sometimes are not what they seem. The false dawn animal *Eozoon canadense* is one such case, a form originally hailed as one of the brightest gems in the scientific crown of the Geological Survey of Canada by Sir William

Opposite page: Neoproterozoic-Cambrian Puncoviscana Formation, at high altitude (over 15,000 feet) in the Andes, Jujuy Province, northwest Argentina (P. Vickers-Rich).

Figure 351. Puncoviscana Formation, the Andes, Jujuy Province, northwest Argentina (P. Vickers-Rich).

Dawson (Hofmann, 1971). *Eozoon* was even noted by both the great geologist Charles Lyell and Charles Darwin as being of great import. Noted in his fourth edition of the *Origin of Species*, published in 1866, Darwin was delighted that a complex fossil organism had been found in the "great void" of the Precambrian where fossils were expected but had not been found. Unfortunately, *Eozoon* turned out to be inorganic, likely formed as minerals separated from one another as rocks were heated and squeezed through metamorphism.

Another case of mistaken identity is a supposed jellyfish fossil from the Neoproterozoic Nankoweap Formation of the Grand Canyon, *Brooksella canyonensis*, originally described by R. S. Bassler in 1941. Glaessner (1969) as well as Runnegar and Fedonkin (1992), among others, noted that *Brooksella* was a pseudofossil, very likely a sedimentary structure produced by compaction of sands. Other structures, like *Rhysonetron* from the Early Proterozoic quartzite north of Lake Huron, were first thought by Hofmann to be some of the oldest metazoan remains, dated at between 2 and 2.5 billion years. Later he suggested that they could be due to compaction of the sands as "a response to varying tensional stress directions associated with the propagation of a crack within a drying mud in ripple troughs" (in Schopf & Klein,

Figure 353. Ring structure, Charnwood Forest, England (R. Jenkins). Diameter of coin about 2.5 cm.

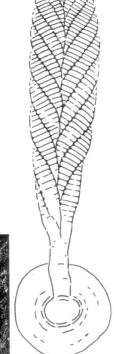

Figure 354. Charnwood Forest Ediacarans: top left, Charnia masoni; *bottom left,* Charniodiscus concentricus *and right, reconstruction of the single living organism that likely produced both fossils (from Ford, 1958).*

1992). Another misinterpreted structure is from the Ranford Formation discovered by mapping geologists in the eastern Kimberley of Australia. Preston Cloud pointed out that these were actually gypsum crystal rosettes, not jellyfish, but unsuspecting collectors and buyers of supposed Ediacaran fossils are sometimes buying these pseudofossils! The buyer, and the serious scientist, must beware.

Charnwood Forest

Roger Mason was a very lucky schoolboy in May 1957, for as he fossicked around in Charnwood Forest of Leicestershire England with three other year-11 schoolboys

Figure 352. Ring Quarry, Charnwood, England, with Trevor Ford inspecting outcrops (R. Jenkins).

(Richard Allen and Richard Balchford) from the Wyggeston School for Boys. They found something unusual. They had peddled their bicycles to a small, disused quarry on the grounds of the Charnwood Forest Golf Club at Woodhouse Eaves to go rock climbing. There was one particularly challenging slab of rocks, rising about seven meters from the ground, tilted off at 45 degrees. As Mason climbed to the top to anchor the safety rope one of the other boys called out that he had found something that looked like a fossil plant. Mason reported this find to his father, a Minister of the Great Meeting Unitarian Chapel and part-time lecturer on philosophy at Vaughan College, which was associated with Leicester University. Father drove out with Roger one evening to examine the find and was convinced that it should be shown to Trevor Ford. At first Ford was not overly enthusiastic to check out the find, but Roger's father finally convinced him and upon seeing it in the field his words were "My God, it is!" – it being a fossil and it turning out to be an Ediacaran that occurred in rocks of a recognized Precambrian age. Later searching of the Woodhouse Beds of the Maplewell Group made up of tuffaceous siltstones and agglomerates turned up a few more similar fossils, this time of two quite distinct forms. Further determined searches of different rock layers, however, failed to produce more.

Trevor Ford announced these finds at a meeting in Sheffield on the 15th of February 1958 and then published two photographs as well as his reconstruction of these frond-like forms that were firmly attached to a rounded disc in the Yorkshire Geology Society journal (Ford, 1958). He named the two different types *Charnia masoni*, of course honoring Roger Mason for his discovery, and *Charniodiscus concentricus*. At the time his paper was published, only twelve of the discs and six of the fronds were known. Those were deposited in collections of the Department of Geology at the University of Leicester.

Ford corresponded with Martin Glaessner in Australia at this time, who noted the Charnwood fossils' striking resemblance to material from the Flinders Ranges of South Australia, which Glaessner had originally reported on as Cambrian in age. Just how much the Charnwood find influenced Glaessner to recognize the Flinders material as Precambrian is controversial (B. McGowran, R. Jenkins, *pers. com.*), but it certainly was the first clear-cut association in the literature of the Ediacarans with a Precambrian age.

As their names suggest, *Charniodiscus* was based on the discs and *Charnia* was based on the plant-like fronds, such as the one found by Roger Mason. Ford, in his 1958 paper certainly noted that at least one specimen suggested that *Charniodiscus* could be the attachment with which the stalk of *Charnia* was connected. Ford also further noted some similarities between these fossils and *Rangea schneiderhohni*, which Gürich (1930) had described from the Nama Formation in Africa. He likewise noted similarity with some of the "medusae" that Reg Sprigg (1947, 1949) had found in Australia. Some who had examined the Australian material before Mason's discovery did not accept that the Charnwood structures were indeed organic remains, claiming that the discs were instead concretions. This, however, was before the frond-like fossils were found (Watts, 1947).

Putting all the information together that he had at the time, Ford's conclusion was that *Charnia masoni* should "most rationally be interpreted as an algal frond and that

Figure 355. Pseudovendia charnwoodensis, Woodhouse Beds, Outwoods, Charnwood Forest, England (from Ford, 1979).

Charniodiscus concentricus may be the basal part of the same alga" and the "only likely alternative is that they represent a primitive coelenterate of unknown affinities."

After the first reports, fossils continued to turn up over the years. Helen Boynton and Trevor Ford (1979) named a form *Pseudovendia charnwoodensis*, which they suggested was some sort of arthropod, a joint-legged animal, and placed it in the Vendomiidae after examining photographs and latex molds of Australian material and consulting with Richard Jenkins and Martin Glaessner, both at that time based in Adelaide. The Australians agreed with Boynton and Ford's idea. The family Vendomiidae itself had been allied with arthropods by Keller and Fedonkin in 1977.

Pseudovendia charnwoodensis was discovered by Helen Boynton on a "loose block in the Outwoods, near Loughborough, in the northeastern part of Charnwood Forest." The rock type it was found in was similar to the Woodhouse Beds where *Charnia masoni* and *Charniodiscus concentricus* had been found years before.

With time, metazoan impressions were found in many different parts of the Charnian Supergroup, which includes the Woodhouse Beds, site of the first discoveries. These impressions include a variety of discoidal forms, some up to 12 cm across, with frondose attachments like *Charnia masoni* and *Charniodiscus concentricus* and the "maybe"

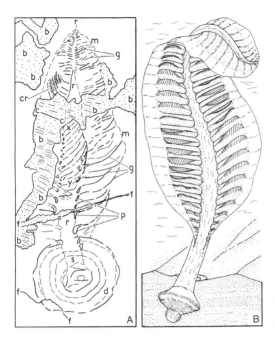

Figure 356. Reconstruction of Charniodiscus concentricus from Charnwood Forest, England (R. Jenkins).

arthropod *Pseudovendia* but in addition a large "bushlike form with multiple fronds" (Jenkins, 1992). As Richard Jenkins points out, and as more material has become known, *Charnia* has been recognized as an elongate, frondose form with its subdivisions or branches aligned in an alternate chevron pattern, clearly quite flexible. Some individuals reach lengths of over 50-60 cm, with breadths in excess of 20 cm. Jenkins noted that some of the discoidal forms resembled *Ediacaria* and suggested that those with significant annulations might represent floats of forms related to the chondrophores.

The Carmarthen Biota

Fossils from the Carmarthen area, Dyfed, southern Wales, were found as the result of a mapping project for students at the University of Swansea. One of the first places the students began to work on was close to a volcanic succession that had been mapped long ago by the British Geological Survey as belonging to the Old Red Sandstone of Devonian age. The students and their teacher soon realized that these rocks looked very different from any others referred to as the Old Red Sandstone. To begin with, they were a dark olive-green and covered with pseudofossils formed of manganese dioxide, which leaves stains that resemble ferns. After a careful and detailed search of these rocks, discoid fossils were located which resembled forms known from Australia that had been called jellyfish, or medusoids by Reg Sprigg (1947) and Martin Glaessner (1959; Cope, 1977, 1982).

The Carmarthen metazoans, all disc-shaped, were preserved in siltstone and fine sandstone, each separated by thin layers of mud – indicative of deposition in quiet seas that were occasionally "intruded" by coarser sediments. The sandwiching of these sediments between volcanic ash and flows called rhyolites, as well as the minerals preserved in the silt, sand and mud clearly signal that these were formed by volcanic ash being deposited out of water.

Fossils are preserved on the underside of beds as convex impressions and have been given such names as *Cyclomedusa* and *Medusinites* (Cope, 1982). Most were collected from a single quarry, but none were actually found in place. The same quarry produced slabs with trace fossils,

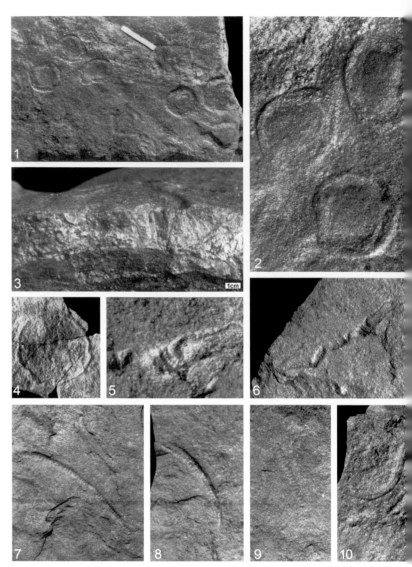

Figure 357 Nimbia (1-4), Archaeichnium (5-6) and Cloudina (7-10) Ediacaran organisms from the Great Basin of Western North America (from Hagadorn & Waggoner, 2000).

some very similar to palaeopascichnids, others gently meandering trails – but all shallow, nearly horizontal trails as one might expect in Neoproterozoic sequences, not in Cambrian or younger rocks.

The Great Basin of California and Nevada and the Grand Canyon of Arizona

A few enigmatic structures, perhaps of body fossils as well as trace fossils, have been reported from silica-rich clastic sediments of the Basin and Range country of eastern California and southwestern Nevada. They have been given such names as *Ernietta* (Horodyski, 1991) and *Swartpuntia*, and are found in the Montgomery Mountains along with the small-shelled fossils of *Cloudina* and other tubular fossils called *Archaeichnium*, *Corumbella* and *Onuphionella*. All associated trace fossils beds – *Planolites*, *Helminthoidichnites* and *Palaeophycus* (Horodyski, 1991; Hagadorn & Waggoner, 2000; Waggoner & Hagadorn, 2002) – lie parallel to the sedimentary thus showing no deep penetration of the sea bed, which is characteristic in younger sediments. James Hagadorn and Ben Waggoner found discoidal forms, which they thought were quite similar to *Nimbia*, in the Wood Canyon Formation

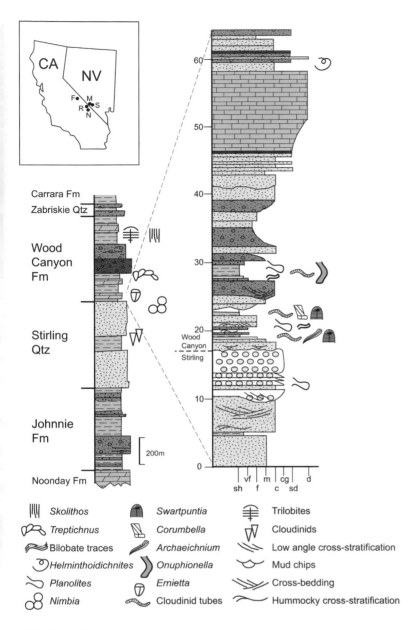

Lithological legend:

sh	shale	c	coarse sandstone
vf	very fine sandstone	cg	conglomerate
f	fine sandstone	sd	sandy dolostone
m	medium sandstone	d	dolostone

Figure 358. Location and stratigraphic sequence of the Wood Canyon Formation, Great Basin of Western North America (from Hagadorn & Waggoner, 2000). F, Funeral Mountains; M, Montgomery Mountains; N, Nopah Range; R, Resting Springs Range and S, Spring Mountains (D. Gelt).

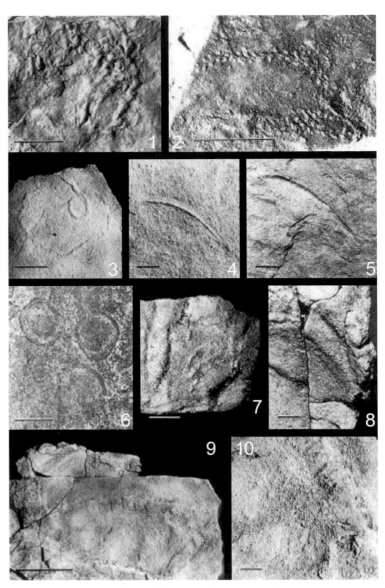

Figure 359. A selection of impressions suggestive of Ediacarans, from the Kelso Mountains, California (courtesy of J. W. Hagadorn from his Ph.D. dissertation, 1998, and accessible on http://www. amherst.edu/~jwhagadorn/publications/Diss.pdf). (Image 1313) All, except 6 (from the Upper Stirling Quartzite), from the lower member of the Wood Canyon Formation: (1-2) Treptichnus pedum; (3) Gordia; (4-5) tubular fossils; (6) similar to Nimbia; (7) possible arthropod-like form; (8-10) Swartpuntia (fossils in the collections of the Los Angeles County Museum of Natural History). Scales = 1 cm.

as well as in the underlying Stirling Quartzite. Above the Wood Canyon Formation were sediments bearing the trace fossil *Treptichnus pedum*, a form used by many biostratigraphers to indicate a Cambrian age.

Although these fossils are tantalizing, more and better-preserved specimens need to be found before their true relationships are clearly understood. In some cases, additional material is needed to clearly establish whether these traces are indeed even the remains of once living animals or simply imprints made by physical, not organic, processes.

North Carolina's "Chimney" Fossils

North Carolina is a place of gentle manners and broadleaf forests that for the most part blanket rock outcrops. Quarries and river cuts are about the only places where rocks are exposed. It is a place that gets quite cold in the winter, and anyone with a good wood fire can comfortably sit inside their abodes, gazing at the snow outside with a cup of hot chocolate or a hot toddy in hand. So, it was that Steve Teeter, now an earth sciences teacher from Oakboro, North Carolina, was building his grandfather's fireplace from local rocks when he noticed a cigar-sized fossil, first described as a trilobite (an extinct group related to the joint-legged arthropods) (St

Jean, 1973). This fossil was not a trilobite but turned out to be a metazoan very similar to *Pteridinium*, a Neoproterozoic form known best from the southern deserts of Namibia, northern Russia and the arid ranges of Australia (Milton, 1985; Gibson, Teeter & Fedonkin, 1984). The rocks bearing this fireplace (or "chimney") fossil are part of the Carolina Slate Belt – metamorphosed or altered rocks that began their "lives" as sediments or those associated with volcanoes.

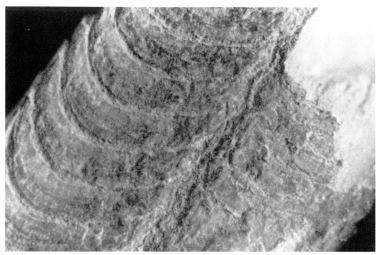

Figure 360. Pteridinium *from Rock Hole Creek, North Carolina, one of the Ediacaran fossils found in the slates used to construct a chimney (from Gibson, Teeter & Fedonkin, 1984).*

Figure 361. Pteridinium *from Island Creek, North Carolina. Holotype on left (3602); paratype on right (3603); width across vanes approximately 3.5 cm (from Gibson, Teeter & Fedonkin, 1984).*

It was not until Teeter earned a degree in science education at the University of North Carolina, followed up by graduate work in geology, that he renewed his interest in the chimney fossil. The chimney had actually been torn down, and the rocks, originally from Little Bear Creek, had been stored for safekeeping and later use behind his father's barn – thank goodness. Steve decided to build another chimney, this time for himself. He split apart one of the stones that had been used in the original chimney, and there before his eyes was another of the cigar-sized and shaped imprints. This time the fossil went on the mantelpiece, not *in* the fireplace, for all to see. But still nothing happened, that is, until 1982. At that time another local resident, Sara Samson, also trained in geology, began finding trilobites in the area, and the "fame" of these discoveries made it all the way to the *New York Times*. Sara found these fossils in place in the rocks where Steve had found his fossil – he knew that the rocks containing his "chimney fossils" had come from Little Bear Creek, but where was anyone's guess.

What was most interesting about Sara Samson's discovery was that the trilobites she found were not closely related to anything else known from North America. They resembled much more closely African forms! Geologists were certain that Africa had at times in the past collided with southeastern North America, and so this explained the similarities of the fossils from this period – which turned out to be early Phanerozoic, Cambrian in age.

Although Steve thought his fossil could be a trilobite too, when he finally got in touch with a geologist who knew something about fossils, Pete Palmer, in Colorado, Pete pointed out this could not possibly be such an animal. There were not the three body lobes present on Steve's fossil that were present in all trilobites. Palmer suggested that Steve show a photograph of the fossil to some other experts who would be attending a meeting of paleontologists in England. It just so happened that Boris Sokolov was at that meeting and gave a presentation on fossils from Russia – and once Palmer saw Sokolov's slides he thought he had the solution to Steve's dilemma. One of Sokolov's fossils looked almost exactly like Steve's. Sokolov suggested that the best person to identify this fossil was Mikhail Fedonkin, a colleague of his in Moscow.

Once Fedonkin saw Steve Samson's fossil, he immediately identified it as *Pteridinium*, pointing out the similarities it had with his own fossils from the White Sea region of northern Russia (Gibson, Teeter & Fedonkin, 1984). Such fossils are also very common in the southern deserts of Namibia in the Nama Group of late Precambrian age. And, so, the chimney dilemma was solved – these fossils were not trilobites but older forms from the late part of the Precambrian.

In 1983, yet another *Pteridinium* fossil was found when Gail Gibson gave a talk to a group of primary school students. One of the students in the class had brought the teacher a fossil from Rock Hole Creek. This fossil added more detail to what was known of the other two specimens already in hand.

Clearly, much more is to be learned from the Carolina Slate Belt and "Steve Teeter does all he can to control a wry satisfied smile when students in his 8th-grade science classes suggest that he tear down his fireplace in search of more fossils" (Milton, 1985).

Figure 362. Guillermo Aceñolaza (far left), Gilberto Aceñolaza (middle) and Mikhail Fedonkin in the Puncoviscana Formation, Jujuy Province, NW Argentina (P. Vickers-Rich).

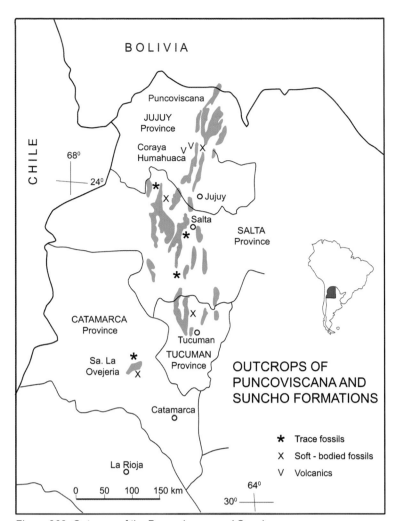

Figure 363. Outcrops of the Puncoviscana and Suncho formations, NW Argentina (from Aceñolaza and Durand, 1986). (D. Gelt)

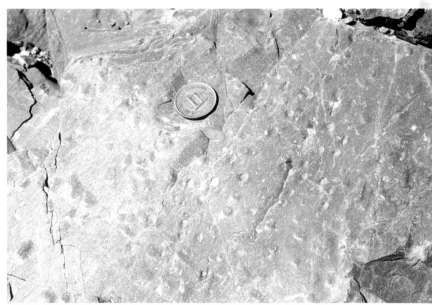

Figure 364. Beltanelloides-like structures, Puncoviscana Formation, Purmamarca, Jujuy Province, NW Argentina (Argentinian peso coin for scale) (P. Vickers-Rich).

The Enigma of South America

For such a large continent, the rocks of South America have not given up many secrets about their Precambrian past. It is one of the great challenges to paleontologists working on the history of animals during the Neoproterozoic. Although there is an abundance of trace fossils of various sorts and kinds in such formations as the Puncoviscana Formation of northwestern Argentina (Salta and Jujuy provinces in particular) and southern Bolivia, studied in detail by Gilberto and Guillermo Aceñolaza and Luis Buatois and Gabriela Mangano, little in the way of definite macroscopic metazoans is known. *Oldhamia* is one of the "trace fossils" that is found at several levels in the late Neoproterozoic to Cambrian Puncoviscana Formation. Perhaps it is a trace, some suggest, made by an arthropod of some sort (Aceñolaza & Durand, 1984). *Oldhamia* consists of a mass of fan-shaped structures – curved ridges and grooves. Some other workers suggest that *Oldhamia* may even be a body fossil, some allying it with *Aspidella* (Runnegar, 1992). The most recent interpretation posed is that of Luis Buatois and Gabriela Mangano (2004). They suggest it is a vermiform, a worm-like organism,

relatively stationary, that fed by mining shallowly under algal and bacterial mats. *Oldhamia* is always found in association with microbial mats. Although not everyone would agree, *Oldhamia* appears to be mainly, if not entirely, Cambrian,

191

Figure 365. Oldhamia, *Puncoviscana Formation, Late Neoproterozoic-Cambrian of northwest Argentina* (P. Vickers-Rich). *Length of fan-shaped structure about 3.5 cm.*

not Neoproterozoic in age, and often dominates the deeper ocean sediments, relicts in the Ediacaran biotas, commonly associated with microbial mats, in Cambrian assemblages (Buatois & Mangano, 2004). Such relict communities were mainly restricted to the depths where the burrowing organisms, the "spear-bearers" of the agronomic revolution of deep burrowing, had not yet colonized.

The Puncoviscana Formation clearly spans both the Neoproterozoic and part of the Cambrian. The few fossils that may be metazoan body fossils come from the lower part of this formation, where traces are restricted to surface or very shallow sediment feeding. Guillermo Aceñolaza (2004) noted an enigmatic tube-like structure nearly 13 cm long and from 1.2-1.4 cm wide, preserved as a mold from the La Higuera locale near the Choromoro River in Tucuman Province, northwestern Argentina. He tentatively referred this to *Sphenothallus*, noting its similarity to the common Early and Middle Cambrian *Selkirkia* – forms that might be related to living pogonophorans. But it was a single specimen. More need to be found before one can be certain of any affinity.

Also from the Puncoviscana Formation are impressions referred to *Beltanelliformis* and *Sekwia* (Aceñolaza & Durand, 1986; Aceñolaza *et al.*, 2005), from the Munano area of Salta, the Humahuaca area of Jujuy, the Sierra de La Ovejeria in Catamarca and the Choromoro area of Tucuman Province. These structures occur in small and large sizes, and many have marginal bands that are distinct from the remaining impression. Some of the smaller, spherical structures appear to occur in lines, reminiscent of the much older *Horodyskia* from Montana. Body fossils and traces alike were emplaced in a continental slope environment, the site of a large submarine fan system, which formed alongside of a reasonably stable continental margin, the edge of the Protopacific of the Gondwana supercontinent. The fan-building river dumped a mass of sediments into a subsiding basin, which continued to subside over a lengthy period, giving rise to the Puncoviscana Formation. By Middle Cambrian times, however, around 500 million years ago, the southern Pampean Mountain Range began to rise, turning this passive margin into an active one (Jezek *et al.*, 1985).

Figure 366. *Puncoviscana Formation outcrops east of Cerro de los Cobres, high Andes, northwest Argentina* (P. Vickers-Rich).

One other intriguing locale in South America is in the Pantanal region of western Brazil. The Tamango Formation of the Corumba Group has yielded "a moderately diverse assemblage of fossils and biological mediated sedimentary structures including stromatolites, putative metazoans and trace fossils" (Babcock *et al.*, 2005). A Neoproterozoic age is suggested by the presence of the small carbonate-shelled *Cloudina*. A minimum age of 569 +/- 20 million years (Cordani *et al.*, 1985) is based on a radiometric (rubidium/strontium) date, determined on the Diamantino Formation that lies well above the Corumba Group. All of these fossils occur above a diamictite that is claimed by some to be of glacial origin, one of the latest glacial events in the Neoproterozoic. But, some geologists have questioned the glacial nature of these deposits, noting that they may simply represent a debris accumulation that should not be attributed to cold conditions (Van Straden & Zimmermann, 2004).

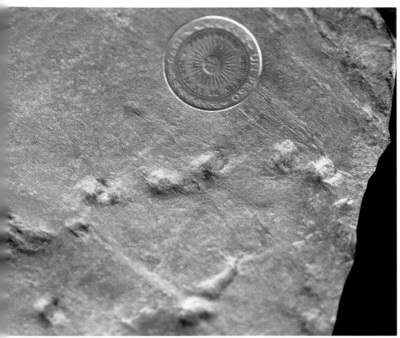

Figure 367. "String of beads" structure that could be body fossils or traces, Puncoviscana Formation, Neoproterozoic to Cambrian, NW Argentina (Argentinian peso coin for scale, 2.5 cm in diameter) (P. Vickers-Rich).

Figure 368. Clay interclast conglomerate in the Puncoviscana Formation on the same bedding planes as, and in contrast to, the spherical Beltanelloides-like and "string of beads" fossils in the Puncoviscana Formation, Purmamarca, Jujuy Province, NW Argentina (P. Vickers-Rich).

Figure 369. Outcrops of the Puncoviscana Formation, near Cerro de los Cobres, Neoproterozoic to Cambrian, NW Argentina (P. Vickers-Rich).

Figure 370. Tubular structure, likely organic, Puncoviscana Formation, La Higuera, Tucuman Province, NW Argentina (Guillermo and Gilberto Aceñolaza). Specimen PIL 14.545, cf. Sphenothallus. Maximum width about 1.5 cm.

Whatever the origin of the diamictite, within it one form is of particular interest, *Corumbella werneri*, which has been allied with the present-day coronate scyphozoan, very similar to the living *Stephanoscyphus*, as well as an extinct group, the conularids (Babcock *et al.*, 2005), already reported from the Vendian rocks of the White Sea in Russia. *Corumbella werneri* was a tubular metazoan, reaching up to 80 mm in length and 25 mm in diameter. It has an accordion-like structure with a midline groove that runs the entire length of the tube, which is surrounded by ringlets. This is not the only South American locale in which *Corumbella* has been recovered – it was first found in the Corumba-Ladario, Mato Grosso do Sul region of southwestern Brazil (Hahn *et al.*, 1982), from the Claudia limestone quarry. Of some interest is the report of this form in the western USA (Hagadorn & Waggoner, 2000).

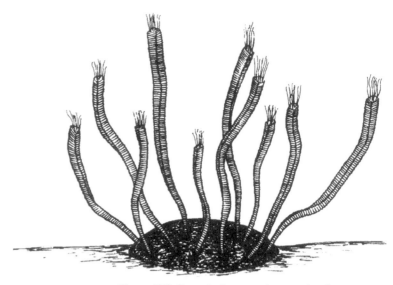

Figure 373. Corumbella werneri, *reconstruction, a possible scyphozoan colonial cnidarian or, perhaps, a conularid-like organism, Ladario, Brasil (from Babcock et al., 2005).*

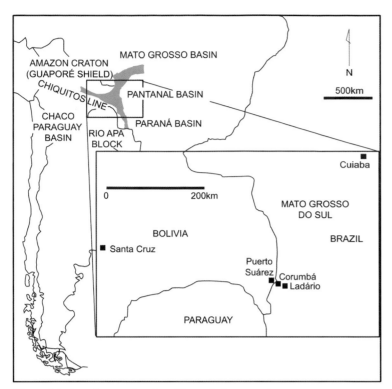

Figure 371. *Locality map for site of* Corumbella werneri, *Corumba region, western Brazil (from Babcock et al., 2005) (D. Gelt).*

Figure 372. Corumbella werneri, *Itau Quarry, Ladario, Brazil (from Babcock et al., 2005). Width of tube about 6 mm.*

China

The most important Ediacaran fossil sites in China are the Yangtze Gorge area in western Hubei Province, Lantian in southern Anhui Province, Wegn'an in central Guizhou Province and Gaojiashan in southern Shaanxi Province. Stratigraphic sections in the Gorge have been used as a yardstick for Chinese stratigraphers to correlate Ediacaran successions elsewhere in China.

Briefly, Ediacaran successions in the Yangtze Gorge area are divided into the Doushantuo Formation and the overlying Dengying Formation. The Doushantuo Formation rests on diamictites of the Nantuo Formation, generally accepted as glacial in origin, perhaps during the Marinoan glaciation. The basal Doushantuo Formation consists of a 5-meter-thick dolostone resting on the Nantuo Diamictite. Similar dolostones ("cap carbonates") have been found overlying most all glacial deposits assigned to the Marinoan globally. Above the cap carbonate lie black shales, with abundant pea-sized chert nodules, followed by thin-bedded dolostones and limestones, with a black shale at the topmost. The entire Doushantuo Formation appears to have been deposited in an intracratonic basin, probably in deep, subtidal environments (Zhang *et al.*, 1998). The overlying Dengying Formation consists of carbonate rocks with little or no muds, silts or sands. It was likely deposited in shallow, subtidal to intertidal environments, as there is evidence for subaerial exposure and paleokarst formation.

Correlation using stable carbon isotope data and limited radiometric dates suggests that the Doushantuo and Dengying formations were deposited during the Ediacaran Period. Recent radiometric dating by Dan Condon and colleagues supports this interpretation (Condon *et al.*, 2005). Condon and his colleagues have suggested that the cap carbonate of the basal Doushantuo was deposited around 635 million years ago, a similar age to some of the glacial deposits in Namibia. They also dated a volcanic ash bed a few meters above the cap carbonate at around 632 million years and another ash bed near the Doushantuo-Denying boundary at about 551

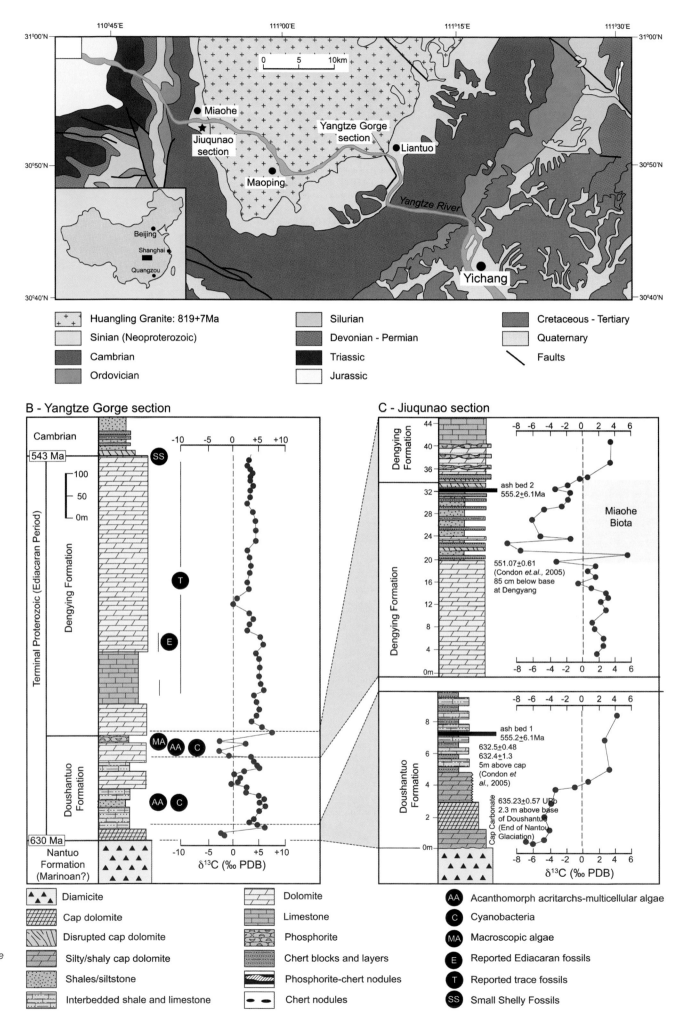

Figure 374. Locality map and stratigraphic/geochemical record of the Yangtze Gorge and Jiuquno sections of south China, area where the rich Doushantuo Assemblage of Ediacaran fossils has been recovered (modified from Zhang et al., 2005) (D. Gelt).

million years. Thus, the Doushantuo Formation, about 260 meters thick in the Yangtze Gorge area, seems to represent about 84 million years of geological history (or roughly 90% of the Ediacaran Period). The Dengying Formation, on the other hand, reaches more than 600 meters in thickness, but represents only 9 million years of time. Stratigraphic thickness can be deceiving, when it is used to estimate time!

Ediacaran rocks in China are known as the Sinian System, a stratigraphic term first introduced in 1882 by the German geographer Baron Ferdinand von Richthofen and later refined in 1922 by the German-American geologist Amadeus William Grabau, then a professor of geology at Peking University. J. S. Lee (Lee J. S.) made the first attempt in the 1920's to systematically document the Sinian stratigraphy in the Yangtze Gorge area (Lee, 1924). In the 1970's and 1980's many Chinese paleontologists, including Cao Ruiji, Chen Meng'e, Ma Guogan, Xing Yusheng, Yin Chongyu, Yin Leiming, Zhang Luyi and Zhu Shixing, carried out systematic paleontological research on the Sinian System in the Yangtze Gorge area. They discovered a rich record of stromatolites, trace fossils, marine algal fossils (*Vendotaenia*) and tubular forms (*Sinotubulites*) in the Dengying Formation, as well as abundant and diverse spiny acritarchs (organic-walled microfossils) preserved in chert nodules of the Doushantuo Formation.

A most important discovery came in 1978 when Cheng Meng'e and Ma Guogan recognized a variety of macroscopic carbonaceous compressions in the black shale of the uppermost Doushantuo Formation near the village of Miaohe. The true significance of these fossils was not appreciated immediately – until the botanist Zhu Weiqing interpreted some as fossil algae (Zhu & Chen, 1984). Subsequent excavations by Chen Meng'e and colleagues recovered more than 20 new species of well-preserved macrofossils (Steiner, 1994; Ding *et al.*, 1996; Xiao *et al.*, 2002). These finds have been interpreted as a variety of different animals – sponges, cnidarians and various sorts of worms – but not all researchers agree with these identifications. A number of the tubular fossils in this assemblage have been variously interpreted as scyphozoan tubes, pogonophoran worm tubes and alga tubes (Xiao *et al.*,

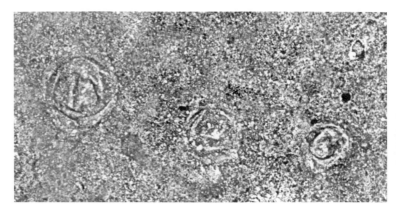

Figure 376. Chuaria circularis *(courtesy of T. Ford). Diameter of individuals about 2 μm.*

2002): *Calyptrina striata, Sinospongia typical, Sinospongia chenjunyani* and *Protoconites minor*.

Since the discovery of the Miaohe Assemblage, similar macroscopic fossils have been found in Ediacaran shales elsewhere in South China. The Lantian Formation in southern Anhui Province contains a dozen macro-algal fossils, some of which are quite similar to forms found at Miaohe (Yuan *et al.*, 1999). Recently, a few of these Miaohe forms have been found in black shales of the Lower Doushantuo Formation in the Yangtze Gorge (Tang *et al.*, 2005) and in mudstones of the Upper Doushantuo Formation in eastern Guizhou Province (Zhao *et al.*, 2005). Clearly the Miaohe-type preservation has a much wider distribution than originally thought

In addition to the typical benthic marine forms identified in the Miaohe Assemblage, shale and mudstones of the Doushantuo Formation have also yielded a number of circular compressions, which are reminiscent of chuarids. Typically they are preserved as two-dimensional compressions (Tang *et al.*, 2005), but occasionally occur as three-dimensionally preserved, internal molds, such as those recovered from the Lantian Formation in southern Anhui Province (Yuan *et al.*, 2001). The Lantian chuarids are preserved as pyrite pseudomorphs, are small (only 2-3.4 mm in diameter) and show a medial split, which appears to be an encystment structure. These are quite similar to eukaryote megacysts. Chuarids, in particular *Chuaria circularis*, have been interpreted in many different ways. Some workers have suggested that they are acritarchs (Evitt, 1963), especially similar to the leiosphaerids. Others have noted their similarity to

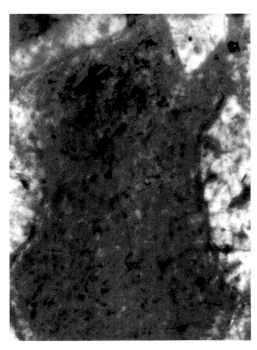

Figure 375. Sponge fossils, Doushantuo Formation, Yangtze Gorge, western Hubei Province, China (courtesy of C.-W Li). Width of base about 390 μm

Figure 377. Calyptrina striata, *Doushantuo Formation, Yangtze Gorge, China (Xiao, et al., 2002). Width of tube about 5 mm.*

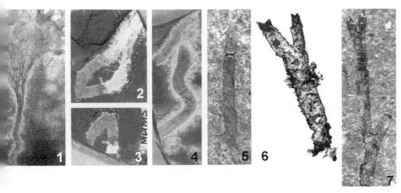

Figure 378. Proconites minor, *Doushantuo Formation, Yangtze Gorge, China (Xiao, et al., 2002).*

Figure 379. *Embryo concentration, slightly etched (embryos in place), Doushantuo Formation, Yangtze Gorges, south China (courtesy of C.-W. Li; photo by S. Morton). Largest embryos abut 500 μm in diameter.*

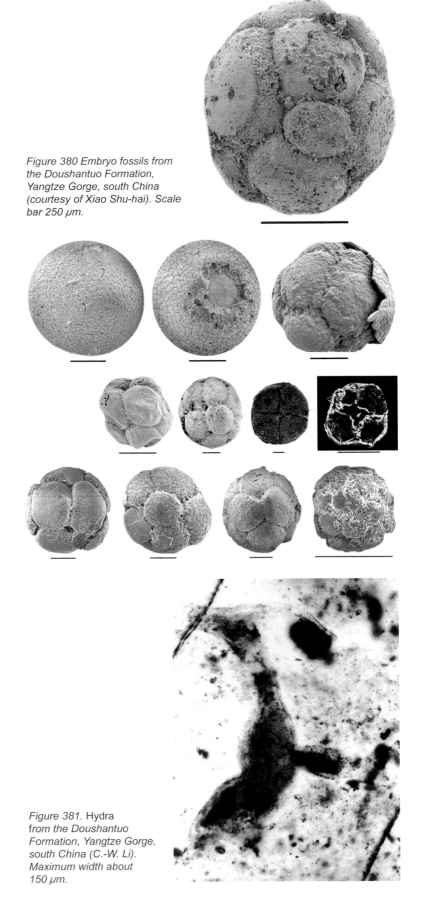

Figure 380 *Embryo fossils from the Doushantuo Formation, Yangtze Gorge, south China (courtesy of Xiao Shu-hai). Scale bar 250 μm.*

Figure 381. *Hydra from the Doushantuo Formation, Yangtze Gorge, south China (C.-W. Li). Maximum width about 150 μm.*

eukaryotic algal cysts, with some affinities to *Tawuia* (Ford & Breed, 1973; Vidal & Ford, 1985; Butterfield *et al.*, 1994). Still others have noted similarities with *Beltanelliformis*, perhaps some sort of green algae (Yuan *et al.*, 2001). There has even been a suggestion that *Chuaria* could have been colonial cyanobacteria, somewhat like *Nostoc*. What cannot yet be resolved is whether they are plant or animal.

Perhaps the most significant discovery in the Ediacaran record of China was made in a phosphorite unit of the Doushantuo Formation at Weng'an, central Guizhou Province. The Doushantuo Formation there is only about 40 meters thick, much thinner than in the Yangtze Gorge. This unit was likely deposited in a shallow, subtidal to peritidal environment, as a subaerial exposure surface has been developed in the Middle Doushantuo at Weng'an. The phosphorites at Weng'an also contain abundant intraclastic grains, further suggesting a shallow-water depositional environment. In 1984, Zhu Shixing discovered microfossils from this locality, interpreting some as cellularly preserved eukaryotic algae, including red algae. Later the fossils were compared with modern coralline red algae (Zhang, 1989; Zhang & Yuan, 1992; Xiao *et al.*, 2004).

The preservational quality of these fossils is superb, reflecting morphology at the cellular and subcellular levels. The fossils are phosphatized and surrounded by dolomitic matrix, making it possible to extract fossils from the rock in three dimensions, using acetic acid digestion methods. Thus both the internal and external morphology of these tiny fossils can be studied.

197

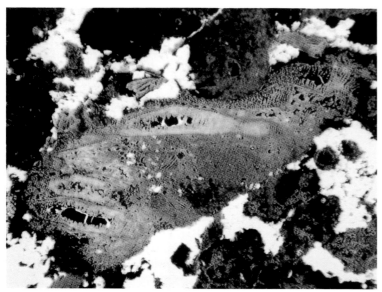

Figure 382. Tabulate coral from the Doushantuo Formation, Yangtze Gorge, south China (courtesy of C.-W. Li). Length about 2 mm.

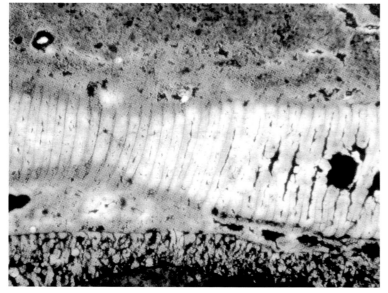

Figure 383. Tabulate coral from the Doushantuo Formation, Yangtze Gorge, central China (courtesy of C.-W. Li).

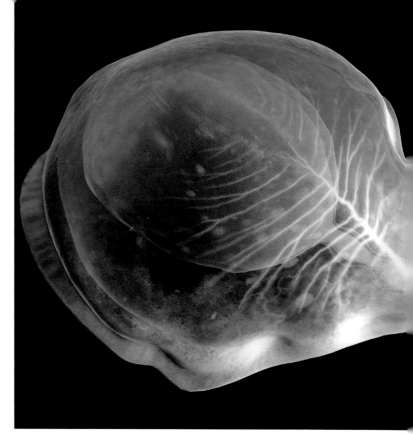

Figure 384. Reconstruction of Vernanimalcula from the Doushantuo Formation near Weng'an, Guizhou Province, south China (courtesy of artist Amadeo Bachar, Scientific American and David Bottjer). Length of specimen about 180 μm.

As these fossils were studied, researchers from the Nanjing Institute (Xue Yaosong) interpreted some as dividing cells of green algae, while the Harvard group (Andrew Knoll, Zhang Yun and Xiao Shuhai) suggested that they were eggs and embryos of animals. Evidence supporting the egg-embryo interpretation includes the consistent embryo size with doubling cell numbers and decreasing cell size, arrangement of cells in such a way that the interior cells cannot be photosynthetic, as well as the presence of an ornamented envelope that is similar to a metazoan egg case. Simultaneous with the work of Knoll (Xiao *et al.*, 1998) and his colleagues, another group led by Li Chia-wei and Chen Jun-Yuan came to much the same conclusion, based on material from the Doushantuo phosphorites at Weng'an (Li *et al.*, 1998), that they were embryos and larvae of sponges. Later work has suggested the presence of cnidarian tubes (Xiao *et al.*, 2000) and bilaterians (Chen *et al.*, 2002; Chen *et al.*, 2004) – one example being *Vernanimalcula guizhouena*. Recent studies by Li Chai-wei and his group using Synchrotron X-ray microscopy

(SXM) have revealed body fossils of parasitic nematode-like bilaterians living in some of the Doushantuo embryos (Li, 2006). The assignment of *Vernanimalcula* to the bilaterians, however, has been questioned by some (Bengtson & Budd, 2004).

It would not be surprising that organisms other than animals and algae could have lived in Doushantuo time. Recently Yuan Xulai and his colleagues have reported the discovery of possible marine lichenoids, an association of algal or cyanobacterial autotrophs and fungal heterotrophs in the Doushantuo phosphorite. It is an indication that the ecological interactions among Doushantuo organisms were likely complex including autotrophy, heterotrophy and symbiosis, at the very least.

More classic Edicacara fossils are also known from the Doushantuo – *Paracharnia dengyingensis*. This form appears in the Middle Dengying Formation in the Yangtze Gorge (Sun, 1986) – originally interpreted as a colonial cnidarian, a sea pen. Recently, similar frond-like forms have been reported from the same area (Xiao *et al.*, 2005), but interpreted as procumbent organisms that lived near the water-sediment interface, rather than erect like sea pens. Unlike other Ediacaran fossils of a similar morphology, preserved as casts and molds in sandstones and siltstones, the Dengying fossils are encased in a fine-grained limestone, thus allowing study in thin section – and it appears that these forms thrived in this carbonate-rich environment.

The Dengying Formation in the Yangtze Gorges area also boasts abundant trace fossils and skeletal tubes known as *Sinotubulites baimatuoensis* (Chen *et al.*, 1981). These are true metazoan fossils of macroscopic size. A promising site currently being investigated by Hua Hong of Northwest University (Xian) is located in Gaojiashan, southern Shaanxi Province. Here the Dengying Formation is known to contain

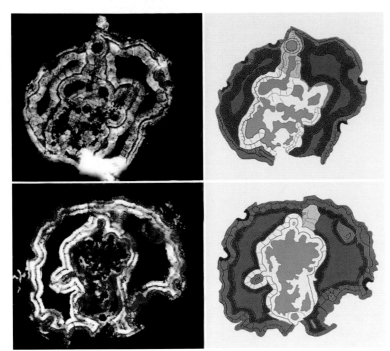

Figure 385. Thin section showing fossil of Vernanimalcula *from the Doushantuo Formation of Weng'an, south China (courtesy of C.-W. Li). Top illustrations holotype (X00305), coronal section passing through ventral mouth, from specimen about 180 μm in length. Bottom illustrations, coronal section passing through dorsal wall of pharynx. Both thin sections about 50 μm thick.*

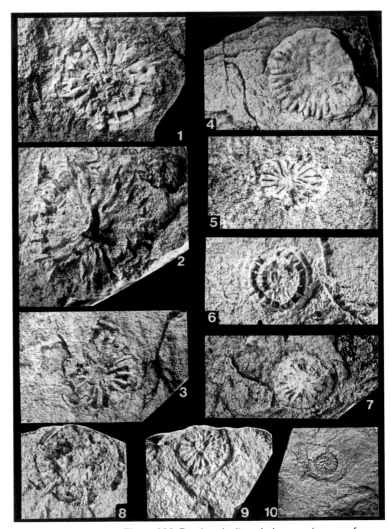

Figure 386. Persimedusites chahgazensis, *page from Hahn & Pflug, 1980. Average diameter of specimens about 1.5 cm. Figures 1-8 from Chah-Mir Mine, north-eastern Tajkuh, Kushk Series, Iran; Figures 9-10 from the Kushk Mine, northeast of Bafgh, Yazd area, Iran.*

exquisitely phosphatized tubes of *Cloudina hartmannae* as well as other tubular fossils, *Conotubus* and *Gaojiashania*. *Cloudina* has a global distribution in Ediacara times, and may be one of the few fossils of this period that can be used as an index fossil (Hua *et al.*, 2003; Hua *et al.*, 2005). *Cloudina* from the Dengying Formation is also characterized by the presence of tiny drill holes into its tubes – perhaps the evidence for early predators (Bengtson & Yue, 1992; Hua *et al.*, 2003), and similar to holes recently observed by Li Chia-wei in some of the embryo fossils from the Dengying Formation (Li, 2006).

From the northeast of China, Liaoning Province, discoidal fossils have been recovered that are distinct from the disc-like forms known from other Ediacaran assemblages. They also do not resemble the carbonaceous compressions recovered from other parts of China. They come from the Xingmincun Formation, whose age is actually not precisely established, leaving some doubt as to its late Precambrian age, but this Jinxian biota needs further study (Zhang, Hua & Reitner, 2005).

A Few Other Spots

Ediacaran fossils and other megascopic evidence of Precambrian life have been reported from a number of other locales around the world. So far, none have been fully substantiated. Some have been or seem likely to be refuted, but further study and the discovery of additional specimens could well substantiate other reports noted below and bring them into the realm of the "definite" occurrences (G. Narbonne, *pers. com.*).

One such possibility occurs on the Digermul Peninsula in northeastern Finnmark, northern Norway, where a varied fauna of discoidal forms have been recovered from the Inverely Member of the Stappogiedde Formation: *Ediacaria,*

Cyclomedusa, Beltanella, Hiemalora and *Nimbia*. Most of these, if not all, may simply be holdfasts, many probably referable to *Aspidella* (G. Narbonne, *pers. com.*), for organisms that lived above in the open water (Farmer *et al.*, 1992).

Other spots include western Scotland, where chains of pellets given the name *Neonereites uniserialis* by Brasier and McIlroy (1998) may have been left by a metazoan in the sediments that became the Bonhaven Formation, which lies just above the Port Askaig Tillite laid down by a glacier. These sediments are possibly older than 590 million years ago (although the date is under revision and the biogenicity of these structures has been questioned (G. Narbonne, *pers. com.*). In this area of Scotland rocks in the Diabaig Formation, dating at around 1000-1200 million years, contain wrinkle structures indicating binding by microbial mats that also show aerial exposure – they could be deep desiccation cracks, for example. This indicates that even at this ancient time, some forms of life were apparently living on land (Prave, 2002).

Farther to the east, a form given the name of *Persimedusites chahgazensis* was collected in the Esfordi Formation of Central Iran near Kushk (Hahn & Pflug, 1980), thought to be some sort of medusa, but Bruce Runnegar (in Schopf & Klein, 1992) was not

Figure 387. Persimedusites chahgazensis, *from Iran (courtesy of B. Hamdi and M. Jafari). Diameter of structure about 1.5 cm.*

convinced of this and called it a structure with uncertain affinities – using the Latin *incertae sedis*! J. J. Dozy (1984) noted that a fossil from the valley of the Lozara River in northwestern Spain had some affinities to forms like *Arborea* and *Rangea*, but nothing more has been discovered from there and the true biogenicity is in question without more material.

From the Krol and Tal formations of the Krol Belt in Uttar Pradesh in the Lesser Himalayas some Ediacarans have been reported: *Beltanella, Cyclomedusa davidi, Conomedusites lobatus, Charniodiscus* cf. *arboreus, Dickinsonia, Irridinitus* sp., *Medusinites, Nimbia occlusa, Pteridinium* cf. *carolinaense, Sekwia* cf. *excentrica, Tirasiana, Zolotytsia biserialis,* perhaps *Beltanelliformis* and *Kimberella* cf. *quadrata* (Shanker & Mathur,

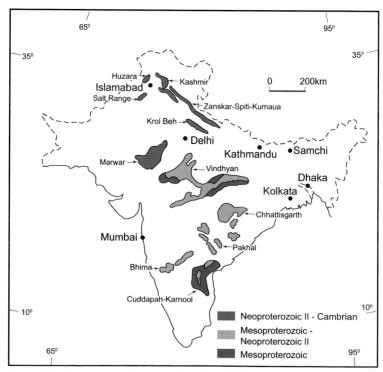

Figure 388. Distribution of Proterozoic sediments in India (courtesy of P. K. Maithy and G. Kumar). (D.Gelt)

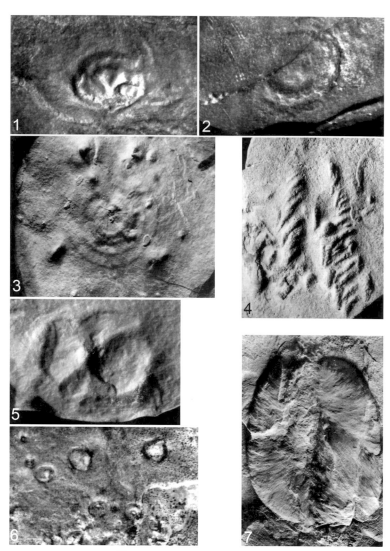

Figure 389. A selection of possible Ediacaran fossils from India – Nainital District, Uttar Pradesh, Krol Formation. Names have been associated with these fossils by G. Kumar and P. K. Maithy. This assemblage offers the opportunity for a better understanding of the Asia macro-metazoans in the future. (1, 2) cf. Cyclomedusa; (3,5) cf. Conomedusites lobatus; (4) cf Charniodiscus sp.; (6) cf. Dickinsonia, but a holdfast identity should be considered (courtesy of G. Kumar & P. K. Maithy)

1992; Maithy & Kumar, 2004). These specimens need review and new collection associated with a detailed record of geological context. These same sediments also preserve Small Shelly Fossils and traces. They offer good possibilities for further studies in the future. Sponge body fossils have also been reported from the Buxa Dolomite, northeastern Lesser Himalayas of India, west Siang District of Arunchal Pradesh (Shukla *et al.*, 2004).

A few Ediacaran-like fossils were also reported by Mark McMenamin from northwestern Sonora, Mexico from the Clemente Formation – a possible *Cyclomedusa plana, Sekwia,* an erniettid and a few trace fossils. McMenamin suggested the age for these fossils was in excess of 600 million years, making them some of the oldest, if not the oldest, Ediacarans. The fossils are rare and not well preserved, and this site needs much further development before the true biological nature of the material can be determined.

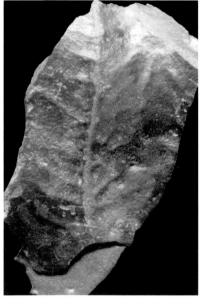

Figures 390-391. Possible Ediacarans from the Nainital District, Uttar Pradesh, Krol Formation of India. Top, cf. Kimberella quadrata; lower right, unnamed form with proarticulatan architecture (courtesy of G. Kumar and P. K. Maithy).

Other forms such as *Baikalina sessilis* from the Kurtun Formation near Lake Baikal (Sokolov, 1973) seem to be truly biogenic, in this case likely related to some of the common Nama fossils of Namibia such as *Pteridinium* or *Ernietta*. Their similarity was supported by Namibian expert Hans Pflug when he visited Novasibirsk and studied the collections there. Runnegar (1992) has noted that such forms as *Brabbinthes* from Alaska appear to be hexactinellid sponge spicules. *Pararenicola huaiyanensis* from the Jiuliqiao Formation of Anhui Province in China may well be a metazoan body fossil. *Redkinia spinosa* from the Redkino Formation of the Nepeitsino bore, and other places, as well as *Sinosabellidites huainanensis* from the Liulaobei Formation of Anhui Province in China are good candidates for metazoans. The black, chitinous comb-like *Redkinia spinosa* from the Redkino Formation of the Nepeitsino bore in European Russia, *Redkinia fedonkini* from the Ukraine, as well as tubular *Sinosabellidites huainanensis* from the Liulaobei Formation of Anhui Province of China are all metazoans. Medusoid fossils have also been reported from sites in Australia other than the Flinders Ranges – from the Hamersley Range in the northwest of Western Australia (Edgell, 1964), dated at around 1600-1800 million years. These are certainly mineral concretions and not organic (Walter, 1972). Fossils from the southwest of Australia in the Stirling Range (Rasmussen *et al.*, 2002), with a date of between 1800 and 2500 million years is a bit more convincing as organic remains, but the dates placed on these fossils are in serious doubt.

Using the biodiverse locales as starting points, as "museum of specimens" with which new finds can be compared, it is simply a matter of time before more diversity is added to the Neoproterozoic metazoan list. One must always be mindful, however, of not over-interpreting the impressions left behind in the sands and clays, for sedimentary and metamorphic structures and traces can oft-times be deceptive.

There are other locales from which Neoproterozoic metazoans have been reported here and there, but most of these need to be further investigated before the real or unreal nature of the supposed fossils is established. Bruce Runnegar and Mikhail Fedonkin (in Schopf & Klein, 1992) highlighted the then known material and evaluated the probability of biogenicity. Forms from the Aksumbinsk Formation of Kazakhstan (*Aksumbensis)* appear to be pseudofossils, a fossil "jellyfish" from the Nimbahera Formation of Madhya Pradesh is classed as *incertae sedis* – of uncertain affinities based on the material available. *Ichnusa cocozzi* from Sardinia appears to be a pseudofossil, not of biogenic origin, *Jixiella capistratus*; several species of *Mashania* from Heilonjiang apparently sedimentary structures; *Kullingia concentrica* from Norway and many other parts of the world a trace fossil; *Majaella verkhojanica* and *Suvorovella* from the Maya River of Yudoma in Siberia of uncertain affinities; *Medusinites simplex* from Liaoning Province in China a pseudofossil; *Sajanella* from the Irkutsk area apparently a deformed microbial mat, not a metazoan body fossil. *Tyrkanispongia tenua* from the Gonam Formation along the Urchur River of Siberia appears to be volcanic glass shards (Runnegar,1992)!

Figure 392. Redkinia cf. spinosa, Ust'Pinega Formation, Redkino Series, Russia (F. Coffa, courtesy of the Paleontological Institute, Moscow). Bottom specimen about 0.7 cm in length.

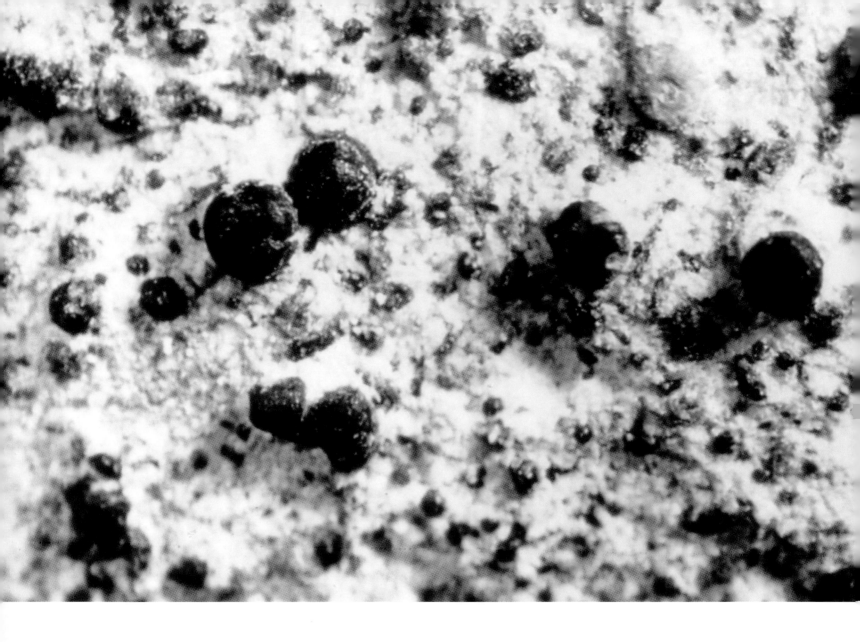

400um

PART 3: OTHER EVIDENCE OF ANIMALIA

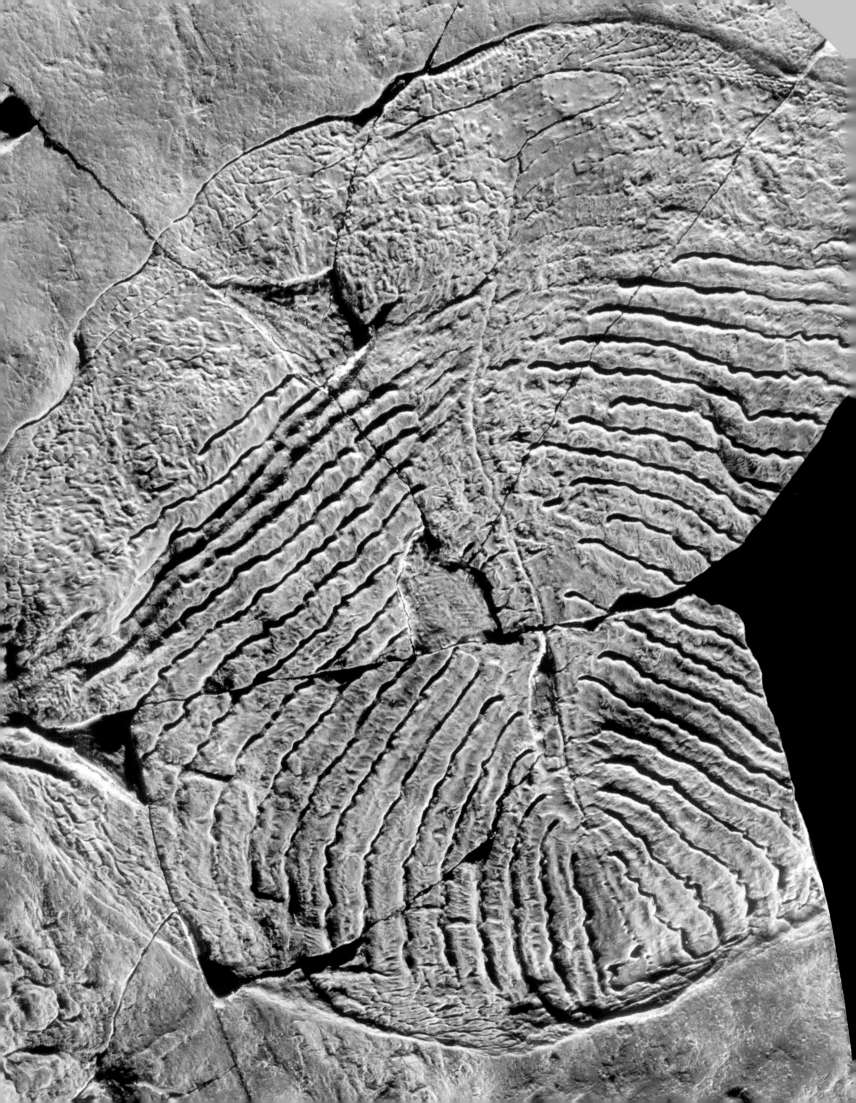

First Trace of Motion

M. A. Fedonkin and P. Vickers-Rich

In practice, speculations about the past,
If they are not to be entirely idle,
Must relate to the traces which
The past has left

A. J. Ayer — *The Central Questions of Philosophy*

Beaches are spectacular places to visit for anyone interested in forensics. They are truly "crime scenes," ripe for analysis, though most visitors are blinded by the beauty and complexity of the oysters, seaweed bulbs and the flotsam loosened by humanity – fishing floats, old sandals and an exultation of fascinating, unidentifiable fragments. But, a closer look at the sand reveals the "footprints" left behind by an army of organisms, which madly and consistently "farm" the sediments for their intertidal treasures before the sea yet again invades, with predictability, controlled by the Moon, with the incoming tides.

The curious beachcomber can spend hours watching the activities of animals that feed on, leave tracks in and burrow, both deeply and shallowly, in search of food or protection into this "mobile" environment. For the most part, such traces are sculpted by metazoans, those animals with many cells and a variety of complex behaviors. Metazoans have been modifying the beaches and sea floors from their beginnings, and these activities can sometimes be fossilized as "traces," morphologically discrete structures that result from the behavior of animals and how that is reflected in the sediments with which they interact. Unfortunately, beach traces are very infrequently preserved in the fossil record, but traces left in less disturbed environments, such as below storm-wave base sea floors or quiet lake floors, are. It is these structures that allow paleontologists to reconstruct past behavior over long periods of time, even into the distant Precambrian when metazoans totally lacked shells or skeletons.

One must be careful in interpreting these fossil clues, however, for some complex traces can be made by organisms that themselves are not so complex. For example, burrowing sea pens, anemones and hydroids can leave complex traces even though they are far less complicated animals than worms and arthropods, which can, at times, leave simpler traces! However, the trace fossils that are preserved are more likely

Figure 393. Feeding bundles left as traces on a modern beach, Four Mine Beach, Port Douglas, Queensland, by living ghost crabs (Family: Ocypodiae) (P. Vickers-Rich). Large spheres about 1 cm in diameter.

to be those made by animals that were full-time burrowers and plowers of the sea floor. But, as Anthony Martin (*pers. com.*) points out, the preservation of trace fossils is dependent on many factors – the very nature of the substrate, the behavior and anatomy of the trace maker, whether the traces are buried quickly enough and whether the sediments are geochemically amenable to their preservation. He notes that borings in bivalve shells, tiny in comparison to deep burrows, such as the 50 cm deep burrowing callianassid shrimp, are more likely preserved as they are in hard shells and not soft sediment. Another caution is that one animal can make many different kinds of traces, depending upon environmental factors. To even further complicate matters, different animals can leave traces that are quite similar. So, paleoichnologists, scientists who study fossil traces, certainly have their work cut out for them!

Opposite page: Yorgia waggoneri, *Vendian sequence, Valdai Series, Zimnii Bereg, White Sea, Russia, part of a large quarry block collected by Andrey Ivantsov (PIN 5024) (F. Coffa).*

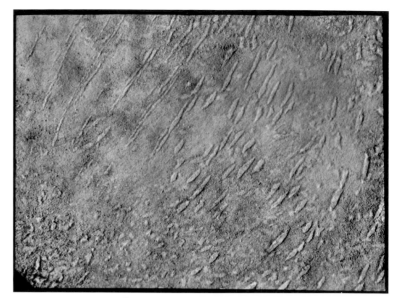

Figure 394. Nenoxites, holotype. Meandering feeding trail. Vendian sequence, Suz'ma River, White Sea region (M. Fedonkln). Trace about 5 mm in width.

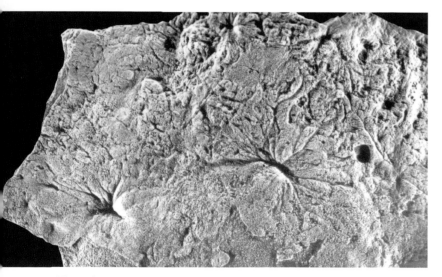

Figure 395. Pseudorhizostomites howchini, pull out structures likely produced when a holdfast was ripped out of the sediment. Vendian sequence, Zimni Bereg, White Sea, Russia (M. Fedonkin). Width of structures about 3.5 cm.

What Is Real, What Is Not?

In real contrast to the detail left by metazoan body fossils or shells, any experienced observer of modern life knows that tracks, trails and burrows are often quite enigmatic. Who made the traces and how? Or, in fact, are they imprints of bodies of living organisms, or is there an inorganic explanation (Jensen *et al.*, 2005)? Unlike in the modern world, a "paleowatcher" cannot wait patiently until the "perpetrator" of the trace appears – and then connect trace with maker. The paleontologist is not so lucky. Association of tracks with track makers in the fossil record is rare – especially in the Precambrian. The mollusk-like *Kimberella* and both *Yorgia* and *Dickinsonia* associations only recently discovered in the sedimentary sequences of Russia and Australia are some of the few examples of direct association in the Precambrian.

Not all imprints that have been called "trace fossils" have been made by living organisms either. These should more correctly be called tool marks, not trace fossils. These are made by inanimate objects, sometimes initiated by water current flow, wind, bouncing rocks or clay fragments on a bottom surface of sea or stream bed, ice or other crystals growing in sediments, gas bubbles escaping from sediments or scratch marks left on the bottoms of rock surfaces during turbidity flows or submarine avalanches or even generated during times when rocks are being metamorphosed (*e.g.* Seilacher & Meschede, 1998). Confusing, too, are living organisms, such as termites or roots of plants, etching their way into sediments of a much older age, but these are true trace fossils – one must be careful in interpreting the timing of their production.

The paleontologist must ask a few questions to help decide whether trace fossils were made by living organisms or by other means, whether they are real trace fossils or mere inorganic pseudofossils or doubtful fossils, dubiofossils (Frey, 1975; Crimes & Harper, 1977). Sometimes a final decision cannot be made. Such decisions must wait for future discoveries or, in fact, may never be possible based on the data at hand.

Question 1. Similarity to Modern Traces. Do the trace fossils closely resemble structures made today by living organisms? Do they resemble burrows in soft sediments, boring in hard surfaces, feeding structures, fecal materials, walking traces or trackways, crawling trails, points of interaction between organisms?

Question 2. Complexity. Are the structures complex? Generally biological structures are far more complex than those left by abiogenic processes.

Question 3. Chronological/Evolutionary Correspondence. Do the structures occur in a sequence from bottom to top that corresponds to the occurrence of the types of organisms that most likely could have made them? Apparent trace fossils found in rocks dating from hundreds of millions of years before body fossil evidence of metazoans must be regarded with suspicion. The chance of metazoans evolving twice, or evolving hundreds of millions of years before their first reliable body fossil record, is so small that the most likely explanation is to doubt the animal origin of such structures.

Question 4. Abundance. Is there just one trace or many? Most organisms do not occur alone, but in populations and communities. One single, odd trace is not convincing for a biological origin. When some new biological process evolves that is successful, it generally becomes very common. Those Ediacaran trace fossils, reflecting this new way of processing the sediments or moving across them, are very common.

Question 5. Geographic Distribution. Is the distribution of the trace fossils very restricted or widespread? Although organisms certainly have environmental preferences, generally there is a wider distribution than just a single locale.

Question 6. Restricted Stratigraphic Range. Does the trace have a limited range in the rock sequence or does it occur again and again? Organisms live for only a certain period of geologic time, then go extinct. Gas-escape structures and crystal growth in sediments do not. However, a complicating fact is that simple patterns left behind by many organisms can easily be convergent and therefore can have a long stratigraphic range (*e.g.* "*Planolites*," which has a range from Ediacaran to recent). Conversely there are sedimentary structures that are more common (or more commonly preserved) in certain periods of time (S. Jensen, *pers. com.*).

Question 7. Diversity. Is there some diversity of trace fossils? Generally, if of biological origin, there is more than one type of trace fossil. This applies for most of the Phanerozoic, but becomes more debatable the closer one approaches the commencement of animal-sediment interactions towards the end of the Ediacaran (*pers. com.*, S. Jensen).

Question 8. Associated Fossils. Is there a direct association of trace with trace maker? In some cases a fossil, either a body fossil or a mineralized skeleton or shell, is directly associated with a trace. Although this is quite rare, it is the most compelling evidence for biogenicity, an Ediacaran example being *Kimberella*. A resting trace (a bioglyph or "bioprint") can provide anatomical details of the trace maker, sometimes additional information given beyond the actual fossil itself (A. Martin, *pers. com.*), and it certainly gives insights into behavior not always given by the body fossil.

Question 9. Sedimentary Context. Are the trace fossils preserved in sedimentary rocks? Traces left by organisms occur in sedimentary sequences. These may be metamorphosed, but their original nature was sedimentary and not igneous (except in very unusual circumstances such as footprints left in volcanic ash or organisms entombed in a lava flow that poured into an ancient lake).

Question 10. Environmental and Ecological Context. Are the trace fossils found in environments and ecological associations that mirror what relates to observed associations in the modern world?

Figure 397. Kimberella quadrata *feeding traces, Zimnii Gory, White Sea, Russia (M. Fedonkin). Body fossil about 2 cm in length.*

Question 11. Contemporaneity. Were the traces formed at the time the sediments were deposited, or have they been superimposed at a later date? Could they be plant roots or burrows made by such animals as termites, bees or barnacles at a much later date (Sando, 1972)?

If "asked" these questions and the trace "answers" yes to most, then there is a good chance that these trace fossils represent imprints made by once living organisms, and they can pinpoint the presence of certain kinds of metazoans. Even more interesting is that trace fossils can give insights into the behavior of now-extinct organisms and how they interacted with their environment.

Late Proterozoic Trace Fossils

The first trace fossils occur in the Proterozoic, but they are nowhere nearly as diverse as they are in younger rocks. These first ichnofossils seem to represent, for the most part, surface-movement trace fossils (called *Repichnia*; Seilacher, 1964), a series of complex patterns, repeated again and again on a horizontal surface (Fedonkin, 1985; Aceñolaza &

Figure 398. Gordia antiquaria, *a horizontal trace fossil, Flinders Ranges, South Australia (South Australian Museum P27977) (S. Morton). Width of slab base about 20 cm.*

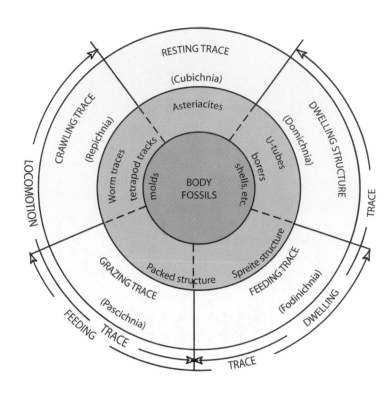

Figure 399. Main behavioral categories describing the formation of trace fossils (modified from Dolf Seilacher's classification). (D.Gelt)

Figure 400. Kimberella quadrata *grazing traces on the sole of the bed, Puttapa Ranges, South of Copley, Flinders Ranges, South Australia, Brent Bowman Collection. Rounded "pellets" represent holes in the microbial mat, perhaps inparted by grazing action of* Kimberella *(S. Morton).*

2cm

Durand, 1986; Aceñolaza, 2004; Crimes & Anderson, 1985). The trace fossils of this time are all nearly two-dimensional and confined to bedding planes (Fedonkin, 1986). Such forms as *Bilinichnus simplex* are trails of two, shallow, parallel

Figure 401. Neonereites, *strings of fecal pellets or perhaps ?protists. Vendian sequence, Suz'ma locale, White Sea region, northern Russia (M. Fedonkin).*

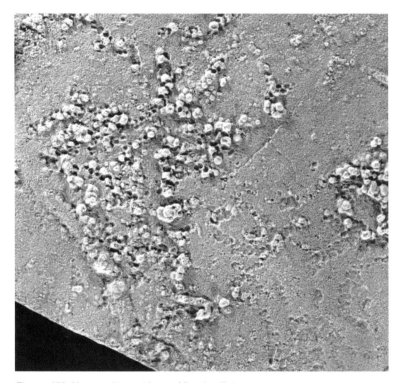

Figure 402. Neonereites, *strings of fecal pellets or perhaps ?protists, distorted. Vendian sequence, Suz'ma locale, White Sea region, northern Russia (M. Fedonkin).*

grooves, which resemble the traces left by living gastropods where the lateral parts of the foot sink more deeply into the sediments than the central part. This sort of trace records peristaltic movement, where waves of contraction passed along the underside of the animal as it moved along. Until relatively recently, it seemed that most trace fossils of the Neoproterozoic were preserved in sediments deposited in relatively shallow water – with only a few places recorded that represented deeper water assemblages – one the volcanic-rich sediments of the Albermarle Group, an Ediacaran sequence in North Carolina, southeastern USA (Gibson *et al.*, 1984), where a variety of trace fossils identified as *Gordia* (*Circulichnis*), *Neonereites* and *Planolites* (perhaps related to *Helminthoidichnites*) have been found – though some of these identifications have been questioned. Deep

Figure 403. Shallow, horizontal trace fossils and Tribrachidium (SAM D244), positive epirelief (natural cast of external molds), Bathtub Gorge, Flinders Ranges, South Australia (S. Morton). Maximum width of slab about 27 cm.

hollow beads may have formed as egg masses, or tiny soft-coral polyps (Gehling *et al.*, 2000), or by bottom-living algae or single-celled animals such as protists (Seilacher *et al.*, 2003).

True grazing trace fossils (*Pascichnia*) are relatively rare in the Neoproterozoic, one excellent example of which is *Kimberella*, considered in detail below. Possible grazing trace fossils are associated with *Yorgia* and *Dickinsonia*, which appear to be a combination of resting trace fossils (*Cubichnia*) and feeding trace fossils – not yet recognized as a common type in younger Phanerozoic sediments (Ivantsov & Malakhovskaya, 2002). Just how these metazoans, known from both trace and body fossils, moved is yet to be determined. They may have glided between feeding stops, where they remained for some time, essentially absorbing nutrients from the bacterial mat below. Perhaps they used rhythmic contractions of their segmented bodies to slowly creep forward, much like a leech or scale worm. Or perhaps they simply decoupled from the mat and leisurely floated near the bottom to the next stop, dropped to the floor, absorbed their fill of food resources – microorganisms and other small animals - then decoupled again and moved on yet again (Ivantsov & Malakhovskaya, 2002; Ivantsov & Fedonkln, 2002).

marine ichnofaunas, though not diverse, are also known from northwest Canada, Central Spain and northwestern Argentina (Seilacher *et al.*, 2005; Narbonne & Aitken, 1990; Vidal & Jensen, 1994; MacNaughton *et al.*, 2000; Crimes, 2001).

Nenoxites curvus are meandering and ribbon-like trace fossils with transverse ridges or rugae – a reflection of an animal moving by contractile waves passing through the underside of the body, again just as in living gastropods – but the ridges may indicate that this metazoan was, like living gastropods, collecting food particles from the mud. In fact, it may have been selective feeding (Fedonkin, 1977). *Palaeopaschichnus* has been interpreted as a feeding meander trace (Glaessner, 1969; Palij, 1976), or one produced by the anterior end of the body and regular muscular contractions, leaving a succession of narrowly arched grooves along the trail, essentially perpendicular to the direction of movement (Fedonkin, 1977). Conversely these structures may have been produced by serial growth on the sea floor of sausage-shaped chambers of a large, single-celled protozoan (Seilacher *et al.*, 2003).

Trace fossils clearly reflecting movement through and feeding within sediments (called *Fodinichnia*), rather than organisms simply passing across the surface, have been given names such as *Planolites*. With its circular cross section and smooth surface, *Planolites* traces were formed by peristaltic motion that involved the whole body, not just the bottom surface. *Neonereites* may reflect a metazoan not only moving through the sediments, but leaving a trail of "fecal pellets," suggesting that this metazoan may have been eating, digesting and excreting mud plus undigested ingredients packed in slimy pellets after removing nutrients. Anthony Martin (*pers. com.*) has also suggested that the "pellets" could be pads of sediment pushed behind the trace maker by its walking legs (certainly observed in such forms as juvenile limulids). However, since no body fossils have been identified which could have left such fecal trails, such serial sets of

Figure 404. Aulichnites, bilobate locomotion trail, Suz'ma locale, White Sea region, northern Russia (M. Fedonkin).

209

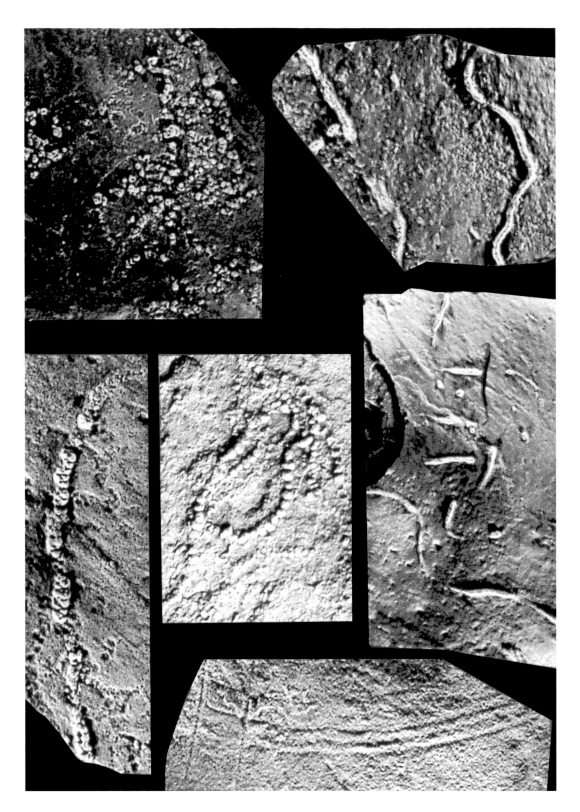

Figure 405. A variety of traces from the *Suz'ma locale, White Sea region, northern Russia. For detail see Figs. 401, 402 and 404 (M. Fedonkin).*

Fedonkin (1977, 1985) and Bengtson (1993), among many others, have noted a number of trends in the trace fossil record during the latter part of the Proterozoic, just before the Cambrian. Trace fossils tend to increase in size, complexity, diversity, depth of substrate penetration and abundance during this time, with a dramatic increase in diversity perhaps at the very end of the Proterozoic. In fact, this was the beginning of three-dimensional colonization of the sediment, providing access to new ecospace, with its new resource base.

However, Mary Droser and her colleagues (2002) considered that there was really only a very limited range of trace fossils in the latest Ediacaran strata. Many apparent trace fossils are tapering tubes or possibly even fragments of marine algae.

It was only at the base of the Cambrian System that a great variety of trace fossils appear, some indicating burrowing into the sea floor and some made by animals with hard appendages (Walter *et al.*, 1989; Landing, 1994; Droser *et al.*, 1999). As noted above, most Ediacaran trace fossils tend to be shallow trace fossils – there are no deep burrowers at this time – and most all are parallel to sea floor surfaces – examples being *Planolites Nenoxites, Aulichnites, Bulinichnus* and *Helminthoidichnites.*

To be specific, the number of different sorts of trace fossils known from the late Proterozoic Ediacaran times is less than 15, which is apparent from published names (Seilacher *et al.*, 2005). Behavioral complexity of these trace fossils is quite limited. Most of these have no clear connection with contemporary body

Figure 406. *Trilobite trackways, MacDonnell Ranges, Northern Territory, Cambrian (Steve Morton). Width of trace about 10 cm.*

Figure 407. *Artist's reconstruction of metazoan that produced* Climactichnites *traces, Cambrian of North America (M. Fedonkin).*

Figure 408. Climactichnites *traces, Cambrian of North America (M. Fedonkin).*

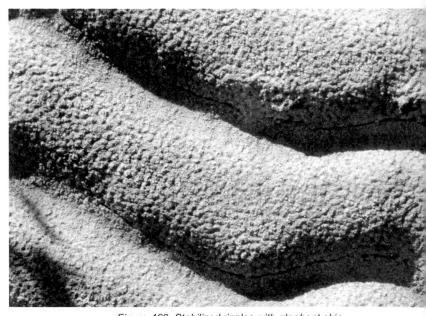

Figure 409. *Stabilized ripples with elephant skin texture. The microbial mats stabilized the ripple marks, likely the cause of their preservation. Distance between ripple crests about 8 cm (P. Vickers-Rich).*

fossils and most were no broader than about 10 mm – they were small. As mentioned before, most trace fossils are found in shallow water sediments. Both diversity and overall size of trace fossils decrease with increasing water depth. This trend is surely related to the decreasing availability of food resources and oxygen needed for respiration with depth as well.

Most of the small sediment processors seem to have fed by selectively removing minute food particles from the sea floor without ingesting large quantities of mud or sand, for they have left no strings of fecal pellets common to this sort of feeding.

The Verdun Syndrome and the Agronomic Revolution

Ediacaran metazoans apparently were not capable of penetrating the ocean bottom sediments any great distance, most "burrows" being no deeper than 2-3 mm. Many suggestions have been made about why burrowing and attendant sediment processing did not occur. One suggestion is that cyanobacterial-laden sediments may have been so anoxic below a certain depth that they were of no real nutritive value. Or perhaps the "burrowers" did not yet have the needed muscular set-up to allow sufficient forces to be developed in order to dig deep into the sediments.

Dzik (2003, 2004) and others have noted that perhaps there was no "need" for deep burrowing due to the lack of active predators with vision. Dzik applied the name "Verdun Syndrome" to the revolution that took place at the Ediacaran-Cambrian boundary, alluding to the World War I battle where armor and trenches were the only protection from enemy attacks. Thus, with the development of vision and active epifaunal predator hunting, there was immense selective pressure to "put on armor" (thus develop shells) or "dig trenches" into the sediments for survival. Perhaps the first deeper burrowing was not related to feeding at all, but to escape from predation. But, as Anthony Martin (*pers. com.*) points out, certain types of predators using other senses than sight (chemotaxis or "smell" and tactile sensations) were also part of the scene at this time and should be further integrated into the predator-prey scenarios developed to explain the development of defenses at the Precambrian-Cambrian changeover.

Seilacher and Pfluger (1994) called this change at the Ediacaran-Cambrian boundary the "Agronomic Revolution" because they noted how trace fossils change from two-dimensional mat scraping to three-dimensional plowing and deep burrowing on the sea floor. Droser and colleagues (2002) demonstrated that the process was more one of mixing. Animal activity progressively stirred up the mud on the sea floor and wiped out the evidence of shallow burrowing as the variety of sediment feeders increased and they experienced competition. However it is viewed, this change in the variety and intensity of burrowing reflects fundamental changes in marine ecosystems. Whether it was the appearance of predators, the "animal arms race" (the Verdun Syndrome), or the advent of "grave robbing" or mining for food (the Agronomic Revolution), the Early Cambrian marks a revolutionary change in the history of life. It is quite possible that intense competition for food resources in two-dimensional and highly populated habitats in shallow marine basins left little choice but colonization of deeper sediments and the ocean depths (Fedonkin, 1987).

Figure 411. Circulichnus trace, top of the Vendian sequence, Olenek River, Arctic Siberia (F. Coffa). Maximum width of slab 12 cm.

Figure 410. Palaeopascichnus sp., growth of serial sets of sausage-shaped tubes on bedding plane (South Australian Museum, 36854), Wonoka Formation, Flinders Ranges, South Australia (S. Morton). Maximum width partial slab 10 cm.

But feeding and retreat in sediments both require one condition – there must be sufficient aeration to allow oxygen-dependent organisms to use this resource. Aeration of the sediments is dependent not only on enough oxygen to be present in the water above the sea floor, but that it penetrates the sediments as well. Bioturbation by organisms, even to a shallow depth, began to open up this environment – not only did it bring oxygen but it also disrupted the bacterial mats that bound the sediments – it disrupted the stability of the substrate necessary for the formation of the mats and other biogenic structures like stromatolites. In addition, organisms that were processing the sediments enabled more oxygen to penetrate ocean floor sediments. Add to that the rise of filter-feeding organisms, such as sponges, in the world's oceans, which removed many of the fine particles and in the process "cleared" the marine waters. The result - light penetrated farther down and more effectively in shallower water alongside sediments increasing in porosity (Fedonkin, 1987, 1992; Logan *et al.*, 1995). This in turn fueled the colonization of many areas by photosynthesizing eukaryotes, which may have "invaded" areas previously dominated by bacteria, where anoxic conditions had prevailed. Thus, in many ways, the sea floors at the end of the Ediacaran were changing, being charged more and more with oxygen. They were definitely becoming more habitable deep down, both in the ocean depths themselves and within the sediments (Crimes & Fedonkin, 1994).

The dramatic increase in size and diversity of animals, and the beginning of their deeper sediment penetration, has been documented at the very end of the Ediacaran and the beginning of the Cambrian in the Rovno Regional Stage of the Eastern European Platform of Russia and in the Nemakit Daldynian horizon of the northern Siberian Platform (Fedonkin, 1990). It is not until the mid-Cambrian when trace fossils show more modern patterns. Interesting too is that, although the diversity of trace fossils and depth of burrows in shallow water environments increase in the Early Cambrian, deep-water communities (*e.g.* the Puncoviscana Formation ichnobiota of NW Argentina) are quite different. Microbial mats are still a part of the ocean floor ecosystem well into the Cambrian with associated typical trace fossils (Buatois & Mangano, 2003).

The Track Makers

Who were the perpetrators of such trace fossils? Were they related to living forms like modern cnidarian sea pens, which we know can quite actively burrow? Or were they trails left by more complex animals? So far, this is a difficult question to answer with any certainty. One fact is clear, however – few, if any, of the trace fossils were left by animals with appendages, for these would have left quite distinct marks. Some rare trace fossils from the White Sea region of Russia may be the rare exception, but they are yet undescribed. Such tracks do show up clearly in the Early Cambrian.

If there were mobile animals living on Ediacaran sea floors, why did they not leave more trace fossils? This seems very strange, as body fossils and trace fossils are often preserved on the same surfaces. It needs to be remembered that Ediacaran sea floors were colonized by biomats (Gehling, 1999; Seilacher & Pfluger, 1994), a combination of dead organic matter, bacterial films and all sorts of simple algae, including the relatives of modern seaweeds. These surfaces are commonly overlooked in the search for the geometrically regular biological shapes of individual animal fossils. But if one looks at an image of an Ediacara fossil, the background upon which it is preserved is almost always criss-crossed with ridges, grooves, humps and bumps with a texture that resembles leather or some sort of a meshwork. This so-called "old elephant skin" or "rhino hide" pattern is common in all Ediacara fossil assemblages. In places it is obvious that these ancient mats had been ripped up and shuffled together like shards of pottery where they had been stiffened by mineral deposits. Some surfaces appear as if they had been covered by a plastic wrap and had been afterwards wrinkled and deformed. It should be noted, however, that preservation of surface traces is the exception, be it with or without the presence of ubiquitous mats (S. Jensen, *pers. com.*).

The point to be made is that tiny Ediacaran animals moving over these mats could not leave trails that would be preserved unless they could break through, furrow into them or even dissolve away these mats. Then, and only then, could traces be preserved as fossils.

Despite the low diversity of the Ediacaran trace fossils, still there is some variety that reflects a number of different behaviors where animals have managed to actively graze on or furrow through or into the mats. *Planolites* trace fossils are rather straight, sometimes crossing one another's paths. And then there are the meandering or sinusoidal patterns of *Nenoxites*, *Aulichnites* and *Helminthoida*. Whereas *Aulichnites* is a "leisurely" meandering, bilobate path, *Yelovichnus* is tightly curved, suggesting a very thorough processing of the sediments. *Gordia*, as its name implies, seems to tie itself in knots, twisting and turning on the flat ocean bottom sediments. There are other patterns like the circular trace fossils of *Circulichnus* or even star-shaped bioturbations, though those are less common. A few other trace fossils even show some depth of burrowing, such as *Bergauria*, which can form either on the horizontal or in more vertical orientation, leaving a series of "rings" perhaps imprinted by a peristaltic type of motion as the trace maker moved upwards or sideways.

Trail's End: Metazoan Body Fossils Associated with Trace Fossils

Only in rare cases are metazoan body fossils so characteristic of late Ediacaran times found in association with traces of their activity. These have only been recognized in the last decade. They are indeed rare exceptions and are best preserved in the clays and silts exposed along the cliffs of the White Sea in northern Russia, to the east of Arkhangel'sk and on the sandy sea floors preserved in the Flinders Ranges of South Australia, where Ediacara animals lived below fair-weather wave reach on the slopes of deltas. Special conditions entombed this assemblage – sand avalanches, caused by storms or seasonal water mass pulses from stream inflow along the coasts, off the shoulder of a shallow marine shelf, emplaced fine sand over a muddy bottom with a framework of cyanobacterial mats. These mucilaginous mats essentially bound the sediments together, preserving many sedimentary structures, such as ripple marks, which would otherwise have been washed away by bottom currents or tides. Grazing on this organic-rich carpet, itself with many layers ranging from oxygen-rich upper to anoxic base, were such forms as *Yorgia*, *Dickinsonia* and *Kimberella*. They left their "absorption" footprints, their "grazing" plow marks and sometimes even a gliding trail.

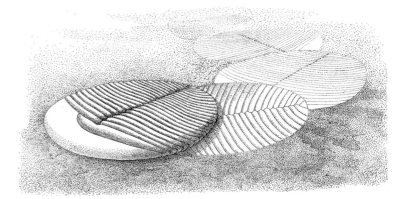

Figure 412. Reconstruction of Yorgia waggoneri leaving a trail of absorption traces in the pervasive microbial mat that covered the seabeds of the time (drawn by A. Besedina and I. Tokareva for M. Fedonkin).

Yorgia waggoneri is a rather large (up to 25 cm), bilaterally symmetric, segmented organism. It appears to have been an animal with a proper body cavity, a coelom – a group called triploblastic, perhaps related to annelid worms or some more complex group (Dzik & Ivantsov, 1999). Along with another form, *Dickinsonia*, it is placed in the family Dickinsoniidae.

Both *Yorgia* and *Dickinsonia* left distinctive "footprints" in the sands and muds of the Ediacaran, and both are similar to the gigantic *Epibaion*, one which reached lengths of over 44 cm (Ivantsov & Malakhovskaya, 2002; Dzik, 2003; Fedonkin, 2003). *Yorgia*'s trace fossils are a series of similar smooth, oval shapes, which outline the body with a distinctly segmented pattern. The oval footprints form a chain or sometimes a cluster and may slightly overlap one another. Each "footprint" or "touch down" platform along the trail reflects the finest detail of the outside of *Yorgia*, an impression of its underside. Molds of *Dickinsonia costata* have been found at the end of a series of overlapping casts of "footprints" that are identical to each other, but just a little larger than the preserved body fossil impression of this metazoan. These imprints are the hollows left where *Dickinsonia* sat and rotted or absorbed the underlying mat before moving on until it and its footprints were smothered by a sand avalanche and both preserved as fossils (Gehling *et*

Figure 414. Yorgia waggoneri, *Vendian sequence, Valdai Series, Zimnii Bereg, White Sea, Russia, part of a large quarry block collected by Andrey Ivantsov (F. Coffa). Body fossil width about 4 cm, but can reach larger sizes, up to 25 cm.*

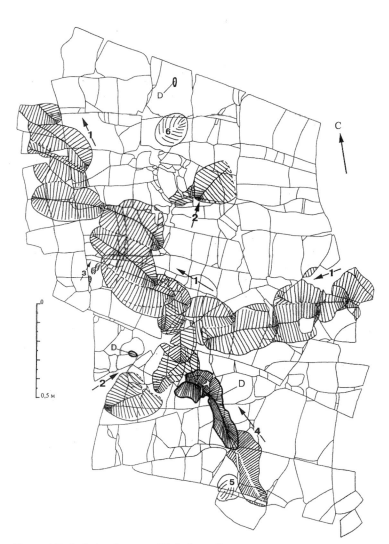

Figure 413. *A chain of traces of* Epibaion axiferus *and a body fossil of a very large* Dickinsonia sp. *at the end of a chain of "footprints" (Ivantsov & Malakhovskaya, 2002) (PIN 5024).*

Figure 415. Yorgia waggoneri, *Vendian sequence, Valdai Series, Zimnii Bereg, White Sea, Russia, part of a large quarry block collected by Andrey Ivantsov (PIN 5024) (F. Coffa). Body fossil width about 4 cm.*

al., 2005). It certainly was an organism that appears to have moved, thus ruling out ideas that it could have been some sort of fungi or lichen (Retallack, 1994).

The smooth and positive relief of the trail, the absence of a sharp boundary between the trail and the surrounding sediments and uniform shape of the "footprints" signal the difference between a trace fossil, a trail and the negative imprint left by a body. Body fossils usually show a wide spectrum of the deformations and preservation styles in what paleontologists call a deep negative hyporelief – which is a concave imprint imposed on the bottom (or sole) of the sediments that covered the animal, bringing about its demise. To the contrary, the "footprints" appear to have formed as positive casts of hollows made into the cyanobacterially impregnated sediments.

What seems to have happened in the case of *Yorgia* and *Epibaion* is that they settled on the microbial or algal mat and somehow absorbed the nutrients below. Once those nutrients had been exhausted, perhaps when the anoxic layers of the sea bottom mats were reached, *Yorgia* loosened its grip, so to say, and moved on to another spot where it began "feeding" again. Perhaps this was by gliding to the next spot, or perhaps even "coming unstuck" and gently floating to the next drop point, where the process was repeated all over again. It might have used some sort of ciliary organ to creep or just expanded and contracted like a worm and moved on to its next spot. Nonetheless, it left no remains of the movement trail, as the motion did not affect the underlying mat as did the longer "feeding" stop. When these "prints" were first uncovered they were certainly not thought to be feeding traces. They were interpreted as impressions of two different animals, then as roll-over prints of the same animal (Dzik & Ivantsov, 1999), then as traces left as resting spots by an organism using mucous cement to stick it to the bottom (Ivantsov, 2001) and finally as feeding traces (Ivantsov & Malakhovskaya, 2002), a type of feeding for large organisms unknown in the Phanerozoic.

Fan Scratchers

Kimberella quadrata was originally described as a cubomedusa (Glaessner & Wade, 1966; also Jenkins, 1984). However, new fossil material from the Vendian of the White Sea region has changed that view. *Kimberella* appears now to have been a mollusk-like, triploblastic animal with a tall,

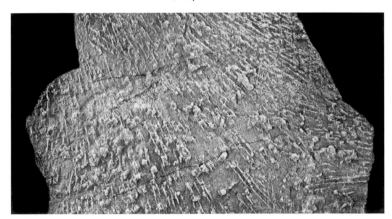

Figure 416. Radulichnus, *the paired scratch marks left by* Kimberella quadrata *as it fed on the microbial mat. Ust'-Pinega Formation, Upper Vendian sequence, White Sea, northern Russia. Block width about 9.5 cm., latex peel of sole of bed, whitened for photography (A. Mazin).*

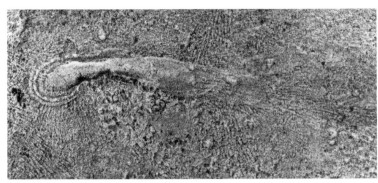

Figure 417. Kimberella quadrata *body fossil and associated trace (PIN 4853-318), Ust'-Pinega Formation, Upper Vendian sequence, White Sea, northern Russia. Width of body fossil @ 5 mm. Latex peel of bed sole, whitened for photography (A. Mazin).*

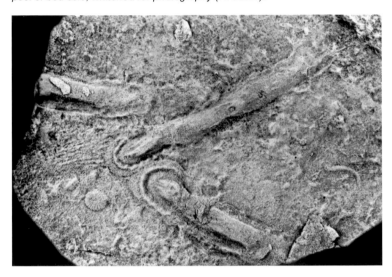

Figure 418. Kimberella quadrata *body fossils and associated traces (PIN 4853-5), Ust'-Pinega Formation, Upper Vendian sequence, White Sea, northern Russia. Length of middle trace and body fossil about 4.5 cm. Latex peel of bed sole whitened for photography (A. Mazin).*

soft and unmineralized shell (Fedonkin & Waggoner, 1997). Recent excavations, systematically carried out by Andrey Ivantsov, on the Solza River of the Onega Peninsula to the west of Arkhangel'sk, exposed rocks representing ancient sea beds with a number of short crawling trails and feeding scratch marks clearly associated with numerous small (0.5 inch long) specimens of *Kimberella*. The scratch marks are arranged in the shape of fans.

Other trace fossils made by *Kimberella* are scratch marks arranged in fan shapes. These scratch marks were originally thought to be spicules, first noticed by Martin Glaessner and Mary Wade (1966) in the Ediacaran rocks of South Australia's Flinders Ranges. Somewhat later Richard Jenkins (1992) suggested that these trace fossils had been produced by some sort of arthropod. In 1991, Jim Gehling described these trace fossils in detail, noting the fan-like array. After the discovery of two specimens of arrays of scratch trace fossils together with footprints of *Kimberella*, he suggested that they were the feeding impressions of some sort of mollusk (Gehling et al., 1996). That same year, Fedonkin and Waggoner (1997) made a strong case for *Kimberella* being a "bilaterally symmetrical, benthic animal with a non-mineralized, univalved shell, resembling a mollusk in many respects." The story began to come together as more and more specimens were collected from the Vendian sequences, especially by Andrey Ivantsov, along the White Sea in northern Russia. Dolf Seilacher then made the connection of such scratch marks with *Kimberella* in a letter to Mikhail Fedonkin. Jim Gehling (Ph.D. thesis, 1996) had independently made the same connection when he noted on two separate slabs collected in 1993 from the Flinders Ranges fans associated with *Kimberella*. The specimen that clinched the association of *Kimberella* with the scratch marks, however, was one palm-sized slab found in an Italian collection repatriated to the Paleontological Institute in Moscow. It had been illegally excavated from the White Sea cliffs, sold on the open market and ended up in a foreign collection. Once the director of the museum where the specimen was housed discovered its history, he returned it to its rightful home. This single specimen provided the best evidence for association. Since then, more associated scratches and body fossils of *Kimberella* have been found. More than 800 specimens of *Kimberella* have been recovered and are part of the collections of the Paleontological Institute, Moscow, giving an in-depth picture of this fascinating, complex metazoan which existed long before animals developed true shells and skeletons.

First described by Fedonkin (1992) as radial arrays of sponge spicules, such fan-shaped scratch marks are common in the Vendian fossil assemblages of the White Sea region. They are preserved in positive relief on the bottoms or soles of sedimentary beds, indicating excavation to depths of 1- 2 mm. The scratches come in pairs and seem to reflect the imposition of paired radula-like elements that may have been borne on the end of a retractable proboscis that was first extended and then drawn back into the mouth region where food was deposited, then the proboscis extended again. Once the region that could be reached by extension of this proboscis (which has been preserved on some of the White Sea specimens) was fully "farmed," *Kimberella* backed up and began to farm a new area, thus moving backwards, rather than forwards. This was a reasonable strategy to avoid wasting time moving over an already exhausted food source.

The Currency of Trace Fossils

Trace fossils are useful indicators of the behavior of long-dead organisms, and certainly occur in some abundance in Ediacaran sediments. They paint a world where most activities took place on the sea floor and not below it. They also reflect that animals were using a variety of feeding styles, even in these early times. They must, however, be used with caution since a single kind of animal can leave a variety of trace fossils and different animals can, of course, leave traces that are indistinguishable. So, it is that, for the most part, trace fossils, though they bear names of genera and species as do body fossils and those with shells and skeletons, the names do not mean the same thing. They do not necessarily represent one particular species, as do the binomial names like *Homo sapiens*, for us humans. They also are not nearly so useful for telling time, for use in biostratigraphy, as are body fossils. But, clearly, they add a dimension to the understanding of the past, unique to them, and when used together with body fossils and composition/structure of the enclosing sediments, yield a better understanding of Ediacaran environments and biodiversity than any one line of evidence ever will, but need to be consistently revised in light of new evidence (Jensen, Droser & Gehling, 2006). Best of all, trace fossils showcase the behavior of once living metazoans better than any other line of fossil evidence (A. Martin, *pers. com.*).

Figure 419. "Treptichnus pedum," *a trace fossil, from the late Neoproterozoic/Early Phanerozoic of Namibia (photo by M. Fedonkin & P. Vickers-Rich). Length of longest trace about 4 cm.*

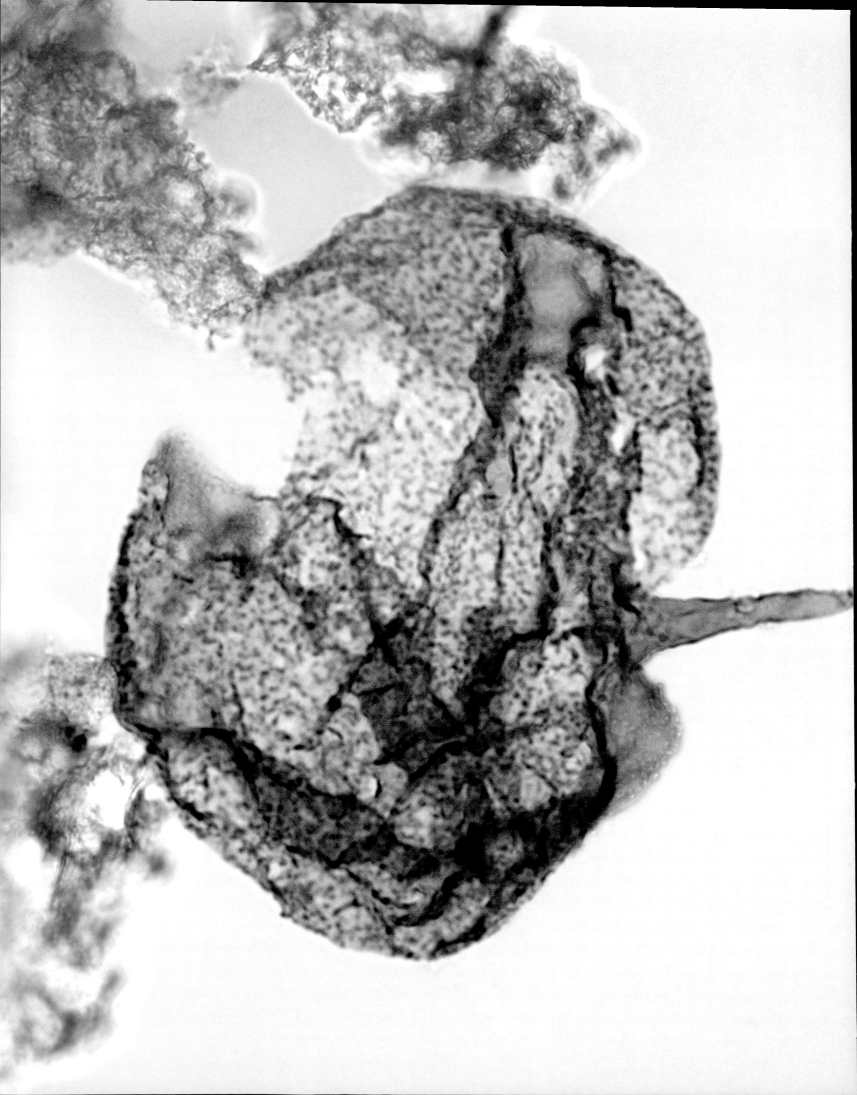

CHAPTER 13

The World of the Very Small:
Fueling the Animalia

K. Grey

It was Darwin (1859) who pointed out a mystery that has only been partially answered today. Although fossils were abundant from the Cambrian period onwards, there was a lack of evidence for life in the preceding Precambrian era. This was despite the fact that living things already showed a high degree of organization by the time of the first known fossils. Darwin's comments spurred researchers to find evidence of this earlier life. We now know that life was abundant throughout much of the long interval before the Cambrian explosion; the problems are mainly ones of size, preservation (most organisms lack hard parts) and knowing what to look for and where to find it.

Many Precambrian fossils belong to the world of the very small. Various microfossils had already been described from the Phanerozoic, and it was not long before reports appeared of Precambrian fossils, most notably those of Walcott (1883, 1895, 1899). Other records were added in the early 1900's, and these included descriptions of the structures now known as stromatolites, which are and were built by microbes, mainly bacteria. Kalkowsky (1908) coined the German term "stromatolithi," which later became "stromatolite," and pioneering studies of these structures were carried out in Australia by Mawson (1925) and Madigan (1932, 1935). However, there was considerable debate about the biogenicity of both the microscopic structures and the edifices they built, a problem that persists even now. The sometimes-heated debates about whether Precambrian structures were really fossils, resulted in a highly skeptical publication by Seward (1931), which, as Schopf and Klein (1992) pointed out, stifled research on Precambrian fossils for many years. One or two reports of Precambrian fossils continued to filter through, but it was not until Reg Sprigg's discoveries (Sprigg, 1947, 1949) of fossil soft-bodied "jellyfish" from South Australia (as described above) and their subsequent confirmation as being older than Cambrian in age, that the study of Precambrian fossils began in earnest. The study of the world of the very small, the microfossils, was to play a major role in this.

Opposite page: Ceratosphaeridium mirabile, *565 million year old acritarch from the Wilari Dolomite Member, Observatory Hill 1 drill hole, Officer Basin, South Australia (Commonwealth Palaeontology Collection 36417). Specimen is 124.5 μm in maximum diameter (K. Grey).*

Figure 420. Charles Darwin (courtesy of J. Stilwell).

Overcoming Barriers

As well as examining the actual remains of metazoans, one needs to understand the food sources that fueled these organisms and thus, this chapter takes a brief look at the world of the very small, the planktonic and benthic microbiota that fueled the rise of animals. These small fossils often reflect the state of environmental conditions as well as serve as time markers that make it possible to correlate between geographically distant sites bearing the earliest of metazoans.

The atmosphere of mistrust of Precambrian fossils prevails to this day, particularly with regard to some of the oldest fossil-like structures. Claims that structures older than 3.0 billion years old, or even 1.9 billion years old, are microfossils or stromatolites have recently been challenged, especially by Martin Brasier and his colleagues at Oxford University (Brasier *et al.*, 2002; Morbath, 2005). The similarity of some Archean microfossils to chemically precipitated structures or biomorphs (Garcia-Ruiz *et al.*, 2003; Hyde *et al.*, 2004) also causes unease. Current interest in the search for life, and more especially for fossil life, on Mars has caused scientists to critically re-assess the criteria used for recognizing fossils. Such problems have limited the use of two significant aspects of Proterozoic paleontology: stromatolite biostratigraphy and palynology. Both methods have the potential to produce substantial contributions to Proterozoic correlations, but have rarely been accepted or applied outside the former Soviet Union, Europe, China, and to a lesser extent, Australia and India. In particular, North American investigators have concentrated on the sedimentological controls of stromatolite morphology, and on the paleobiology of associated chert microfossils. However, the possibility of using the fossils for telling time (biostratigraphy) (Hill *et al.*, 2000) has largely been ignored. These factors have influenced the direction of research in Proterozoic paleobiology and the aspects of study that have received greatest attention.

Zonation			Assemblage zone
Ediacaran Complex Acanthomorph Palynoflora (ECAP)		4	*Ceratosphaeridium mirabile / Distosphaera australica / Apodastoides verobturatum*
		3	*Tanarium irregulare / Ceratosphaeridium glaberosum / Multifronsphaeridium pelorium*
		2	*Tanarium conoideum / Schizofusa risoria / Variomargosphaeridium litoschum*
		1	*Appendisphaera barbata / Alicesphaeridium medusoidum / Gyalosphaeridium pulchrum*
Ediacaran Leiospere Palynoflora (ELP)		L	*Leiosphaeridia crassa / Leiosphaeridia jacutica*

Figure 421. Australian Ediacaran acritarch zones (after Grey, 2005). (D.Gelt)

What is Palynology?

Palynology is the study of acid-insoluble microfossils. These minute fossils can be present in great abundance in small samples of rock, especially of drill core. They include groups such as spores, pollen, acritarchs, dinoflagellates, diatoms and chitinozoans. Many of these groups are found only in rocks younger than Precambrian, but the acritarchs are very significant in later Proterozoic rocks. Fossils are extracted by dissolving about 25 g of ground-up rock (preferably shale or siltstone) in a succession of acids. Hydrochloric acid dissolves any carbonate, and hydrofluoric acid (highly dangerous, thus requiring very strict safety precautions) dissolves silicates (the major constituent of most rocks). The residue consists mostly of organic material and a few heavy minerals. The latter can be removed by the use of heavy-liquid separation in a centrifuge, or if the fossils are too delicate, by a swirling technique similar to gold panning. Some fossils may have a coating of coal-like material or of pyrite, and they can be cleaned up by the use of fuming nitric acid. Residues are filtered to remove very fine particles of organic matter that could obscure the view of the fossils, and the fossils may be separated into different-sized fractions for easier observation. Finally the residue is strewn evenly on a microscope slide, and the coverslip cemented in place. Fossils obtained in this way are of great value for oil and gas exploration. They are usually abundant, are widespread regardless of local environments and show rapid evolution. They allow an exploration geologist to determine precisely where a sample came from in the geological column – that is, its age.

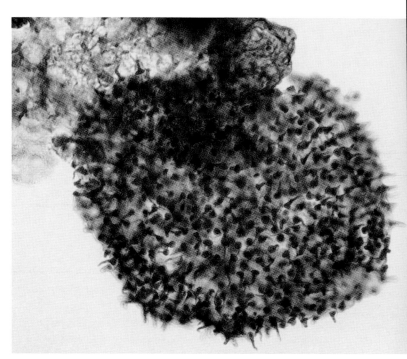

Figure 422. Taedigerasphaera lappacea, *565 million year old acritarch from the Wilari Dolomite Member, Observatory Hill 1 drill hole, Officer Basin, South Australia (Commonwealth Palaeontological Collections 36506). Maximum diameter, 170 μm (K. Grey).*

In the 1930's, paleontologists began to study microfossils extracted from coal and to use pollen in examining climatic changes associated with the Quaternary ice age. These studies rapidly developed into the use of palynology as a biostratigraphic tool throughout much of the Phanerozoic. Major advances in the classification of acid insoluble microfossils were made in the early 1960's, when dinoflagellates and acritarchs were recognized as separate groups of organisms (Downie *et al.*, 1963; Evitt, 1963). Studies of early Paleozoic acritarch zonation advanced rapidly. A few pioneering studies of Proterozoic palynomorphs took place, but many of the species described are now regarded as abiogenic, or at the best as dubiofossils. Soviet palynologists began to document Proterozoic palynomorphs in the early 1950's (Naumova, 1951) but

detailed studies began in the 1960's, and numerous papers reporting Precambrian microfossils were published. The work was championed by Boris Vasilievich Timofeev, who examined Proterozoic material from a variety of geographical areas (Jankauskas & Sarjeant, 2001). Timofeev and his Russian colleagues published more than 30 papers in rapid succession, documenting fossils in the latest Proterozoic, as well as carrying out studies on even older microfossils. However, few of these publications received attention outside the USSR, largely because of the difficulty of obtaining and translating the literature. Western scientists were discouraged by the poor-quality illustrations, poor stratigraphic constraints, and the perception that the material contained many artifacts and contaminants.

In the 1950's and mid-1960's, Western scientists rediscovered Precambrian paleontology. Interest was rekindled by finds of unequivocal and exceptionally well-preserved chert microfossils in the Gunflint Chert in North America (Tyler & Barghoorn, 1954; Barghoorn & Tyler, 1965) and in the Bitter Springs Formation of Australia (Barghoorn & Schopf, 1965). In particular, the Gunflint Chert provided evidence that stromatolites were formed by the activities of bacteria, because abundant, well-preserved examples of their fossilized remains are found forming the stromatolitic lamina. At about the same time, descriptions of extant stromatolites still forming in Shark Bay, Western Australia, were published (Logan, 1961), further establishing the role of microbial mats in the construction of stromatolites. Recognition that many chert microfossils were remains of benthic microbial communities (BMC's) opened up the discipline of Proterozoic paleobiology, and renewed interest in interpreting paleoenvironments through comparative studies of sedimentology and microbiology. Moreover, the fact that the fossils were embedded in chert removed doubts about the provenance of the fossils; they were clearly present at the time the chert crystallized. North American researchers, in particular, pounced on this new approach for studying early life because organic material was preserved in superb detail in fine-grained chert.

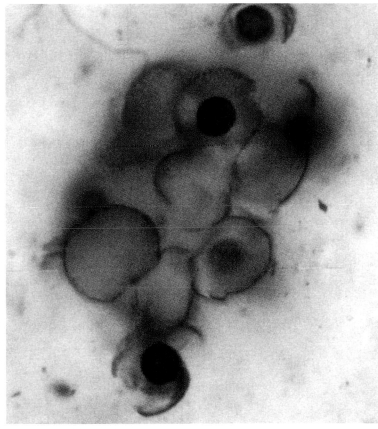

Figure 423. Microfossils in Neoproterozoic Bitter Springs Formations chert, Macdonnell Ranges, Northern Territory, Australia (J. Warren and I. Stewart). Individual cells about 10 μm in diameter.

The downside of the concentrated research on chert microfossils was that it directed attention away from other types of microfossils, particularly those that were preserved in mudstone, siltstone and shale, and which required palynological extraction. The systematic documentation of biostratigraphic ranges, vital to Phanerozoic biostratigraphy, was neglected. A dichotomy occurred,

with North American paleobiologists concentrating on Precambrian paleobiology, taxonomy and paleoenvironment, and European, USSR and, to a lesser extent, Chinese geologists studying palynology and biostratigraphy. Chert microfossils continue to provide valuable insights into early evolution, but they present a biased representation of Proterozoic environments because chert is mainly formed in shallow water. It is also difficult to use chert microfossils for systematic stratigraphic analysis, because chert horizons have a patchy distribution. Fossil descriptions tend to be of individual specimens occurring in stratigraphic and geographic isolation, and the surface of the microscope slide usually cuts through specimens so they are viewed in cross-section. This contrasts with palynological specimens, which are usually complete and have been laterally compressed during burial. These different ways of observing fossils have inevitably led to problems with how the fossils are named, and recent studies have had to resolve problems of duplication or misidentification.

The Problems of Contaminants and Degradation
Precambrian micropaleontologists must worry about the problems of artifacts and contaminants. Some abiogenic structures can resemble microfossils and form in the type of environment where the original organisms lived. They may have formed when the sediment was buried and/or during diagenesis. If fossils are present in thin section, you have to be sure they are in the rock matrix, and not in veins or cracks that may be much younger in age. They can also be introduced during laboratory preparation. This means that great care must be taken in the laboratory to ensure that no cross-contamination takes place between samples or by atmospheric pollution. It is important for the researcher to be familiar with possible contaminants and to dismiss anything that looks suspicious. One way of monitoring possible contaminants is to leave a slide exposed in the lab with a thin coating of mounting medium and to periodically check for contaminants. It is also worth checking small amounts of the chemical reagents used, because as they get older, bacteria and fungi may grow in them. Another way of checking is to process a rock such as a piece of granite with the batch of normal samples. If you find microfossils in the granite control sample, you know there are contamination problems!

Another factor that must be considered when trying to identify species is the taphonomy or postmortem degradation of the specimens. Not every specimen present on a slide will be a textbook example of what the species looks like. Specimens can decompose after they die, and features such as spines become twisted or broken as the specimen is buried. Micropaleontologists must learn to recognize degraded specimens as well as ones that are superbly preserved.

Species Identification
Identifying a specimen means the paleontologist must be familiar with the species already described – worldwide. Each specimen is carefully matched to a described species to make sure that it is similar in morphology and dimensions. Where possible, size ranges are plotted as scatter diagrams to show the range of variation. Specimens that fall outside known patterns are described as new species. Their affinities can be determined by matching them to modern analogues and known species. For Proterozoic acritarchs, a major reference work is an extensive review of fossil protists and their modern counterparts by Tappan (1980). The paleontologist then needs to understand the biology of modern organisms and their distribution patterns in order to interpret the fossil ones.

The task of describing microfossils and palynomorphs from a range of Precambrian stratigraphic successions has continued with increasing momentum, and both the stromatolite and microfossil record are now well documented. Interest has focused on two areas. One is the search for the earliest records of life (Schopf, 1983; Schopf & Klein, 1992).

Figure 424. Modern environments where microbialites (stromatolites and thrombolites) form today (upper left, Lake Clifton, W.A.; upper right, Lake Walyungup, W.A.; lower left, Lake Richmond, W.A.; lower right, hot springs in Yellowstone National Park, Wyoming, North America) (K. Grey).

Such studies have been invigorated in recent years by the search for fossil life on Mars. The second major focus has been on the documentation of biotas in the latest Proterozoic. Microfossil assemblages associated with the evolution of the Ediacaran biota have become the center of attention for several reasons. Palynological studies have played an important role in defining the new Ediacaran system. The development of a preliminary acritarch zonal scheme holds promise of better dating and correlation throughout this interval. Additionally, acritarchs were most probably the food source for the earliest metazoans, so it is important to understand how they evolved and responded to environmental changes and how changes in phytoplankton populations may have resulted in corresponding changes in animal populations. Such studies are in their infancy, but are already producing exciting results (Grey, 2005).

Microfossils Through Time

Uncertain still are the ranges and precise ages of many Precambrian microfossil assemblages or of the stromatolites constructed by certain forms of microbial activity. But, it is clear that life on Earth evolved slowly and steadily for about 2500 million years before diversification rates changed drastically some 600 million years ago. The earliest organisms were probably Archaea, followed by simple Bacteria. These organisms likely used chemical pathways to provide their energy source. The oldest widely accepted direct evidence for life is found in the 2.8 billion year old Fortescue Group of the Pilbara, and in similar-age rocks from southern Africa, as noted

Evidence for Life on Another Planet

Exploration on Mars has not yet produced convincing evidence of life, although the discovery of atmospheric methane raises interesting questions about how this gas could be produced at constant levels without a living source. While debate continues about the possibility of extant Martian life, scientists are also searching for extinct life. Earth and Mars had similar early histories. If life evolved on Earth, could it have also evolved on our sister planet before it lost most of its atmosphere? The answer is to search for fossil evidence. The Mars Rover missions have found similar sedimentary environments to those containing Earth's ancient fossils. Investigations continue, and chemical analysis of rock samples for traces of biogenic carbon plays a vital role in these studies. There is also a search for body fossils, the morphological structures that were once part of a living organism.

The search for Martian fossils has provoked enormous debate about how to identify fossils in Earth's oldest rocks. First, of course, there needs to be confidence in the biogenicity of Earth structures before making comparisons with any similar structures found on Mars. Some scientists, Martin Brasier among them, now claim that structures from the Pilbara in Western Australia, previously thought to be the oldest microfossils from rocks ranging in age from 3.496 to 3.465 billion years, are simply artifacts, and not indicators of biogenicity. Similarly, they are unconvinced that domes in the Dresser Formation and cones in the Strelley Pool Chert (3.496 and 3.388 billion years old, respectively), also in Western Australia, are stromatolites, preferring to interpret them as ripple marks or chemical precipitates. As debate rages, researchers are re-evaluating criteria that define microfossils and stromatolites. This involves reviewing fundamental principles, including descriptive terminology, and environmental context.

The search for life on other planets is an exciting new frontier of science known as astrobiology. It is a multidisciplinary study, and astrobiologists will need a firm grounding in paleontology and its techniques to evaluate any evidence of fossil life obtained from our neighboring planets.

in Chapter 1. As previously mentioned above, well-developed stromatolite-like structures in the Warrawoona (Walter *et al*., 1980) and Kelly Groups of the Pilbara, ranging in age from 3.5 to 3.4 billion years (Hofmann *et al*., 1999), are not currently accepted as stromatolites by some researchers, although others point to their similarity to much younger forms and regard them as evidence for the continuity of the fossil record back to the earliest Archean. The presence of microfossils has also been disputed.

Bacteria appear to have been the predominant form of life for at least 2 billion years. They formed mats on the substrate, the benthic microbial communities (BMC's), and some of them induced precipitation, producing structures known as stromatolites. It is still uncertain when bacterial organisms began to use photosynthesis to extract energy from sunlight. There is some chemical evidence to suggest that organisms which produced oxygen as a by-product of photosynthesis were likely preserved in the Fortescue Group. They may have evolved earlier than this, but convincing evidence is lacking. The atmosphere must have contained abundant free oxygen by the time the large areas of banded iron formations (BIF's) were formed (see Chap. 1).

The earliest eukaryotes, in the group Algae, may have appeared in the later Archean, but evidence for this is sparse. Small spheres and filaments have a record extending back

Figure 425. Different types of fossil stromatolites: Top, from left to right: Baicalia burra, Tungussia wilkatanna, both branching stomatolites on cut faces of drill core (core about 40 mm in width), about 750 milion years in age; Conophyton "new form" from the Stag Arrow Formation, aged about 1050 million years (maximum width of illustrated specimen abut 70 mm, but cones can be up to 50 cm in diameter and Ephyaltes edingunnensis, from the Yerrida Group, 2100 million years old (maximum diameter about 80 cm. Bottom, left: Acaciella australica from the Skates Hills Formation, about 800 million years old. The domes are up to 2 m in diameter, the smaller columns about 1 cm in diameter. Ian Williams – geologist with the Geological Survey of Western Australia (in photo) discovered the "string of beads" Horodyskia fossils in Western Australia. Bottom, right: Earheedia kuleliensis, a large dome of the type illustrated in Figure 20, about 1750 million years old. Malcolm Walter, a major contributor to Archean and Proterozic research in Australia, provides the scale (K. Grey).

into the Archean, but it is not until about 1850 million years ago that spheres with larger diameters and classed as acritarchs first appear in the rock record. They are generally considered to be of algal affinity, although the evidence for this is still inconclusive.

Stratigraphic-distribution patterns of organisms for much of the Late Mesoproterozoic and early Neoproterozoic remain unclear, and additional studies are needed before

Figure 426. Chuaria circularis, *a megascopic algae,* Neoproterozoic, Puza Shale, Spain (F. Coffa).

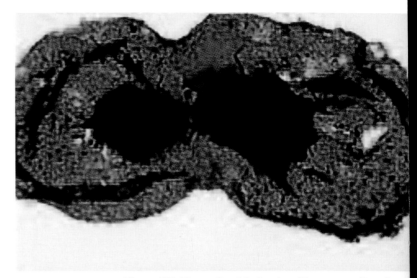

Figure 427. The acritarch Leiosphaerida crassa *with cell division underway, ABC Range Quartzite, SCYW1a drill hole, Stuart Shelf, South Australia dated at 600 million years. Specimen width 312.5 μm wide (K. Grey).*

time spans for different fossils are certain. Better dating is needed for much of the rock succession. For the most part, pre-Ediacaran microfossils from around the globe seem to consist of large populations of colonial coccoid bacteria, abundant filaments, other morphologically simple spheres of probable bacterial origin, together with simple sphaeromorph acritarchs, sparse, often large, sphaeromorph acritarchs, some ornamented forms, and rare spiny acritarchs, all of probable algal origin (Knoll, 1996). Some middle Mesoproterozoic rocks, such as the Bangemall Group of Western Australia, contain forms such as *Crassicorium pendjariensis* and *Pterospermopsimorpha pileiformis*, which have complex, multilayered structures. Other Mesoproterozoic and early Neoproterozoic acritarchs are characterized by the formation of one or more membranes, such as *Trachyhystrichosphaera aimika*. This species has hollow, cylindrical, processes of variable length, that are scattered and irregularly arranged, and which support a thin, outer membrane. *Valeria lophostriata* is a large form characterized by thin, parallel, closely spaced ribs or costae that looks like a pale fingerprint. There are some rare acritarchs (acanthomorphs) with spines or processes (relatively major morphological features that project from the vesicle surface and are typically more than about 10 % as long as the diameter of the vesicle or body of the acritarch).

Typical process-bearing taxa include *Tappania* and *Vandalosphaeridium*. *Tappania plana* is present in the @1400 million year old Roper Group of the Northern Territory (Javaux et al., 2001, 2003) and in the Baicaoping Formation of China (Yin & Gao, 1996). The processes in many of these pre-Ediacaran acritarchs appear to be different from those seen in the Ediacaran (prompting Nick Butterfield to suggest recently that many may be fungi – Butterfield, 2005). They are variable in length, suggesting that they grew outwards from the vesicle at different times (sometimes they consist only of a bud-like projection, whereas others on the same vesicle may be long, tubular structures). They tend to be cylindrical in shape and have a hollow, tubular center, and are sometimes open-ended.

An important and diverse assemblage of acritarchs was described from the Lakhanda and Miroedikha formations of Siberia (Timofeev, Herman & Mikhailova, 1976; Jankauskas et

al., 1989; Herman, 1990), which for many years were thought to be early Neoproterozoic in age, but which are more likely older, Mesoproterozoic in age (Rainbird et al., 1988). Similar assemblages are present in the Avzyan Formation in the Urals (Sergeev, 1992, 1994), and the Changlongshan Formation in the western Yan Shan region of China (Luo, 1991).

Add to these the enigmatic "strings of beads" (see Chap. 2), which some consider to be metaphytes (ancient plants similar to some modern seaweeds) (Grey & Williams, 1990) and others suggest may be animals, such as hydrozoans (Yochelson & Fedonkin, 2000).

By the early Neoproterozoic or Tonian (1000-850 million years ago), acritarch assemblages were beginning to show considerable diversity. Coccoid bacteria and filaments were still common, as were the simple spherical leiospheres, but the Tonian was characterized by the appearance of a broader range of morphology and abundant large sphaeromorph acritarchs.

Ornamented forms, such as *Valeria lophostriata*, became more numerous, and groups such as the prismatomorphs, typified by *Octoedryxium truncatum*, are recorded for the first time. Forms with equatorial flanges, the pteromorphs, are common, although these may have a record dating back to @1600 Ma (Buick & Knoll, 2001). The large spherical acritarch, *Chuaria circularis*, is a distinctive species that can be abundant in this part of the succession. Several species of process-bearing acritarchs were reported from the 900–800 Mt Wynniatt Formation of the Shaler Group, Victoria Island, northwest Canada (Butterfield & Rainbird, 1998), although Butterfield now considers some of these forms to be types of fungi rather than acritarchs. Ornamented forms, such as *Trachyhystrichosphaeridium laminaritum*, characterized by alveolate walls; *Trachysphaeridium laufeldii*, which has echinate sculpture and a rimmed, circular opening (one of the few probable excystment structures recorded from the Neoproterozoic); *Kildinosphaera verrucata*, which has bulbous, verrucate sculpture; and *Cymatiosphaeroides kullingii*, with numerous, thin, solid, cylindrical processes that support a multilaminate membrane, are present in the later Tonian, and many of these forms extend into the Cryogenian

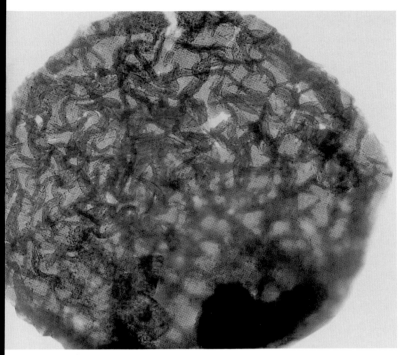

Figure 428. Cerebrosphaera buickii, *an acritarch from the Hussar Formation, Empress 1A drill hole, Officer Basin, Western Australia, 750 million years old. Specimen is 150 µm in diameter (K. Grey).*

(Knoll, 1996). *Trachyhystrichosphaera aimika* is common and is wide-ranging both geographically and stratigraphically, and is known from at least 15 successions (Butterfield *et al.*, 1994), including the Svanbergfjellet Formation in Spitsbergen and the Wynniatt Formation of the Shaler Group in Arctic Canada (Butterfield & Rainbird, 1988).

Polygonomorphs, such as *Octoedryxium truncatum*, occur in large numbers, but only through short stratigraphic intervals. Complex colonial forms, such as *Eosaccharomyces ramosus*, complex filamentous forms, vase-shaped microfossils, and macroscopic eukaryotes (probably small seaweeds) are also present (Knoll, 1996). Vase-shaped microfossils are known from Neoproterozoic marine sediments worldwide. They are usually preserved as chert or phosphate and have a rounded base and a truncated narrow neck with a terminal opening. They have complex internal structures, and have been interpreted as testate amoebae (Martí Mus & Moczydlowska, 2000; Porter & Knoll, 2000).

Cryogenian (the time of major ice ages, thus the name, between 850 and 630 million years ago) assemblages occur in Australia in the Centralian Superbasin and Adelaide Rift Complex (Preiss, 2000), especially in the Bitter Springs Formation of the Amadeus Basin (Barghoorn & Schopf, 1965; Schopf, 1968; Zang & Walter, 1992), the Alinya Formation in the eastern Officer Basin (Zang, 1995; Gravestock *et al.*,

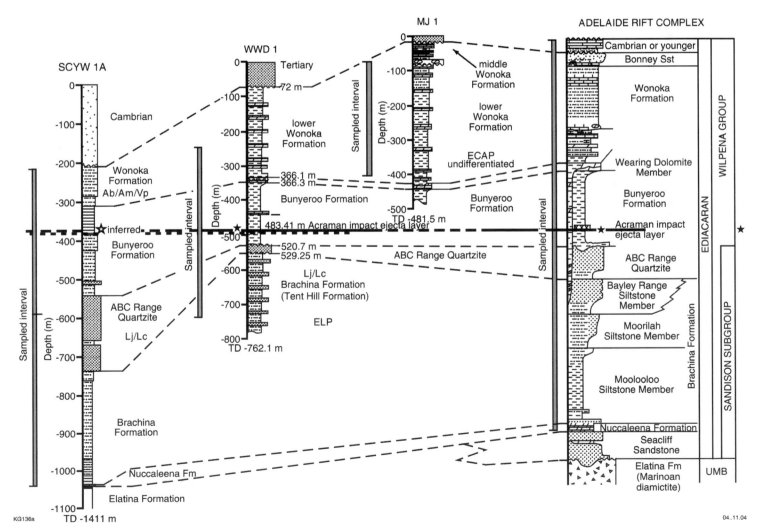

Figure 429. Correlation of sediments in the Ediacaran type section in the Adelaide Rift Complex, Flinders Ranges, Australia, with drill core sediments from the Stuart Shelf containing acritarchs (K. Grey).

225

1997), the Buldya Group of the western Officer Basin (Grey & Cotter, 1996; Cotter, 1999; Hill *et al.*, 2000), the Burra Group in the Adelaide Rift Complex (Schopf & Barghoorn, 1969; Knoll *et al.*, 1975), and the Albinia Formation of the Ngalia Basin (Walter & Cloud, 1983), all in Australia.

Direct dating is lacking for much of the Cryogenian in Australia, but it is likely younger than the Gairdner Dyke Swarm, dated at around 825 million years. Successions can be correlated using stromatolites, microfossils and isotope chemostratigraphy (Hill *et al.*, 2000). An older assemblage, dominated by various species of *Leiosphaeridia, Satka colonialica*, the colonial form *Synsphaeridium* and numerous filaments (many belonging to *Siphonophycus*) has been identified from the Alinya, Browne and Skates Hills formations, the Tarcunyah Group, and the Gillen and (possibly) the Loves Creek Members of the Bitter Springs Formation. A younger assemblage is found in the upper Buldya Group of Western Australia, the Burra Group of South Australia (Grey *et al.*, 2005) and very recently in the Finke Beds in the Amadeus Basin. It contains *Chuaria circularis* and *Cerebrosphaera buickii* and *C. ananguae*. The distinctive species *Cerebrosphaera buickii* appears to be restricted to this interval and in the western Officer Basin first appears in the middle Hussar Formation (Cotter, 1999; Hill *et al.*, 2000; Grey *et al.*, 2005). It has also been recorded from the Svanbergfjellet Formation of Spitsbergen in rocks that are about the same age (Butterfield *et al.*, 1994).

The integration of data gathered from the rocks, isotope chemostratigraphy, stromatolites and microfossils (Hill & Walter, 2000; Hill *et al.*, 2000) has produced consistent results and consequently has improved understanding of the sequence of events, both biotic and environmental, in the Centralian Basin of central Australia.

Unfortunately, there is very little microfossil or stromatolite information allowing reconstruction of assemblages from the time interval between two significant Proterozoic glaciations, the older Sturtian to the younger Marinoan. Few

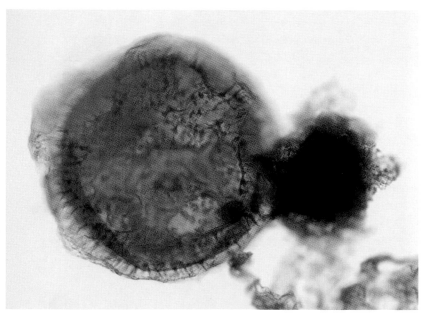

Figure 430. Distosphaera australica *(Commonwealth Palaeontological Collections 36449), an Australian acritarch. Specimen 210 μm maximum diameter; age 560-570 million years (K. Grey).*

specimens have been recorded globally. Lithologies are generally unsuitable for microfossil preservation, and the few assemblages recorded are poorly preserved, very sparse and dominated by simple leiosphere acritarchs (Vidal & Moczydowska-Vidal, 1997; Knoll, 2000). At present, little is known about the composition of these rare assemblages or their distributions. The situation changes considerably after the glaciations, when acritarchs became increasingly diverse and complex (Grey, 2005).

Ice or Impact?

One of the more interesting questions confronting paleontologists is why, after nearly 2 billion years of conservative evolution, relatively simple organisms were replaced in a very short period by the diverse biota that contains the ancestors of most phyla known today, as pointed out in Chapter 2.

In Australia, there is an almost continuous fossil record extending from rocks 3.4 billion years old in the Pilbara region to the Proterozoic-Cambrian boundary. These earliest fossils (still the subject of considerable controversy) consist of simple bacterial spheres and filaments and the laminated structures, stromatolites, which they built. Life on Earth remained predominantly the province of bacteria for nearly 2 billion years, and apart from the introduction of the Algae, showed only a slow and gradual change in diversity.

Between about 0.6 and 0.55 billion years ago, there was a major change in the biota globally (Grey, 2005). This is clearly demonstrated by detailed studies of continuous drillcore from petroleum and mineral exploration from three sedimentary basins covering much of central and southern Australia (Zang & Walter, 1992; Grey 2005). More than 1000 samples were studied by dissolving the rock in various acids to extract organic material. The slides produced in these studies contain bacterial spheres and filaments, fragments of benthic microbial mats and planktonic green algae (acritarchs).

The acritarch assemblage consists mostly of simple spheres or large spiny forms. Time range charts for individual acritarch species through this interval reflect some interesting distribution patterns. Beginning in the early Neoproterozoic, assemblages consist mainly of cyanobacteria and planktonic single-celled, smooth-walled, spherical Algae (leiospheres). These species are mostly long ranging and conservative, not changing much over long periods of time. There are a few species with ornament, and the ornament on most forms consists only of short spines or thickened ridges. Even these tend to have shorter time ranges, they are few in number.

One explanation that has been put forward for the rapid species diversification is the "Snowball Earth" hypothesis (Hoffman & Schrag, 2000, 2002; Hoffman *et al.*, 1998), discussed in Chapter 2. Aware that glacial deposits are found on all continents and seem to be clustered into at least two global ice ages, known as the Sturtian and Marinoan glaciations, these researchers suggested that the ice probably covered the whole Earth, causing a crisis for many organisms that preferred balmier conditions. At such a time, the Earth may have resembled the ice-covered moon of Jupiter, Europa, though many researchers disagree with this reconstruction. Hoffman and Schrag suggested that under these severe conditions a few species survived in specialized environments like hot springs, or in the ice itself, or in open channels in the ocean-ice cover. Species that lived

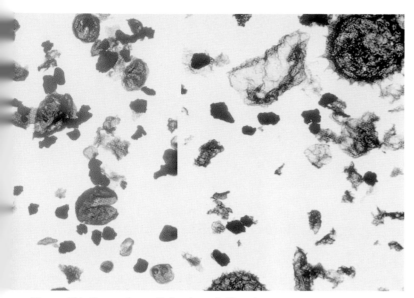

Figure 431. Comparison of leiosphere (left) and acanthomorph (right) acritarchs of Ediacaran age, South Australian drill core material. Scale bar 10 μm (=0.01 mm) (K. Grey).

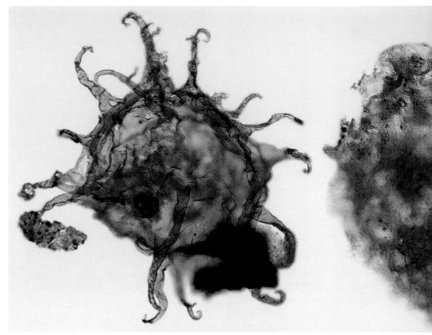

Figure 432. The acanthomorphic acritarch Tanarium conoideum (Commonwealth Palaeontological Collections 36508) from the Officer Basin drill core material of Ediacaran age, about 560-570 million years. Maximum diameter about 150 μm, without spines. Spines may be up to 200 μm in length. (K. Grey).

in these particular environments, or refugia, were likely quite specialized. Once the ice melted and temperatures rose, perhaps these specialized forms were able to emerge, take advantage of rapidly changing conditions and give rise to an explosion of new life forms.

Assuming that scenario, the fossil record from Australia suggests that glaciation may have been only part of the story. Although the glaciations were severe (glacial rocks were deposited close to the Equator in Australia), they do not seem to have affected the biota as severely as predicted by the Snowball Earth hypothesis. Australian assemblages are certainly sparse during and between the Sturtian and Marinoan glaciations. Samples taken just above the Marinoan glaciation levels are barren, indicating that the species diversity and biomass was probably depleted by the glaciation. But, rather than reflecting a rapid diversification as predicted by the Snowball Earth theory, species diversity remains low throughout the time of marked sea-level rise that resulted from melting of the ice cap. The leiosphere acritarch-dominated biota, similar in composition to the pre-glacial assemblages, persists through several hundred meters of drillcore. Post-glacial species show essentially no change from pre-glacial forms. There is no evidence to suggest either colonization by rapidly diversifying species or invasion of species from refugia such as hot springs (Grey, 2005).

Moreover, Australia had at least three, if not four, glacial episodes, and data currently emerging suggests that the Australian glaciations may not be precisely of the same age as glaciations elsewhere in the world. In Australia, the oldest glaciation, the Sturtian, is younger than 725 million years (Grey et al., 2005). Glacial rocks in King Island (which lies in Bass Strait south of mainland Australia) and Tasmania were recently dated at about 580 million years (Calver et al., 2004), and they probably correlate with the Elatina Formation, the sediments deposited during the Marinoan glaciation on the mainland of Australia. In the Kimberley region of Western Australia, sediments and indicators of an even younger glaciation, the Egan Formation, overlies the Marinoan glacial rocks and is probably about 560 million years old. The Elatina Formation has traditionally been correlated with glacial units on other

continents, such as the Nantuo Formation in China. However, the age of the Chinese succession was recently established as about 630 million years old (Condon et al., 2005). Either the Elatina Formation is older than the glaciation rocks on King Island and Tasmania (and they represent a fourth Australian glaciation), or the idea of correlating glaciations globally must be questioned. At present, it seems that the idea of only two major global glaciations in the Neoproterozoic is probably too simplistic. Further dating is necessary before there will be any realistic assessment of the number of glacial events for the Neoproterozoic. This is, in fact, the task of a recently approved UNESCO International Geological Correlation Project (IGCP518).

There is, however, striking change in the acritarch assemblages at one point in time. This change takes place during a post-glacial sea-level rise (Grey, 2005), and is rapid and dramatic. Nearly 60 new species, apparently belonging to a completely new group of large, green algae, appear and diversify in the time represented by less than 150 m of drillcore taken from an area to the northwest of the Flinders Ranges in South Australia. The smooth-walled leiosphere acritarchs, so dominant during the Sturtian post-glacial sea-level rise, become scarce. The transition from the leiosphere-dominated palynoflora to the younger complex spiny acritarchs (acanthomorph) was abrupt and apparently happened simultaneously, as recorded in all the drillholes studied in Australia. There is a dramatic change in size, complexity and taxonomic diversity, indicating significant evolutionary change over a very short period of time. Not only did the numbers of acanthomorph species increase in the Ediacaran, but there are also significant differences in both the spinosity and in the structure of the skeletal wall between pre-Ediacaran and Ediacaran process-bearing acritarchs. In general, the walls in Ediacaran acritarchs are smooth and translucent, whereas those in older forms tend to be thick, opaque and rather grainy

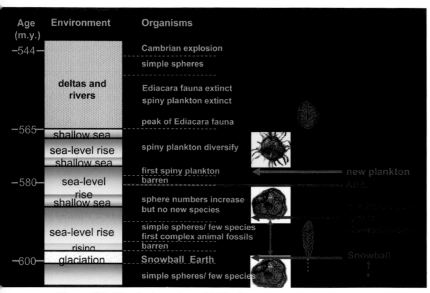

Figure 433. Physical events and biologic response reflected during the Ediacaran in Australia. AIEL indicates the Acraman impact level (K. Grey).

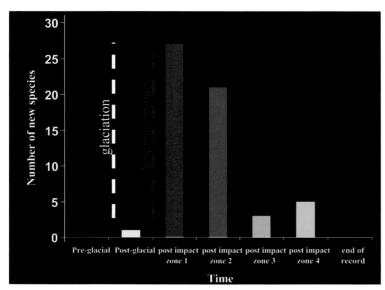

Figure 434. Effect of the rise of new species relative to the Acraman bolide event in Australia, most spectacularly demonstrated in the Officer Basin faunas (K. Grey).

Figure 435. Acraman event layer in Bunyeroo Formation, Ediacaran succession of the Flinders Ranges National Park, South Australia (P. Vickers-Rich).

Figure 436. Acraman event layer in Ediacaran sequence of the Flinders Ranges, South Australia (P. Vickers-Rich).

in appearance. The spines of pre-Ediacaran acritarchs are hollow, cylindrical tubes of variable length whereas Ediacaran assemblages are dominated by acanthomorphic acritarchs with tapering, hollow processes. The mode of process development seems to set the group of tubular-process-bearing acritarchs apart from tapering-process types that develop later. There are a few records of acanthomorphs with highly complex process morphologies before the base of the Ediacaran, but the Ediacaran forms show significant differences from the older ones. The diversification in acanthomorphs allows definition of four distinct zones of spiny acanthomorphic acritarchs (Grey, 2005).

It is significant that the change occurs just above a distinctive marker horizon, known from both field outcrops and drill core (Grey et al., 2003; Hill et al., 2004; Grey, 2005), the Acraman impact ejecta layer. The Acraman layer is a blanket of angular fragments of volcanic rocks and shattered quartz crystals, which were dumped instantaneously into mudstone and siltstone accumulating on a marine shelf. The debris has been traced to a 90 km-diameter impact crater, at the site of Lake Acraman in South

Australia. An asteroid about 4.7 km in diameter likely formed the crater and scattered ejecta across an area 1000 km in diameter. The impact was followed by a massive tsunami, possibly as devastating as the tsunami which hit southeast Asia on 26 December 2004, killing more than 300,000 people. The age of the impact has not been precisely determined, but a best estimate is about 570 million years ago. At this time Earth had no vegetation cover to protect land surface, and such an impact would have produced an enormous dust cloud, far larger than one associated with younger impacts on a vegetated earth. The cloud probably spread to the stratosphere, a cloud large enough to block sunlight for a period long enough to shut down most photosynthesis and bring about global biotic crisis (Grey et al., 2003; Grey, 2005).

Figure 437 (opposite page). Effect of Acraman event at approximately 570 million years ago on the overall composition of the acritarch faunas of Australia, a dramatic change in the dominance of leiospheres to that of the spiny acanthomorphs (K. Grey).

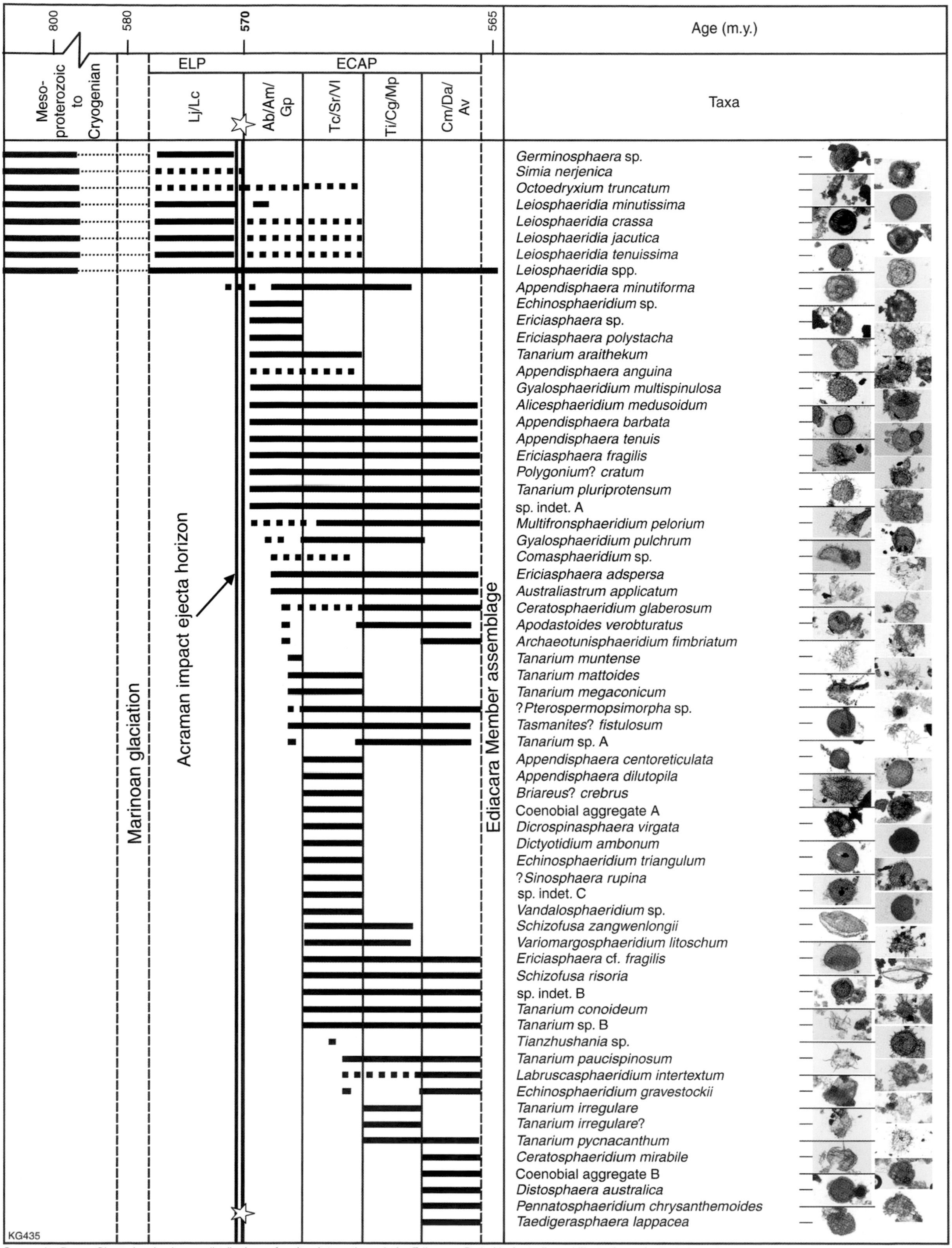

Age (m.y.)

800 | 580 | 570 | 565

Meso-proterozoic to Cryogenian

Marinoan glaciation

Acraman impact ejecta horizon

ELP — Lj/Lc

ECAP — Ab/Am/Gp | Tc/Sr/Vl | Ti/Cg/Mp | Cm/Da/Av

Ediacara Member assemblage

Taxa

Germinosphaera sp.
Simia nerjenica
Octoedryxium truncatum
Leiosphaeridia minutissima
Leiosphaeridia crassa
Leiosphaeridia jacutica
Leiosphaeridia tenuissima
Leiosphaeridia spp.
Appendisphaera minutiforma
Echinosphaeridium sp.
Ericiasphaera sp.
Ericiasphaera polystacha
Tanarium araithekum
Appendisphaera anguina
Gyalosphaeridium multispinulosa
Alicesphaeridium medusoidum
Appendisphaera barbata
Appendisphaera tenuis
Ericiasphaera fragilis
Polygonium? *cratum*
Tanarium pluriprotensum
sp. indet. A
Multifronsphaeridium pelorium
Gyalosphaeridium pulchrum
Comasphaeridium sp.
Ericiasphaera adspersa
Australiastrum applicatum
Ceratosphaeridium glaberosum
Apodastoides verobturatus
Archaeotunisphaeridium fimbriatum
Tanarium muntense
Tanarium mattoides
Tanarium megaconicum
?*Pterospermopsimorpha* sp.
Tasmanites? *fistulosum*
Tanarium sp. A
Appendisphaera centoreticulata
Appendisphaera dilutopila
Briareus? *crebrus*
Coenobial aggregate A
Dicrospinasphaera virgata
Dictyotidium ambonum
Echinosphaeridium triangulum
?*Sinosphaera rupina*
sp. indet. C
Vandalosphaeridium sp.
Schizofusa zangwenlongii
Variomargosphaeridium litoschum
Ericiasphaera cf. *fragilis*
Schizofusa risoria
sp. indet. B
Tanarium conoideum
Tanarium sp. B
Tianzhushania sp.
Tanarium paucispinosum
Labruscasphaeridium intertextum
Echinosphaeridium gravestockii
Tanarium irregulare
Tanarium irregulare?
Tanarium pycnacanthum
Ceratosphaeridium mirabile
Coenobial aggregate B
Distosphaera australica
Pennatosphaeridium chrysanthemoides
Taedigerasphaera lappacea

Composite Range Chart showing known distributions of acritarch taxa through the Ediacaran Period in Australia and illustrations of selected species

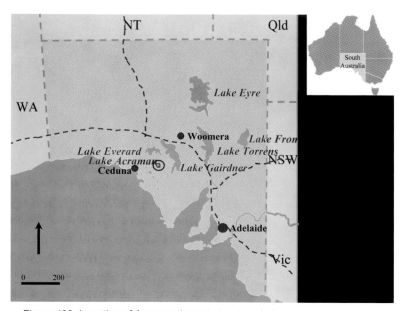

Figure 438. Location of Acraman impact structure, Lake Acraman, South Australia (K. Grey).

declines in biomass are associated with mass-extinction events in younger parts of the geological record. Changes in the biomarker record have also been associated with these biotic crises (McKirdy *et al.*, 2003, 2006).

Why would an impact cause a shift in population that allowed a new type of organism to become dominant? Organisms that depend on photosynthesis, such as benthic mats and leiospheres, would be devastated by the effects of the post-impact cosmic winter, but it probably had less effect on the small population of spiny acritarchs that was around just before the impact, and was never a major component of the population. These particular spiny acritarchs had the capacity to produce protective shells or cysts when confronted by adverse conditions. In this, they appear to be similar to some modern dinoflagellates, and some of these can encyst and sink out of the water column within 20 minutes of a significant drop in light and/ or temperature. Resting cysts can remain dormant, even for years, until conditions improve sufficiently for them to hatch (or excyst) and diversify rapidly. This would give

The fossil record is not the only evidence for the effect and recovery of the biota from a crisis. A major, short-lived decrease in the ratio of carbon isotopes indicates a sudden decline in organic carbon, and then there is a rise that mirrors species diversification (Calver & Lindsay, 1998). Similar

Figure 439. Location of Acraman impact structure, Lake Acraman, South Australia. Present Lake Acraman shown in (red). Final crater size (collapse crater) shown in green (modified from NASA Shuttle image).

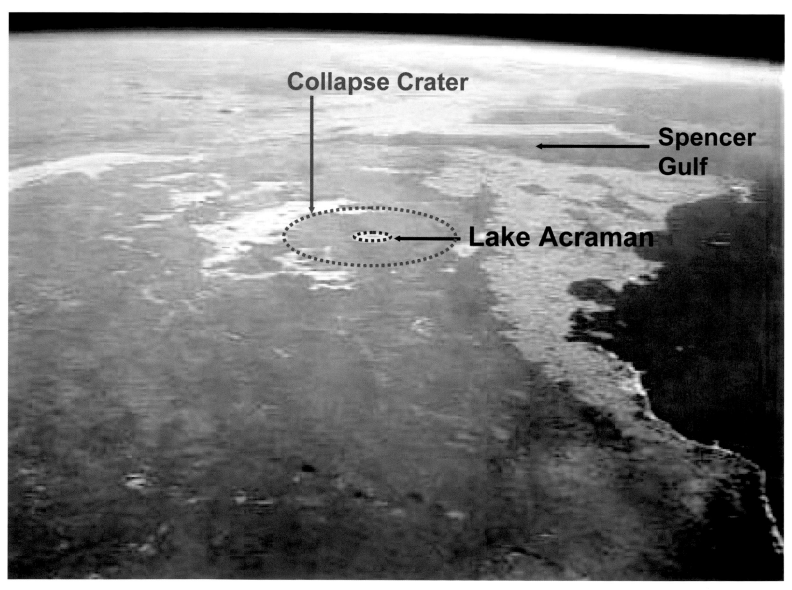

Collapse Crater

Spencer Gulf

Lake Acraman

them a competitive advantage over the leiospheres and mat communities, which would be slow to recover from an impact event, and which probably never recovered their former status. Other organisms, such as seaweeds, also began to diversify at this time. The evolution of more diverse communities must have provided an ideal food source for an evolving fauna, encouraging diversification of the Ediacarans. Eventually, overgrazing by new animal species probably led to the extinction of both Ediacaran acanthomorphic acritarchs and fauna.

It is early days yet in these investigations, and further studies are needed to thoroughly test the ideas outlined above. The spiny acritarch assemblage has been recognized outside Australia, but detailed biostratigraphy is lacking in other parts of the world. Comparable studies, with data on stratigraphic ranges, are needed for China and east Siberia to establish whether the patterns are indeed global ones and associated with the asteroid impact in Australia (Grey, 2005).

The rapid changes that took place in the biota before the Cambrian diversification probably resulted from the cumulative effect of several closely spaced crises: the Marinoan glaciation, the Acraman impact event and the evolution of filter feeders, grazers and predators. The aftermath of a severe glaciation followed by a large asteroid impact may have impacted on the opportunities for living systems on Earth about 570 million years ago and helped set the scene for the Cambrian diversification.

Figure 440. Glaciation, Antarctica (T. Rich).

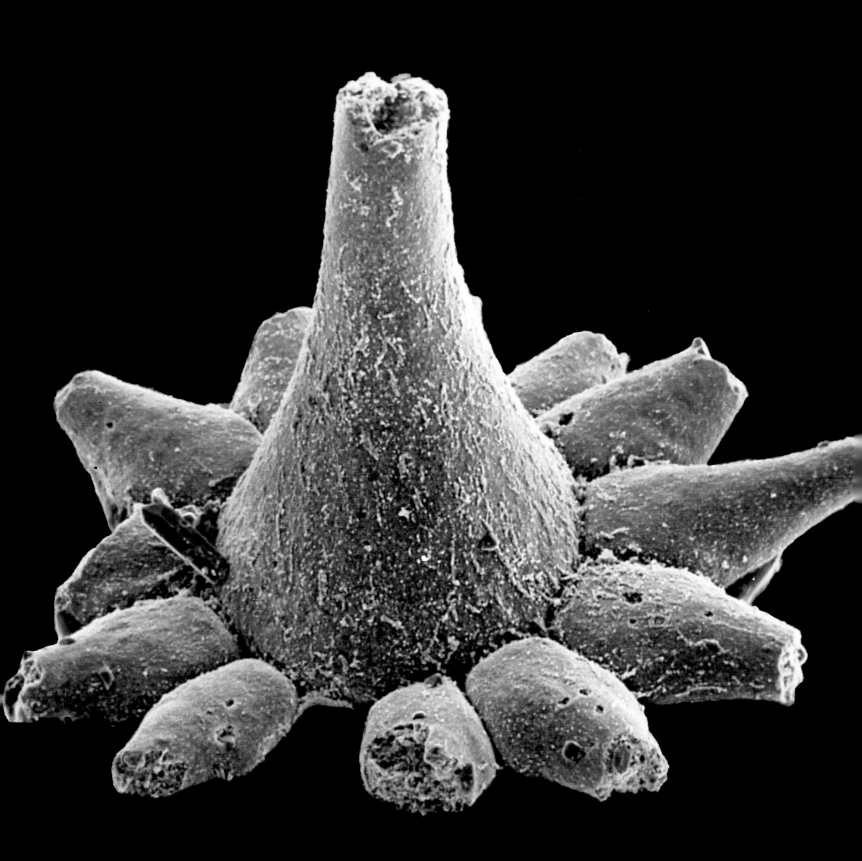

PART 4: A DRAMATIC CROSSROADS — THE CAMBRIAN "EXPLOSION"?

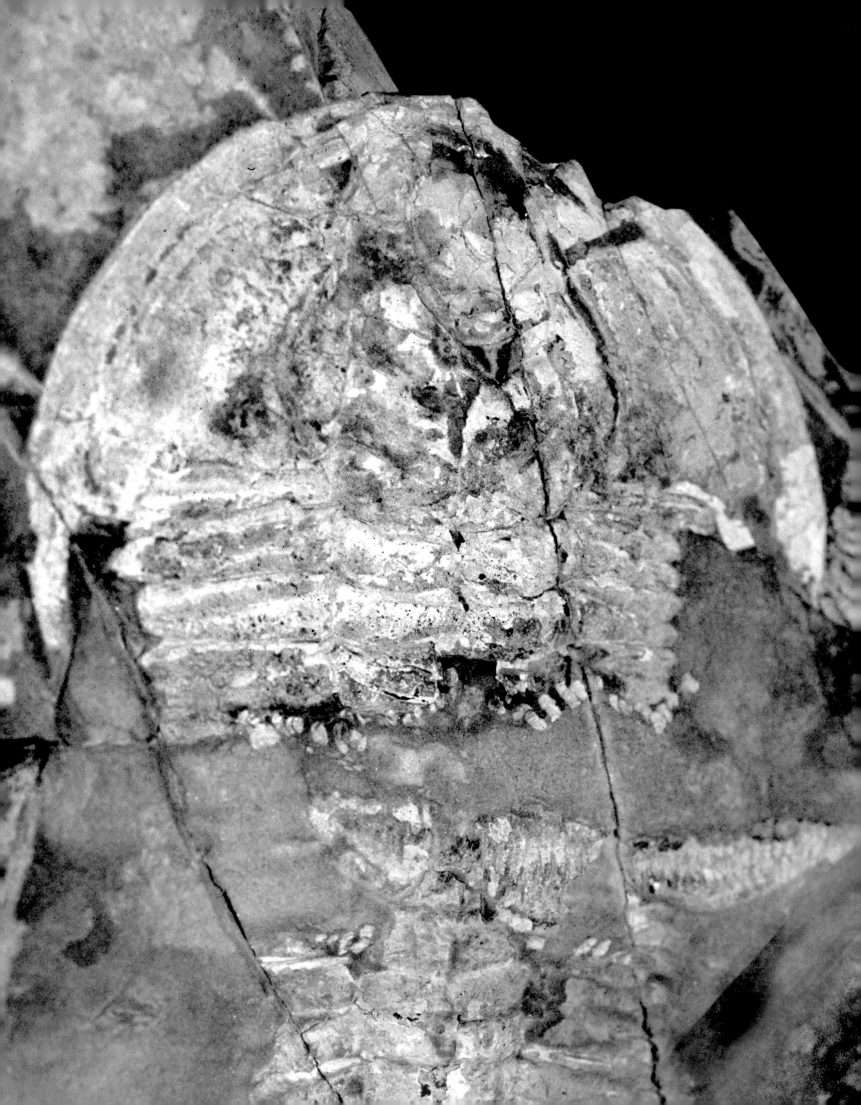

Body Plans, Strange and Familiar, and the Enigma of 542

P. Vickers-Rich

"Fossils and Egyptian hieroglyphs share daunting similarities: Both consist of arcane geometric glyphs in rock that conceal deeper meanings from the rude enquirer, and are capable of false translation. Remember Shelley's fakery 'And on the pedestal the words appear: My name is Ozymandias, King of Kings: Look on my works ye Mighty and despair!' and take note of fossil feuds."

Martin Brasier and Jonathan Antcliffe (2004)

The Ediacarans are indeed the enigma to which Martin Brasier and Jonathan Antcliffe (2004) refer. Elements of this strange assemblage of metazoans have been known since the early 20th-century discoveries in the deserts of Namibia, likely even before. But it was not until Reg Sprigg, a young mapping geologist with a penchant for fossils, found his first "jellyfish" in the coarse sandstones of the Flinders Ranges of South Australia that their true ancientness was realized and the puzzlement over their place in life's family tree began in earnest.

The Origin of Body Plans

Bob Brain began his career in the caves of southern Africa, meticulously searching, as forensic "artists" do, for the clues to the origins of humankind. But in retirement, his mind turned to older rocks, much older rocks. He has combed the low hills that rise above the game-graced Savannah grasslands of Etosha Pan in the north of Namibia for ancestors of the first animals. His search has not been disappointing.

He began his hunt for microfossils of single-celled organisms, but found something far more interesting, some of the oldest, if not the oldest, sponges – not just their spiky spicules, but beautiful whole sponges, exquisitely recorded in calcium phosphate, a proven preserver of soft-bodied organisms from many other parts of the world (Brain *et al.*, 2004). The rocks of this region, limestones of the Otavi Group, crop out in the Otavi Mountainland and to the west in the Kaokoveld, and date somewhere between 720 and 590 million years ago. These fossils seem to signal that some phyla, still alive today, had differentiated long before they acquired hard parts in the Cambrian or later, and thus show up in the fossil record.

But just what is a poriferan, and how did Brain recognize the structures that he observed in the rocks of Etosha as signifying a particular animal? Wallace Arthur has written a book that clarifies this called *The Origin of Animal Body Plans* (1997) and at the beginning makes the statement: "Body plans are easy to exemplify but difficult to define. It is generally accepted that higher taxa such as phyla and classes are characterized by unique body plans, while lower ones, such as genera and species are not. Frequent references are made, for example, to the vertebrate body plan, the insect body plan, the molluscan body plan, and so on." Arthur goes on to point out the sorts of characteristics that scientists use in defining body plans and the ways in which these characters can be manifested. His Table 2-1 (slightly modified here) nicely summarizes these characters and their manifestations:

Character	Manifestation (or Character State)
1. Skeleton	Hydrostatic, internal, external or absent entirely
2. Symmetry	Bilateral, radial, asymmetric
3. Paired Appendages	0, 2, 3, 4, many
4. Body Cavity	Acoel (*i.e.* none), pseudocoel, hemocoel, coelom
5. Cleavage Pattern	Spiral, radial, syncytial
6. Segmentation	Segmented, unsegmented

Opposite page: Redlichid trilobite from the Early Cambrian of Big Gully, Emu Bay of Kangaroo Island, South Australia (P. Vickers-Rich, courtesy of the South Australian Museum).

Figure 441. Etosha Pan, northern Namibia, where the Otavi Group crops out (M. Fedonkin).

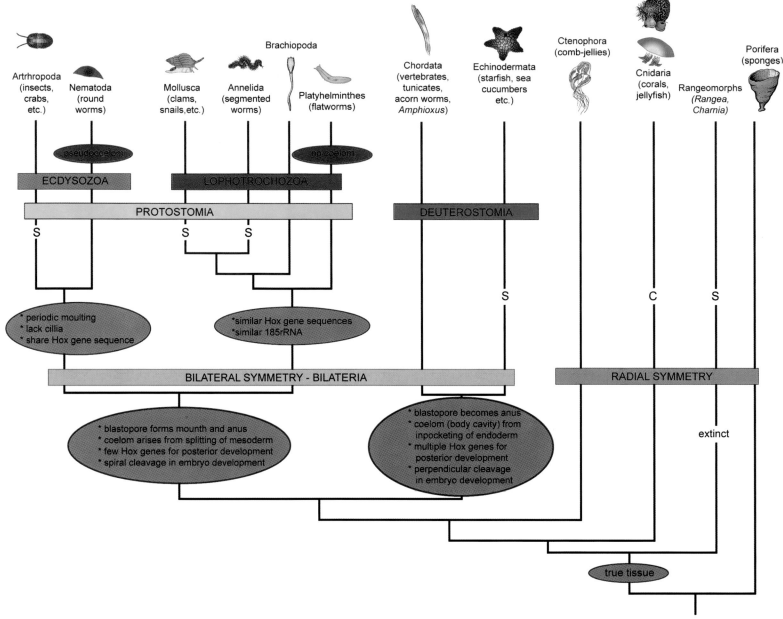

Figure 442. Family tree of the metazoans based on DNA, embryology and morphology (many sources including Valentine, 2004; drafting by D. Gelt).

Underlying these body plans, of course, are the genetic instructions producing them, and sometimes a quite different genetic background can lead to a very similar body plan. Body plans very similar to one another can come from very different sources – they can converge. This appears to be the case with the sponges, for example, noted long ago and pointed out by Kevin Peterson and Nicholas Butterfield (2005). They noted that many sponges may look very similar to each other in body plan, but based on studies of their DNA, calcareous sponges seem to be more closely related to what are called the eumetazoans (the ctenophores, cnidarians and triploblasts – animals with three embryonic tissues that develop into specific adult structures) - than to other sponge groups with siliceous skeletons. So, when analyzing the gross morphology of any animal, be it sponge, coral or mammal, in order to understand true relationships, it is quite necessary to expand your database to include not only the obvious morphological body plan, but also the genetic and developmental information as well. One needs to be able to separate body plans that have developed from very different

backgrounds that might have converged on one another from those that have a common ancestor somewhere back in time. And this is not always possible when working with fossils of once living forms, which have left only ghostly imprints in the ancient sands of an ancient sea.

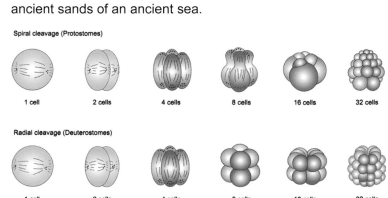

Figure 443. Embryonic development in the early stages of cell division illustrating the differences of protostomes from deuterostomes (D. Gelt).

Wallace Arthur (1997) is a positive man when he notes that he will concentrate on "what we know rather than what we do not." He does point out, however, "It is easy to be despondent about our ability to reconstruct phylogenetic history [that is, the family trees of the different major groups of animals]. The evolutionary literature of the past 150 years contains hundreds of different proposed patterns of interrelationship." Arthur pointedly comments that not only gross morphological data can sometimes be rather ambiguous, but so too molecular data, for rates of evolution vary and "molecules, like morphologies . . . are subject to parallel and convergent evolution." There is no perfect measure of who is related to whom, and thus every bit of data that can be brought to bear on working out family trees, is the most lucrative way of getting at reality.

The basic puzzle of the Ediacaran biota is simple – to what are the organisms that make it up related? Are animals that make up this bizarre collection related closely to groups that we know well in rocks younger than 542 million years ago to present, or not? Or are Ediacarans members of some groups that have left no post-Precambrian record? The answer to both of these questions seems to be "yes." Some forms, like *Kimberella*, may have Phanerozoic relatives, while others, such as *Rangea*, may not. And as research proceeds on the Ediacarans, more and more seem to have some post-Precambrian relatives, but certainly not all.

Bruce Runnegar (1992) rightly pointed out that it was Reg Sprigg's discovery of his "jellyfish" fossils in the Flinders Ranges of South Australia that set the scene for classification of Ediacaran metazoans. For some time after that, the circular discs from there and elsewhere were classed as scyphozoans, fossil jellyfish. When Roger Mason found his frond-like fossils in Leicestershire in 1957, they were first thought to be fossil algae. Glaessner (1959) modified those views and interpreted frond-like forms as pennatulate cnidarians, fossil soft corals or sea pens. Subsequent to that many new forms, with various interpretations, surfaced. Namibian fossils entered the scientific scene, even though they had been known since the early 20th century, with Hans Pflug's work (1970). Pflug thought that forms like *Pteridinium* and *Ernietta* were unique,

neither sponges nor coelenterates to which they had been initially assigned. He suggested that they belonged to a distinct, extinct group, which he named the Petalonamae. The Petalonamae, according to Pflug, was a complex, monophyletic group of colonial metazoans. Martin Glaessner and Mary Wade in Australia continued to ally the Ediacarans with modern phyla. Mikhail Fedonkin (Runnegar, 1992) took another approach – he suggested viewing this group of intriguing metazoans as leftovers from an older Precambrian radiation that was primarily obscure (1985) and thus set up a classification based on this assumption, erecting names of phyla such as the Proarticulata, Trilobozoa and Cyclozoa and the Class Vendiomorpha, where symmetry of growth was used to sort out different groups. In 1983, Adolf Seilacher proposed yet another idea. He suggested that nearly all the Vendian/Ediacaran taxa belonged to a separate animal kingdom, the Vendiobionta. He later modified this concept, admitting that some forms did seem to have relatives in the Phanerozoic (Buss & Seilacher, 1994; Seilacher *et al.*, 2003) (see below). Certainly, however, as noted by Guy Narbonne (2004), there were forms with rangeomorph architecture, so common among many of the Ediacarans, that did not seem to survive beyond the end of the Precambrian, but not all Ediacarans shared this internal structure. The rangeomorphs seemed to possess a "forgotten architecture," which disappeared after the beginning of the Cambrian, some 542 million years ago, a construction that had been dominant in the late Neoproterozoic.

Figure 444. Pteridinium *sp., Nama Group, Farm Aar, Aus region, southern Namibia (M. Fedonkin). Width of specimen on right about 5 cm.*

Others have suggested that certain Ediacarans were related to fungi (Peterson, Waggoner & Hagadorn, 2003) or even lichens (Retallack, 1994). Specifically, these researchers noted that forms, such as *Aspidella, Charnia* and *Charniodiscus* had fungal similarities. They pointed out that these forms were "multicellular or multinuclear, lived below the photic zone, could not move or defoul themselves, did not exhibit taphonomic shrinkage and were not transported or moved." They single out *Aspidella* as a form that had indeterminate growth, with no maximum size limits and growth zones similar to modern mycelia. Other Ediacarans exhibit fractal growth patterns. Importantly, Peterson and his colleagues pointed out that just because some of the Ediacarans exhibited these characteristics of algae and

Figure 445. Aspidella terranovica, *holdfast and stalk, Ediacara Hills, Flinders Ranges, South Australia (S. Morton). Diameter of holdfast about 9 cm.*

Leo Buss and Adolf Seilacher (1994), following on Seilacher's (1984, 1988, 1992) original idea, agreed that Ediacarans belonged to a single Kingdom, the Vendobionta, quite separate from all other life, but related to the Eumetazoa. The vendobionts were viewed as cnidarian-like organisms, which lacked cnidae (nematocysts – the stinging cells that stun prey that can then be consumed) – a definition that rather fits the living combjellies (Ctenophora). The vendobionts, according to Buss and Seilacher, were nourished by a symbiotic association with photosynthetic or chemosynthetic organisms. In this paper, Buss and Seilacher characterized the vendobionts as elongate, sac-like forms, lacking a mouth and any sort of structures like stinging cells, which characterize the cnidarians. This architecture, called by Seilacher a "quilted" construction, resembled a series of inflated tubes. Seilacher thought that the vendiobionts were infaunal metazoans, forms that lived entirely enclosed in the sediments. Buss and Seilacher did note that some late Neoproterozoic trace fossils (such as *Bergaueria* and sand skeletons, the Psammocorallia, described by Seilacher in 1992) could represent true cnidarians, but maintained that forms in the Vendobionta were not themselves cnidarians. Seilacher and colleagues (2003) accepted the presence of sponges (Porifera) in the Neoproterozoic, but not of coelenterates – pointing out that what were once considered "medusoids" should be considered pseudofossils, such as *Pseudorhizostomites,* now thought to be gas escape structures or "pull out" features for attachment discs of sessile, frond-like organisms, one example being *Aspidella.* Seilacher and his colleagues later accepted *Kimberella* as a mollusk (although Valentine, 2004 would question this), but not forms like *Spriggina, Parvancorina, Mialsemia, Vendia* and *Vendomia* as arthropods. They noted the lack of arthropod trackways or burrows in Ediacaran sediments and suggested that none of the Ediacarans seemed to have developed paired appendages. *Arkarua*, on the other hand, might have been an echinoderm, a trilobozoan or even a vendobiont. In addition to these more familiar forms, Seilacher and others (Narbonne, 2005) thought that some forms previously thought to be trace fossils, might, in fact,

lichens, not all possessed this body plan. They noted that "the hypothesis that these fossils were functionally fungus-like need not imply that the organisms were members of the crown-group Fungi."

Adolph Seilacher, Dimitri Grazhdankin and Anton Legouta (2003) suggested that some Ediacarans, such as *Aspidella*, may have been giant, single-celled protists, analogues being the living xenophyophores (Legouta & Seilacher, 2001). However, Kevin Peterson and colleagues (2003) countered that, noting there were no traces of any agglutinate, mineralized or otherwise supportive outer wall that would favor this idea, leaves only "the slime moulds, complex multicellular algae, metaphytes, fungi, and animals among the eukaryotes, as the only modern analogues which can be reasonably considered." They then rule out most of these: myxobacteria, actinomycetes and slime moulds, as they do not reach such large size or complexity as the frond-like Ediacarans. Living below the photic zone would further rule out metaphytes, algae and lichens. Peterson further excluded the metazoans, based on modes of preservation and growth. Nicholas Butterfield, in another paper (2005), suggested that some fossils assigned to the acanthomorphic acritarchs from the Wynniatt Formation of Victoria Island in northwestern Canada and the Roper Group of Australia (dated at 1430 million years) – *Tappania* – could also be related to fungi.

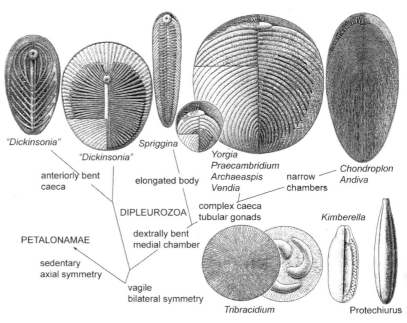

Figure 446. Family tree of the Dipleurozoa as proposed by Dzik (2003).

be body fossils – *Neonereites*, *Intrites* and *Yelovichnus* – the remains of xenophyophores, a group of living marine protists that are today restricted to very deep waters – 500 m to 7000 m or more – giant single-celled organisms. The vendobiont was challenged in the early 1990's by Gehling (Gehling, 1991; Runnegar, 1995) and is yet a most controversial concept, engendering lively scientific debate.

Jerzy Dzik (2003) suggested other possible relationships for the Ediacarans, defining a dipleurozoan body plan for such forms as *Yorgia*, *Dickinsonia*, *Marywadea* and *Spriggina*, originally classed as polychete worms, which have surprising similarity to that of chordates. He pointed to the Early Cambrian chordate *Yunnanozoon*, with its segmented muscle units situated above the notochord, and suggested that "Perhaps this functional anomaly has a historical explanation." On the other hand, he outlined aspects of the dipleurozoans in their serial, laterally placed "caeca" and "gonads" that are also shared with nemerteans, with the dorsal quilt of the dipleurozoans being a homologue (but actually perhaps only an analogue) to the "rhynchocoel" (which may be a true coelom; Turbeville, 1986) in the nemerteans. Nemerteans, the ribbon worms, are primarily marine animals with a variety of lifestyles – some live under rocks, some within algal mats, mud or sand – a few have mucus-lined burrows or tubes. Some even live in deep water, some in symbiotic arrangements with tunicates, crabs and bivalve mollusks. They normally are less than 20 cm or so in length, but *Lineus longissimus* has been reported to reach up to 30 meters (Great Barrier Reef Marine Park Authority, on line data, 2003). Nemerteans have true guts, blood vessels, brains, but no true body cavity (coelom) and, interestingly, a retractable proboscis with which they spear, grab and poison prey. They are truly active predators.

Dzik (2002, 2003) proposed that some of the petalonameans share similarities with living ctenophores, the combjellies – such forms as *Rangea*, *Charniodiscus*, *Bomakellia* – again noting similarities of Ediacaran forms with the Cambrian ctenophores *Fasciculus* and *Maotianoascus*. Other researchers have pointed out similarities of forms such as *Charniodiscus*, *Vaizitsinia*, *Glaessneria* and *Khatyspytia* with the living sea pens, the Pennatulacea. However, G. C. Williams (1997), an expert on this living group, firmly disputes this alliance – he notes that the lateral branches of these frond-like forms are not of the same architecture as the leafy sea pens, in fact so different that they are not even functional homologues.

Martin Brasier and Jonathan Antcliffe (2004) suggest that heterochrony might have played a role as well, with changes in the process of development of such forms as the rangeomorphs producing different morphologies that could change with changing conditions – juvenile forms could have survived to "adulthood," carrying with them their juvenile morphology. Clearly, understanding of the morphology and relationships of many Ediacarans is still very much under debate.

A Conundrum of Classifications

It is enlightening to examine the ways numerous researchers have tried to place the Ediacaran biota into some order. As Wallace Arthur noted, there are literally hundreds of classifications – a true conundrum. Those presented below are constructions proposed by the paleontologists who have worked closely with the late Neoproterozoic metazoans, up

close and personal. They and their close colleagues over the years have worked with the detailed gross morphology, both internal and external, left behind by the most enigmatic of metazoans and the oldest. Quite clearly, there is no general agreement on the placement of many forms. There are a variety of hierarchies with different rules of placement (Budd & Jensen, 2000). A good example is *Kimberella*. Originally, it was classified as a jellyfish (a scyphozoan) by Durham (1978). Later Sepkoski (1992) classed it as a cnidarian of uncertain affinities. Then Mikhail Fedonkin and Ben Waggoner (1997) presented a case for molluscan affinities. *Kimberella* had a soft shell, a structure that appears to have been a foot, another a possible proboscis with paired hard parts

The Ordering of Metazoans

Before examining classification of the Ediacarans, it is helpful to understand the basic subdivisions of Phanerozoic animal phyla, with emphasis on the living forms we know. Many different arrangements of members of the animal kingdom exist, but a classical one is that of Raven and Johnson (1986).

Kingdom: Animalia

Subkingdom: Parazoa (animals that lack symmetry)

    Phyla: Placozoa (minute, flattened marine organisms with two cell layers) and Porifera (the sponges, which may have developed more than once)

Subkingdom: Eumetazoa (animals that have definite shape and symmetry, and usually tissues organized into organs and organ systems)
A. *Radially Symmetric Animals*:

    Phyla: Cnidaria (corals, sea anemones, etc.) and Ctenophora (combjellies) – which may not be closely related. These two phyla are considered relicts of a former highly diverse assemblage

B. *Bilaterally Symmetric Animals (Bilaterians)*

(1) Acoelomates (animals lacking a body cavity)

    Phyla: Mesozoa (minute worm-like parasites), Platyhelminthes (flatworms), Rhynchocoela (ribbon worms)

(2) Pseudocoelomates (animals with a pseudocoel)

    Phyla: Nematoda (eelworms, roundworms), Nematomorpha (horsehair worms), Rotifera (rotifers), Loricifera (tiny animals, only microns across, which live in spaces between sand grains)

(3) Coelomates (animals with a coelom – a true body cavity)

    (a) *Protostomes* **(coelomates with a mouth that develops from or near the blastopore, that is, a hole in a sac that develops in the growing embryo)**

    Phyla: Mollusca (clams, snails, squids, etc.), Annelida (segmented worms such as earthworms and leeches), Pogonophora (beard worms), Onychophora (velvet "worms") and Arthropoda (joint-legged animals such as insects, spiders, crabs, etc.)

    (b) *Deuterostomes* **(coelomates with the anus forming from or near the blastopore)**

    Phyla: Echinodermata (starfish, sand dollars, sea cucumbers), Chaetognatha (arrow worms), Hemichordata (acorn "worms"), Chordata (tunicates, lancelets and the vertebrates)

similar to a radula of some living mollusks. Body fossils of this form are associated with movement trails and feeding patterns. Thus, with the accumulation of more data over time, ideas concerning the relationship of *Kimberella* have rather dramatically changed.

What is important to remember is that classifications are a construct to reflect current knowledge about relationships of the organic world. Classifications will always be dynamic, changing, and the more data that researchers bring to the table for consideration, the closer they will likely get to solving the mystery of relationships, just as a forensic expert closes in on the "crime" and the "criminal." And some "crimes" are solved, some never. Scientists and non-scientists need to step up to the challenges to find out more, to examine all aspects of the data. In the end, there will always be differences of opinion, but most satisfying is when all of the data is brought to the table by a variety of different specialists, and one paradigm shines above all others.

Table of Selected Classifications Through Time

Classification of Martin Glaessner (Robison & Teichert, 1979)

?Phylum: Porifera
    Class, Order, Family uncertain *Tyrkanispongia*, **Vologdin & Drozdova 1970**
Phylum: Coelenterata
    Class: Hydrozoa
        Family: Chondroplidae
            Chondroplon, **Wade 1971**
            Ovatoscutum, **Glaessner & Wade 1966**
        Family: Porpitidae
            Eoporpita, **Wade 1972**
    Class: Scyphozoa
        Family: Uncertain
            Albumares, **Fedonkin in Keller & Fedonkin 1976**
            Brachina, **Wade 1972**
            Hallidaya, **Wade 1969**
            Kimberella, **Wade 1972**
            Skinnera, **Wade 1969**
    Class: Conulata
        Family: Conchopeltidae
            Conomedusites, **Glaessner & Wade 1966**
        Medusae of Uncertain Affinities
            Cyclomedusa, **Sprigg 1947**
            Ediacaria, **Sprigg 1947**
            Lorenzinites, **Glaessner & Wade 1966**
            Mawsonites, **Glaessner & Wade 1966**
            Medusinites, **Glaessner & Wade 1966**
            Planomedusites, **Sokolov 1972**
            Pseudorhizostomites, **Sprigg 1949**
            Rugoconites, **Glaessner & Wade 1966**
    Problematical Coelenterata ("Petalonamae" of Pflug 1970)
        Family: Pteridiniidae
            Pteridinium, **Gürich 1933**
        Family: Rangeidae
            Rangea, **Gürich 1930**
         Family: Charniidae
            Charnia, **Ford 1958**
            Charniodiscus, **Ford 1958**
            Glaessneria, **Germs 1973**
        Family: Erniettidae
            Ernietta, **Pflug 1966**
            Erniofossa, **Pflug 1966**
            Ernionorma, **Pflug 1972**
            Erniobeta, **Pflug 1972**
            Erniograndis, **Pflug 1972**

Unrecognizable and rejected genera assigned to the "Erniettomorpha"
        Erniocarpus, **Pflug 1972**
        Erniocentris, **Pflug 1972**
        Erniocoris, **Pflug 1972**
        Erniopelta, **Pflug 1972**
        Erniotaxis, **Pflug 1972**
    Family: Uncertain
        Arumberia, **Glaessner & Walter 1975**
        Baikalina, **Sokolov 1972**
        Namalia, **Germs 1968**
        Nasepia, **Germs 1972**
Phylum: Annelida
    Class: Polychaeta
        Family: Vologdinophyllidae
            Cloudina, **Germs 1972**
        Family: Dickinsoniidae
            Dickinsonia, **Sprigg 1947**
        Family: Sprigginidae
            Spriggina, **Glaessner 1958**
            Marywadea, **Glaessner 1976**
        Family: Anabaritidae
            Anabarites, **Missarzhevshi in Voronova & Missarzhevsky 1969**
Phylum: Arthropoda
    Superclass: Trilobitomorpha (but Class Uncertain)
        Family: Vendomiidae
            Vendomia **(Keller in Keller & Fedonkin 1976)**
            Onega, **Fedonkin in Keller & Fedonkin 1976**
            Praecambridium, **Glaessner & Wade 1966**
    Class: Branchiopoda
        Family: Parvancorinidae
            Parvancorina, **Glaessner in Glaessner & Daily 1959)**
            Doubtful Arthropoda
            Velancorina, **Pflug 1966**
Phylum: Pogonophora ?
        Family: Saarinidae
            Calyptrina, **Sokolov 1965**
        Family: Sabelliditidae
            Paleolina, **Sokolov 1965**
Phylum: Uncertain
    Class, Order and Family Uncertain
        Redkinia, **Sokolov 1976**
        Tribrachidium, **Glaessner in Glaessner & Daily 1959**
        Vermiforma, **Cloud in Cloud *et al.*, 1976**

Taxa with Doubtful Invertebrate Affinities
    Family: Suvorovellidae
        Suvorovella, **Vologdin & Maslov 1960**
        Majeella, **Vologdin & Maslov 1960**
    Family: Uncertain
        Petalostroma, **Pflug 1973**

Classification of J. Wyatt Durham (1978)

Medusoids
    Albumares, **Fedonkin 1976**
    Asterosoma (for *Brooksella*), **Bassler 1854**
    Beltanella, **Sprigg 1947**
    Bronicella, **Sokolov 1973**
    Charniodiscus, **Ford 1958**
    Chondroplon, **Wade 1971**
    Conomedusites, **Glaessner & Wade 1966**
    Cyclomedusa, **Sprigg 1947**
    Ediacaria, **Sprigg 1947**
    Hallidaya, **Wade 1969**
    Kimberella, **Glaessner & Wade 1966**
    Lorenzites, **Glaessner & Wade 1966**
    Mawsonites, **Glaessner & Wade 1966**
    Ovatoscutum, **Glaessner & Wade 1966**
    Planomedusites, **Sokolov 1972**
    Rugoconites, **Glaessner & Wade 1966**

Skinnera, **Wade 1969**
Suvorella, **Vologdin & Maslov 1960**
Tirasiana, **Palij 1969**

Pennatulacean
    Arborea, **Glaessner & Wade 1966**
    Charnia, **Ford 1958**
    Pteridinium, **Gürich 1930**

Platyhelminthes
    Brabbinithes, **Allison 1975**

Annelida
    Dickinsonia, **Sprigg 1947**
    Spriggina, **Glaessner 1948**

Mollusca
    Cloudina, **Germs 1972**
    Wyattia, **Taylor 1966**

Arthropoda
    Onega, **Fedonkin 1976**
    Praecambridium, **Glaessner & Wade 1966**
    Vendia, **Keller 1969**
    Vendomia, **Keller 1969**

Pogonophora
    Paleolina, **Sokolov 1967**
    Sabellidites, **Yanishevsky 1926**

Echinodermata
    Tribrachidium, **Glaessner 1959**

Incertae Sedis
    Anabarites, **Missarzhevsky 1969**
    Baikalina, **Sokolov 1972**
    Nemiana, **Palij 1969**
    Parvancorina, **Glaessner 1958**
    "17" genera, Anderson 1976

Classification of Mikhail A. Fedonkin (1985, 1987, 1990)

Phylum: Coelenterata
  Class: Cyclozoa
    Family: Unknown
      Nemiana simplex
      Beltanelliformis brunsae
      Cyclomedusa radiata
      Cyclomedusa davidi
      Cyclomedusa plana
      Cyclomedusa minuta
      Cyclomedusa delicata
      Tirisiana disciformis
      Palliella patelliformis
      Ediacaria flindersi
      Nimbia occlusa
      Nimbia dniesteri
      Protodpleurosoma rugulosum
      Eoporpita medusa
      Kaisalia mensae
      Veprina undosa
    Family: Chondroplidae
      Ovatoscutum concentricum
      Chondroplon **sp.**
  Class: Inordozoa
      Irridinitus multiradiatus
      Armillifera parva
      Elasenia aseevae
    Family: Bonatiidae
      Bonata septata
    Family: Hiemaloriidae
      Hiemalora stellaris
      Evmiaksia aksionovi
  Class: Trilobozoa

    Family: Albumaresidae
      Albumares brunsae
      Anfesta stankovskii
    Family: Incertae sedis
      Tribrachidium heraldicum
      Skinnera
  Class: Conulata
    Family: Conchopeltidae
      Conomedusites lobatus
  Class: Scyphozoa
    Family: Incerta sedis
      Stauridinia crucicula
    Family: Pomoriidae
      Pomoria corolliformis
Phylum: ? Platyhelminthes
      Platypholinia pholiata
      Vladimissa missarzhenskii
Phylum: Proarticulata
  Class: Dipleurozoa
    Family: Dickinsoniidae
      Palaeoplatoda segmentata
      Dickinsonia costata
      Dickinsonia tenuis
      Dickinsonia lissa
  Class: Vendomorpha
    Family: Vendomiidae
      Onega stepanovi
      Vendomia menneri
      Vendia sokolovi
Phylum: Arthropoda
  Class: Paratrilobita
      Bomakellia kelleri
      Mialsemia semichatovi
    Family: Sprigginidae
      Spriggina borealis
      Spriggina floundersi
Phylum: Petalonamae
    Family: Charniidae
      Charnia masoni
    Family: Pteridiniidae
      Pteridinium nenoxa
      Pteridinium simplex
      Inkrylovia lata
      Archangelia valdaica
      Podolimirus mirus
      Valdainia plumose
    Family: Incertae sedis
      Ramellina pennata
      Vaveliksia velikanovi
      Zolotytsia biserialis
      Vaizitsina sophia
      Lomozovis malus

Classification of J. John Sepkoski (Schopf & Klein, 1992)

Phylum: Trilobozoa
    Family: Albumaresidae
    ?Family: Anabaritidae
    Family: Uncertain
      Tribrachidium
Phylum: Cnidaria
  ?Class Petalonamae
    Order Erniettamorpha
      ?Family: Bomakellidae
      ?Family: Dickinsonidae
      Family: Erniettidae
      Family: Pteridiniidae
    Order: Rangeomorpha
      Family: Charniidae
      Family: Rangeidae
  Class: Cyclozoa
    Family: Cyclomedusidae
    ?Family: Medusinitidae
  Class: Hydrozoa

In Sepkoski's classification he notes the number of genera in the following: Trilobozoa (4), Petalonamae (22), Cyclozoa (6), Hydrozoa (3), Scyphozoa (4), Anthozoa (1), Cribricyathida (1) and Paratrilobita (3).

Since the 1980's, ideas about the relationships of metazoans have been in a state of radical change, and today, traditional views, such as those expressed by Raven and Johnson have been largely replaced as phylogenetic constructs based on a mass of information from molecular biologists, developmental biologists and morphologists which has accumulated and been integrated and analyzed (*e.g.* Eernisse & Peterson, 2004). Current opinion is that the ancestry to the metazoans appears to lie among the choanoflagellates. Another insight, noted above, based on new data is that the Phylum Porifera, the sponges, appears to contain forms which came from different ancestries – it is not a monophyletic group. The calcareous sponges

seem closely related to the eumetazoans, and the siliceous sponges form a separate group. Eernisse and Peterson ally a number of groups together as Lophotrochozoa –including 1) the brachiopods and phoronid worms, 2) the rotifers, gnathostomulids and platyhelminth worms and 3) the bryozoans, nemerteans, annelids, mollusks and echiuran and sipunculan worms. Two excellent summaries of recent ideas concerning metazoan relationships are Giribet (2003) and Glenner *et al.* (2004) as well as the in-depth book by Valentine (2004) – these give some idea of the diversity of opinions.

Summarizing many new studies, Eernisse and Peterson (2004) noted that the nematode worms were actually primitive bilaterians and lump them with such groups as priapulids and loriciferans. Martindale and Henry (1998) suggested that the combjellies (Ctenophora) may be the missing link between radially symmetric metazoans and bilaterally symmetric forms. It will require further work in the future to gain any sort of consensus. Eernisse and Peterson sensibly advocate that "a total-evidence approach with several different types of data derived from numerous taxa . . . can only lead to a better understanding of the interrelationships among the major animal lineages and, of course, to animal evolution itself."

The same diversity of opinions is expressed with regard to classification of the Ediacarans, some present below. It will be some time, however, and will require considerable cooperation between paleontologists, developmental biologists, molecular biologists and morphological neontologists (and likely many others!) before a reasonably agreed upon family tree of the metazoans, both living and extinct, will be arrived at – a wonderful nexus of scientific inquiry for the immediate future.

Bruce Runnegar (1992) pointed out that "the lack of a secure taxonomy of the Ediacara fauna is a clear symptom of the poor condition of our knowledge" of these puzzling animals. It is also related to the present state of flux of classification based on primarily living metazoans (again see Giribet, 2004 and Valentine, 2003). Runnegar further noted some important points. Some of the Ediacarans were certainly animals – there is no doubt about that. They were not all giant protists, if indeed any were. Their trace fossils (tracks, trails, grazing patterns, burrows) clearly demonstrate that some were mobile – *Kimberella, Yorgia, Dickinsonia*, among others. Some were clearly feeding on and within the microbial mats that covered much of the sea floor at the time, *Kimberella* with its fan-shaped grazing patterns and *Yorgia* with its "footprints" likely left by its chemoprocessing of that mat structure. *Kimberella*, in fact, appears related to mollusks, one of the few groups which most researchers would agree has a clear connection to Phanerozoic metazoans. *Yorgia* and *Dickinsonia* have no uncontested Phanerozoic relatives, and just how they fed has intrigued many. Runnegar drew a comparison of these forms to *Riftia*, a "gutless," highly derived annelid that today lives near hydrothermal vents. *Riftia* lacks a mouth, digestive tract and anus – that is, as an adult, for its larvae have both a mouth and a gut and are free swimmers, using cilia to propel themselves. As an adult it settles to the bottom, assembles chemosynthesizing sulfide-oxidizing bacteria in a large internal organ, reduces its mouth and gut and gets on with its new life (Gardiner & Jones, 1993). The hydrogen sulfide that is pumped out of the vents is absorbed at the head end of this giant "tube worm," transported to the organ with its awaiting bacteria by the blood. The bacteria then use the sulfide to produce energy for the generation of organic molecules that serve as the "worm's" food. Ingenious!

Runnegar goes further to note that since *Yorgia* and *Dickinsonia* did not live near vents that something a bit different might be happening – they still likely depended on chemosymbiosis for their "food" and could have contained bacteria that used the local chemistry. What might have supplied that local energy was the interface between the oxic and the anoxic layers in a living microbial mat. Hydrogen sulfide would have been present in the mat, while oxygen needed to complete reactions could have been absorbed by these forms from the overlying seawater – oxygen supplied by organisms that photosynthesized in the waters above. The footprints left by *Yorgia* and *Dickinsonia* could be due to absorption of the microbial mat down to the level where no further hydrogen sulfide was extractable. At least that is one scenario, proposed here, for these early metazoans.

Other forms were not mobile, they were tethered, sedentary. One classification places them in the Petalonamae – *Charnia, Charniodiscus, Aspidella, Glaessnerina, Glaessneria, Rangea* and others. They may have no Phanerozoic progeny (Narbonne, 2005) – but some researchers still look to the living sea pens as possible relatives, though Williams (1997) makes a strong case against this alliance. Still other groups may be related to Phanerozoic forms and may not – the jury is still out on this case – *Vendia* and *Praecambridium* as trilobite relatives, *Tribrachidium* and *Arkarua* as possible echinoderms, *Ausia* and a new form from the White Sea of Russia as possible protochordates (Fedonkin, Vickers-Rich & Swalla, *in prep.*), *Spriggina* and *Parvancorina* as possible arthropods (the latter with Cambrian forms, such as *Skania, Primicaris* and *Parvancorina larviformis*, of marked similarity (Lin *et al.*, 2006), and so on. Narbonne (2005) aptly points out that "in little more than a decade, the affinities of the Ediacara biota went from being a well-established 'fact' to one of the greatest controversies in paleontology."

Embryonic Revelations

Understanding of where animals came from and just when they developed is not always spotlighted by large specimens. In fact, spectacular fossils that shed considerable light on the early metazoans come in very small sizes. The Doushantuo phosphorites, Doushantuo Formation of the Yangtze Platform of south China, are dated at 580 +/- 1 million years (Zhang, Yin, Xiao & Knoll, 1998; Zhang & Yuan, 1992; Li *et al.*, 1998; Zhang *et al*, 1998; Xiao & Knoll, 2000; Condon *et al.*, 2005). These phosphorites were deposited after the Nantuo Tillite was laid down, which mirrors one of the major glaciations in the Neoproterozoic just prior to the appearance of the majority of Ediacaran metazoans. The Doushantuo Formation is a succession of shales, some with significant phosphatic layers, and carbonates. Above lie the Dengying sediments with rare Ediacaran fossils and at the very top, early Cambrian Small Shelly Fossils. About 600 km to the southwest of the Yangtze Gorge, in the Weng'an region of Guizhou Province, the Doushantuo sediments contain tiny fossils, beautifully preserved, in phosphate-rich shales. These minute forms are the three-dimensional, microscopic remains of both plants and animals. The metazoans preserved are the oldest metazoan body fossils that most researchers would agree are truly metazoans (Valentine, Jablonski & Erwin, 1999).

The Doushantuo animal embryos are many times (3-5 x) larger than the red and green algae also preserved as phosphatized fossils. The embryos seem to be encased in a protective envelope during divisions or what embryologists call cleavage – from the time the first cell divided into two and then two into four and so on. The geometry of the developing embryo from these first divisions – four-cell stage, 16-cell stage, etc. – are the sorts of configurations typical of animals and not plants. In contrast to the precisely controlled cell division and obstruction of cell movement by rigid cell walls in the early development of plants, early embryonic differentiation of animals is not so constrained. The eventual differentiation of embryonic tissues in animals, as can be observed in the Doushantuo fossils, was achieved by the "programmed cell migration unobstructed by cell walls" (Xiao, 2002).

Early considerations of the Doushantuo embryos (Xiao & Knoll, 2000; Xiao, Yuan & Knoll, 2000) noted that embryo size (just visible to the naked eye), the constant presence of an enveloping membrane and the precise cleavage pattern of the cells in the early stages of development were all consistent with an animal identity. First opinions were not specific as to whether these were diploblastic or triploblastic animals, that is, animals with two or three embryonic body layers (Valentine, Jablonski & Erwin, 1999). Shuhai Xiao (2002), however, advanced the idea that many of the Doushantuo embryos represented triploblastic animals – that is, organisms with three embryonic body layers – one layer being the ectoderm that develops into nerves, skin and many other organs; another the mesoderm, which forms reproductive organs, muscles and more; and finally the endoderm, which forms the basis of internal organs, like the gut. Xiao advocated that many of the embryos were not of sponges or jellyfish, which have only two embryonic body layers – diploblasts – but represented more complicated animals. It still remains to be established, however, if they were truly triploblastic animals (Valentine, *pers. com.*, 2006). But, without doubt, these small fossils establish that quite complicated animals were in existence before the majority of the larger, soft-bodied Ediacarans appeared in the fossil record. It should be noted that some larger metazoans, for example, *Charnia wardi* from Newfoundland, have a similar antiquity to the Doushantuo embryos (Narbonne & Gehling, 2003; Bowring & Schmitz, 2003). This is certainly in agreement with estimates made by those working on the very structure of the genes, molecular data, that predicts a rather "deep time" origin for many metazoan groups. "Even the most conservative molecular clock estimates from the time of the basal split of metazoans predate the classical Cambrian radiation" (Wills & Fortey, 2000). Conservative molecular estimates give a 670 million year date for the split between protostomes and deuterostomes, a 600 million year date for the split between chordates and echinoderms and the split between the sponges and the rest of the metazoans, the eumetazoans, at around 630 to 604 (Erwin, 1999; Peterson & Butterfield, 2005) or even older at 900 million years (Wills & Fortey, 2000). Could some of the Chinese embryos have been embryos of the Ediacarans themselves?

The Enigma of 542 (Million Years before Present)

Just what happened on Earth about 542 million years ago is still a bit of a mystery. Over a short space of time, but not instantaneously (Knoll & Carroll, 1999), many different life forms acquired hard parts; this was a significant proportion of these hard tissue was constructed of calcium phosphate

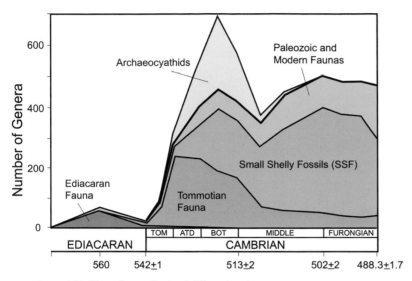

Figure 447. Diversity gradients of different metazoan groups from the Ediacaran through the Cambrian, reflecting the major faunal turnover, where generic diversity is plotted through time (Sepkoski in Schopf and Klein, 1992). (D. Gelt)

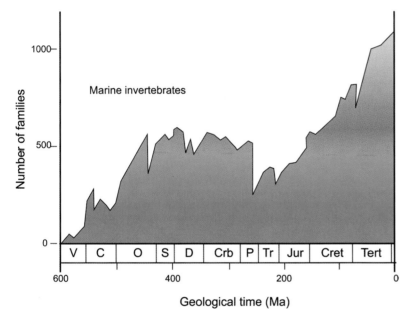

Figure 448. Number of families of marine invertebrates plotted through time (modified from Briggs & Crowthers, 2001). (D. Gelt)

(Cook & Shergold, 1984). It is not as if hard tissues were not being deposited before that significant date – but they were not widespread in the Ediacaran and for the most part laid down only by small organisms, such as *Cloudina and Namacalatus* (although in the case of *Namacalatus* crystallization of the calcite could have been post-depositional) and a variety of tiny Small Shelly Fossils (SSF's), which were diverse in the very earliest Cambrian. But, large animals did not seem to have had hard parts before that date. Ediacarans left their soft-bodied imprints in the sands and clays, bodies that, however, could have been reinforced by a resistant tissue, such as cellulose, which would have assisted in their preservation (Vickers-Rich, 2006). Tunicates are the only known animals to generate cellulose, but they seem to have developed the ability to deposit this supportive structure by horizontal gene transfer at some time in the past (Nakashima *et al.*, 2004).

Correlated with the widespread appearance of hard parts, metazoans increased in biodiversity over a period of 10 to 20 million years, beginning about 542 million years ago – commonly termed the "Cambrian explosion." The literature on this biodiversification event is enormous (see Zhuravlev & Riding, 2001 as well as Schopf & Klein, 1992 for extensive bibliographies; Budd & Jensen, 2000 for an up-to-date summary; Shu *et al.*, 1999, and Chen *et al.*, 1997 for summary of the earliest Cambrian faunas). What lit the fuse that led to such an explosion? And was the fuse long or short; was this really a biotic explosion, or simply a reflection of a better fossil record once hard parts made it possible to record the presence of certain body plans? All such questions are currently energetically being debated in the pages of scientific journals. There are many theories, but one thing is for certain – it was truly the end of most of the Ediacarans.

Figure 449. Tubular organism from the Early Cambrian Big Gully, Emu Bay of Kangaroo Island, South Australia (P. Vickers-Rich, courtesy of the South Australian Museum).

What was the cause of the seemingly rapid initiation of hard part deposition? Was it due to increase in levels of oxygen in the oceans brought about by sequestering of organic carbon (Kirschvink & Raub, 2003 and many others), to the lowering of marine salinity (Kanuth, 2004; Vickers-Rich, 2006), to a turnover of ocean waters and the rise of unoxygenated bottom waters rich in phosphates in the surface seas (Cook & Shergold, 1984), to the massive influx of weathered continental sediments, filled with clay minerals, nutrients and high levels of calcium as the great Transgondwanan Mountain Range shed its freshly exposed and weathered "skin" (Squire *et al.*, 2006; Kennedy *et al.*, 2006) into the extensive continental shelf seas that developed with marine transgressions of the Cambrian that certainly provided more habitable environments, new ecospace that had not existed before (Brasier & Lindsay, 2001)? Perhaps the rapid movement of tectonic plates characteristic of this time brought buried carbon- and methane-rich sediments into the tropics where their release initiated several periods of Greenhouse conditions (Kirschvink & Raub, 2003). Was this Cambrian explosion partly due to the growing cephalization and concentration of chemo- and photoreceptors (*e.g.* eyes) into organs critical to the precise identification of prey and thus the rise of efficient and large predators? A

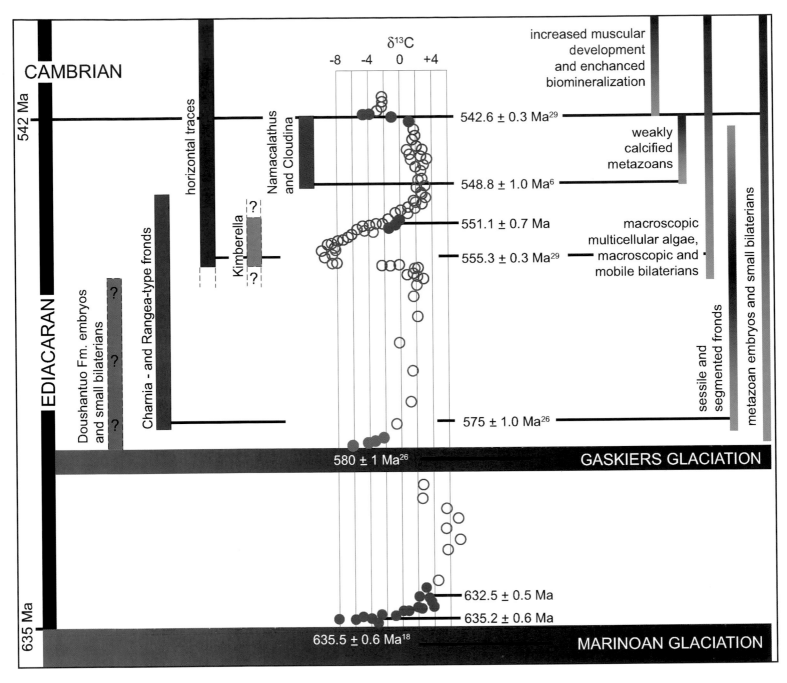

Figure 450. Fauna, flora, carbon isotope signatures
and radiometric dates for sequences of Ediacaran age
globally (modified from Condon et al., 2005) (D. Gelt).

proliferation of hard parts would have had many advantages to their possessors: protection from predators both in providing a barrier but in also allowing more efficient burrowing and greater efficiency of muscular control. There are many ideas swirling around the scientific community currently. Many or all may be correct, but none stands out as the single, overriding answer.

And the question could be asked yet again, how are the Ediacarans related to the metazoans of the Cambrian

– assemblages made up of both shelled and soft-bodied forms from the Middle Cambrian Burgess Shale of North America (Gould, 1989; Collins, Briggs & Conway Morris, 1983), the Early Cambrian of Emu Bay of southern Australia and the Chengjiang of south China (Chen et al., 2004)? Some Ediacarans, as noted in previous text, appear to have relatives that survived into the earliest Cambrian (Jensen, Gehling & Droser, 1998), including forms which closely resemble Ediacaran taxa, such a Parvancorina, thought to be an early

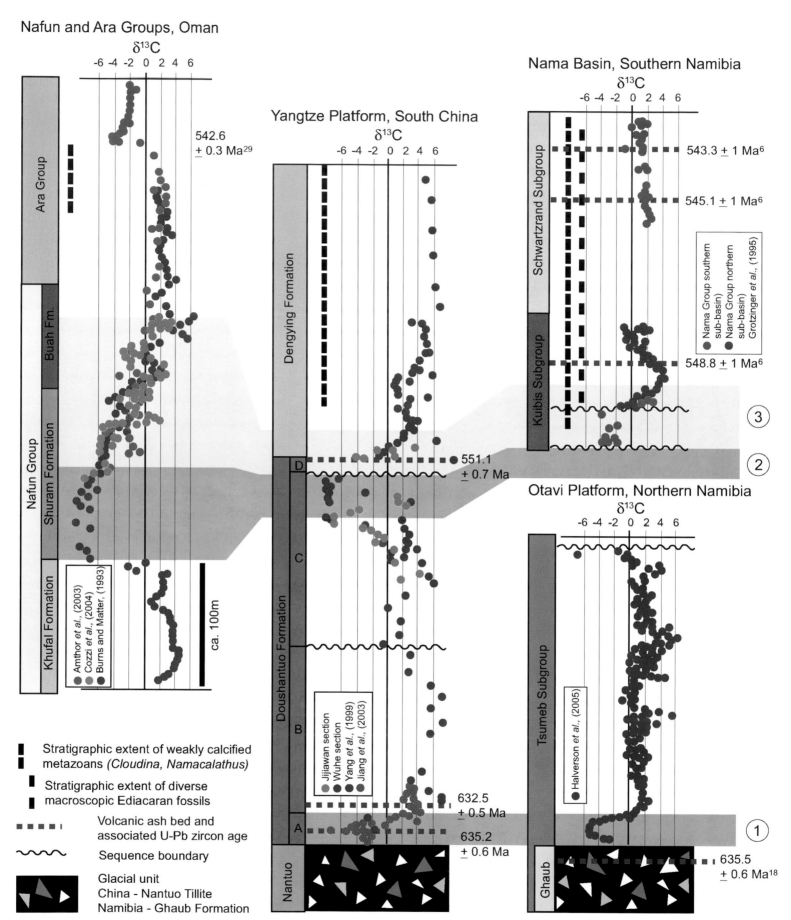

Nafun and Ara Groups, Oman
δ¹³C

Yangtze Platform, South China
δ¹³C

Nama Basin, Southern Namibia
δ¹³C

Otavi Platform, Northern Namibia
δ¹³C

542.6 ± 0.3 Ma[29]

543.3 ± 1 Ma[6]

545.1 ± 1 Ma[6]

548.8 ± 1 Ma[6]

Nama Group southern sub-basin)
Nama Group northern sub-basin)
Grotzinger *et al.*, (1995)

551.1 ± 0.7 Ma

632.5 ± 0.5 Ma

635.2 ± 0.6 Ma

635.5 ± 0.6 Ma[18]

Amthor *et al.*, (2003)
Cozzi *et al.*, (2004)
Burns and Matter, (1993)

Jijiawan section
Wuhe section
Yang *et al.*, (1999)
Jiang *et al.*, (2003)

Halverson *et al.*, (2005)

ca. 100m

■ Stratigraphic extent of weakly calcified metazoans (*Cloudina, Namacalathus*)

■ Stratigraphic extent of diverse macroscopic Ediacaran fossils

▰ ▰ ▰ Volcanic ash bed and associated U-Pb zircon age

〰〰 Sequence boundary

Glacial unit
China - Nantuo Tillite
Namibia - Ghaub Formation

Figure 451. Carbon isotope patterns in several Ediacaran sequences globally – Oman, China and Namibia (modified from Condon et al., *2005; drafted by D. Gelt). 1) Correlation of cap carbonate excursion (δ¹³C starting at -2%, going to a nadir of -5%; 2) invariant negative δ³C excursion in the range of -7%; 3) transition from negative to positive (@1% values after Upper Doushantuo/Shuram/Kuibis anomaly.*

arthropod (Lin *et al.*, 2006). But, for the most part it may have been that the demise of most Ediacarans opened up niches, for new forms that flourished in the Phanerozoic, giving rise to the animals we are so familiar with today – similar to how the dinosaur extinctions at the end of the Mesozoic paved the way for the Cenozoic mammalian and avian diversification. The

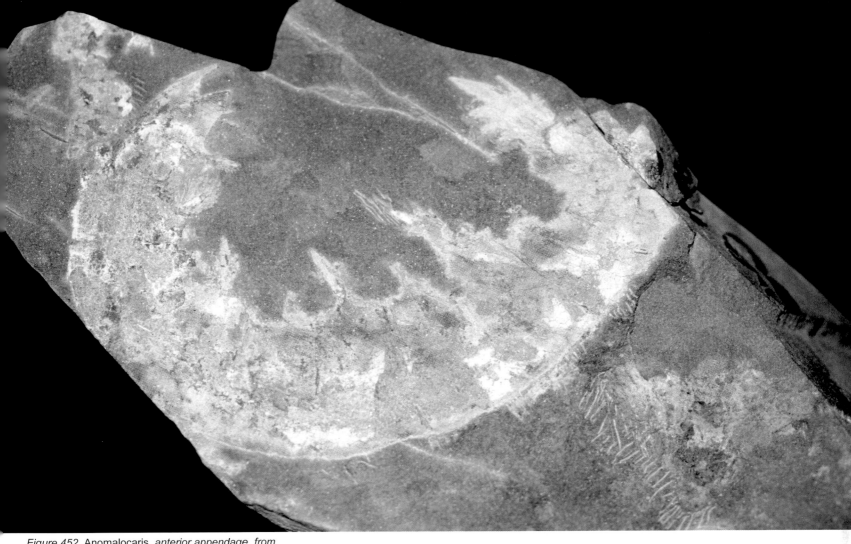

Figure 452. Anomalocaris, anterior appendage, from the Early Cambrian Big Gully, Emu Bay of Kangaroo Island, South Australia (P. Vickers-Rich, courtesy of the South Australian Museum).

Ediacarans were probably, for the most part, the prelude to the "Cambrian explosion" (Knoll & Carroll, 1999).

Was the Cambrian explosion an event that happened rapidly or was it spread over 10-20 million years – as Richard Fortey has called it, "an explosion with a long fuse?" Alan Cooper and Richard Fortey (1998; and Fortey, 2001) note that even the most conservative molecular evidence suggests that there were long periods of evolutionary innovation and development of body plans before the fossils record presence. Molecular evidence, noted before, suggests that diversification of bilaterian animals took place as long ago as 1000-900 million years (Erwin, 1999; Droser, Jensen & Gehling, 2002; Hedges *et al.*, 2004; Blair & Hedges, 2005). Discoveries of Ediacarans, exemplified by *Kimberella* (Fedonkin & Waggoner, 1997) and *Vendoconularia* (Ivantsov & Fedonkin, 2002), perhaps related to Phanerozoic taxa, strengthen the case for pre-Cambrian innovation and earlier development of some modern body plans, long before they appear full blown in the fossil record. As noted below, animals with two embryonic germ layers (diploblastic) were perhaps present by 600-610 million years ago, those with three embryonic germ layers (triploblastic) may have appeared by 570 million years or so, and that it was not until the earliest Cambrian that unambiguous deuterostomes were around (Erwin, 1999). Recent finds in Russia may soon alter this idea. But others (Budd & Jensen, 2000) would question much of this evidence. They view the Cambrian explosion as quite real.

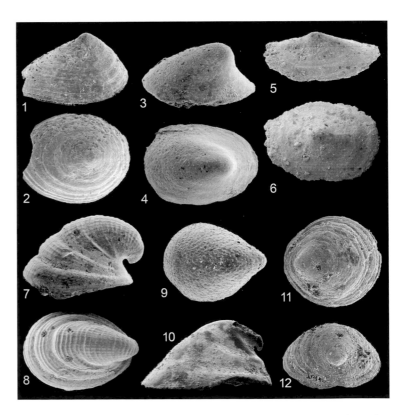

Figure 453. Small Shelly Fossils (SSF's) from Australia, Siberia and China, representative of the family Helcionellidae, Tommotian, Early Cambrian in age (P. Yu. Parkhaev).

247

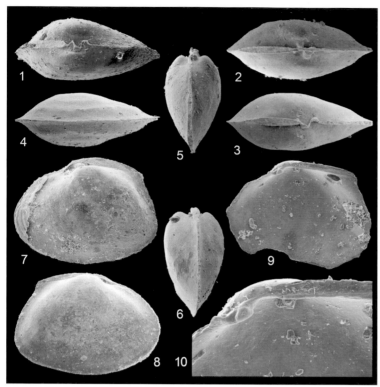

Figure 454. *Small Shelly Fossils (SSF's)*, Pojeta runnegari, *Early Cambrian, Yorke Peninsula, South Australia (P. Yu. Parkhaev).*

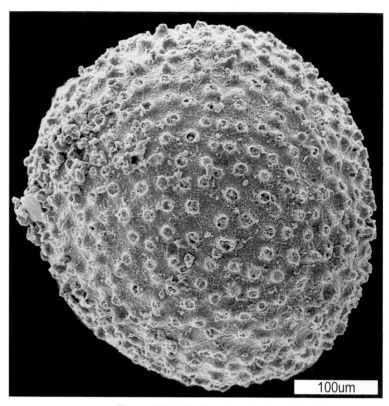

Figure 456. *Aetholicopalla adnata, a Small Shelly Fossil (SSF) from the Early Cambrian (Botomian) borehole SYC-101, Parara Limestone, York Peninsula, South Australia (P. Yu. Parkhaev).*

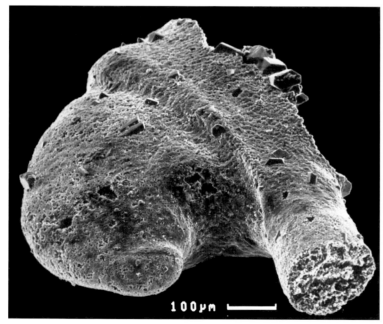

Figure 455. Runnegarella americana, *Forteau Formation, Yorke Peninsula, South Australia, South Australia, a Small Shelly Fossil. (P. Yu. Parkhaev)*

Figure 457. Skolithos, *vertical burrows, Cambrian, MacDonnell Ranges, Northern Territory (S. Morton). Width of specimen about 22 cm.*

During the very earliest part of the Cambrian, a number of spicular, tubular and conical forms, known as Small Shelly Fossils (SSF's) developed, some of which appear to be small versions with relations to certain molluscan groups. Some sclerites of fossil SSF's are very similar to the pyrite armor plating present in the "foot" of some living oceanic vent-dwelling gastropods (Waren *et al.*, 2003). Other SSF, seem to have either distant (Parkhaev, 1999) or no clear relationships to either modern or even Paleozoic forms. It was not until sometime in the Early Cambrian, between 540 and 530 million years ago, that

Figure 458. Early Paleozoic sea floor community reconstruction (courtesy of the Paleontological Institute, Moscow).

the truly rapid development of body plans occurred (Erwin, 1999), one particularly illustrative fossil site produced the Chengjiang Fauna of south China, dated at @530 million. A possible vertebrate, *Haikouella* (Chen, Huang & Li, 1999), is part of this assemblage. Hard parts certainly come on strong at this point, in a number of different groups – but did the underlying architecture of metazoan bodies suddenly develop at this time, or was it only the precipitation of hard parts that was the premier event?

In addition to large body fossils and microfossils, tracks and trails of animals are known by 555 million years ago (Droser, Jensen & Gehling, 2002). Earliest traces were quite simple – unbranched. They were small, less than a few mm in

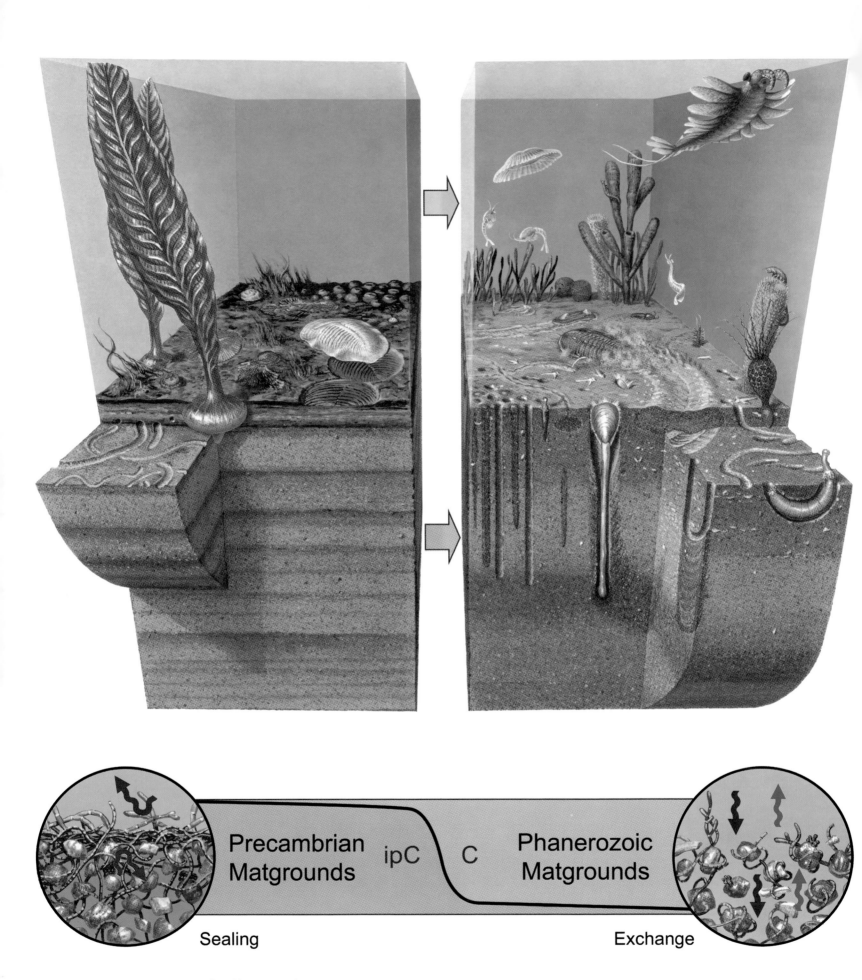

Precambrian Matgrounds   ipC   C   Phanerozoic Matgrounds

Sealing

Exchange

Figure 459. Changes in the sea floor biota across the
Ediacaran-Cambrian boundary (artist Peter Trusler).

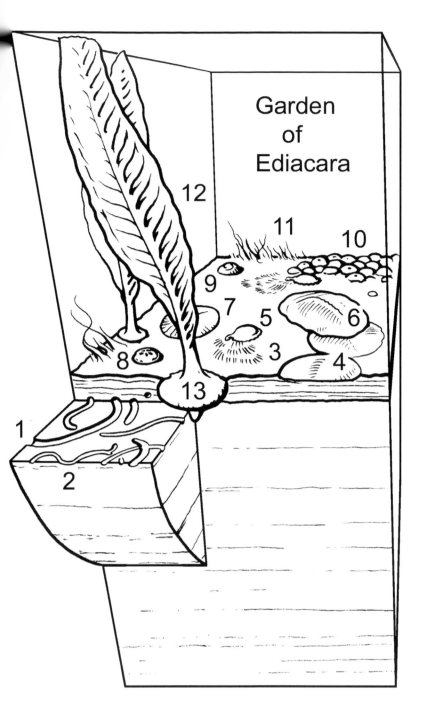

Garden
of
Ediacara

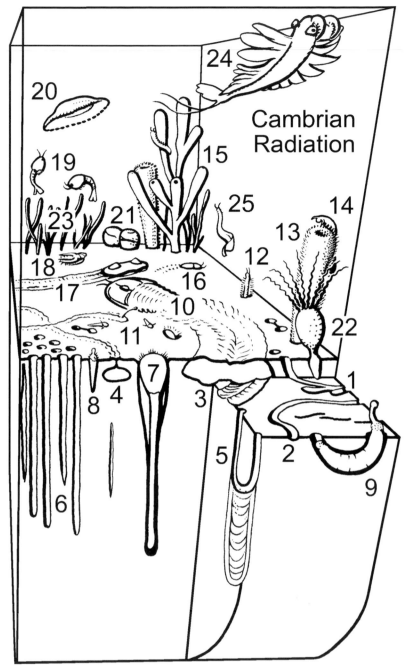

Cambrian
Radiation

Figure 460. Key for The Garden of Ediacara. Extensive marine microbial mat grounds covered the ocean bed. In the photic zones these were probably dominated by cyanobacteria and in the absence of vertical bioturbation, the mats became layered, essentially sealing the sediment below from gas and nutrient exchange. The mat structure greatly influenced the style of preservation of the Precambrian "metazoans" and was responsible, in part, for the remarkable "death mask" soft tissue impressions so characteristic of the Ediacaran fossil record. The biota consisted of: Under Mat Miners – 1, Planolites; 2, Archaeonassa – both trace fossils. Mat Scratchers – 3, Radula scratchings made by 5, Kimberella. Mat Etchers – 4, Trace fossil "footprints" or body impressions etched into the mat by 6, Yorgia; 7, Dickinsonia also left such records of its resting positions. Mat Encrusters – 8, Arkarua; 9, Tribrachidium; 10, Nemiana; 11, filamentous algae; 12, Charniodiscus and a diversity of 13, holdfasts or disc-shaped forms, probably "anchors" belonging to a variety of frond-like organisms. Reconstruction by Peter Trusler.

Figure 461. Key for the Agronomic Revolution. The Cambrian Radiation. With the Cambrian came the advent of skeletal hard parts and a major escalation of trophic structure. Vertical bioturbation of the sediments increased with various burrowing lifestyles. Predation increasingly became a driving force for ecological and evolutionary change and by the mid-Cambrian, the Agronomic Revolution was complete: the patterns for the remainder of the Phanerozoic ecosystems were in place. Trace fossil remains (ichnotaxa): 1, Planolites; 2, Treptichnus pedeum; 3, Cruziana (trilobites produce a variety of burrowing, furrowing, walking and resting traces, each with given names); 4, Psammichnites; 5, Diplocraterion; 6, Skolithos. Organic fossils remains, fauna: 7, Lingulella (inarticulate brachiopod in burrow); 8, Selkirkia and 9, Ottoia, both priapulid worms in vertical and U-shaped burrows respectively; 11, hyolithids; 12, Burgessochaeta, a polychaete worm; sponges Vauxia (13) and Chancelloria (15) with 14, Onycophoran Aysheaia crawling over them; 17, Halkieria; 20, Eldonia, possibly related to holothurians; 21, helicoplacoids and 22, Gogia spiralis, both echinoderms; 25, Pikaia, a chordate; and a variety of "arthropods" – 10, Redlichia, a trilobite; 16, Naraoia, a soft-bodied trilobite; Marrella; 19, Waptia; and 24, Amplectobelua, an anomalocarid. Flora: 23, Marpolia, an alga.

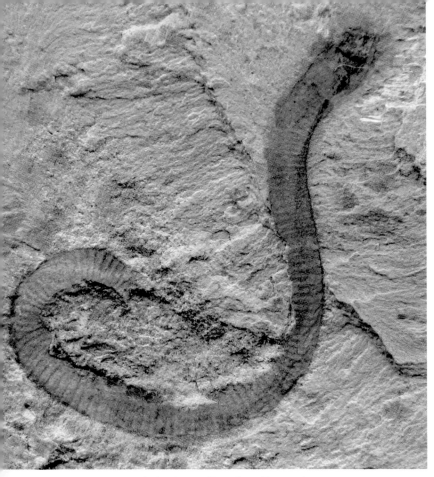

Figure 462. Maotainshania *sp., Chengjiang Fauna of Yunnan Province, southern China dated at around 530 million years, Early Cambrian (courtesy of Chia-wei Li). Length of specimen about 35 mm.*

Figure 463. Microdictyon sinicum, *a lobopodian or velvet worm (worms with legs), Chengjiang Fauna of southern China dated at around 530 million years, Early Cambrian (courtesy of Chia-wei Li). Length of specimen about 25 mm.*

diameter, and were made close to the sediment-water interface, horizontally oriented. Some are clearly similar to those known to be generated by bilaterian animals. This sort of trace fossil type continued into the Early Cambrian. It was not until nearly mid-Cambrian times, however, that animals began to burrow deeper and bioturbate sediments well below the sediment-water interface. This changed the preservational world forever. Just what brought this about is also yet unresolved – predation, greater oxygenation of the sediments, competition for food resources on the sea floor or water column, demise of widespread microbial mats"…"the list of possibilities is long.

As George Williams, a specialist on the pennatulaceans, the sea pens, so aptly points out: "the problematic nature of interpreting the nature of the Ediacaran fossils [or for that matter the nature of Neoproterozoic and Early Cambrian trace fossil makers] is in part due to the specialized state of science in which workers in different fields concerned with similar issues are unaware of each other's work and findings. It would be more productive if neontologists (*i.e.* systematists) and Precambrian paleontologists worked more closely together to decipher the affinities of the organisms in the Ediacaran and Burgess Shale faunas" (Williams, 1995) and this certainly seems to be happening in a big way at present.

Future progress in understanding all aspects of the late Proterozoic – its biota, its environments and their interactions – truly lies in interdisciplinary science and the sharing and mutual understanding of one another's databases and insights. Maybe then, we will truly begin to fathom the Ediacarans and the world in which they lived, as well as the potential they hold for the future.

Figure 464. The trilobite Yunnanocephalus yunnanensis, *Chengjiang Fauna of southern China dated at around 530 million years, Early Cambrian (courtesy of Chia-wei Li). Length of specimen about 15 mm.*

Figure 465. *The brachiopod* Lingulella chengjiangensis, *Chengjiang Fauna of southern China dated at around 530 million years, Early Cambrian (courtesy of Chia-wei Li). Length of specimen about 4.5 cm.*

Figure 467. *The arthropod* Fuxianhuia protensa, *Chengjiang Fauna of southern China dated at around 530 million years, Early Cambrian (courtesy of Chia-wei Li).*

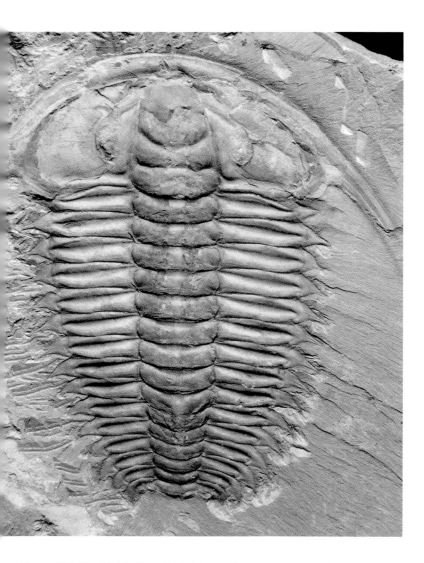

Figure 466. *The trilobite* Eoredlichia intermedia, *Chengjiang Fauna of southern China dated at around 530 million years, Early Cambrian (courtesy of Chia-wei Li). Length of specimen about 4.2 cm.*

Figure 468. *The trilobite* Naraoia spinosa, *Chengjiang Fauna of southern China dated at around 530 million years, Early Cambrian (courtesy of Chia-wei Li). Length of specimen about 27 mm.*

Figure 469. Isoxys *sp., a common element in many Cambrian faunas. It is a bivalved phyllopod crustacean, possibly related to ostracods. Chengjiang Fauna of southern China dated at around 530 million years, Early Cambrian (courtesy of Chia-wei Li). Length of specimen about 4 cm.*

Figure 471. Vetulicola cuneatus, *thought to be a deuterostome and in its own phylum, Chengjiang Fauna of southern China dated at around 530 million years, Early Cambrian (courtesy of Chia-wei Li).*

Figure 472. Algae, Chengjiang Fauna of southern China dated at around 530 million years, Early Cambrian (courtesy of Chia-wei Li). Width of block about 17 cm.

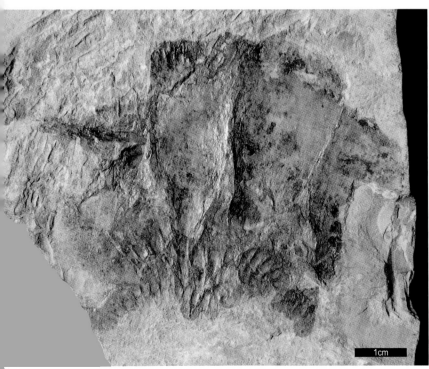

Figure 470. Peytoia *sp., thought to be the mouth parts of the large predator Anomalocaris, Chengjiang Fauna of southern China dated at around 530 million years, Early Cambrian (courtesy of Chia-wei Li). This form was originally interpreted as a jellyfish.*

Figure 473. Yuknessia *sp., a fossil algae, Chengjiang Fauna of southern China dated at around 530 million years, Early Cambrian (courtesy of Chia-wei Li). Width of block about 10 cm.*

Figure 474. Haikouella lanceolata, *Early Cambrian Maotianshan Shale, Ercai Village, Haikou, Kunming, southern China, Chengjiang Fauna (courtesy of Chia-wei Li and from Chen et. al., 1999).*

Figure 477. Trace fossils, *Early Cambrian Maotianshan Shale, Ercai Village, Haikou, Kunming, southern China, Chengjiang Fauna (courtesy of Chia-wei Li and from Chen et. al., 1999). Width of block about 25 cm.*

Figure 475. Reconstruction of Haikouella lanceolata, *Early Cambrian Maotianshan Shale, Ercai Village, Haikou, Kunming, southern China, Chengjiang Fauna (courtesy of Chia-wei Li and from Chen et. al., 1999).*

Figure 478. Trace fossils, *Early Cambrian, Meishcun, Jinning, Yunnan, southern China (courtesy of Chia-wei Li). Width of traces about 0.5 cm or less.*

Haikouella lanceolata

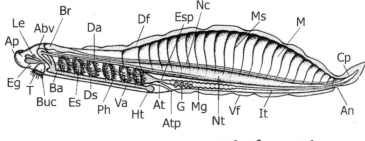

Haikou, Yannan

Figure 476. Haikouella lanceolata, *Early Cambrian Maotianshan Shale, Ercai Village, Haikou, Kunming, southern China, Chengjiang Fauna (courtesy of Chia-wei Li and from Chen et. al., 1999).*

Figure 479. Modern cnidarians, Seto Marine Labs,
Shirahama, Japan (E. Savazzi)

Figure 480. Living bivalve, with numerous blue "eyes"
– light-sensitive areas, Seto Marine Labs,
Shirahama, Japan (E. Savazzi)

ATLAS OF
PRECAMBRIAN METAZOANS

Patricia Vickers-Rich

Mikhail A. Fedonkin

James G. Gehling

Maxim V. Leonov

Andrey Yu. Ivantsov

Patricia Komarower

Margaret Fuller

with sincere thanks for significant assistance to

D. Gelt, S. Jensen, N. Hunt, E. Vnukovskaya,

E. Serezhnikova, G. Narbonne,

M. Laflamme, D. Rice, and J. Stilwell

Note: *Many descriptions presented in this* Atlas *are updated on previously published information.*

Classification of Ediacaran fossils is in a very fluid state at the moment as noted by Narbonne (2005), Gehling et al. (2005) and many others. In this Atlas *we have used the Phyla Proarticulata and Petalonamae (which may include forms not closely related, thus may be polyphyletic groupings), but are cognizant that these are controversial and that the taxonomy of Ediacarans may well change dramatically in the years to come.*

We would welcome comments from those with differing ideas, and if those comments could be forwarded to one of us (PVR at pat.rich@sci.monash.edu.au), they will be incorporated into the revised edition of this book.

Specimen numbers for illustrations in the Atlas *are either those of the holotype or where holotype is not illustrated, the number of the specimen shown is given in the title.*

Opposite page: Pteridinium simplex *in field on Farm Aar, southern Namibia. Nama Group rocks of late Neoproterozoic age (M. Fedonkin). Geology pick head for scale, @ 3 cm long.*

The Ediacarans. This material has been reproduced with permission of the Australian Postal Corporation. The original work is held in the National Philatelic Collection (art by Peter Trusler).

Albumares brunsae (holotype)

Occurrence: Rare
Locality: Suz'ma (5 km upstream from Suz'ma River mouth), White Sea, Russia; Reaphook Hill, Flinders Ranges, South Australia
Rock Unit: Ust'Pinega Formation, Valdai Group, Verkovka Formation of Grazhdankin, 2004; Ediacara Member, Rawnsley Quartzite, Wilpena Group
Description: Small discoid organisms, with more than 100 small marginal tentacles. Three-fold symmetry. Umbrella-like with three ridges radiating from center, narrowing and dividing lobes of the umbrella. From center of each of three lobes, three canals arise and split at least four times into smaller canals as they approach outer margins of organism. Diameter of umbrella, 13 mm; length of lobes, 5 mm; thickness of tentacles, 0.15 mm
Reference: Keller & Fedonkin, 1977
Type Specimen: Paleontological Institute, Moscow, GIN N4464/14
Classification: Phylum: Cnidaria; Class: Trilobozoa; Family: Albumaresidae. This group has also been allied to echinoderms and sponges by other workers

Anabylia improvisa (holotype)

Occurrence: Uncommon
Locality: Khorbusuonka River, tributary of the Olenek River, Yakutia, Olenek Uplift, Arctic Siberia
Rock Unit: Khatyspyt Formation, Khorbusuonka Group, Yudomian
Other Names: Taxon could be a synonym of *Inaria karli* (Grazhdankin, in press)
Description: Low relief cast on the sole of bed; rounded triangular in shape but poor preservation; up to nine fine, radial ribs; ribs narrow towards apex; ribs separated by fine furrows; three concentric zones are present at opposite ends of apex, with increasing relief distally from apex; maximum width, 9 mm; ribs 0.3-0.8 mm wide
Reference: Vodanjuk, 1989
Type Specimen: Geological Museum, Novosibirsk No. 913/7 (TSSGM)
Classification: Phylum: Petalonamae, attachment disc for frond-like organism

Andiva ivantsovi (PIN 3993/5110)

Occurrence: Uncommon in South Australia; not rare in Zimnii Bereg Assemblage
Locality: Zimnii Bereg, White Sea, Russia; Flinders Ranges, South Australia
Rock Unit: Ust'Pinega Formation, Valdai Group (Verkova, Zimnii Gory and Yorga formations of Grazhdankin, 1994); Ediacara Member, Rawnsley Quartzite, Wilpena Group
Description: Mop-head-like organism resembling a looped thumb print. Body oval to elongate and constructed of many fine, U-shaped segments that sweep backwards from a paired midline, ending in marginal lappets. Deformed impressions indicate this was a pliable organism, much like *Dickinsonia*. Resembles posterior end of *Dickinsonia tenuis*
Reference: Fedonkin, 2002; Narbonne, 2005
Type Specimen: Paleontological Institute, Moscow, PIN 3993/5108
Classification: Phylum Proarticulata; Class: Cephalozoa (Ivantsov, in press); Family: Yorgiidae; Narbonne (2005) places this taxon with the dickinsoniomorphs, while others have placed it in the Hydrozoa

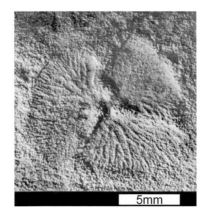
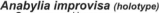

Albumares brunsae

Anabylia improvisa

Andiva ivantsovi

Anfesta stankovskii (PIN 3993/5706)

Occurrence: Rare
Locality: Zimnii Bereg and Kharakta River, White Sea, Russia; South Australia
Rock Unit: Ust'Pinega Formation, Valdai Group (Verkovka and Yorga formations of Grazhdankin, 2004)
Description: Small, hemispherical form, flattened, with three-fold symmetry; three elongate ridges radiate from center, each separated by angles of 120 degrees. These sausage-like ridges have rounded edges at both ends, proximal and distal. Whole organism divided into a number of narrow, radial lobes, always divisible by three – these bifurcate twice near their ends. In some specimens, tentacles present. Similar to *Albumares*, *Skinneria* in overall shape as well as *Albumares*, but more discoidal in shape and not dominantly tri-lobate in overall form. Diameter up to 18 mm; length of ridges up to 5 mm; width of ridges, up to 1.3 mm
Reference: Fedonkin, 1984
Type Specimen: Paleontological Institute, Moscow, PIN3993/260-A
Classification: Phylum: Cnidaria; Class: Trilobozoa; Family Albumaresidae; group has also been allied with echinoderms and sponges

"Arborea arborea" (holotype; See Charniodiscus arboreus)

Occurrence: Uncommon
Locality: Ediacara Hills, Flinders Ranges, South Australia
Rock Unit: Rawnsley Quartzite
Other Names: See *Charniodiscus arborea, C. concentricus, Rangea arborea* and others
Reference: Glaessner, 1959; Glaessner & Daily, 1959; Glaessner & Wade, 1966
Type Specimen: South Australian Museum SAM P12891

Archaeaspinus fedonkini (holotype)

Occurrence: Uncommon in South Australia
Locality: Zimnii Bereg and Kharakhta, White Sea, Russia; Chace Range of Flinders Ranges, South Australia
Rock Unit: Ust'-Pinega Formation, Valdai Group (Verkovka and Yorga formations of Grazhdankin, 2004); Ediacara Member, Rawnsley Quartzite, Wilpena Group
Other Names: "*Archaeaspis fedonkini*" (Ivantsov, 2001), pre-occupied genus name (a trilobite)
Description: Small bilateral animal of proarticulatan body plan with rounded shape (in top view); head area non-segmented and present in all stages of growth. Segments (isomeres) wide, short, with rounded ends, bent posteriorly; number of isomeres no more than 15 pairs. Head area bears non-paired lobe that is elongated mediolaterally and slightly asymmetric – with left side better developed. Dorsal surface covered with small knobs
Reference: Ivantsov, 2001
Type Specimen: Paleontological Institute, Moscow, PIN 3993/5053
Classification: Phylum: Arthropoda (stem group or primitive arthropod) or Phylum: Proarticulata (Ivantsov, 2004)

Anfesta stankovskii

Arborea arborea

Archaeaspinus fedonkini

"Archaeaspis" fedonkini
(See Archaeaspinus fedonkini)

Archangelia valdaica *(holotype)*
Occurrence: Rare
Locality: Suz'ma (5 km upstream from Suz'ma River mouth), White Sea, Russia
Rock Unit: Ust'Pinega Formation, Valdai Group (Verkovka Formation of Grazhdankin, 2004)
Other Names: Junior synonym of *Onegia nenoxa* (according to Grazhdankin, Ph.D. ms.)
Description: Bilaterally symmetric organism with ovoid shape, and midline axial zone. Axial zone made up of succession of short, broad segments that form zigzag line in middle. Segments decrease in size from one end to other. Lateral parts display gentle, transverse undulation. Width, 27 mm; length of segments at midline ranges from 2.5-4.5 mm; width 5-7 mm
Reference: Palij, Posti & Fedonkin, 1979
Type Specimen: Paleontological Institute, Moscow, GIN 4464/50 (original number; now PIN 3992/59)
Classification: Phylum: Petalonamae; Family: Pteridiniidae

Arkarua adami *(holotype)*
Occurrence: Rare, but common in restricted, deeper water sediments
Locality: Chace Range and Devil's Peak, Flinders Ranges, South Australia
Rock Unit: Ediacara Member, Rawnsley Quartzite, Wilpena Group
Description: Disc-shaped to hemispherical organism with five-pointed, star-shaped grooves forming opening. Flattened specimens almost polygonal in shape with segmented rim. Rim formed by shallow depression with outer ridge. Original body shape probably domed, being about 1-2 mm in height
Reference: Gehling, 1987
Type Specimen: South Australian Museum SAM P26768
Classification: Phylum: Echinodermata

Armillifera parva *(holotype)*
Occurrence: Rare
Locality: Zimnii Bereg and Solza River, White Sea, Russia
Rock Unit: Ust'Pinega Formation, Valdai Group (Verkovka and Yorga formations of Grazhdankin, 2004)
Description: Small, medium-sized, with oval shaped body consisting of convex central region and flattened periphery; central dome complicated by series of "comma-shaped" depressions arranged in two rows in alternating order ("gliding reflection") – at least four of these depressions, increasing in size from narrow to broad end of organism; peripheral zone bears fine radial furrows
Reference: Fedonkin, 1980
Type Specimen: Geological Institute, Moscow, GIN 4482/109; now PIN 3993/5213
Classification: Phylum uncertain, but Ivantsov (2007) placed in the Phylum Proarticulata based on its "symmetry of gliding reflection," where segments meet along the midline in a zigzag fashion. Sometimes also placed in the Phylum: Cnidaria; Class: Inordoza

Arumberia banksii *(GSM Geol. Surv. London 49163, from Longmyndian)*
Occurrence: Common in certain facies
Locality: Northern Territory, South Australia; Flinders Ranges (both Ediacaran and Early Cambrian); Longmyndian of England and Wales; White Sea, Russia; Urals; Podolia, Ukraine; northern France and the Channel Islands (probably Early Cambrian); Avalon Peninsula, Newfoundland; southern Namibia
Rock Unit: Arumbera Sandstone and Central Mount Stuart Formation of the Northern Territory; Pound Subgroup and Billy Creek Formation of South Australia; Valdai Series and Sylvitsa Series; Series Rouges du Golfe Normanno-Breton; Signal Hill and Musgravetown Groups; Nama Group
Description: Organism with a series

Archangelia valdaica

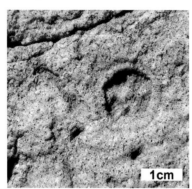
Arkarua adami

Armillifera parva

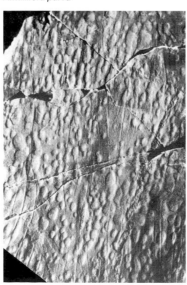
Arumberia banksii

of rectilinear ribs, radiating from a gently sloping elevation and separated by shallow grooves with both inner and outer walls. Bland (1984) noted that the impressions are present on both the tops and undersides of beds, often as part and counterpart. These impressions "exhibit a low relief of mounds and hollows." These lie between arrays of straight to gently curving parallel to subparallel fine ridges and in broad shallow grooves that are between 5 and 90 cm in length and maintain about the same width throughout
Reference: Glaessner & Walter, 1975; Bland, 1984; McIlroy & Walter, 1997
Type Specimen: Bureau of Mineral Resources, Canberra, Commonwealth Palaeontological Collection CPC 14948
Classification: Affinities of this structure are unclear. Originally described as a colonial cnidarian, later interpreted as a sedimentary structure produced by high-flow regime bottom currents acting on microbial-mat bound sediments, yielding sharp ribs and smooth hollows. Previously, comparisons were drawn with such forms as *Ernietta*

Askinica dimerus *(holotype)*
Occurrence: Rare, only two specimens
Locality: Basin of the rivers Askyn and Karanyurt and basin of Zelem River, Bashkeria, southern Urals
Rock Unit: Ashynskaya Sequence, Zeganskaya Series
Description: Oval-shaped imprint with an unevern undulating surface, which is dominated by a central raised disc. On one side of the central disc there are two raised spheres, while on the other side there are concentric bands; the central part has more topography than the periphery; there is a peripheral zone that is about 2-3 mm wide and sharply defined; diameter of imprint, 19 mm; diameter of cental disc, 2.5 mm; height of the imprint, 3-4 mm
Reference: Bekker, 1996
Type Specimen: Geological Museum Novosibersk, TSGM 1/11406 (316/89)
Classification: Likely holdfast of a frond-like organism although originally assigned to the Cnidaria

Aspidella costata *(Geol. Mus. Novisibersk GMN 673-51)*
Occurrence: Uncommon
Locality: Khorbusuonka River, tributary of the Olenek River, Yakutia, Olenek Uplift, Arctic Siberia; Urals
Rock Unit: Khatyspyt Formation, Khorbusuonka Group, Yudomian; Sylvitsa Series
Other Names: Also probably *Aspidella hatyspytia*; redescription as *Inaria hatyspytia* (Grazhdankin, *in press*)
Description: Large (diameter of holotype 74 mm), round or ellipsoidal form with wide marginal area, ribbed zone between marginal area and central area; ribbed zone wide, convex (preserved on sole of bed) ribbed zone separated from marginal zone by furrow; middle zone depression separated from ribbed zone by fine ridge; ribbed zone with lobes possessing very fine ribs, often merging
Reference: Vodanjuk, 1989
Type Specimen: Geological Museum, Novosibirsk, TSGM 913/5
Classification: Phylum: Petalonamae, attachment disc for frond-like organism; considered by many as a form-taxon for holdfasts that could represent many different taxa

Aspidella hatyspytia *(holotype)*
Occurrence: Common
Locality: Khorbusuonka River, tributary of the Olenek River, Yakutia, Olenek Uplift, Arctic Siberia
Rock Unit: Khatyspyt Formation, Khorbusuonka Group, Yudomian
Other Names: Probably *Aspidella costata;* redescription as *Inaria hatyspytia* (Grazhdankin, *in press*)
Description: Ellipsoidal shape, marginal rim well developed; ribbed zone separated from margin by furrow; middle part of organism deeply depressed; preserved on sole of bed; ribbed zone contains up to 10 broad, inflated ribs

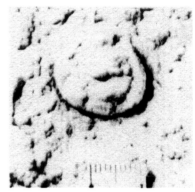
Askinica dimerus

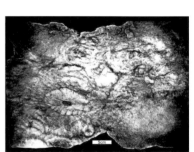
Aspidella costata

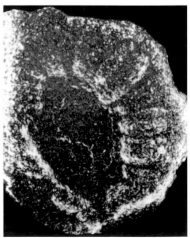
Aspidella hatyspytia

or spokes that are separated by deep furrows; best-preserved ribs divided by two fine furrows, which essentially divide these ribs into three zones; holotype maximum diameter 18 mm
Reference: Vodanjuk, 1989
Type Specimen: Geological Museum, Novosibirsk TSGM 913/4
Classification: Phylum: Petalonamae, anchor disc for frond-like organism

Aspidella terranovica (field specimen from Flinders Ranges, South Australia)
Occurrence: Very common
Locality: Flinders Ranges, South Australia; Avalon Peninsula, Newfoundland
Rock Unit: Rawnsley Quartzite, Wilpena Group; Fermeuse Formation, St John's Group
Other Names: *Beltanella gilesi*, (possibly *Cyclomedusa radiata, C. davidi, C. gigantea, C. minima, C. delicata, Ediacara flindersi* although not all researchers would agree with these synonymies, [Serezhnikova, pers. com., 2006.]), *Glaessneria imperfecta, Irridinitus multiradiatus, Jampolium wyrshykoowski, Madigania annulata, Medusinites patellaris, Paliella patelliformis, Paramedusium patellaris, Planomedusites patellaris, Protodipleurosoma rugulosum, Spriggia annulata, Spriggia wadea, Tateana inflata, Tirasiana coniformis, Tirasiana concentrallis, Tirasiana disciformis, ?Vendella larini*
Description: Disc-shaped and preserved in convex relief on the bases of beds. Most specimens have concentric rings and radial grooves. Originally described as several species of separate organisms; studies of thousands of specimens suggest that they simply represent different preservation styles of the same structure, perhaps for a few different organisms. When compared to the disc that is attached to the stalk of the frond-like form *Charniodiscus* – it is clear that these fossils were all probably attachment discs for such fronds. Obviously, from the name *Cyclomedusa* these forms in the past were considered to be body imprints of jellyfish – but jellyfishes have four-fold symmetry and mouthparts, features which are not found on these fossils. The variety of patterns of concentric rings and grooves probably simply due to different modes of compaction and the degree of sedimentary infilling of these bulbous holdfasts. Because they were embedded in the substrate, they had a much better chance of being preserved and this explains their commonness in the fossil record
Reference: Billings, 1872; Gehling *et al.*, 2000 and references therein
Type Specimen: Geological Survey of Canada GSC221
Classification: Phylum: Petalonamae, attachment disc for frond-like organism

Ausia fenestrata (holotype)
Occurrence: Extremely rare (one hand specimen with several "individuals")
Locality: Farm Plateau, Aus area, Namibia, southern Africa
Rock Unit: Dabis Formation, Nama Group
Description: A sponge-like form, bag-like with rows of large pores that occur in well-defined strips. Strips appear to wrap around the body of organism as diagonals while pores themselves form lines that lead from base to top of organism. The pores are large, relative to those found in poriferans. Organism was certainly compressable, for in only specimen known, which preserves at least two individuals, one shows the fenestrae or pores at the base as slits only, compressed, whereas those nearer what appears to be the dorsal end are open and uncompressed. On same individual, fenestrated part of organism attaches to narrowed base, that, although showing vague outlines of strips, which are more vertically oriented, presents an unfenestrated surface with a granular texture – quite reminiscent of *Ernietta*

Aspidella terranovica

Ausia fenestrata

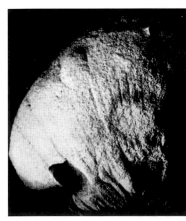

Baikalina sessilis

Reference: Hahn & Pflug, 1985
Type Specimen: Geological Survey of Namibia GSN 393 (F542)
Classification: Uncertain but may have relationships with Urochordates (Fedonkin, Vickers-Rich & Swalla, *in prep.*, 2007)

Baikalina sessilis
Occurrence: Rare
Locality: Lake Baikal
Rock Unit: Kurtun Formation
Other Names: Probably belongs in *Pteridinium* or *Ernietta*
Description: Organism bag-like
Reference: Sokolov, 1972
Type Specimen: Not specified
Classification: ?Phylum: Petalonamae

Barmia lobatus (from Bekker, 1996, pl. 3, no. 4)
Occurrence: Rare, two specimens only
Locality: Askyn River Basin, to the east of the village of Solontsy, Bashkeria, southern Urals
Rock Unit: Ashynskaya Sequence, Zeganskaya Series
Description: Oval shaped imprint with clearly defined concentric zonation, best defined in the central disc which is nearly flattened center; surface is deformed into folds; similar to *Conomedusites*, but does not have a four-fold symmetry; disc definitely deformed indicating that the surface was elastic. Diameter 20-28 mm; central zone, 3-6 mm
Reference: Bekker, 1996
Type Specimen: Geologlcal Museum, Novosibersk, TSGM 1/11406 (307/89)
Classification: Likely holdfast of a frond-like organism although originally assigned to the Cnidaria

"Beltanella gilesi" (holotype; See *Aspidella terranovica*)
Occurrence: Rare
Locality: Ediacara Hills, Flinders Ranges, South Australia; Mackenzie Mountains, Northwest Territories, Canada
Rock Unit: Rawnsley Quartzite, Wilpena Group; Sheepbed Formation, Windermere Supergroup
Other Names: Synonymised with *Aspidella*: described originally as an anemone
Reference: Sprigg, 1947; Gehling *et al.*, 2000 and references therein
Type Specimen: South Australian Museum SAM T3-2056

"Beltanella podolica" (holotype; same specimen as *Bronicella podolica*)
Occurrence: Rare
Locality: Dniester River, Russia
Rock Unit: Yarysev Formation
Type Specimen: Same specimen as *Bronicella podolica*

Beltanella zimimica (TSGM 9/87)
Occurrence: Rare
Locality: Zelem River, Bashkeria, Ural Mountains
Rock Unit: Bakyeievskaya Formation
Description: Inflated oval imprint with slightly displaced central disc. The surface of the imprint is divided into a few sectors, which are defined by radial ribs that originate at the central disc. Along the external border of one of these sectors there are a series of concentric furrows present. Surface of the imprint has some three dimensions. External border is somewhat deformed
Reference: Bekker, 1992
Type Specimen: TSGM/TSNEGR Leningrad Museum (St Petersburg); N1/11406 (9/57)
Classification: Very likely a holdfast for a frond-like organism

Beltanelliformis brunsae (PIN 4464/115)
Occurrence: Relatively common
Locality: Suz'ma and Zimnii Bereg, White Sea, Russia; Lorino Borehole, Russia; Aus Area, Namibia; South Australia; Wernecke Mountains and Mackenzie Mountians, NW Canada; Southern China; Nainital district, Uttar

Barmia lobatus

Beltanella gilesi

Beltanella podolica

Beltanella zimimica

Beltanelliformis brunsae

Pradesh, India
Rock Unit: Ust'-Pinega Formation, Valdai Group; Redkino Series; Dabis Formation, Nama Group; Rawnsley Quartzite, Wilpena Group; Sheepbed Formation, Windermere Supergroup; Doushantuo Formation; Krol Formation
Other Names: Perhaps = *Nemiana* Palij, 1976; *Hagenetta*
Description: Round imprints with high positive hyporelief. Concentric wrinkles distributed on peripheral edges. Central area smooth and flat (Menner, 1974 in Keller *et al.*,1974)
Reference: Menner, 1974 in Keller *et al.*, 1974
Type Specimen: Geological Institute GIN 4310/10
Classification: Probably an algae, but specific classification not established

Beltanelloides sorichevae *(PIN 3992/501)*

Occurrence: Common
Locality: Yarnema Borehole (depth of 118.4 m), Lyamtsa, Zimnii Bereg, White Sea, Russia; Ukraine; Podolia; possibly Urals, Russia; Spain
Rock Unit: Ust'Pinega Formation; Valdai Series (Lyamtsa, Verkovka and Zimnii Gory formations, Grazhdankin, 2004); Lomozov Beds, Mogilev Formation, Redkino Series; Cherny Kamen Formation, Sylvitsa Series; Pusa Shales and Axorejo Sandstone
Other Names: May be synonym with *Beltanelliformis brunsae* (Fedonkin, *pers. com.*); Narbonne & Hofmann, 1987); some specimens have been called *Nemiana* but the two taxa are distinct (M. Leonov, *in press*)
Description: Simple rounded organism, typically preserved as negative, rounded impression with thin irregular concentric wrinkles on periphery of an organic film. Fossils of this organism lay in beds of massive, structureless clays, and imprints have the same structure on both the upper and lower surfaces of the rock. The enclosing clay deforms both above and below fossils. *Beltanelloides* was a spherical organism with thin, but durable envelope around it and cavity in its center. Perhaps it was a planktonic autotroph, an idea originally proposed by Sokolov in 1976
Reference: Sokolov, 1965, 1972
Type Specimen: Not established
Classification: Possibly planktonic algae, but some researchers have suggested affinities with the Cnidaria

Bergaueria sp.

Occurrence: Rare
Locality: Stirling Ranges, Western Australia
Rock Unit: Stirling Formation
Description: Small (10 mm in length) elliptical discoid impressions
Reference: Cruse, *et al.*, 1993
Type Specimen: ?
Classification: Uncertain, with similarity to *Intrites*, placed by some in the metazoa, identified by others as a trace fossil. The great supposed antiquity of these (2016-1215 million years) requires further investigation – both as to the age and as to the classification

Blackbrookia oaksi *(holotype)*

Occurrence: Rare
Rock Unit: Lub Cloud Greywacke Member, Ives Head Formation, Blackbrook Group
Description: "The single impression consists of two subquadrangular impressions lying side by side with traces of a third alongside; each of the two measures 160 mm in diameter. The margin of the impression is an irregularly raised ridge up to 5 mm wide. Within this are very irregular lobate ridges comparable with those of *Ivesia lobata*." Boynton & Ford, 1995
Reference: Boynton & Ford, 1995
Type Specimen: Holotype *in situ* on Ives Head, Shepshed, Charnwood Forest Leicestershire Museum, UK, SK477 170. Casts in Leicester University Geology Department, Collections Accession No. 115576 and in Leicestershire Museum Geology Collection Accession No. G36/1994
Classification: Phylum: ?Cnidaria

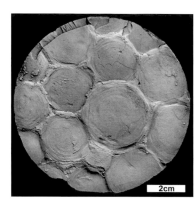

Beltanelloides sorichevae

Blackbrookia oaksi

Bomakellia kelleri

Bomakellia kelleri *(holotype)*

Occurrence: Rare, a single specimen
Locality: Suz'ma (Syuzma) (5 km upstream from the Suz'ma River mouth), White Sea, Russia
Rock Unit: Ust'-Pinega Formation, Valdai Group (Verkovka Formation of Grazhdankin, 2004)
Other Names: Dzik (2002) has suggested that *Bomakellia* might be congeneric with *Rangea* and both with *Mialsemia*; some similarities shared with *Paracharnia*
Description: Large organism with crescent-shaped front region and elongated body with numerous lateral appendages. Broad front end with convex ovate, axial lobe. Sides of front end and posterior part of central lobe flatter. The posterior part of axial lobe with rounded process, itself with two notches on both sides. Join between body and front lobe forms a sinuous boundary. Body narrow at front and gradually broadens towards middle, then narrows significantly towards posterior end. Margins of flat dorsal side with two rows of tubercles, all about same size. Closely spaced lateral "appendages" present that are oval in shape and split into two branches at end, increasing in size posteriorly. *Mialsemia* and *Bomakellia* are similar in many ways, but even though *Mialsemia* is elongate, it still has an ovate shape. *Bomakellia* has a larger anterior region. Length about 95 mm; length of thorax about 80 mm; maximum width of thorax not including the appendages, 22 mm; width of head, 30 mm; length of appendages from 7-18 mm; width from 4 to 8 mm; diameter of dorsal tubercles, 1-2 mm. Dzik (2002) interprets the "appendages" as branches of petaloids, with a central axis, with 4 vanes preserved
Reference: Fedonkin, 1985; Dzik, 2002
Type Specimen: Paleontological Institute, Moscow, PIN 3992/508
Classification: Originally classified in Phylum: Arthropoda, Class: Paratrilobita, but now serious consideration should be given to possible rangeomorph or ctenophore (Dzik, 2002) relationships

Bonata septata

Occurrence: Rare
Locality: Zimnii Bereg, White Sea, Russia
Rock Unit: Ust'-Pinega Formation, Valdai Group (Verkovka Formation of Grazhdankin, 2004)
Other Names: Central part of *Bonata* similar to mid-Cambrian *Peytoia nathorsti*
Description: Round or ovate impressions made up of two concentric zones – outer zone relatively broad and flat with rare concentric ridges. Inner zone has more relief and shows organism had some thickness. This zone is marked by numerous radial, wedge-shaped lobes with rounded ends that do not reach centre of the organism. Diameter of body up to 40 mm; diameter of central zone 8-10 mm; length of wedge-like lobes up to 5 mm or more; number of lobes ranges from 11-19
Reference: Fedonkin, 1980
Type Specimen: Geological Institute, Moscow, GIN 4482/52-7
Classification: Phylum: Cnidaria; Class: Cyclozoa; Family: Bonatiidae

Brachina delicata *(holotype)*

Occurrence: Uncommon
Locality: Brachina Gorge, Flinders Ranges, South Australia
Rock Unit: Rawnsley Quartzite, Wilpena Group
Description: Ovoid to oblong form with punctate hood and outside rim consisting of a smooth band, onto which small, ridge-like structures radiate outward. Not oral side of medusoid as originally interpreted. Hence, paratypes representing convex side are likely referred to *Aspidella* or a related form.
Reference: Wade, 1972
Type Specimen: South Australian Museum SAM F17343/P40933
Classification: Unknown

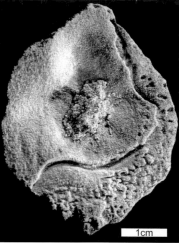

Bonata septata

Bonata septata

Bonata septata

Brachina delicata

Bradgatia linfordensis

Bradgatia linfordensis (no number, field specimen)
Occurrence: ?
Locality: Leicestershire, UK; Conception Bay, Avalon Peninsula, Newfoundland
Rock Unit: Charnian Supergroup; Mistaken Point Formation
Description: Bush-shaped organism consisting of multiple branches bearing fractal divisions. Attached to sea floor and flattened by currents before burial; reached 20 cm diameter
Reference: Boynton & Ford, 1995; Narbonne et al., 2001; Narbonne, 2005
Type Specimen: Specimen remains in the field and thus no type number given
Classification: Unknown, but related to other fractal branching rangeomorphs of Ediacaran age in Leicester and Newfoundland

Bronicella podolica (holotype)
Occurrence: Common
Locality: Dniester River, Ukraine
Rock Unit: Bronnitsy Member, Yaryshev Formation
Other Names: Same as "Beltanella podolica"
Description: Small (2- 5 mm diameter), circular convex imprints preserved on soles of beds in positive hyporelief, or flattened "with a hardened enamel-like surface either smooth or in deformed specimens sculpted with thin layer of wrinkles and folds" (Fedonkin et al., 1983)
Reference: Zaika-Novatsky, 1965; Palij, 1976; Palij et al., 1979
Type Specimen: Kiev State University KGU 1811
Classification: Cnidaria, incertae sedis (Palij, 1976)

"Charnia grandis" (See Chrania masoni)
Occurrence: Not common
Locality: Charnwood Forest, England
Rock Unit: Woodhouse Formation, Maplewood Group
Other Names: Charnia masoni
Description: See Charnia masoni
Reference: Boynton & Ford, 1995; length about 60 cm
Classification: Phylum: Petalonamae; Family: Charniidae, but this frond-like form may have been a colonial animal or even a large alga, but see Charnia masoni discussion
Type Specimen: ?

Charnia masoni (holotype from Ford publication)
Occurrence: Relatively common
Locality: Charnwood Forest, UK; Newfoundland; White Sea, Russia (Solza River, Zimnii Bereg); Olenek Uplift, Siberia, Russia; Flinders Ranges, South Australia; Mistaken Point Reserve, Newfoundland; Northwest Canada and other places
Rock Unit: Woodhouse Formation, Maplewood Group; Ust'Pingea Formation, Valdai Group (Verkovkaand Zimnii Gory Formations of Grazhdankin, 2004); Khatyspyt Formation; Ediacara Member, Rawnsley Quartzite, Wilpena Group; Sheepbed Formation; Mistaken Point and Trepassey formations, Conception Group, Newfoundland
Other Names: Glaessnerina grandis; Glaessnerina siberica; Rangea grandis and Rangea siberica; Charnia grandis
Description: Frond-shaped organism with curved midline ridge, forming zigzag pattern; secondary branches divided by deep grooves, with maximum number of 15. Reaches up to 170 mm in length of frond with width up to 80 mm. Secondary branches up to 6 mm in width and 13 mm length; parallel-sided to ovate
Reference: Ford, 1958; Narbonne & Aitken, 1995; Nedin & Jenkins, 1998; Narbonne, 2005
Type Specimen: Leicester University, UK, LEIUG 2382
Classification: Phylum: Petalonamae; Family: Charniidae, but this frond-like rangeomorph may have been a colonial animal; some researchers have suggested it to be a large alga but occurrence of charniids in the subphotic zones rules this unlikely

Bronicella podolica

Charnia grandis (Geology Today, Nov-Dec 1999)

5cm
Charnia masoni

Charnia siberica

"Charnia siberica" (holotype; same specimen as Rangea siberica; See Charnia masoni)
Occurrence: Rare
Locality: Olenek Uplift, Siberia
Rock Unit: Khatyspyt Formation
Other Names: Charnia masoni, Rangea siberica
Description: Taxon could be synonymous with Charnia masoni – needs further study
Reference: Sokolov, 1972, 1973; Glaessner, 1984
Type Specimen: Same specimen as Rangea siberica

Charnia wardi (holotype)
Occurrence: ?
Locality: Newfoundland
Rock Unit: Brook and basal Mistaken Point Formation, Conception Group
Description: Long (up to 2 m) and narrow (less than 6 cm) charniid with small holdfast (less than 6 cm).
Reference: Narbonne & Gehling, 2003; Narbonne, 2005
Type Specimen: Royal Ontario Museum, Toronto, ROM 38628
Classification: Phylum: Petalonamae; Family: Charniidae; but this frond-like rangeomorph may have been a colonial animal or even a large alga, but see discussion for Charnia masoni

Charniodiscus arboreus (holotype)
Occurrence: Uncommon
Locality: Ediacara Hills, Flinders Ranges, South Australia; Newfoundland; ?Nainital District, Uttar Pradesh, India
Rock Unit: Rawnsley Quartzite, Wilpena Group; probably Briscal Formation; Krol Formation
Other Names: Arborea arborea, Rangea arborea
Description: Frond-shaped organism, smooth on one side, with branches alternating along central stalk on the other. Tube-like structures arranged along each flap-like branch. Fronds uncommon in Ediacara Fauna of South Australia, most preserved as damaged specimens with little more than the central stalk and stubs of the side branches. Original specimen of related form found in Leicester, UK, has disc attached to stem, proving association. Large numbers of fronds with discs in place known in the Conception Group of Newfoundland, but preservation poor and does not allow detailed description or refined identification. In the Flinders Ranges rocks, Ediacara Member, fronds have apparently been torn away from their anchoring holdfasts and transported prior to burial, so both parts of organism are preserved separately. One exceptional specimen, however, South Australian Museum (SAM P19690), demonstrates association of holdfast and frond. This species of Charniodiscus normally has more than 20 primary branches, approximately same width throughout frond, with distal branches only slightly narrower than proximal equivalents
Reference: Glaessner & Daily, 1959; Glaessner & Wade, 1966; Jenkins & Gehling, 1978; Laflamme, Narbonne, & Anderson, 2004; Narbonne, 2005
Type Specimen: South Australian Museum hypotypes SAM F16718/ P40592
Classification: Phylum: Petalonamae; Family: Charniidae; but this frond-like rangeomorph may have been a colonial animal or even a large alga; but see discussion for Charnia masoni

Charniodiscus concentricus (holotype)
Occurrence: Rare
Locality: Charnwood Forest, England; Flinders Ranges, South Australia
Rock Unit: Woodhouse Formation, Maplewood Group; Rawnsley Quartzite, Wilpena Group
Other Names: South Australian specimens may be junior synonyms of Arborea arborea, Rangea arborea
Description: Nearly circular disc, type measuring 6.4 cm, others as small as 5 cm diameter, with irregular raised boss in center and faint concentric

Charnia wardi

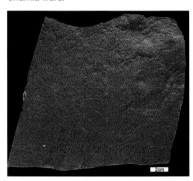
Charniodiscus arboreus

Charniodiscus concentricus

265

corrugations around it, flat border, no radial or quadripartite markings present
Reference: Ford, 1958; Jenkins & Gehling, 1978; Narbonne, 2005
Type Specimen: University of Leicester, UK, Department of Geology, No. 2383/3
Classification: Phylum: Petalonamae; likely a holdfast for a frond-like rangeomorph

Charniodiscus longus (holotype)
Occurrence: Uncommon
Locality: Ediacara Hills, Flinders Ranges, South Australia and possibly in the Zimnii Bereg sequence of the White Sea, Russia
Rock Unit: Rawnsley Quartzite, Wilpena Group
Other Names: Same specimen as *Rangea longa* and *Glaessnerina longa*
Description: Very elongate frond with closely spaced divisions on each primary branch. Flap-like covers on primary branches that clearly show they belong to *Charniodiscus*. Shape of frond similar to type specimen of *Charnia wardi* from the Avalon Peninsula of Newfoundland. However, divisions on secondary branches quite small, and shape is not so clearly fusiform. Most specimens preserved as casts on tops of beds, often with depressions at base of stalk, where frond was attached to buried holdfast
Reference: Glaessner & Wade, 1966; Jenkins & Gehling, 1978; Sun, 1986; Narbonne, 2005
Type Specimen: South Australian Museum SAM P13777 (preserved on the sole or bottom of a bed)
Classification: Phylum: Petalonamae; Family: Charniidae; but this frond-like rangeomorph may have been a colonial animal or even a large alga, but see discussion for *Charnia masoni*

Charniodiscus longus

Charniodiscus oppositus (SAM P82190)
Occurrence: Rare
Locality: Ediacara Hills, Flinders Ranges, South Australia; Zimnii Bereg, White Sea, Russia; Dniester River Basin, Podolia, Ukraine
Rock Unit: Rawnsley Quartzite, Wilpena Group; Ust'-Pinega Formation, Valdai Group (Yorga Formation of Grazhdankin, 2004); Mogilev Formation
Other Names: May be junior synonym of *Rangea arborea*
Description: Very similar to *C. arboreus*. As with all *Charniodiscus*, species fronds are much more rarely preserved. Organism stood erect in water above sea floor, vulnerable to wave action and storm surges, while their anchors or holdfasts lay buried in sediment and were more easily and frequently preserved
Reference: Jenkins & Gehling, 1978; Narbonne, 2005
Type Specimen: South Australian Museum SAM F17337/P40312
Classification: Phylum: Petalonamae; Family: Charniidae; but this frond-like rangeomorph may have been a colonial animal or even a large alga, but see discussion for *Charnia masoni*

Charniodiscus oppositus

"*Charniodiscus planus*"
(holotype; See **Aspidella terranovica**)
Occurrence: Common
Locality: Dniester River, Podolia, Ukraine
Other Names: "*Cyclomedusa plana*" and *Aspidella terranovica*
Rock Unit: Yaryshev Formation
Reference: Sokolov, 1972, 1973
Type Specimen: ?
Classification: Phylum: Petalonamae; Family: Charniidae

Charniodiscus planus

Charniodiscus procerus (field image)
Occurrence: Common
Locality: Mistaken Point, Newfoundland
Rock Unit: Upper Mistaken Point Formation, Conception Group and the Trepassey Formation, St John's Group
Description: *Charniodiscus* with less than 15 primary branches. Prominent, well-defined stem representing greatest portion of length. Frond lanceolate and

Charniodiscus procerus

typically bent to one side of stem. Basal attachment disc circular, 16-54 mm in diameter, unornamented, and consisting of smooth outer region lacking concentric rings and central boss to which stem attaches. Stem 31-136 mm long and joins high-relief central boss of disc to foliate main body. Frond complex and ornamented, 47-154 mm long and 15-57 mm wide, composed of two identical half leaf-shaped foliate structures joined together along central stalk. From stalk extend 8-14 subparallel, laterally directed primary branches, situated along the stalk and diverging at 45-90 degrees. Primary branches parallel to subparallel-sided and all of about same width throughout frond
Reference: Laflamme, Narbonne & Anderson. 2004; Narbonne, 2005
Type Specimen: Royal Ontario Museum, Toronto, ROM 36046
Classification: Phylum: Petalonamae; Family: Charniidae; but this frond-like rangeomorph may have been a colonial animal or even a large alga, but see discussion for *Charnia masoni*

Charniodiscus spinosus (field image)
Occurrence: Common
Locality: Mistaken Point, Newfoundland
Rock Unit: Mistaken Point Formation, Conception Group
Description: *Charniodiscus* with less than 15 branches. Front ovate, aligned with stem and terminated by prominent spine. Ovate frond, 39-164 mm long and 18-98 mm wide and similar to *C. procerus* in complexity and ornamentation. Primary branches separated by distinct furrows and composed of multiple secondary lobes. Stem joins disc to foliate main body.
Reference: Laflamme, Narbonne & Anderson, 2004; Narbonne, 2005
Type Specimen: Royal Ontario Museum, Toronto, ROM 36047;
Classification: Phylum: Petalonamae; Family: Charniidae; but this frond-like rangeomorph may have been a colonial animal or even a large alga, but see discussion for *Charnia masoni*

Charniodiscus spinosus

Chondroplon bilobatum (holotype)
Occurrence: Rare
Locality: Flinders Ranges, South Australia; Zimnii Bereg, White Sea, Russia
Rock Unit: Rawnsley Quartzite, Wilpena Group; Ust'Pinega Formation, Valdai Group (Zimnii Gory and Yorga formations of Grazhdankin, 2004)
Other Names: Form from White Sea succession may be a distinct form, but specimen not well enough preserved to allow more to be said than it belongs in this group (the Family Chondroplidae of Wade, 1971
Description: Part and counterpart known. Chambered, bilaterally symmetric form with narrow central axis, preserved as groove. Set of crescentic chambers joins central axis and curve concentrically around proximal end. Earlier chambers are broader, first chamber being large and elongate, while successive individual chambers decrease in breadth towards posterior end and thin towards central groove. Chambers curve away from central axis and end in sawtooth margin
Reference: Wade, 1971; Fedonkin, 2002
Type Specimen: South Australian Museum, SAM F17335a and b/P40473a and b
Classification: Originally placed in the Phylum Cnidaria; Class: Cyclozoa; Family: Chondroplidae; Fedonkin suggested that it was likely related to bilaterians

Chondroplon bilobatum

Cloudina sp. (field specimen)
Occurrence: Common, reef-formers
Locality: Namibia, China, Siberia, Canada, global
Other Names: Several other possibly related taxa, including *Conotubus* and *Sinotubulites*
Rock Unit: Nama Group; Dengying Formation; and others

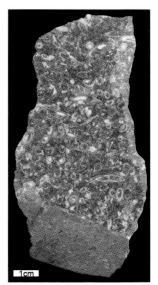
Cloudina sp.

Description: Small (but visible to the naked eye, *e.g. Cloudina hartmannae* reaches 2.5-6.5 mm), calcite-impregnated skeletons, built of multilayered tubular structures that close off at base. Tubes can have wedge-like protuberances, consist of nested funnels; they lack any internal transverse cross-walls. Walls made up of randomly oriented micron sized crystals. Some have daughter funnels that split within parental funnels. Thought to have evolved specialized skeletal secretion organs that could be retracted within parental funnels and generate new funnels. Capable of asexual reproduction. Similar in the growth patterns with living serpulid annelids
Reference: Germs, 1972; Glaessner, 1984; Grant, 1990; Hua *et al.*, 2005
Classification: Uncertain, but suggestions are annelid worms, algae, large foraminifera or cnidarian grade of organization. Another form, *Conotubus*, from the Gaojiashan Member of the Dengying Formation in China is thought to be a more primitive relative of *Cloudina*

Conomedusites lobatus
(holotype)
Occurrence: Common
Locality: Flinders Ranges, South Australia; Dniester River Basin near Novodnestrovskaya Power Station, Podolia, Ukraine; Zimnii Bereg, White Sea, Russia; Nainital District, Uttar Pradesh, India
Rock Unit: Rawnsley Quartzite, Wilpena Group; Lomozov Beds, Mogilev Formation; Ust'-Pinega Formation, Valdai Group (Verkovka Formation of Grazhdankin, 2004); Krol Formation
Description: Cone-shaped form divided by four deep radial grooves, thus producing peripheral boundary that is tetralobate, giving this organism almost square shape. Each lobe further divided by smaller grooves, and along peripheral boundary there are sometimes a number of thick tentacles
Reference: Glaessner & Wade, 1966; Fedonkin, 1985
Type Specimen: South Australian Museum SAM P13789
Classification: Phylum: Cnidaria; Class: Conulata; Family: Conchopeltidae

Corumbella werneri
(Ohio State University OSU 46401)
Occurrence: Common
Locality: Pantanal, Western Mato Grosso do Sul, southwestern Brazil; genus also reported from the western USA (Hagadorn & Waggoner, 2000)
Rock Unit: Tamengo Formation, Corumba Group
Description: Narrow, ringed, annulated, mostly parallel-sided tube having four-fold radial symmetry; annular rings meet at longitudinal grooves making the midlines; internally midlines are marked by carinae or ridges; width of tubes about 5 to 6 mm
Reference: Hahn *et al.*, 1982; Babcock *et al.*, 2005
Classification: Phylum: Cnidaria; Class: Scyphozoa (coronate); Family: Corumbellidae, or perhaps a conularid

Cyanorus singularis (holotype)
Occurrence: ?
Locality: Kharakhta, Solza, Zimnii Bereg, White Sea, Russia
Rock Unit: Ust'Pinega Formation, Valdai Group (Verkhovka, Zimnii Gory and Yorga formations of Grazhdankin, 2004)
Description: Small, elongate bilateral animals; long unsegmented head, isomeres with posterior flexure, tapering distally; digestive distributive system consists of axial channel and lateral diverticula, which in head region branch a few times; in trunk, each isomere contains one lateral tubular outgrowth, perhaps part of a digestive system
Reference: Ivantsov, 2004
Type Specimen: Paleontological Institute, Moscow, PIN 4853/83
Classification: Bilaterian, placed by

Conomedusites lobatus

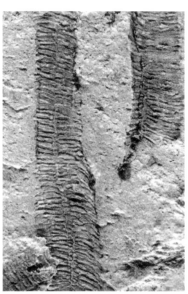
Corumbella werneri

Cyanorus singularis

some in the Phylum: Proarticulata; Class: Cephalozoa; Family: Sprigginidae (Ivantsov, 2004 and *in press*)

"Cyclomedusa cliffi" (holotype;
See *Aspidella terranovica*)
Occurrence: Rare
Locality: Cliffe Hill Quarry, Markfield, Leicestershire, UK
Rock Unit: Hallgate Member, Bradgate Formation, Maplewell Group
Other Names: See *Aspidella terranovica*, but not all researchers would agree with this synonymy (Serezhnikova, *pers. com.*)
Description: "A poorly preserved incomplete ovoid disc 150 x 120 mm (the latter estimated from the broken block) with a slightly undulating margin. Within this is a flat area 20 mm in diameter surrounding a slightly depressed area 20 mm wide. In the centre is a raised convex boss estimated at 30 x 22 mm." Boynton & Ford, 1995
Reference: Boynton & Ford, 1995
Type Specimen: Leicestershire Museum, UK, Accession Number G730/1993
Classification: Phylum: ?Cnidaria; attachment disc for frond-like organism or perhaps; Class: Cyclozoa; Family: unknown or, perhaps, Phylum: Petalonamae

"Cyclomedusa davidi" (SAM
P82195; See *Aspidella terranovica*)
Occurrence: Common in the Australian sequence but rarer in Russia
Locality: Ediacara Hills, Flinders Ranges, South Australia; Zimnii Bereg, Suz'ma, Yarnema, White Sea, Russia; Podolia, Ukraine; Charnwood Forest, England; Yellowhead Platform, southwestern Canada; Central Urals; Newfoundland; Nainital District, Uttar Pradesh, India
Rock Unit: Rawnsley Quartzite, Wilpena Group; Ust'-Pinega Formation, Valdai Group; Mogilev Formation (Ukraine); McManus; Sheepbed Formation, Hadrynian Miette Group, Windermere Supergroup; Albemarle Group; Kamen Formation, Sylvitsa Series; Conception Series; Krol Formation
Other Names: *Aspidella terranovica*, although not all agree with this synonymy (Serezhnikova, *pers. com.*, 2006); *Madigania annulata*
Description: Circular fossil with numerous concentric grooves, slightly separating elevated areas. Some *Cyclomedusa* species have thin, straight radiating grooves, interpreted by some as gastrointestinal imprints. As name implies, these fossils were originally thought to be jellyfish imprints, but more recent suggestions are that they are the holdfasts or anchors for a number of different organisms. Some have been found attached to such forms as *Charniodiscus arboreus*. *C. davidi* itself is characterized by having low central tubercle (positive hyporelief). Spacings between concentric grooves sometimes inconsistent around outside of organism and range between 0.5 and 4 mm. Diameter of organisms ranges from 10 to 50 mm. Shelf, separating the inner and outer zones, has more distinct relief in large specimens and lacks radial grooves
Reference: Sprigg, 1947, 1949; Jenkins, 1984; Sun, 1986; Gehling *et al.*, 2000
Type Specimen: Holotype poorly preserved; South Australian Museum SAM T5-2057
Classification: Phylum: Petalonamae

"Cyclomedusa delicata" (PIN
3992/355); See *Aspidella terranovica*)
Occurrence: Rare
Locality: Suz'ma, White Sea, Russia
Rock Unit: Ust'-Pinega Formation, Valdai Group (Verkovka Formation of Grazhdankin, 2004)
Other Names: See *Aspidella terranovica*, but not all researchers would agree with this synonymy (Serezhnikova, *pers. com.*, 2006)
Description: Large, ovate and

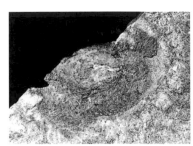

Cyclomedusa cliffi

Cyclomedusa davidi

Cyclomedusa delicata

flattened form with conical central part and many concentric ridges. These cover entire body in regular fashion and are separated by fine furrows. Ridges in central part of *C. delicata* narrower than on flattened peripheral part. Largest specimen with diameter of 116 mm, width of fine concentric ridges ranges from 0.5 to 1.2 mm, while width of broader ridges reaches 2.5 mm. Number of ridges can reach 40.
Reference: Fedonkin, 1981; Gehling *et al.*, 2000
Type Specimen: Geological Institute GIN, Moscow, 4464/336
Classification: Phylum: ?Cnidaria; attachment disc for frond-like organism or perhaps; Class: Cyclozoa; Family: Unknown or perhaps Phylum: Petalonamae

"Cyclomedusa gigantea"
(holotype; See *Aspidella terranovica*)
Occurrence:
Locality: Ediacara Hills, Flinders Ranges, South Australia
Rock Unit: Rawnsley Quartzite, Wilpena Group
Other Names: Synonomized with *Aspidella terranovica,* but not all researchers would agree with this synonymy (Serezhnikova, *pers. com.*)
Reference: Sprigg, 1949
Type Specimen: South Australian Museum SAM TI8 2035

"Cyclomedusa minuta"
(holotype; See *Aspidella terranovica*)
Occurrence: Rare (one specimen)
Locality: Suz'ma, White Sea, Russia
Rock Unit: Ust'Pinega Formation, Valdai Group (Verkhovka Formation, Grazhdankin, 2004)
Other Names: Not a valid taxon. Juvenile form of other larger discs (Gehling, *pers. com.*, 2005), *Aspidella terranovica,* but not all researchers would agree with this synonymy (Serezhnikova, *pers. com.*)
Description: A small discoidal form preserved as flattened cast with slightly indented central tubercle, surrounded by regular concentric ridges (six) of uniform width. Diameter about 7 mm, with central tubercle about 1 mm and ridges around 0.4 to 0.6 mm. Rather similar to *C. davidi,* but smaller and with ridges more regularly spaced and more flattened overall. Differs from *C. plana* in lacking smooth and broad outer body and from *C. radiata* in being much smaller and lacking radial grooves present in *C. radiata.* Some features might be due to deformation after death and different stages of growth
Reference: Fedonkin, 1981
Type Specimen: Geological Institute, Moscow, GIN 4464/112

"Cyclomedusa plana" (See *Aspidella terranovica*)
Occurrence: ?
Locality: Ediacara Hills, Flinders Ranges, South Australia; Zimnii Bereg, White Sea, Russia; Dniester River Basin, Podolia, Ukraine; Mackenzie Mountains, Northwest Territories, Canada
Rock Unit: Rawnsley Quartzite, Wilpena Group; Ust'-Pinega, Valdai Group (Verkovka Formation of Grazhdankin, 2004); Mogilev and Yaryshev formations; Sheepbed Formation, Windermere Supergroup
Other Names: *Aspidella terranovica,* but not all researchers agree with this synonymy (Serezhnikova, *pers. com.*, 2006.); possibly *Glaessneria imperfecta;* see *Glaessneria plana*
Description: Broad, flat disc, sometimes divided by irregular radial grooves, a feature characteristic of large specimens with diameters greater than 30 mm. Small specimens usually quite smooth. Diameters range from 15 to 300 mm See *C. davidi* for more detail of genus
Reference: Glaessner & Wade, 1966; Gehling *et al.*, 2000
Type Specimen: South Australian Museum SAMP13778

Cyclomedusa gigantea

Cyclomedusa minuta

"Cyclomedusa radiata" (SAM 16729; See *Aspidella terranovica*)
Occurrence: Rare in the White Sea sequence
Locality: Flinders Ranges, South Australia; Zimnii Bereg, White Sea, Russia; Dniester RIver, Podolia, Ukraine
Rock Unit: Rawnsley Quartzite, Wilpena Group; Ust'-Pinega Formation, Valdai Group (Yorga Formation of Grazhdankin, 2004); Mogilev Formation
Other Names: *Aspidella terranovica,* but not all researchers agree with this synonymy (Serezhnikova, *pers. com,* 2006.); *Glaessneria radiata*
Description: Flat central dome separated by deep groove from relatively narrow outer zone with numerous, fine radial grooves also visible in central section. Central dome with fine concentric grooves. Total diameter reaches 36 mm, diameter of central dome, 24 mm. Radial grooves on outer part spaced at most 1.5 mm but often much closer together
Reference: Sprigg, 1947; Jenkins, 1984; Gehling *et al.*, 2000
Type Specimen: South Australian Museum SAM T23-2037

Cyclomedusa radiata

"Cyclomedusa serebrina"
(holotype same as *"Pollukia"*; See *Aspidella terranovica*)
Occurrence: Rare
Locality: Dniester River, Podolia, Ukraine
Rock Unit: Mogilev Formation (Ukraine)
Other Names: *Aspidella terranovica,* but not all researchers agree with this synonymy (Serezhnikova, *pers. com,* 2006.); *"Pollukia serebrina"*
Reference: Palij, 1976
Type Specimen: Kiev State University KGU 1735

Cyclomedusa serebrina

Dickinsonia brachina (holotype)
Occurrence: ?
Locality: Brachina Gorge, Flinders Ranges, South Australia
Rock Unit: Rawnsley Quartzite, Wilpena Group
Other Names: Difficult to distinguish from *D. lissa* and *D. tenuis* in early stages. May be synonym of *D. tenuis.*
Description: See *D. tenuis*
Reference: Wade, 1972
Type Specimen: South Australian Museum SAM P17998
Classification: Bilaterian, but Phylum uncertain; in the past has been placed in the Phylum Proarticulata based on its "symmetry of gliding reflection" where segments meet along the midline in a zigzag fashion. Class: Dipleurozoa; Family: Dickinsonidae. Gehling *et al.* (2005) do not agree that the segments are offset.

Dickinsonia brachina

Dickinsonia costata (SAM 13760)
Occurrence: Very common in Australia and in Russia
Locality: Ediacara Hills, Flinders Ranges, South Australia; Kharakhta and Solza rivers, Zimnii Bereg, White Sea, Russia; Dniester River Basin, Podolia, Ukraine
Rock Unit: Rawnsley Quartzite, Wilpena Group; Ust'-Pinega Formation, Valdai Group (Verkhovka, Zimnii Gory and Yorga formations of Grazhdankin, 2004); Mogilev Formation
Other Names: *Dickinsonia minima, Dickinsonia spriggi, Papillionata eyrie*
Description: Form has ovoid, sometimes almost circular body, with low relief. Body segmented, divided down middle with ridge, possibly outline of gut. Head end defined by location of largest and broadest of segments, which gradually increase in size towards head; tail defined by end where finer segments located. In some larger specimens strong, sharp ridges occur (often with secondary faint ridge) along trailing margins of each segment. Faint outlines of soft body tissue that had been extended beyond main body mass, like those left by extension of foot in living mollusks, often preserved around outside of higher relief main body impression. Rarely seen are branching structures faintly preserved that cut across segments,

Dickinsonia costata

possibly digestive caecae. Postmortem structure, such as overfolds, tears and indentations frequently observed on larger specimens, assumed to be damage or deformation imparted during burial. Smallest specimens less than 1 cm long with only 6-10 segments, largest more than 20 cm with more than 50 segments in largest forms. Species probably "tactophobic," for they never seem to touch – on one bedding surface alone with over 200 specimens, not a single one in contact with another. All *Dickinsonia* species had large surface areas, perhaps for use in respiration or gathering nutrition, or both
Reference: Sprigg, 1947; Glaessner & Wade, 1966
Type Specimen: South Australian Museum SAM T6-2055
Classification: Bilaterian, but Phylum uncertain; in the past has been placed in the Phylum Proarticulata based on its "symmetry of gliding reflection" where segments meet along the midline in a zigzag fashion. Class: Dipleurozoa; Family: Dickinsonidae. Gehling *et al.* (2005) do not agree that the segments are offset and note that this form has been assigned to the dickinsonids, cnidarians, polychaete worms, vendobionts and stem-group chordates

"Dickinsonia elongata" (SAM P13760; See *Dickinsonia rex*)
Occurrence: ?
Locality: Ediacara Hills, Flinders Ranges, South Australia
Rock Unit: Rawnsley Quartzite, Wilpena Group
Other Names: Type specimen is probably damaged specimen of *Dickinsonia costata*. Other material referred to *Dickinsonia rex*
Reference: Glaessner & Wade, 1966; Wade, 1972
Type Specimen: South Australian Museum SAM P13367

Dickinsonia lissa (PIN 3993/5512)
Occurrence: Uncommon in South Australia
Locality: Flinders Ranges, South Australia; White Sea, Russia
Rock Unit: Ediacara Member, Rawnsley Quartzite, Wilpena Group; Ust'-Pinega Formation, Valdai Group (Yorga Formation of Grazhdankin, 2004)
Description: In general, similar to *D. costata* but very delicate, with broad "gut"(midline division and segments that taper parallel to tail end; segments very thin. Mid-ridge quite tall and does not thin towards posterior or back end
Reference: Wade, 1972
Type Specimen: South Australian Museum SAM F17466/P40134
Classification: Bilaterian, but Phylum uncertain; in the past has been placed in the Phylum Proarticulata based on its "symmetry of gliding reflection" where segments meet along the midline in a zigzag fashion. Class: Dipleurozoa; Family: Dickinsonidae. Gehling *et al.* (2005) do not agree that the segments are offset (see *D. costata* classification discussion)

"Dickinsonia minima" (holotype; See *Dickinsonia costata*)
Occurrence: ?
Locality: Flinders Ranges, South Australia
Rock Unit: Ediacara Member, Rawnsley Quartzite, Wilpena Group
Other Names: Synonym of *Dickinsonia costata*
Reference: Sprigg, 1949; Wade, 1972
Type Specimen: South Australian Museum SAM T51-2000

Dickinsonia rex (holotype)
Occurrence: Rare
Locality: Flinders Ranges, South Australia
Rock Unit: Rawnsley Quartzite, Wilpena Group
Other Names: *Dickinsonia elongata*
Description: Similar to *D. costata*, but with many more segments. Segments change little in thickness for two-thirds of body length. First anterior segment length/axial length ratio is smaller than in *D. costata*. Segments generally tend

Dickinsonia elongata

Dickinsonia lissa

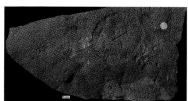
Dickinsonia minima

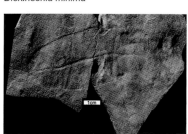
Dickinsonia rex

to be similar in width from outer edge of body to mid-ridge. However, the very large specimen (SAM P40167) has segments that tend to narrow by less than half their width and curve quite sharply anteriorly and then posteriorly when they reach mid-ridge, perhaps suggesting tall mid-ridge. With overall length of 81 cm and width of 65 cm, this *Dickinsonia* is the largest to be discovered
Reference: Jenkins, 1992
Type Specimen: South Australian Museum, SAM P40167
Classification: Bilaterian, but Phylum uncertain; in the past has been placed in the Phylum Proarticulata based on its "symmetry of gliding reflection" where segments meet along the midline in a zigzag fashion. Class: Dipleurozoa Family: Dickinsonidae. Gehling *et al.* (2005) do not agree that the segments are offset (see *D. costata* classification discussion)

"Dickinsonia spriggi" (holotype; See *Dickinsonia costata*)
Occurrence: ?
Locality: Ediacara Hills, Flinders Ranges, South Australia
Rock Unit: Rawnsley Quartzite, Wilpena Group
Other Names: *Dickinsonia costata*
Reference: Glaessner & Wade, 1966
Type Specimen: South Australian Museum SAM T49-2007

Dickinsonia tenuis
Occurrence: ?
Locality: Ediacara Hills, Flinders Ranges, South Australia; Kharakhta and Solza rivers, Zimnii Berg (*D. cf. tenuis*), White Sea, Russia; Dniester River Basin, Podolia, Ukraine
Rock Unit: Rawnsley Quartzite, Wilpena Group; Ust'-Pinega Formation, Valdai Group (Verkhovka, Zimnii Gory and Yorga formations of Grazhdankin, 2004); Mogilev Formation
Other Names: May be same as *Dickinsonia brachina*
Description: Similar to *D. costata*, but broad (length being twice that of width) and with very fine segments. Some Individuals have more than 100 delicate segments (isomeres).
Reference: Glaessner & Wade, 1966
Type Specimen: South Australian Museum SAM P13791
Classification: Bilaterian, but Phylum uncertain; in the past has been placed in the Phylum Proarticulata based on its "symmetry of gliding reflection" where segments meet along the midline in a zigzag fashion. Class: Dipleurozoa; Family: Dickinsonidae. Gehling *et al.* (2005) do not agree that the segments are offset (see *D. costata* classification discussion)

"Ediacaria" booleyi (field specimen)
Occurrence: Common
Locality: South side of Booley Bay, near Duncannon, County Wexford, Ireland (Eire)
Rock Unit: Booley Bay Formation, Upper Cambrian
Description: Large circular disc-like structures with the dorsal side divided into 3 concentric zones, with additional concentric lines and numerous thin radial features that become more prominent towards the periphery, showing a ribbed or tubercular nature. Central zone nearly flat, passing outwards into a more steeply sloping middle zone, which leads to a gently sloping, flange-like outer zone. Ventral side has alternate areas of coarse and fine concentric markings and numerous fine radial lines, some thicker and more prominent at the edges; specimens vary in size but a typical specimen has a total radius of about 5 cm with a central disk radius of 1.5 cm. (Crimes *et al.*, 1995; MacGabhann, 2006)
Reference: MacGabhann, *in press*
Type Specimen: Geological Survey of Ireland, Dublin; GSI F12410.
Classification: Crimes *et al.*,1995 suggested that this form was related to Some primitive animals that left no

Dickinsonia spriggi

Dickinsonia tenuis

Ediacaria booleyi

Phanerozoic offspring and were part of the Ediacaran fauna that survived into the Late Cambrian in deep marine environments. MacGabhann has recently suggested that these fossils do not belong in the genus *Ediacaria* or any known Ediacaran taxon due to their apparent high density and tough integument. He did not rule out the possibility that they could be inorganic

"Ediacaria flindersi" (See

Aspidella terranovica)
Occurrence: Common
Locality: Flinders Ranges, South Australia; Zimnii Bereg, White Sea, Russia; Dniester River Basin, Podolia, Ukraine; Khorbusuonka Basin, Olenek Uplift, Siberia; Wernecke Mountains and Mackenzie Mountains, Northwest Canada
Rock Unit: Rawnsley Quartzite, Wilpena Group; Ust'-Pinega Formation, Valdai Group (Verkovka Formation of Grazhdankin, 2004); Yaryshev Formation; Khatysput Formation; Sheepbed Formation and Blue Flower Formation, Windermere Supergroup
Other Names: *Aspidella terranovica*, but not all researchers agree with this synonymy (Serezhnikova, *pers. com.*, 2006.) *Ediacaria* fossils are rounded and slightly convex imprints preserved on the soles of the layers, preserved in three dimensions
Description: Benthic organisms whose basal parts have radial symmetry preserved as flat to low-relief imprints of three-zonal attachment surface with radial structures. Specimens have central disc and outer ring separated by deep annular groove; distinct tubercle in the central disc. Radial grooves mainly confined to outer ring. Surfaces of specimens generally smooth; sometimes internal cast of stem can be preserved. Diameters of specimens range from 40 to more than 450 mm. Fossils may be holdfasts of forms like *Charnia* and *Charniodiscus*
Reference: Sprigg, 1947, 1949; Fedonkin, 1985; Serezhnikova, 2005
Type Specimen: South Australian Museum SAM T1-2058

Ediacaria flindersi

Elasenia aseevae (holotype)
Occurrence: Rare
Locality: Dniester Valley near the Novodnestrvskaya Hydroelectric Power Station, Podolia, Ukraine
Rock Unit: Lomozov Beds, Mogilev Formation
Other Names: Perhaps a synonym of *Ediacaria* and thus *Aspidella terranovica*
Description: Small discoidal organism, preserved as semi-spherical casts. Inner zone relatively large central disc. Outer zone contains several small, round tubercles that adjoin central disc. Central disc flat, taking up about one-third area of whole organism. Diameter of body, 11 mm; diameter of central disc, 4 mm; diameter of tubercles in outer zone about 2 mm
Reference: Fedonkin, 1983; Velikanov, Aseeva & Fedonkin, 1983
Type Specimen: Paleontological Institute, Moscow, PIN 3994/338
Classification: Phylum: Cnidaria; Class: Inordozoa; Family: uncertain

Elasenia aseevae

Elasenia uralica (holotype)
Occurrence: One holotype and a few fragments
Locality: Basin of the rivers Askyn and Karanyurt, Bashkeria, southern Urals
Rock Unit: Ashynskaya Sequence, Zeganskaya Series
Description: Small disc-shaped imprints preserved in positive hyporelief as a half-sphere, with large raised, central disc with a central structure; in the peripheral zone there are a number of raised oval-shaped areas; diameter of single specimen, 15 mm; diameter of central disc, 4 mm; width of peripheral band, 3 mm; diameter of central structure, 1.5 mm. Show some similarity to *Beltanella zilimica*
Reference: Bekker, 1996
Type Specimen: Geological Museum,

Elasenia uralica

Novosibersk,TSGM 1/11406 (316/89)
Classification: Likely holdfast of a frond-like organism although originally assigned to the Cnidaria

Eoporpita medusa (SAM D-159)
Occurrence: Common in Russia
Locality: Ediacara Hills, Flinders Ranges, South Australia; Zimnii Bereg, White Sea, Russia, perhaps also the Urals; Mackenzie Mountains, Northwest Territories, Canada
Rock Unit: Rawnsley Quartzite, Wilpena Group; Ust'-Pinega Formation, Valdai Group (Yorga Formation of Grazhdankin, 2004); Cherny Kamen Formation, Sylvitsa Series; Sheepbed Formation, Windermere Supergroup
Description: Rounded or ovate in outline with raised central disc or cone, surrounded by concentric rings, usually lower in relief. Outer rings sometimes have fine, radiating striations, but generally smooth. Tentacle-like structures appear to be attached to outer rim of central zone and radiate at approximately right angles outward, sometimes observed as single tentacles but often as mass
Reference: Wade, 1972; Fedonkin, 1985
Type Specimen: South Australian Museum SAM T27-2019
Classification: Once interpreted to be in the Phylum: Cnidaria; a chondrophorine; now thought to represent a benthic form, perhaps a xenophophoran protist (Seilacher *et al.*, 2003) or even a complex holdfast (Serezhnikova, *pers. com.*, 2007)

Epibaion axiferus (holotype)
Occurrence: Common in Russia
Locality: Lyamtsa, Solza River, Zimnii Bereg, White Sea, Russia
Rock Unit: Ust'-Pinega Formation, Valdai Group (Lyamtsa, Verkhovka, Zimnii Gory and Yorga formations of Grazhdankin, 2004).
Description: Large bilateral organisms with proarticulate body plan. Body elongated-oval and completely subdivided into isomeres. Isomere orientation is, in general, radial, but 2/3 of segments in posterior end turned backwards. Isomeres narrow and equally thick. Only in posterior part of body and immediately near anterior end does thickness of isomeres decrease, number exceeds 120 pairs. Isomere surface smooth. Longitudinal axis of ventral side marked by narrow lobe tapered from both ends and connecting inner ends of all isomeres. Width of axial lobe almost equal in different-size specimens. Resting tracks demonstrate long imprints of isomeres and axial lobe.
Reference: Ivantsov & Malakhovskaya, 2002
Type Specimen: Paleontological Institute, Moscow, PIN 3993/5199
Classification: Bilaterian, but Phylum uncertain; in the past has been placed in the Phylum Proarticulata based on its "symmetry of gliding reflection" where segments meet along the midline in a zigzag fashion. Class: Dipleurozoa; Family: Dickinsonidae (see *D. costata* classification discussion)

"Erniaster apertus" (See *Fig. 130,*
this volume, Pflug Plate 28) See
Ernietta plateauensis
Locality: Farm Aar or Plateau, Aus area, Namibia
Rock Unit: Nama Group
Reference: Pflug, 1972
Type Specimen: Geological Survey of Namibia GSN 230 (F465-H) (Pflug No 238)

"Erniaster patellus" (See *Fig. 130,*
Pflug Plate 29) See *Ernietta*
plateauensis
Locality: Farm Aar, Aus area, Namibia
Rock Unit: Dabis Formation, Nama Group
Reference: Pflug, 1972
Type Specimen: Geological Survey of Namibia GSN 293 (F484-H) (Pflug No 238)

Eoporpita medusa

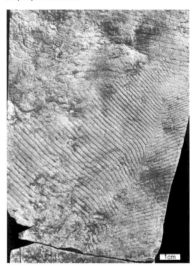

Epibaion axiferus

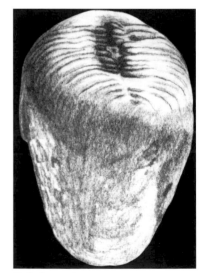

Erniaster apertus

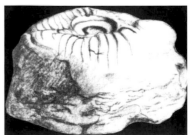

Erniaster patellus

"Ernietta" sp.
Occurrence:
Locality: Suz'ma, Zimnii Bereg, Winter Coast, White Sea, Russia
Rock Unit: Ust'-Pinega Formation, Valdai Group
Reference: Not yet described
Type Specimen: None designated

"Ernietta aarensis" (See Fig. 130, Pflug Plate 34) See Ernietta plateauensis
Locality: Farm Aar or Plateau, Aus area, Namibia
Rock Unit: Dabis Formation, Nama Group
Reference: Pflug, 1972
Type Specimen: Geological Survey of Namibia GSN 36 (F407-H) (Pflug No 36)

Ernietta plateauensis (See Fig. 130, Pflug Plate 34)
Occurrence: Common
Locality: Farm Plateau, Aus area, Namibia, southern Africa
Rock Unit: Dabis Formation, Nama Group
Other Names: Erniaster, Erniobaris, Erniobeta, Erniocarpis, Erniocentris, Erniocoris, Erniodiscus, Erniofossa, Erniograndis, Ernionorma, Erniopelta, Erniotaxis (all Pflug, 1972), and perhaps Baikalina (Sokolov, 1972)
Description: The body bilaterally symmetric and segmented. Body axis bent into shape of U. Zigzag median dorsal suture line, at which the segments of sides meet alternatingly" (Pickford, 1995, after Pflug, 1966). Jenkins (in Lipps & Signor, 1992) noted that this form was rather sac-shaped
Reference: Pflug, 1966, 1972; Jenkins et al., 1981, Narbonne, 2005
Type Specimen: Geological Survey of Namibia GSN 283 (F429-H) (Pflug No 277)
Classification: Phylum: Petalonamae; Family: Erniettidae; Narbonne has recently allied with dickinsoniomorphs, but Gehling et al. (2005) do not agree

"Ernietta tschanabis" (See Fig.130, Pflug Plate 34) See Ernietta plateauensis
Locality: Farm Aar or Plateau, Aus area, Namibia
Rock Unit: Dabis Formation, Nama Group
Reference: Pflug, 1972
Type Specimen: Geological Survey of Namibia GSN 12 (F464-H) (Pflug No 12)

"Erniobaris baroides" (See Fig. 130, Pflug Plate 32) See Ernietta plateauensis
Locality: Farm Aar or Plateau, Aus area, Namibia
Rock Unit: Dabis Formation, Nama Group
Reference: Pflug, 1972
Type Specimen: Geological Survey of Namibia GSN (F451-H) (Pflug No 282)

"Erniobaris epistula" (See Fig. 130, Pflug Plate 31) See Ernietta plateauensis
Locality: Farm Aar or Plateau, Aus area, Namibia
Rock Unit: Dabis Formation, Nama Group
Reference: Pflug, 1972
Type Specimen: Geological Survey of Namibia GSN 268 (F449-H) (Pflug No 268)

"Erniobaris gula" (See Fig. 130 Pflug Plate 33) See Ernietta plateauensis
Locality: Farm Aar or Plateau, Aus area, Namibia
Rock Unit: Dabis Formation, Nama Group
Reference: Pflug, 1972
Type Specimen: Geological Survey of Namibia GSN 410 (F427-H) (Pflug No 410)

Ernietta aarensis

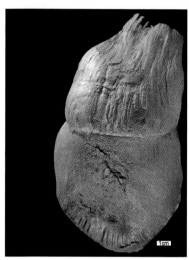

Ernietta plateauensis

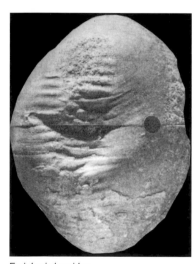

Erniobaris baroides

Erniobaris parietalis

"Erniobaris parietalis" (See Fig. 130, Pflug Plate 33) See Ernietta plateauensis
Locality: Farm Aar or Plateau, Aus area, Namibia
Rock Unit: Dabis Formation, Nama Group
Reference: Pflug, 1972
Type Specimen: Geological Survey of Namibia GSN 267 (F450-H) (Pflug No 267)

"Erniobeta forensis" (See Fig. 130, Pflug Plate 39) See Ernietta plateauensis
Locality: Farm Aar or Plateau, Aus area, Namibia
Rock Unit: Dabis Formation, Nama Group
Reference: Pflug, 1972
Type Specimen: Geological Survey of Namibia GSN 287 (F435-H) (Pflug No 287)

"Erniobeta scapulosa" (See Fig. 130, Pflug Plate 39) See Ernietta plateauensis
Locality: Farm Aar or Plateau, Aus area, Namibia
Rock Unit: Dabis Formation, Nama Group
Reference: Pflug, 1972
Type Specimen: Geological Survey of Namibia ?GSN 285 (F426) (Pflug No 286)

"Erniocarpis carpoides" (See Fig. 130, Pflug Plate 35) See Ernietta plateauensis
Locality: Farm Aar or Plateau, Aus area, Namibia
Rock Unit: Dabis Formation, Nama Group
Reference: Pflug, 1972
Type Specimen: Geological Survey of Namibia GSN 290 (F448-H) (Pflug No 290)

"Erniocarpis sermo" (see Fig. 130, Pflug Plates 35, 36) See Ernietta plateauensis
Locality: Farm Aar or Plateau, Aus area, Namibia
Rock Unit: Dabis Formation, Nama Group
Reference: Pflug, 1972
Type Specimen: Geological Survey of Namibia GSN 289 (F425-H) (Pflug No 289)

"Erniocentris centriformis" (See Fig. 130, Pflug Plate 28) See Ernietta plateauensis
Locality: Farm Aar or Plateau, Aus area, Namibia
Rock Unit: Dabis Formation, Nama Group
Reference: Pflug, 1972
Type Specimen: Geological Survey of Namibia GSN 279 (Pflug No 279), but missing from collections in October 2006

"Erniocoris orbitiformis" (See Fig. 130, Pflug Plate 36) See Ernietta plateauensis
Locality: Farm Aar or Plateau, Aus area, Namibia
Rock Unit: Dabis Formation, Nama Group
Reference: Pflug, 1972
Type Specimen: Geological Survey of Namibia GSN 388 (F417-H) (Pflug No 388)

"Erniodiscus clypeus" (See Fig. 130, Pflug Plate 27) See Ernietta plateauensis
Locality: Farm Aar or Plateau, Aus area, Namibia
Rock Unit: Dabis Formation, Nama Group
Reference: Pflug, 1972
Type Specimen: Geological Survey of Namibia GSN 278 (F486-H) (Pflug No 278)

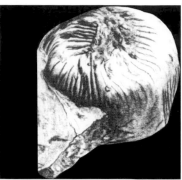

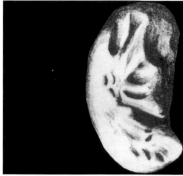

Erniobeta forensis

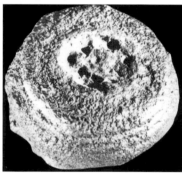

Erniocarpis carpoides

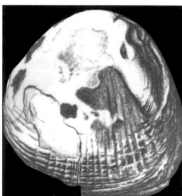

Erniocentris centriformis

Erniocoris orbitiformis

Erniodiscus rutilus

"Erniodiscus rutilus" (See
Fig. 130, Pflug Plate 27) See *Ernietta
plateauensis
Locality: Farm Aar or Plateau, Aus
area, Namibia
Rock Unit: Dabis Formation, Nama
Group
Reference: Pflug, 1972
Type Specimen: Geological Survey
of Namibia GSN 275 (F471-H) (Pflug
No 275)

"Erniofossa prognatha" (See
Fig. 130, Pflug Plate 27) See *Ernietta
plateauensis
Locality: Farm Aar or Plateau, Aus
area, Namibia
Rock Unit: Dabis Formation, Nama
Group
Reference: Pflug, 1972
Type Specimen: Geological Survey
of Namibia GSN 409 (F488-H) (Pflug
No 409)

"Erniograndis paraglossa" (See
Fig. 130, Pflug Plate 38) See *Ernietta
plateauensis
Locality: Farm Aar or Plateau, Aus
area, Namibia
Rock Unit: Dabis Formation, Nama
Group
Reference: Pflug, 1972
Type Specimen: Geological Survey
of Namibia GSN 384 (F221-H) (Pflug
No 384)

"Erniograndis sandalix" (See
Fig. 130, Pflug Plate 38) See *Ernietta*
plateauensis
Locality: Farm Aar or Plateau, Aus
area, Namibia
Rock Unit: Dabis Formation, Nama
Group
Reference: Pflug, 1972
Type Specimen: Geological Survey
of Namibia GSN 192 (F389-H) (Pflug
No 192)

"Ernionorma abyssoides" (See
Fig. 130, Pflug Plate 29) See *Ernietta*
plateauensis
Locality: Farm Aar or Plateau, Aus
area, Namibia
Rock Unit: Dabis Formation, Nama
Group
Reference: Pflug, 1972
Type Specimen: Geological Survey of
Namibia GSN 280 (Pflug No 280)

"Ernionorma clausula" (See
Fig. 130, Pflug Plate 31) See *Ernietta*
plateauensis
Locality: Farm Aar or Plateau, Aus
area, Namibia
Rock Unit: Dabis Formation, Nama
Group
Reference: Pflug, 1972
Type Specimen: Geological Survey of
Namibia GSN 371 (F476-H) (Pflug 371)

"Ernionorma corrector" (See
Fig. 130, Pflug Plate 32) See *Ernietta*
plateauensis
Locality: Farm Aar or Plateau, Aus
area, Namibia
Rock Unit: Dabis Formation, Nama
Group
Reference: Pflug, 1972
Type Specimen: Geological Survey
of Namibia GSN 488 (F481-H) (Pflug
No 488)

"Ernionorma peltis" (See **Fig.
130, Pflug Plates 29, 30) See *Ernietta***
plateauensis
Locality: Farm Aar or Plateau, Aus
area, Namibia
Rock Unit: Dabis Formation, Nama
Group
Reference: Pflug, 1972
Type Specimen: Geological Survey
of Namibia GSN 380 (F483-H) (Pflug
No 380)

"Ernionorma rector" (See **Fig. 130,
Pflug Plate 32) See *Ernietta***
plateauensis
Locality: Farm Aar or Plateau, Aus

Erniofossa prognatha

Erniograndis paraglossa

Ernionorma abyssoides

Ernionorma corrector

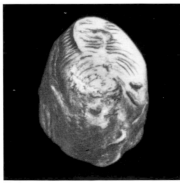

Ernionorma rector

area, Namibia
Rock Unit: Dabis Formation, Nama
Group
Reference: Pflug, 1972
Type Specimen: Geological Survey
of Namibia GSN 257 (F482-H) (Pflug
No 257)

"Ernionorma tribunalis" (See
Fig. 130, Pflug Plate 31) See *Ernietta*
plateauensis
Locality: Farm Aar or Plateau, Aus
area, Namibia
Rock Unit: Dabis Formation, Nama
Group
Reference: Pflug, 1972
Type Specimen: Geological Survey
of Namibia GSN 797 (F487-H) (Pflug
No 268)

"Erniopelta scrupula" (See
Fig. 130, Pflug Plate 33) See *Ernietta*
plateauensis
Locality: Farm Aar or Plateau, Aus
area, Namibia
Rock Unit: Dabis Formation, Nama
Group
Reference: Pflug, 1972
Type Specimen: Geological Survey
of Namibia GSN 236 (F415-H) (Pflug
No 236)

"Erniotaxis segmentrix" (See
Fig. 130, Pflug Plate 37) See *Ernietta*
plateauensis
Locality: Farm Aar or Plateau, Aus
area, Namibia
Rock Unit: Dabis Formation, Nama
Group
Reference: Pflug, 1972
Type Specimen: Geological Survey
of Namibia GSN 396 (Pflug No 396),
but missing from collections in October
2006

Evmiaksia aksionovi (holotype)
Occurrence: Rare (one specimen)
Locality: Zimnii Bereg, White Sea,
Russia
Rock Unit: Ust'Pinega Formation,
Valdai Group (Verkovka Formation of
Grazhdankin, 2004)
Description: Organisms preserved as
flat, discoidal impression within which
are two broad concentric zones. Inner
zone with elevated central tubercle
and numerous ovate structures with
radially oriented long axis. Outer zone
smooth, separated from inner zone by
narrow ridge and lacking any traces
of radial structures but has numerous
irregular concentric ridges. Bears some
resemblance to *Elasenia*, but differs in
the shape of the ovate structures and
their arrangement, a broader outer zone
and the presence of a central tubercle.
No hint of tentacles. Diameter reaches
56 mm; diameter of outer zone, 24 mm;
central tubercle, 10 mm; length of ovate
structure about 5 mm; their width up to
2 mm; width of the ridge between inner
and outer zones, 1 mm
Reference: Fedonkin, 1984
Type Specimen: Paleontological
Institute, Moscow, PIN 3993/657
Classification: ?Phylum: Cnidaria;
Class: Inordozoa; Family: Hiemaloridae
but relationships not well understood

Garania petali (from **Bekker, 1996, pl.
3, no 5**)
Occurrence: Holotype and a few,
incomplete fragments
Locality: Askyn River Basin, to the
east of the village of Sonsnovka,
southern Urals
Rock Unit: Ashynskaya Sequence,
Zeganskaya Series
Description: Small disc-ahaped
specimen with a wide central zone
that Is subdivided by a series of radial
furrows into a few sections and a narrow
peripheral zone with concentric bands;
diameter of imprint, 26 mm; diameter of
central zone, 20 mm
Reference: Bekker, 1996
Type Specimen: Geological Museum
Novosibers**k,**TSGM 1/11406 (307/89)
Classification: Uncertain, perhaps a
holdfast

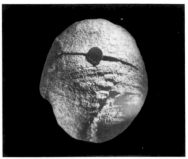

Erniopelta scrupula

Erniotaxis segmentrix

Evmiaksia aksionovi

Garania petali

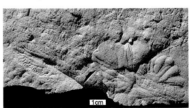

Gehlingia dibrachida

272

Gehlingia dibrachida (holotype)

Occurrence: ?
Locality: Central Flinders Ranges, South Australia
Rock Unit: Ediacara Member, Rawnsley Quartzite (Pound Subgroup)
Description: "A bilaterally symmetric frond-shaped fossil. Each half of the frond is identical and a mirror image to the other half. Each half consists of a swollen axis on the inner edge of the half-frond. This axis bifurcates once, and the bifurcation is directed towards the outer edge of the frond. Numerous tubular structures emanate from the outer edge of the frond axis. These tubules are straight to slightly curved and bifurcate twice before ending abruptly, forming a smooth edge to the frond. A deep groove, as wide as single axis, separates the paired axes. Frond was at least 8 cm in length and 3.1 cm in width." McMenamin & McMenamin, web site
Reference: Gehling, 1988; McMenamin, 1998; www.earthscape.org/r3/mcm02/mcm02b.pdf
Type Specimen: South Australian Museum, SAM P27927
Classification: Phylum Petalonamae

Glaessneria imperfecta (holotype; See Aspidella terranovica)

Occurrence: Rare, two specimens
Locality: Mogilev-Podolski, Derlo River, a tributary of the Dniester River, Ukraine
Rock Unit: Yaryshev Formation, Mogilev-Podolsk Group
Other Names: Aspidella terranovica suggested by Gehling et al., 2000; also perhaps Cyclomedusa plana
Description: Large discoidal form preserved on sole of beds in positive hyporelief; with narrow concentric central zone occupying up to one-third width of fossil, arranged in step-like, conical fashion with regard to elevation; outer zone flat and smooth with radially arranged furrows
Reference: Zaika-Novatski et al., 1968; Güreev, 1987
Type Specimen: Geological Museum, Institute of Geological Sciences, Kiev No. 2127/56

Glaessneria imperfecta

"Glaessneria" plana

(See Cyclomedusa plana)
Locality: Cosmopolitan
Reference: Gureev (1987) reassigned C. plana to a new genus based on the material from both Australia and the Ukraine. The systematics of this group certainly needs reassessment.
Type Specimen: See Cyclomedusa plana

"Glaessneria" radiata

(See Cyclomedusa radiata)
Locality: Cosmopolitan
Reference: Gureev (1987) reassigned C. radiata to new genus based on material from both Australia and the Ukraine. Systematics of this group certainly needs reassessment.
Type Specimen: See Cyclomedusa radiata

Glaessneria siberica (See Charnia masoni)

"Glaessnerina" grandis (holotype; See Chrnia masoni)

Occurrence: Very rare
Locality: Ediacara Hills, Flinders Ranges, South Australia
Rock Unit: Rawnsley Quartzite, Wilpena Group
Other Names: Same specimen as Rangea grandis; see Charnia grandis and Charnia masoni
Description: Frond-shaped fossil, primary branches of which apparently attached to ventral side. Primary branches appear to alternate when joining midline, forming zigzag pattern, which may be result of preservation. They taper to tips at outer margin. Secondary branches cross primary branches at almost right angles

Glaessnerina grandis

forming blunt, rounded edges at margin of primary branches
Reference: Glaessner & Wade, 1966; Germs, 1972, 1973; Jenkins & Gehling, 1978; Runnegar, 1992
Type Specimen: South Australian Museum SAM 12897
Classification: Phylum: Petalonamae, attachment disc for frond-like organism

"Glaessnerina longa" (holotype; same specimen as Charniodiscus longus)

Occurrence: ?
Locality: Ediacara Hills, Flinders Ranges, South Australia; and possibly Dniester River Basin, Podilia, Ukraine and Newfoundland
Rock Unit: Rawnsley Quartzite, Wilpena Group
Other Names: Rangea longa, Charniodiscus longus
Reference: Glaessner & Wade, 1966; Germs, 1972, 1973; Runnegar, 1992
Type Specimen: Same specimen as Charniodiscus longus

Glaessnerina longa

"Hagenetta aarensis" (See Beltanelliformis brunsae)

Locality: Farm Aar or Plateau, Aus area, Namibia
Rock Unit: Dabis Formation, Nama Group
Reference: Hahn & Pflug, 1988
Type Specimen: ?

Hallidaya brueri (holotype)

Occurrence: Uncommon
Locality: Mt Skinner, Northern Territory, Australia
Rock Unit: Mount Stuart Formation
Description: Disc-shaped body, usually measuring between 10 and 30 mm, slightly elevated above three bullae (bulbous structures) arranged in center. Inflated rim surrounds central domed structure with branching ribs radiating outward from central bullae, to outer rim
Reference: Wade, 1969
Type Specimen: South Australian Museum SAM F16464a
Classification: Uncertain or possibly Phylum: Cnidaria; Class: Trilobozoa

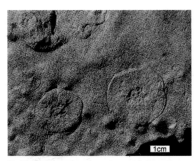

Hallidaya brueri

Hiemalora sp.

Occurrence: Rare
Locality: Avalon Peninsula, Newfoundland
Rock Unit: Mistaken Point Assemblage, Fermeuse Assemblage in the Conception and St John's groups
Reference: Fedonkin, 1992; Gehling et al., 2000
Type Specimen: None
Classification: Family: Hiemaloriidae – relationships not presently understood; in the past was placed in the Phylum: Cnidaria; Class: Inordozoa. Hofmann (2005) noted rare specimens were attached to stems and fronds and thus their likelihood as holdfasts

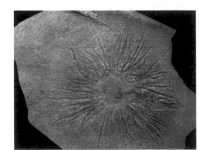

Hiemalora pleiomorpha

Hiemalora pleiomorpha (PIN 3995/251)

Occurrence: Common
Locality: Khorbusuonka River, tributary of the Olenek River, Yakutia, Olenek Uplift, Arctic Siberia; Sekwi Brook area (Mackenzie Mountains, Canada)
Rock Unit: Khatyspyt Formation, Khorbusuonka Group, Yudomian; Blueflower Formation, Windermere Supergroup
Other Names: Mawsonites pleiomorphus (Grazhdankin, in press)
Description: Puzzling benthic organism characterized by radial symmetry with distinct top and bottom that are different in shape; preserved usually in low-relief but sometimes leaving high-relief imprints with rounded central part showing marginal set of rod-like, radial elements projecting beyond main body of the organism
Reference: Vodanjuk, 1989; Serezhnikova, 2006 (in press).
Type Specimen: Geological Museum, Novosibirsk Nr. 913/3
Classification: Family: Hiemaloriidae; relationships not presently understood; in some classifications placed in the Phylum: Cnidaria; Class: Inordozoa. Serezhnikova (2006) interprets these

Hiemalora stellaris

structures as holdfasts that resemble those of living algae and sponges

Hiemalora stellaris *(PIN 3993/309)*
Occurrence: Common
Locality: Lyamtsa, Zimnii Bereg, White Sea, Russia; Dniester River Basin, Podolia, Ukraine; Olenek Uplift, Siberia; Flinders Ranges, South Australia
Rock Unit: Ust'-Pinega Formation, Valdai Group (Lyamtsa, Verkovkaand Zimnii Gory formations of Grazhdankin, 2004); Mogilev Formation; Khatyspyt Formation; Ediacara Member, Rawnsley Quartzite, Wilpena Group
Other Names: Renamed from *Pinegia stellaris*
Description: Bowl-shaped body with numerous projections, radiating from outer margin of body. Morphologically similar form was described by R. Sprigg as *Medusina filamentis*, and this name was later synonomized with *Pseudorhizostomites howchini*. Diameter without "tentacles" ranges from 3 to 40 mm; thickness of tentacles from 0.3 to 2 mm; length of tentacles from 3 to15 mm; max. preserved height, 4 mm
Reference: Fedonkin, 1980, 1982
Type Specimen: Paleontological Institute (originally Geological Institute, Moscow, GIN4482/25)
Classification: Family: Hiemaloriidae – relationships not presently understood; in the past was placed in the Phylum: Cnidaria; Class: Inordozoa; see *H. pleiomorpha* discussion

Ichnusa cocozzi
Occurrence: ?
Locality: Sarrabus area, northeast of Cagliari, Sardinia
Rock Unit: San Vito Formation
Description: Described as resembling *Albumares*, with four- or eight-fold symmetry
Reference: Debrenne & Naud, 1981
Type Specimen: ?
Classification: ? Cnidaria

Inaria karli *(holotype)*
Occurrence: Common
Locality: Chace Range, Flinders Ranges, South Australia
Rock Unit: Rawnsley Quartzite, Wilpena Group
Description: Shaped like garlic bulb, with lobate outline and hollow center. Bulb's diameter equal in size to the stalk's height (but stalk may be incomplete). Radiating internal partitions (?mesenteries) separate lobes and coalesce in central cavity. Number of lobes tends to increase with size, larger specimens having between 10 and 18. Most specimens external and composite molds on top surface of beds, while some specimens have basal surface cast by infilling sand
Reference: Gehling, 1988
Type Specimen: South Australian Museum SAM P27913
Classification: Some researchers have interpreted *Inaria* as a primitive actinian (anemone) with a tubular opening and no tentacles. Others have suggested that *Inaria* could be the remains of a sack-like holdfast of one of the frond-like animals that has had the frond broken away and lost during deposition and subsequent burial, and thus Phylum: Petalonamae?

Inaria n. sp. *(species manuscript name, unpublished)*
Occurrence: Uncommon
Locality: Lyamtsa, White Sea, Russia
Rock Unit: Ust'-Pinega Formation, Valdai Group (Lyamtsa Formation of Grazhdankin, 2004)
Description: See *Inaria karli* for description
Reference: Grazhdankin, 2000, 2004
Type Specimen: Paleontological Institute, Moscow, PIN 4716/3
Classification: Phylum: Petalonamae?

"Inkrylovia lata" *(PIN 3992/507;* See *Pteridinium simplex)*
Occurrence: Uncommon, only fifteen specimens known
Locality: Suz'ma, White Sea, Russia;

Ichnusa cocozzi

Inaria karli

Inkrylovia lata

Mackenzie Mountains, Northwest Territories, Canada
Rock Unit: Ust'-Pinega Formation, Valdai Group (Verkovka Formation of Grazhdankin, 2004); Blue Flower Formation, Windermere Supergroup
Other Names: May be synonymous with *Pteridinium latum* and *P. nenoxa*; junior synonym for *Onegia nenoxa* (Grazhdankin, 2004, Ph.D.)
Description: Large sac-like, bilaterally symmetric organism regularly segmented along its length. Flat segments curved and divided by fine grooves. Arrangement of segments not directly opposite. Body appears to have been very flexible. Width ranges from 40 to 60 mm and observable length up to 70 mm. Judging by fragments, length may reach up to 100 mm. Length of segments ranges from 4.5 to 6.5 mm, remaining constant on individuals.
Reference: Fedonkin, 1979; Grazhdankin, 2004 (Ph.D.)
Type Specimen: Geological Institute, GIN, Moscow, 4464/147
Classification: Although placed by many workers in the Petaloname, Chistyakov *et al.*, 1984 suggested it might be a tunicate

Intrites punctatus *(TSGM 9/87)*
Occurrence: ?
Locality: Zelem River, Bashkeria, Ural Mountains; Winter Coast, White Sea, Russia
Rock Unit: Badyeievskaya Formation; Ust'Pinega Formation
Description: Individuals of circular shape, with a wavy surface with a depression in the middle, preserved on the sole of beds; represent forms that were cylindrical in shape and slightly tilted in orientation; diameter ranges from 1 to 5 mm with smaller forms predominating; height ranges from 1 to 2 mm and diameter of central depression is 0.5
Reference: Becker, 1992
Type Specimen: TSGM/TSNEGR Leningrad Museum (St Petersburg); 9/87

"Irridinitus multiradiatus"
(holotype; See Aspidella terranovica)
Occurrence:
Locality: Dniester River, Podolia, Ukraine; White Sea Region both from outcrop and boreholes; ?Mackenzie Mountains, northwestern Canada; ?Yellowhead Platform, southwestern Canada; Nainital District, Uttar Pradesh, India
Rock Unit: Mogilev Formation; Ust'-Pinega Formation, Valdai Group; Twitya Formation, Windermere Supergroup; Hadrynian Miette Group, Windermere Supergroup; Krol Formation
Other Names: *Aspidella terranovica* suggested by Gehling *et al.*, 2000 but not all researchers agree with this synonymy (Serezhnikova, *pers. com*, 2006)
Description: Small rounded casts of basal part of what may have been polyp, preserved in positive, but low, hyporelief. Central part of fossils commonly caved in. Numerous ridges radiate from center of fossil out to margins where they intersect a narrow marginal ridge. Fossils preserved on top of beds. Diameters reach up to 20 mm; diameter of the central depression 2-3 mm; thickness of radial canals, 0.5 mm
Reference: Fedonkin, 1983; Velikanov, Aseeva & Fedonkin, 1983; Fedonkin, 1985
Type Specimen: Paleontological Institute, Moscow, PIN 3994/524

Irridinitus multiradiatus

Ivesheadia lobata *(field specimen)*
Occurrence: Common
Locality: Leicestershire, UK.; Avalon Peninsula, Newfoundland; South Australia
Rock Unit: Conception and St. John's Groups, Newfoundland; Rawnsley Quartzite, Wilpena Group
Other Names: "Ivesia" lobata was a preoccupied name and thus *Ivesheadia* was proposed to replace it by Boynton & Ford, 1995
Description: Irregularly ornamented, discoidal impressions (5 to 60 cm in

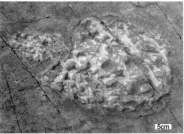
Ivesheadia lobata

diameter). *Ivesheadia* preserved in both positive and negative relief, varying with size. Smaller "lobate disc" morphs of *Ivesheadia* (10-14 cm in diameter) feature V-shaped ridges, pointing inward, and outlining three to five dumbbell–shaped, circular depressions. Sometimes ridges merge with smooth raised and relatively entire margin. Relief of this morph of *Ivesheadia* exceeds that of thickness of crystal tuff that surrounds it, and is greater than any other fossil forms in the assemblage. Larger "bubble disc" forms of *Ivesheadia* (to 60 cm in diameter) are ornamented by rounded to oval craters that are largest near the outside margin. Lobe-shaped hollows lacking in specimens more than 25 cm in diameter, and arrangement of circular hollows becomes more regular with increasing size. Where forms overlap spindle-shaped fossils, it seems that *Ivesheadia* rested over spindles at moment of burial. Thus, it is likely that "morphs" of *Ivesheadia* were result of collapse of chambered, three-dimensional body. The "pizza-disc" form of *Ivesheadia* preserved in positive relief; surface of pizza-disc covered with closely spaced "pustules," each about a centimeter in diameter. Disc bordered by low-relief, sharp ridge. However, discovery of a stalk extending away from the disc, in a few specimens, suggests that *Ivesheadia* could have been large holdfast onto which another structure attached. There is no evidence to support the idea that these were jellyfish, contradicting this idea of Boynton & Ford (1995)
Reference: Boynton & Ford, 1995; Narbonne *et al.*, 2001
Type Specimen: Holotype *in situ* on Iva Head, Shepshed, Charnwood Forest, Leicestershire SK 477 170; casts in Leichester University Geology Department collections Accession No. 115577 and in Leicestershire Museums Geology Collections Accession No. G32/1994, UK
Classification: Phylum: Petalonamae

"Ivesia" lobata (See *Ivesheadia lobata*)

Ivovicia rugulosa (holotype)
Occurrence: Rare
Locality: Zimnii Bereg, White Sea, Russia
Rock Unit: Ust'-Pinega Formation, Valdai Group (Yorga Formation of Grazhdankin, 2004)
Description: Small bilateral animal of proarticulaten plan with oval shape; body not deep; broad, unsegmented head region and long trunk subdivided into long, narrow segments or isomeres whose ends are rounded; most isomeres narrow and flexed posteriorly, but in anterior part of trunk, distal ends flexed anteriorly instead; axial structure present in trunk and consists of two parallel ridges which continue entire length of trunk
Reference: Ivantsov, *in press*
Type Specimen: Paleontological Institute, PIN 3993/5504
Classification: Phylum uncertain, but in the past has been placed in the Phylum Proarticulata based on its "symmetry of gliding reflection," where segments meet along the midline in a zigzag fashion

"Jampolium wyrzhykoowskii"
(See **Aspidella terranovica Jurtia paliji**; from Bekker, 1996, pl. 3, no. 2)
Occurrence: One specimen
Locality: Karanyurt River, Askyn River Basin, Bashkeria, southern Urals
Rock Unit: Ashynskaya Sequence, Zeganskaya Series
Description: Disc-shaped imprint, ovoid, with clearly defined concentric zonation; well-defined peripheral band and raised central disc; reminiscent of some specimens of *Medusinites*, but lacks well-defined radial ribs; perhaps also some related material from the Dneister River Basin, Ukraine; diameter of imprint, 9 mm; diameter of central disc, 3 mm; width of peripheral band, 1-2 mm

Ivovicia rugulosa

Jampolium wyrzhykoowskii

Reference: Bekker, 1996
Type Specimen: Geological Museum, Novosiberisk, TSGM1/11406 (305/89)
Classification: Uncertain, but perhaps a holdfast

Kaisalia levis (holotype)
Occurrence: Rare, one specimen
Locality: Vinozh, Vinnitsa District, Ukraine
Rock Unit: Mogilev Formation
Other Names: See *Kaisalia mensae*
Description: Discoidal fossil preserved in positive hyporelief with convex middle and central zones, while external zone is flattened; weak, concentric wrinkles occur in outer zone; total diameter 150 mm; diameter of inner zone diameter, 37-40 mm; width of middle zone 30 mm; width of external zone 25 mm.
Reference: Gureev, 1987
Type Specimen: Geological Museum, Institute of Geological Sciences, Kiev No. 2127/71
Classification: Phylum: Cnidaria; Class: Cyclozoa or perhaps Phylum: Petalonamae

Kaisalia mensae (holotype)
Occurrence: Rare (eight specimens)
Locality: Zimnii Bereg, White Sea, Russia
Rock Unit: Ust'-Pinega Formation, Valdai Group
Other Names: Perhaps a synonym of *Kaisalia levis*
Description: Large discoidal organism with rounded outline. Regular, flat concentric rings cover entire disc. Rings separated by fine grooves. Divided into three concentric zones. Inner part makes up about one-fifth of area of disc and is bounded by one pair of ridges. The middle, broader zone makes up about half of disc. This region has seven similar-sized concentric rings all of similar width. The inner and middle regions are elevated with respect to outer – this is where organism was thickest. Outer zone much thinner. Diameter of disc up to 160 mm; diameter of central part about 60 mm; diameter of outer zone about 56 mm. Individual rings in outer zone vary in width from 3 to 6 mm
Reference: Fedonkin, 1984
Type Specimen: Paleontological Institute, Moscow, PIN 4482/284
Classification: Phylum: Cnidaria; Class Cyclozoa or possibly Phylum: Petalonamae, probably holdfasts of frond-like organism

Kharakhtia nessovi (holotype)
Occurrence: Rare
Locality: Kharakhta River, Russia
Rock Unit: Ust'-Pinega Formation, Valdai Group (Verkovka Formation, Grazhdankin, 2004)
Description: Medium to large-sized bilaterally symmetric metazoan, body rounded in dorsal view, divided into two rows of transverse segments (isomeres) positioned in alternating order ("gliding reflection") along its longitudinal axis. Segments decrease in size from one end to other, their lateral ends recurved posteriorly. Normally five pairs of segments present. Margin of body covered extensively with coarse radial folds, probably present in living organism and not due to postmortem distortion. Shows similarity to *Yorgia* and *Vendia* Ivantsov, Malakhovskaya & Serezhnikova, 2004
Type Specimen: Paleontological Institute, Moscow, PIN4852/250
Classification: Bilaterian but Phylum uncertain, but in past has been placed in the Phylum Proarticulata based on its "symmetry of gliding reflection" where segments meet along the midline in a zigzag fashion. Class: Vendiamorpha; Family: Uncertain; but see Gehling *et al.* (2005) and Narbonne (2005) for discussion of dickinosoniomorph affinities

Kaisalia levis

Kaisalia mensae

Kharakhtia nessovi

Khatyspytia grandis

Khatyspytia grandis (holotype)

Occurrence: Rare
Locality: Khorbosuonka River, Olenek Uplift, Siberia
Rock Unit: Khatyspyt Formation
Other Names: Some researchers have suggested that there are similarities to *Charniodiscus*
Description: Colony with large attachment organ; thick base and the remainder of the feather-like colony tapers dorsally to apex; growth from axial stem; central axis extends from base to dorsal (distal) apex; branches basally thick, tapering distally and flexed towards apex
Reference: Fedonkin, 1985
Type Specimen: Paleontological Institute, Moscow, PIN 3995/132
Classification: Phylum: Petalonamae

Kimberella quadrata (holotype and PIN 3993-5542)

Occurrence: Rare in Australia; more common in Russia
Locality: Ediacara Hills, Flinders Ranges, South Australia; Suz'ma, Karakhta and Solza rivers, Zimnii Bereg, White Sea, Russia; Nainital Distict, Uttar Pradesh, India
Rock Unit: Rawnsley Quartzite, Wilpena Group; Ust'-Pinega Formation, Valdai Group (Verkhovka, Zimnii Gory and Yorga formations, Grazhdankin, 2004); ?Krol Formation
Other Names: ?*Solza*, ?*Brachina*, *Zolotytsia*
Description: Elongate, boat-shaped form with complex internal structure; tall dorsal, flexible yet stiff, unmineralized shell; frill-like rim around outside of shell – which may have been circular tubular structure with protruding "finger-like structures" rather than true frill. Shell sometimes reached up to 15 cm in length and 5-7 cm in width with height of 3-4 cm. Many specimens, however, were much smaller. Beyond shell and tubular ring, may have extended a "foot" as in modern mollusks that could be extended or contracted. Associated with body fossils are radiating, fan-shaped sets of paired scratch marks (*Radulichnus*), feeding traces of an animal that fed by extracting nutrients from sea floor sediments – has been reconstructed with paired "scolecodont"-like structures embedded within a retractable proboscis. More than 800 specimens known from the White Sea region of Russia, preserving significant information about this form, including not only feeding traces but crawling traces as well. Originally described as a siphonophore, then later as pelagic medusa, still later as box jellyfish
Reference: Glaessner, 1959; Glaessner & Wade, 1966; Fedonkin & Waggoner, 1997
Type Specimen: South Australian Museum SAM P12734
Classification: Phylum: Mollusca; Class: may be related to living chitons (amphineurans) and monoplacophorans

Kimberella quadrata (holotype)

Kimberella quadrata White Sea, Russia (latex peel; PIN 3993-5542),

Kuibisia glabra (holotype)

Occurrence: Rare
Locality: Farm Plateau, Aus Region, Namibia
Rock Unit: Dabis Formation, Nama Group
Other Names: May be the same as *Ernietta plateauensis;* also could be the external covering of *Ausia*
Description: Pickford (1995) paraphrases Hahn & Pflug: "a solitary 'polyp' with approximately conical shape about 10 cm long and 3.5 cm broad at its widest." "Co-axial ribs cover the surface of the polyp. These ribs are all approximately the same shape and size"
Reference: Hahn & Pflug, 1985
Type Specimen: Geological Survey of Namibia GSN 2 (F545)
Classification: Uncertain, but perhaps *Ernietta plateauensis* (Runnegar, 1992) or external surface of *Ausia*. Terminology used in description implies relationship to a polyp-bearing organism – which currently is unwarranted

Kuibisia glabra

Kullingia concentrica (tool mark; PIN 3995/1)

Occurrence: Rare
Locality: Sweden and possibly Khorbusuonka Basin, Olenek Uplift, Siberia; Mackenzie Mountains, Northwest Territories, Canada
Rock Unit: Dividalen Group; Khatyspyt Formation; Sheepbed Formation, Windermere Supergroup
Other Names: A tether, spin or swing mark; bio-mechanical trace fossil
Description: A series of concentric rings
Reference: Foyn & Glaessner 1979; Jensen, Gehling & Droser, 1998; Jensen, Gehling, Droser & Grant, 2002
Type Specimen: Swedish Geological Survey, SGU Type 22, Uppsala
Classification: Once thought to belong to the Chondrophorina (a free swimming medusoid hydrozoan), but it is now thought to be a scratch circle, a current or wave-induced rotation tool mark of an anchored, tubular organism, possibly a sabelliditid (Jensen *et al.*, 2002)

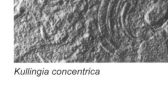

Kullingia concentrica

Lomosovis malus (holotype)

Occurrence: Uncommon
Locality: Novodnestrovsky Quarry, Dniester River, Podolia, Ukraine; Flinders Ranges,South Australia.
Rock Unit: Mogilev Formation; Ediacara Member, Rawnsley Quartzite, Wilpena Group
Description: Large, dendritic, colonial organism. Basal, small discoid attachment. Upper part of stem with fine, tubular processes and ends in crown of fine, long hair-like structures. Such processes can also diverge from sides of basal stem and often bifurcate or split in two. Surface of organism flat. Upper part of basal stem has two semicircular processes, both containing long, tubular growths or shorter brush-like ends. Near base, stem possesses two or three processes of differing lengths. Processes are commonly thick proximally and then narrow. Surface of processes and base smooth, although processes are mainly covered by narrow, elongate ridges. Narrow, double grooves, nearly straight, incise stem from side to side. Based on amount of contortion and crumpling of organism, it appears to have been fairly soft and flexible. Ends of processes and offshoots commonly brush-like. Length of basal stem 60-80 mm; maximum width 25 to 40 mm; length of processes 60-100 mm; thickness, 6 to 11 mm
Reference: Fedonkin, 1983; Velikanov, Aseeva & Fedonkin, 1983
Type Specimen: Paleontological Institute, Moscow, PIN 3994/418
Classification: Phylum: Petalonamae; Family: Uncertain

Lomosovis malus

"Lorenzinites rarus" (See Rugoconites enigmaticus)

Lossina lissetskii (PIN 3993/5057)

Occurrence: Relatively rare
Locality: Solza River and Zimnii Bereg, White Sea, Russia
Rock Unit: Ust'-Pinega Formation, Valdai Group (Verkovkaand Yorga formations of Grazhdankin, 2004)
Description: Very small, bilateral metazoan with proarticulate body plan, elongated body, consisting of head and trunk, separated by narrow furrow; head. Semicircular in outline, its length 30% of total length of animal; head contacts only first pair of isomeres; trunk segmented into short paired isomeres; isomeres with tapering ends; isomeres lie oblique to long axis of body; along axial zone there is no segmentation observed, so individual isomere pairs do not meet along midline; midline axis with medial, longitudinal depression; surface of head region and midline zone of trunk is covered by numerous small knobs
Reference: Ivantsov, *in press*
Type Specimen: Paleontological Institute, Moscow, PIN 3993/5057
Classification: Phylum uncertain, but in the past has been placed in the Phylum Proarticulata based on its "symmetry of gliding reflection" where segments meet along the midline in a zigzag fashion; but see discussion in

Lossina lissetskii

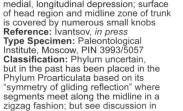

Gehling *et al.* (2005) and Narbonne (2005) concerning dickinsoniomorphs

"Madigania annulata" (holotype; *See Aspidella terranovica*)
Occurrence: ?
Locality: Ediacara Hills, Flinders Ranges, South Australia
Rock Unit: Rawnsley Quartzite, Wilpena Group
Other Names: Synonymized with *Aspidella terranovica* and *?Spriggia annulata, Cyclomedusa davidi*
Reference: Sprigg, 1949
Type Specimen: South Australian Museum SAM T31-2031

Madigania annulata

Majaella verkhojanica (holotype)
Occurrence: ?
Locality: Maya River, Eastern Siberia
Rock Unit: Yuodoma Formation
Description: Skeleton irregularly discoidal/saucer-shaped, made up of two flat walls often with irregular concentric crinkles (Glaessner *in* Robison & Teichert, 1979)
Reference: Vologdin & Maslov, 1960
Type Specimen: ?
Classification: Family: Suvorovellidae; with doubtful invertebrate affinities

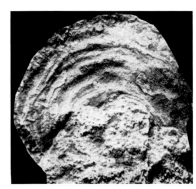

Majaella verkhojanica

Marywadea ovata (SAM P13754; holotype)
Occurrence: Rare (six specimens only)
Locality: Ediacara Hills, Flinders Ranges, South Australia
Rock Unit: Rawnsley Quartzite, Wilpena Group
Other Names: *Spriggina ovata*
Description: Segmented form with head distinct from rest of body and composed of regular angular segmental elements that meet along midline groove. Angular segments and head shield resemble those of arthropod. Has blunt head with sets of ramifying branches extending from center of head to frontal margin. Only two well-preserved specimens (one large and one juvenile) and four poorly preserved specimens of this form are known. Cephalon wide and possesses structures that look very like the digestive caecae in arthropods. Median lobe flattened and broader than in *Spriggina*. No indication that these forms had ability to flex side to side. Although originally thought to be related to polychaete worms, with further study, they seem most similar to some kind of arthropod
Reference: Glaessner & Wade, 1966
Type Specimen: South Australian Museum SAM P13754
Classification: Phylum: Arthropoda; Family: Sprigginidae or Phylum: Proarticulata; Class: Cephalozoa; Family: Sprigginidae (Ivantsov, *in press*)

Marywadea ovata

Mawsonites randellensis (SAM 24595)
Occurrence: Uncommon
Locality: Flinders Ranges, South Australia
Rock Unit: Ediacara Member, Rawnsley Quartzite, Wilpena Group
Other Names: Possibly *Mawsonites spriggi*, maybe a form of *Aspidella*
Description: Disc-shaped with two distinct concentric zones of equal width delineated by sharp furrows and surrounding raised center. Central, middle and outer concentric zones marked by fine striae, which radiate outward to edge of organism. Outer margin scalloped due to raised lobes of unequal width
Reference: Sun, 1986; Jenkins, 1992
Type Specimen: South Australian Museum SAM P24594
Classification: Uncertain, see *Mawsonites spriggi*. But, *M. randellensis* tends to have only marginal lobes, unlike *M. spriggi*, and has a raised central disc, characteristic of some forms of *Aspidella*

Mawsonites randellensis

Mawsonites spriggi (holotype)
Occurrence: Rare
Locality: Ediacara Hills, Flinders Ranges, South Australia
Rock Unit: Rawnsley Quartzite, Wilpena Group

Mawsonites spriggi

Other Names: Possibly *Mawsonites randellensis*, and may be a form of *Aspidella*
Description: Distinctive, disc-shaped organism whose surface is sculptured by many lobes, which unevenly radiate outward from central disc. Inner lobes small and increase in size toward outer margin, where clefts between the lobes indent outer, circular edge
Reference: Glaessner & Wade, 1966; Sun, 1986; Droser *et al.*, 2005
Type Specimen: South Australian Museum SAM P17009/P40943
Classification: *Mawsonites* may have been a jellyfish and later a trace fossil, but it is more likely to be a holdfast

Medusina asteroides

"Medusina asteroides" (PIN 3994/263; *See Medusinites asteroides*)
Occurrence: ?
Locality: Ediacara Hills, Flinders Ranges, South Australia; Zimnii Gory, White Sea, Russia
Rock Unit: Ediacara Member, Rawnsley Quartzite, Wilpenja Group; Ust'-Pinega Formation, Valdai Group
Reference: Sprigg, 1949
Type Specimen: South Australian Museum SAM T40-2021

"Medusina filamentus" (*See Medusinites asteroides*)
Locality: Flinders Ranges, South Australia
Rock Unit: Ediacara Member, Rawnsley Quartzite, Wilpena Group
Reference: Sprigg, 1949
Type Specimen: South Australian Museum SAM T68

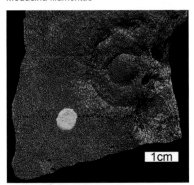

Medusina filamentus

"Medusina mawsoni" (holotype; *See Medusinites asteroides*)
Locality: Flinders Ranges, South Australia
Rock Unit: Ediacara Member, Rawnsley Quartzite, Wilpena Group
Reference: Sprigg, 1949
Type Specimen: South Australian Museum SAM T39

Medusinites applanatus (holotype?)
Occurrence: Two imprints and counterparts
Locality: Ryauzyak River, southern Urals
Rock Unit: Ashynskaya Sequence, Zeganskaya Series
Description: Somewhat deformed ellipse-shaped imprint with broad peripheral area that terminates with finely divided edges; surface of disc characterized by a number of pustules of varying sizes, which may be due to deformation; central zone takes up about one-fourth of total surface area; preserved in negative epirelief; diameter of imprint varies from 16 to 24 mm.; central depression varies from 4 to 6 mm
Reference: Bekker, 1996
Type Specimen: Geological Museum Novosibersk, TSGM 1/11406 (312a/87)
Classification: Uncertain, but perhaps holdfasts

Medusina mawsoni

Medusinites applanatus

Medusinites asteroides (SAM P13785/6)
Occurrence: Common
Locality: Ediacara Hills, Flinders Ranges, South Australia; Zimnii Coast, White Sea, Russia; Dniester River Basin, Podolia, Ukraine; Ryauzyak Basin, South Urals and other places; Wernecke Mountains and Mackenzie Mountains, Northwest Canada; Carmarthen, South Wales
Rock Unit: Rawnsley Quartzite, Wilpena Group; Ust'-Pinega Formation; Mogilev Formation; Basa Formation, Asha Series; Sheepbed Formation, Windermere Supergroup
Other Names: *Medusina asteroides, M. filamentus, M. mawsoni, Medusina mawsoni, Protolyella asteroides*
Description: Small, rounded form with two zones, outer and central. Conical surface of outer zone incised by deep radial grooves
Reference: Sprigg, 1949; Glaessner & Wade, 1966

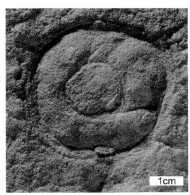

Medusinites asteroides

Type Specimen: South Australian Museum SAM T39
Classification: Phylum: Cnidaria; Class: Cyclozoa; Family: Medusinitidae

Medusinites palijii (holotype)
Occurrence: ?
Locality: Nemia River, a tributary of the Dniester River, Ukraine
Rock Unit: Mogilev Formation
Other Names: *Nemiana simplex*
Description: Originally described as *B. simplex* but later found to be distinct form. Discoidal shape with variable size of central knob relative to size of body; form of central knob irregular; surface of outer zone in large specimens often bears many concentric lines; small forms are similar to members of family Beltanelloididae.
Reference: Gureev, 1987
Type Specimen: Geological Museum, Institute of Geological Sciences, Kiev No. 2127/19
Classification: Phylum: Cnidaria; Class: Cyclozoa; Family: Medusinitidae

Medusinites patellaris (holotype; same specimen as "Planomedusites patellaris" and "Paramedusium patellaris;" See *Aspidella terranovica*)
Occurrence: ?
Locality: Dniester River, Ukraine
Rock Unit: Mogilev Formation
Other Names: *Aspidella terranovica, Planomedusites patellaris, Paramedusium patellaris*
Reference: Sokolov, 1972, 1973; Gehling *et al.*, 2000
Type Specimen: Same as *Planomedusinites patellaris, Paramedusium patellaris*

"*Medusinites*" *sokolovi*
(See *Vendella sokolovi*)
Locality: Ladova and Nemia rivers, tributaries of the Dniester River, Ukraine
Rock Unit: Mogilev Formation
Other Names: *Vendella sokolovi*
Reference: Gureev, 1987
Type Specimen: Geological Museum, Institute of Geological Sciences, Kiev. No number given in Gureev, 1987. Same specimen as *Vendella sokolovi*

Mialsemia semichatovi (holotype)
Occurrence: Rare (one specimen)
Locality: Zimnii Coast, White Sea, Russia
Rock Unit: Ust'-Pinega Formation, Valdai Group (Yorga Formation, Grazhdankin, 2004)
Other Names: Dzik (2002) has suggested that *Mialsemia* could be congeneric with *Rangea* and *Bomakellia*. Fedonkin (*pers. obs.*, 2006) assigns to *Rangea*
Description: Cigar-shaped organism with narrow fusiform body and long, flat appendages on both sides, rounded at their ends. These double pennate "appendages" are largest in central part of the body and decrease in size posteriorly and anteriorly. Body is elongate, smooth and not divided into segments. What appears to be posterior end tapers to abrupt point. Narrow median ridge on body has two short furrows, one on either side. Two short, narrow, ridges radiate away from smooth surface of posterior end of "thorax." Anterior end of body smoothly rounded. What appears to be dorsal end of organism crescent-shaped. Closely spaced tubercles on lateral margin of body at base of each appendage, of which there are seven pairs. Although similar to *Bomakellia*, there are differences in overall proportions and shape and growth pattern of the "appendages"
Reference: Fedonkin, 1985; Dzik, 2002
Type Specimen: Paleontological Institute, Moscow, PIN3993/401
Classification: Originally classified as belonging to Phylum: Arthropoda. Class Paratrilobita; now serious consideration should be given to possible rangeomorph relationships (Dzik, 2002)

Medusinites palijii

Medusinites patellaris

Mialsemia semichatovi

Nadalina yukonensis

Nadalina yukonensis (holotype)
Occurrence: ?
Locality: Yukon, Canada
Rock Unit: Siltstone Unit 1, Windermere Group
Reference: Narbonne & Hofmann, 1987
Type Specimen: Geological Survey of Canada GSC 83022
Classification: ?

Namacalathus hermanastes
Occurrence: Not common
Locality: Driedornvlagte Pinacle Reef Complex, southern Namibia and elsewhere
Rock Unit: Nama Group
Description: Stem topped by outward flaring cup with circular opening at top. Cup perforated by six to seven openings. Small, with maximum diameter of 2.5 cm
Reference: Grotzinger *et al.*, 2000
Type Specimen: Museum of the Geological Survey of Namibia F314
Classification: Uncertain

Namalia villersiensis (See *Ernietta plateauensis; field specimen, 6715*)
Occurrence: Rare
Locality: Farms Buchholzbrunn and Vrede, Namibia
Rock Unit: Dabis Formation, Nama Group
Reference: Germs 1968, 1972; Runnegar, 1992
Type Specimen: No number given in Germs, 1968

Namapoikia rietoogensis
Occurrence: Rare
Locality: Driedoornvlagte, southern Namibia
Rock Unit: Omkyk Member, Zaris Formation, Nama Group (ash overlying this rock unit dated at 548.8 +/-1 million years)
Description: Fully mineralized modules of jointed tubules, about same width throughout (about 1 mm in width). Similar to early Cambrian coralomorphs like *Yaworipkia* from Siberia. Occur on pinnacle reef
Reference: Wood, Grotzinger & Dickson, 2002
Type Specimen: Geological Survey of Namibia, GSN F623
Classification: Relative of corals or sponges

Nasepia altae (holotype)
Occurrence: Rare
Locality: Farms Vrede and Chamis, Namibia
Rock Unit: Kuibis and Schwarzrand subgroups, Nama Group
Description: Leaf-like bodies with fine ribs subparallel to long axis and with clearly marked margins (Glaessner in Robison & Teichert, 1979). Like *Rangea schneiderhoehni* and *Pteridinium simplex* consists of a bundle of spindle-shaped bodies. *Nasepia* differs from both of these taxa in having petaloids of smaller size and in configuration of ribs usally subparallel to long axis of petaloids, not angled. Ribs 0.1 to 1.0 mm wide; petaloids about 10.5 cm across
Reference: Germs, 1973
Type Specimen: South African Museum, SAM K1088.
Classification: Uncertain

Nemiana bakeevi (holotype)
Occurrence: ?
Locality: Zelem River, Bashkeria, Ural Mountains
Rock Unit: Badyeievskaya Formation
Description: Preserved in hyporelief and negative epirelief; individuals preserved as convex, oval-shaped individuals with a smooth surface, usually preserved in groups and rarely as solitary individuals or two or three individuals together. Lack clearly expressed concentric zonation; diameter ranges from 11 to 22 mm
Reference: Becker, 1992
Type Specimen: TSGM/TSNEGR Leningrad Museum, St Petersburg; N1/11406 (318/87)
Classification: Possibly algae or even a colonial cnidarian (possibly an anemone analogue)

Namacalathus hermanastes

Namalia villersiensis

Namapoikia rietoogensis

Nasepia altae

Nemiana bakeevi

Nemiana simplex *(holotype)*

Occurrence: Common in some localities
Locality: Dniester River, Podolia, Ukraine; Zimnii Bereg, White Sea, Russia; Khorbusuonka Basin, Olenek Uplift, Siberia; Aus region of Namibia; genus recognized in South Australia
Rock Unit: Mogilev Formation (Yampol Beds), Yaryshev Formation (Bernasheva Beds) and the Nagoryany Formation (Dzhurzhevka Beds); Ust'-Pinega Formation (Yorga Formation of Grazhdankin, 2004); Khatyspyt Formation; Dabis Formation; Rawnsley Quartzite, Wilpena Group
Other Names: *Tirasiana disciformis* of Palij?; *Beltanelliformis brunsae*; some authors consider *Beltanelloides* as a junior synonym of *Nemiana* but recent studies by Leonov (2006 and in press) show them to be two quite distinct groups with different morphology, ecologies and have produced quite different taphocenoses – *i.e.* preservation styles; *Sekwia kaptarenkoe* perhaps *Vendella*
Description: Impressions of *Nemiana* form positive hyporelief or negative epirelief. Casts with smooth-surface, convex discs, usually arranged in groups, rarely as isolated individuals. In some better-preserved specimens and larger ones, fine concentric grooves preserved, perhaps muscle impressions? In some cases, where it is possible to separate cast from overlying rocks, there appears to be small, rounded depression on relative flat, upper surface, perhaps oral or mouth opening. *Nemiana* individuals often occur in masses, sometimes covering entire bedding plans for metres – apparently a colonial organism. *Nemiana* evidently a benthic, sac-like organism, round as viewed from above, preferring muddy sea floors, and positioned partially submerged in the mud, often in huge concentrations. *Nemiana* concentrations were buried in place by incoming sand avalanches, with sand filling their inner cavities.
Reference: Palij, 1976; Fedonkin, 1981
Type Specimen: Kiev State University No 1818
Classification: Possibly algae or even a colonial cnidarian (possibly an anemone analogue)

Nemiana simplex

Nimbia dniesteri *(PIN 17384B)*

Occurrence: Not common
Locality: Dniester River near Novodnestrovskaya Hydroelectric Power Station, Ukraine
Rock Unit: Lomosov Beds, Mogilev Formation
Description: Small, discoidal organism with flat and smooth central part and clearly protruding marginal ridge that has a trapezoidal cross-section, a structure not present on *N. occlusa*. Shallow, rounded depression in the centre of the disc (diameter 2-3 mm). Diameter of 10 to 30 mm, width of marginal ridge 3 to 6 mm
Reference: Fedonkin, 1983
Type Specimen: Paleontological Institute, Moscow, PIN 3994/384-A
Classification: Phylum: ?Cnidaria; Class Cyclozoa or Phylum: Petalonamae, attachment disc for frond-like organism

Nimbia dniesteri

Nimbia occlusa *(?holotype)*

Occurrence: Common
Locality: Onega Peninsula and Zimnii Bereg, White Sea, Russia; Dniester River Basin, Podolia, Ukraine; Mackenzie Mountains, northwestern Canada
Rock Unit: Ust'-Pinega Formation, Valdai Group (Verkovka Formation of Grazhdankin, 2004); Yampol and Lomozov Beds, Mogilev Formation; Twitya Formation, Windermere Supergroup
Description: Flat, discoidal organisms preserved as annular or ovate structures at base of siltstone bands. Outer part of discs preserved in form of marginal ridge. More extensive central area, three to five times the width of marginal area, commonly smooth, rarely with small tubercle in center. Diameter ranges from 4 to 15 mm

Nimbia occlusa

Reference: Fedonkin, 1981
Type Specimen: Geological Institute, Moscow, GIN 4482/199
Classification: Phylum: ?Cnidaria; Class Cyclozoa or Phylum: Petalonamae, attachment disc for frond-like organism

Nimbia paula *(holotype)*

Occurrence: One specimen only
Locality: Pilipy, Danilov Creek a tributary of the Dniester River, Ukraine
Rock Unit: Mogilev Formation
Description: Discoidal organism, doughnut-shaped with central depression; preserved on soles of beds in positive hyporelief; surface smooth; central zone is knob-like; diameter of disc 5 mm; diameter of circular ridge forming the dounut – 1 to 1.5 mm
Reference: Gureev, 1987
Type Specimen: Geological Museum, Institute of Geological Sciences, Kiev – No number
Classification: Phylum: ?Cnidaria; Class Cyclozoa or Phylum: Petalonamae, attachment disc for frond-like organism

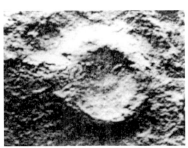

Nimbia paula

Onega stepanovi *(PIN 3992/5005)*

Occurrence: Common
Locality: Suz'ma (5 km from the mouth of the Suz'ma River), Karakhta and Solza rivers, Zimnii Bereg, White Sea, Russia
Rock Unit: Ust'-Pinega Formation (Verkhovka, Zimnii Gory and Yorga formations), Valdai Group
Description: Small organism with flat ovate body, except for centrally raised segmented zone that is shifted towards narrower end of body. Central zone with five paired, longitudinally arranged lobes, each divided by deep, broad groove. Length of lobes gradually decreases toward narrow end of the body. Length 4 to 7.5 mm; width, 2.5 to 3.8 mm; width of lobes about 0.3 mm
Reference: Fedonkin 1977
Type Specimen: Geological Institute, Moscow, GIN 4464/57b originally (now part of the collections of the Paleontological Institute, Moscow, PIN 3992/5049)
Classification: Phylum: Arthropoda; Family: ?Vendomiidae or another variation is Phylum: Proarticulata; Class: Vendiomorpha; Family Vendomiidae

Onega stepanovi (latex peel)

"Onegia nenoxa" *(PIN 3992/400; PIN 3993-5542); Same specimen as* Pteridinium nenoxa; *See* Pteridinium simplex)

Occurrence: Common
Locality: Suz'ma, White Sea, Russia
Rock Unit: Ust-Pinega Formation, Valdai Group (Verkovka Formation, Grazhdankin, 2004)
Other Names: = *Pteridinium simplex*, "*P. nenoxa.*" Junior synonyms include *Archangelia valdaica*, *Inkrylovia lata*, *Suzmites* (according to Grazhdankin, 2004 Ph.D. thesis)
Reference: Keller, 1974; Sokolov, 1976; Grazhdankin & Seilacher, 2002
Type Specimen: Paleontological Institute, Moscow, PIN 3992/1a

Onegia nenoxa

Orthogonium parallelum

(Specimen Lost during WW II)
Occurrence: Rare
Locality: Farm Kuibis, Namibia
Rock Unit: Dabis Formation, Nama Group
Description: Pickford (1995) paraphrases Gürich's (1930) description: "*Orthogonium parallelum* consists of 8 parallel rows of elongated tubes which are square in section lying on a bedding plane of sediment. These tubes are divided into sections, the longest preserved tube being 58 mm long and having 28 mesh-like sections, each of which is 2 mm high and 3 mm wide, each section being separated from its neighbor by a more or less sharply defined groove. Rows 5 + 6 and 7 + 8 of the holotype are closer to each other than they are to their neighbours, thus forming pairs of tubes which are further apart from each other than are the rows in each pair. These square-

section tubes could represent original pneu structure that did not collapse during fossilisation, but were filled with sediment, thereby preserving the three dimensional form of the pneu. At the base of the fossil, the organism bends down into the sediment, thereby forming a bulge. This suggests perhaps that during life, part of the organism was embedded in the sediment, and part of it extended upwards into the overlying water"
Reference: Gürich, 1930, 1933
Type Specimen: Specimen lost in World War II
Classification: Phylum: Petalonamae; Glaessner (in Robison & Teichert, 1979, rejected this as a valid taxon)

Ovatoscutum concentricum
(holotype)
Occurrence: Relatively rare
Locality: Ediacara Hills, Flinders Ranges, South Australia; Zimnii Bereg, White Sea, Russia; Dniester River Basin, Podolia, Ukraine
Rock Unit: Rawnsley Quartzite, Wilpena Group; Ust'-Pinega Formation, Valdai Group (Verkovka and Zimnii Gory formations of Grazhdankin, 2004); Mogilev Formation
Description: Irregularly rounded-oval in shape and sculptured by strong concentric grooves. Rounded shape interrupted by triangular notch on one side and a suture-line (the axis) that arises at top of notch, then continues to opposite margin. Individual grooves broadest near axis and narrow on either side of notch
Reference: Glaessner & Wade, 1966; Wade, 1971; Fedonkin, 1985, 2002
Type Specimen: South Australian Museum SAM P13770
Classification: Thought to belong to the Phylum: Cnidaria; Family: Chondroplidae originally but more recently suggested that it may be related to the dickinsoniomorphs

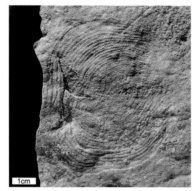
Ovatoscutum concentricum

Palaeopascichnus delicatus
(field specimen)
Occurrence: Uncommon in South Australia; common in White Sea section
Locality: Suz'ma, Zimnii Bereg, White Sea, Russia; Dniester Basin, Podolia; South Australia; Newfoundland; Wales; Namibia
Rock Unit: Ust'-Pinega Formation, Valdai Group (Lyamtsa, Verkhovka, Zimnii Gory and Yorga formations; Kanilov Formation of Grazhdankin, 2004), Mogilev Formation; Ediacara Member, Rawnsley Quartzite, Wilpena Group; Fermeuse Formation; ?Nama Group
Description: First described as a meander trace, Form A, by Glaessner. Consists of arcuate to looped ribs and grooves. However, many specimens show bifurcation of the sets of arcuate loops and appear to represent growth series of serial sets of sausage-shaped tubes on bedding plane
Reference: Palij, 1976; Gehling et al., 2000; Seilacher et al., 2003; Droser et al., 2005
Type Specimen: Ukranian Academy of Sciences, Institute of Geological Sciences, Kiev, IGN1907/7
Classification: Probably some form of collapsed serial beads or tubes growing on the sea floor. Regarded as xenophyophores, giant protozoans that reach up to 25 cm today, by Dolf Seilacher. By others regarded as a trace fossil, but there now seems to be a case for considering it some sort of body fossil which needs further study and clarification. Droser et al. (2005) suggested that other taxa described originally as trace fossil might also be much the same sort of organism: *Yelovichnus*, Intrites, some *Neonereites*, *Orbisiana*, and possibly *Harlaniella*

Palaeopascichnus delicatus

Palaeophragmodictya reticulata
(holotype)
Occurrence: Uncommon
Locality: Flinders Ranges, South Australia
Rock Unit: Ediacara Member, Rawnsley Quartzite, Wilpena Group
Description: Often preserved as part

and counterpart with "spicular" mesh sometimes preserved on at least two fine sandstone layers, indicating sand was trapped within the folded mesh. Specimens generally flattened, sometimes with opening pushed to one side, in down current direction, indicating that these forms were convex domes rather than cup-shaped bodies. A hemispherical, disc-shaped fossil with small disc at apex, spiny rim and central cavity. Most specimens show irregular wrinkling on surface, and in some specimens, perhaps in part due to decay before final burial, very fine network of mesh-like spicules preserved. Radial spicules protrude from rim at base of the disc
Reference: Gehling & Rigby, 1996
Type Specimen: South Australian Museum SAM P32324
Classification: Phylum: Porifera. They may be the oldest known Porifera

Palaeoplatoda segmentata (GIN 4464/103)
Occurrence: Rare
Locality: Suz'ma, (5 km upstream from the river mouth), White Sea, Russia
Rock Unit: Ust'-Pinega Formation, Valdai Group (Verkovka Formation, Grazhdankin 2004)
Other Names: Could be a *Dickinsonia*?
Description: Organism has flattened, leaf-shaped body with undulating margin, indicating elasticity during life of the organism. Body narrow in middle. What appears to be bottom or ventral side of body covered with fine, peripheral, slightly curved ridges. Segmented into fine, transverse ridges present on one side (bottom or ventral surface?) but only weakly on other. Length, more than 70 mm; width, 30 mm; length of individual segments, 0.6 mm; width of median ridge, 2 mm
Reference: Keller & Fedonkin, 1977
Type Specimen: Geological Institute GIN 4464/101
Classification: Phylum Proarticulata; Class: Dipleurozoa; Family: Dickinsoniidae (Fedonkin, 1979); Phylum: Petalonamae (Ivantsov, pers. com., 2006, but classification not clear); see discussion for *Dickinsonia costata*

"*Paliella patelliformis*" (Ukraine specimen; See *Aspidella terranovica*)
Occurence: Common in Podolia but rarer in the White Sea sequence
Locality: Zimnii Bereg, White Sea, Russia; Dniester River Basin, Podolia, Ukraine; Ryauzyak River, South Urals and Central Urals
Rock Unit: Ust'-Pinega Formation, Valdai Group (Verkovka Formation of Grazhdankin, 2004); Lomozov Beds, Mogilev Formation; Basa Formation, Asha Series; Cherny Kamen Formation, Sylvista Series
Other Names: *Aspidella terranovica* suggested by Gehling et al., 2000, but not all researchers agree with synonymy (Serezhnikova, pers. com.)
Description: Flat, circular body composed of central tubercle with flattened tip and broad, smooth outer zone separated from distinct inner zone. These are separated by narrow annular groove. Radial lines present in only largest specimens (diameters from 5 to 45 mm with fine, paired grooves spaced at about 0.5 mm) in outer, but not inner zone. *Paliella* is similar to *Cyclomedusa* but differs in not having concentric groove over entire structure or in central part. Small specimens of *Paliella* similar to *Medusinites asteroides* but differ in not having distinct radial grooves that cut outer zone. Specimens occur in groups or in pairs, where they are pressed close to each other, often rimmed by common narrow zone – suggesting that this form may have reproduced by simple division. Another type of branching reproduction is suggested by some specimens that have long, stem-like chord which extends from central part of larger form to smaller form at end of chord, presumably the offspring
Reference: Fedonkin, 1980
Type Specimen: Geological Institute, Moscow, GIN 4482/51

Palaeoplatoda segmentata

Paliella patelliformis

Palaeophragmodictya reticulata

"Papillionata eyrie" (holotype; See Dickinsonia costata)
Occurrence: ?
Locality: Ediacara Hills, Flinders Ranges, South Australia
Rock Unit: Ediacara Member, Rawnsley Quartzite, Wilpena Group
Other names: Dickinsonia costata
Reference: Sprigg, 1947, 1949
Type Specimen: South Australian Museum SAM T8-2060

Paracharnia dengyingensis
Occurrence: ?
Locallty: Yangtze Gorge, China
Rock Unit: Denying Formation
Description: Frond-like structure partially covered with crust of dark mineral (?pyrite); frond with fusiform depressions separated from each other by elevated ridges that extend towards an axis; frond is much more elongated than in *Rangea,* more like *Bomakellia* (Dzik, 2002); wider stalk than *Charnia*
Reference: Sun, 1986; Steiner 1994
Type Specimen: Wuhan Geological Museum, China, WGM ZnF0011
Classification: Phylum: Petalonamae; Family: Uncertain

Paramedusium africanum
(Specimen Lost)
Occurrence: A single specimen only
Locality: Near railway station at Ham River, Groendoorn, Namibia
Rock Unit: ?Nasep Formation, Nama Group
Description: Pickford (1995) paraphrased Gürich's (1933) original description: "The fossil is broken, with large parts missing, but surviving portion is approximately semi-circular and is flat, with diameter of about 170 mm. Three concentric furrows and ridges (or bulges) along margin which differ in width, the outermost ridge being weakest, innermost one being the widest and lowest....The innermost ridge (or bulge) is covered by radiating ribs 2 to 3 mm from each other which occasionally show net-like connections to each other." Pickford further noted that based on Gürich's description this form was "probably a flat recliner – a mat-like organism that lay sessile on the sea bed. It appears to have had radial growth and serial quilting"
Reference: Gürich, 1930, 1933; Germs, 1972. Specimen lost in World War II
Type Specimen: Lost
Classification: Phylum: Petalonamae

"Paramedusium patellaris"
(Same specimen as "Planomedusites patellaris" and "Medusinites patellaris")

Paravendia janae (holotype)
Occurrence: Rare
Locality: Zimnii Bereg, White Sea, Russia
Rock Unit: Ust'-Pinega Formation, Valdai Group (Yorga Formation, Grazhdankin, 2004)
Other Names: Vendia janae, Aspidella terranovica
Description: Small bilateral form with elongated, oval body (top view) subdivided completely by segments. Lateral edge of body smooth. Four pairs of isomeres curve towards posterior, join in single point. Largest segments of the first pair make up lateral margin of body. Tubular structures, possibly a digestive-distributive system, consist of axial channel, and non-branching, short, lateral outgrowths, one per each isomere
Reference: Ivantsov, 2001 (Vendia janae)
Type Specimen: Paleontological Institute, Moscow, PIN 3993/5070
Classification: Phylum uncertain, but in the past has been placed in the Phylum Proarticulata based on its "symmetry of gliding reflection" where segments meet along the midline in a zigzag fashion. Class: Vendiamorpha; Family: Vendomiidae; possibly a dickinsoniomorph

Papillionata eyrie

Paracharnia dengyingensis

"Paramedusium patellaris"

Paravendia janae (latex peel)

Parvancorina minchami (SAM 12887)
Occurrence: Moderately common
Locality: Flinders Ranges, South Australia; Russia; Zimnii Bereg, White Sea, Russia
Rock Unit: Ediacara Member, Rawnsley Quartzite, Wilpena Group; Ust'-Pinega Formation, Valdai Group (Verkhovka, Zimnii Gory, and Yorga formations of Grazhdankin, 2004)
Description: Small anchor-shaped form, next most common bilateral form in Australia after *Dickinsonia*. Organism with strong arcuate end, just inside margin regarded as "head." Median ridge joining head arc, extends all the way to tail. In smallest specimens, ridge keel-like and highest at tail end. In some specimens delicate paired ridges extend out on either side of midline ridge. Some researchers have suggested these are legs, but they are not jointed, so probably not appendages. Specimens of *Parvancorina* vary in ratio of length to width – most commonly medial-ridge longer than body width. Some specimens are foreshortened, which suggests that tail end was able to contract. In collections from Flinders Ranges in South Australia there are a variety of preservational styles, but some small specimens that seem least deformed suggest that the body was about as deep as was wide. Thus, it appears that there has been significant compression in a large number of specimens. Some forms in younger Chengjiang Fauna of China (*Naraoia*) and also in Canada (Burgess Shale) (*Skania*) show some resemblance to *Parvancorina* and may be surviving relatives
Reference: Glaessner, 1958, 1980; Lin et al., 2006
Type Specimen: South Australian Museum SAM P12774
Classification: Phylum: Arthropoda

Parvancorina saggita (holotype)
Occurrence: Not common
Locality: Solza River, White Sea, Russia
Rock Type: Ust'-Pinega Formation, Valdai Group (Verkovka Formation)
Description: Body elongated oval with wider, probably anterior end. A narrow band, slightly widened anteriorly and posteriorly, extends along margin of body. Central part of body evenly convex in small specimens, while in larger specimens, area occupied by anchor-shaped ridge. In cross-section structure is arched, concave dorsally, width of proximal arched cross-bar or beam about half of its length. Lengthwise structure is straight and relatively wide, cross-beam being much narrower than that in *P. minchami*, in which length of cross-beam nearly equal to longitudinal beam. It further differs from *P. minchami* in having more elongated body, wider marginal band and broader longitudinal beam
Reference: Ivantsov, Malakhovskaya & Serezhnikova, 2004
Type Specimen: Paleontological Institute, Moscow, PIN 4853/89
Classification: Phylum: Arthropoda

Persimedusites chahgazensis
(field specimen)
Occurrence: Not common
Locality: Chah-Mir Mine northeastern Tajkuh; Kushk Mine, northeast of Bafgh, Yazd area, Iran
Rock Unit: Estfordi Formation, Kushk Series
Description: Diameter of specimens approximately 15 mm
Reference: Hahn & Pflug, 1980
Type Specimen: ?
Classification: Hahn & Pflug suggested this form was a medusa, Phylum: Cnidaria. It should be restudied to ascertain its organic nature

Parvancorina minchami

Parvancorina saggita (latex peel)

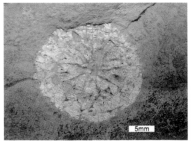

Persimedusites chahgazensis

"Petalostroma kuibis" (holotype)

Nomen nudum
Occurrence: Rare
Locality: Farm Aar, Aus area, Namibia, southern Africa
Rock Unit: Dabis Formation, Nama Group
Description: Saucer-shaped structures up to tens of cm in diameter lacking any indication of internal cavities; irregularly wrinkled surface
Reference: Pflug, 1973
Type Specimen: Pflug, 1973 gave the number of the holotype as 3601 with no indication of institutional collection
Classification: Rejected as a valid taxon by Glaessner (in Robinson & Teichert, 1979)

Phyllozoon hanseni (holotype)

Occurrence: Common
Locality: Flinders Ranges, South Australia
Rock Unit: Ediacara Member, Rawnsley Quartzite, Wilpena Group
Description: Belt-shaped fossil composed of many simple tubular elements that alternate along mid-line. Side elements decrease in size toward tapered end. Usually found in groups with individual pairs in close contact. No evidence of a stem or stalk for attachment. *Phyllozoon* often found in clusters with many specimens lying side-by-side or overlapping, and it has been suggested that possibly these were large, marine algae – like seaweeds. Length of specimens vary from 8-33 cm and width from 2.5 to 4 cm
Reference: Jenkins & Gehling, 1978, Gehling *et al.*, 2005; Narbonne, 2005
Type Specimen: South Australian Museum SAM P19508A (DP51)
Classification: Large organic tubes, thought to be filled with fluid, not muscular tissue (Gehling, *pers. com.*, 2006; possibly marine algae. Ivantsov (2004) placed this taxon in the Phylum: Proarticulata; Narbonne suggested it could be a dickinsoniomorph body fossil; others suggested this taxon to be dickinsoniomprh traces, but Gehling *et al.* (2005) noted that resting traces do not show the overlap that *Phyllozoan* displays; specimens all preserved in positive hyporelief; elements meet alternatively along midline; none show shrinkage (J. Gehling, *pers. com.*, 2007)

"Pinegia stellaris" (holotype

(see *Hiemalora stellaris*, reassigned name)

Occurrence: ?
Locality: Zimnii Bereg, White Sea, Russia (Verkovka Formation, Grazhdankin, 2004)
Rock Unit: Ust'-Pinega Formation, Valdai Group
Other Names: *Pinegia* is a pre-occupied generic name of a Permian insect described by A.V. Martynov in 1928, a junior synonym of *Hiemalora stellaris*, reassigned by Fedonkin in 1982
Reference: Fedonkin, 1980, 1982
Type Specimen: Geological Institute, Moscow, GIN 4482/25

Planomedusites grandis

(Geological Museum GM 2127-70)
Occurrence: Not common, 10 specimens only
Locality: Dniester River, Podolia, Ukraine
Rock Unit: Mogilev Formation
Description: Discoidal form, preserved both as hemisphere and as flattened fossil on sole of beds in positive hyporelief; flattened forms show weakly preserved peripheral ridges; central knob normally conical; diameter of whole form ranges from 22 to 125 mm; diameter of central knob ranges from 5 to 8 mm
Reference: Sokolov, 1972, 1973
Type Specimen: Geological Museum, Institute of Geological Sciences, Kiev – no number available
Classification: Phylum: Cnidaria; Class: Cyclozoa or Phylum: Petalonamae, a holdfast

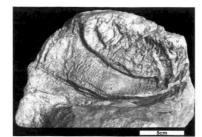

Petalostroma kuibis

Phyllozoon hanseni

Planomedusites grandis

"Planomedusites patellaris"

(Kiev State University; KGU 2127-34; See *Aspidella terranovica*)
Occurrence: Uncommon, eight specimens
Locality: Vinozh, Vinnitsa Region, Dniester River, Ukraine
Rock Unit: Mogilev Formation
Other Names: ?*Aspidella terranovica* (suggested by Gehling *et al.*, 2000); *"Paramedusium patellarism," "Medusinites patellaris"*
Description: Fossil preserved as flattened hemisphere on sole of beds in positive hyporelief; central knob broad; largest specimens show concentric zonation; diameter from 30 to 150 mm; central knob diameter at base 10 to 35 mm
Reference: Sokolov, 1972, 1973; Gureev, 1987; Gehling *et al.*, 2000
Type Specimen: Geological Museum, Institute of Geological Sciences, Russia, see Sokolov,1972

Platypholinia pholiata (PIN 3993/191)

Occurrence: Rare
Locality: Zimnii Bereg, White Sea, Russia (5 km north of Zimnegorsky lighthouse)
Rock Unit: Redkino Horizon, Bed 1, Ust'-Pinega Formation, Valdai Group (Yorga Formation of Grazhdankin, 2004)
Description: Flat, leaf-shaped organisms with acutely narrowed anterior end and smoothly rounded posterior. Midline ridge present on front half of body but disappears posteriorly. On either side of ridge and subparallel to it, finer ridges and furrows. No traces of anus or internal organs. Surface of body flat and smooth, except for midline ridge, where linear structures present, which extend from front of organism to where midline ridge ends. Small depression at very front of body may be terminal mouth – just posterior to this, midline ridge arises. Largest specimen about 60 mm long, up to 28 mm wide, with length of midline ridge varying between 1.5 and 2 mm along its length
Reference: Fedonkin, 1985
Type Specimen: Paleontological Institute, Moscow, PIN 3993/195
Classification: Many similarities with worms in the Platyhelminthes

Podolimirus mirus (holotype;

worth close comparison to *Pteridinium simplex*)
Occurrence: Rare
Locality: Novodnestrovskaya Hydroelectric Station, Dniester River, Ukraine
Rock Unit: Lomosov Beds, Mogilev Formation
Other Names: Similarities to *Valdaina plumosa*, ?*Pteridinium latum* and other *Pteridinium* "species", *Inkrylovia lata*
Description: Large, bilaterally symmetric organism with proarticulate body plan, composed of paired, posteriorly flexed segments (isomeres), wide proximally and narrowing distally. Segments diverge from median axis at about 80-85 degrees and then curve abruptly, running parallel to long axis of organism; length of organism @ 170 mm, width at least 100 mm; width of first proximal segment, 22 mm; nine or fewer pairs of isomeres present
Reference: Fedonkin, 1983; Velikanov, Aseeva & Fedonkin, 1983
Type Specimen: Paleontological Institute, Moscow, PIN 3994/417
Classification: Phylum uncertain, but in the past has been placed in the Phylum Proarticulata based on its "symmetry of gliding reflection" where segments meet along the midline in a zigzag fashion. Class: Vendiomorpha; Family: Vendomiidae but in past classification was placed in the Petalonamae

"Pollukia serebrina" (holotype;

same as *"Cyclomedusa serebrina"*; See *Aspidella terranovica*)
Occurrence: Rare, 1 specimen
Locality: Mogilev-Podolski, Dniester River, Ukraine
Rock Unit: Yaryshev Formation

"Planomedusites patellaris"

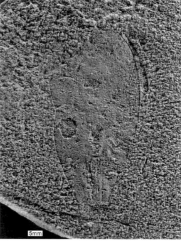

Platypholinia pholiata

Podolimirus mirus

Pollukia serebrina

Other Names: First described as *Cyclomedusa serbrina* by Palij (1969) but Gureev placed it in a separate genus, *Tirasiana disciformis, Aspidella terranovica*
Description: Flattened cast preserved in positive hyporelief on sole of bed with two concentric zones with central knob forming third; appears very similar to *Tirasiania disciformis* and is probably conspecific; diameter 54 mm of body; diameter of internal zone 36-40 mm; central knob diameter 14 mm.
Reference: Palij, 1969; Güreev, 1987
Type Specimen: Kiev State University, KGU 1735

"Pollukia shulgae" (holotype; See Aspidella terranovica)
Occurrence: Rare, 1 specimen
Locality: Novoselka, Ternopol District Borehole 3642 (depth 353.8 m), Ukraine
Rock Unit: Lontova Horizon, Baltic Series, Zburch Formation (Cambrian)
Other Names: First described as "*Cyclomedusa serebrina*" by Palij (1969) but Gureev, 1987, placed it in a separate genus; *Tirasiana* "species;" *Cyclomedusa, Aspidella terranovica*
Description: Small cast preserved in positive hyporelief on the sole of bed with two concentric zones with a small central knob forming third; appears very similar to *Tirasinia disciformis* and is probably conspecific (Fedonkin, *pers. com.*, 2005); outer zone has irregular surface; edge of fossil is irregular; diameter 21-25 mm of body; diameter of internal zone 12-14 mm; central knob diameter 1 mm
Reference: Gureev, 1987
Type Specimen: Geological Museum, Institute of Geological Sciences, Kiev, No. 2127/55

Pollukia shulgae

Pomoria corolliformis(holotype)
Occurrence: Rare
Locality: Zimnii Bereg, White Sea, Russia
Rock Unit: Bed 11, Ust'-Pinega Formation, Valdai Group (Yorga Formation of Grazhdankin, 2004)
Description: Small, discoidal form with two rows of closely spaced tentacles or rootlike projections, one located towards middle of disc and one along margin. Complex, infolded tissue in central part, with at least two long, bent processes which radiate from center, their thickness two to three times that of tentacle-like processes. Diameter ranges from 16 to 20 mm; length of interior "tentacles" 4-6 mm; width of interior "tentacles," 0.3 to 0.4 mm; width of the marginal tentacles up to 0.7 to 0.8 mm; length of central processes up to 10 to 12 mm, width up to 1.5 mm; diameter of central depression or aperture, 2 mm
Reference: Fedonkin, 1980, 1985
Type Specimen: Paleontological Institute, Moscow, PIN 3993/203
Classification: Phylum: ?Cnidaria; Class: Scyphozoa; Family: Pomoriidae or perhaps holdfast of member of Phylum: Petalonamae

Pomoria corolliformis

"Praecambridium" sigillum (on block with other taxa)
Occurrence: Uncommon, being too small to be seen on most surfaces
Locality: Ediacara Hills, Flinders Ranges, South Australia
Rock Unit: Ediacara Member, Rawnsley Quartzite, Wilpena Group
Other Names: Runnegar and Fedonkin (1992) suggested possible synonymy with *Spriggina;* others have noted similarities with *Dickinsonia brachina;* Gehling (*pers. com.*, 2006) notes that there is "no evidence that *Praecambridium* is juvenile *Dickinsonia* or juvenile *Spriggina* (or *Marywadea*)." He points out that juvenile forms of all these taxa are known in South Australia and are not *Praecambridium*
Description: Tiny, bilateral, oval-shaped form with 3-6 arcuate segments/ appendages and head shield bearing paired, branching ribs. Ribs on either side coalesce near midline, then taper to points at outer edge, ends pointing towards the back of animal
Reference: Glaessner & Wade, 1966

Praecambridium sigillum

Type Specimen: South Australian Museum SAM 13798
Classification: Uncertain, see Other Names above

Protechiurus edmondsi (holotype)
Occurrence: Only one specimen known
Locality: Farm Plateau, Aus area, Namibia, southern Africa
Rock Unit: Dabis Formation, Nama Group
Description: Pickford (1995) paraphrases Glaessner's description of this form as "cigar shaped… with a broad based, spatulate" anterior end "and eight more or less prominent longitudinal muscle bands." Length 74 mm, maximum width 19 mm. Fossil with one end tapering to point and other end with broadly rounded, flattened morphology
Reference: Glaessner, 1979; Pickford, 1995
Type Specimen: Geological Survey of Namibia GSN P44
Classification: Disputed identity; Glaessner suggested *Protoechiurus* belonged to the worm Phylum Echiura, Family: Echiuridae. Runnegar (1992) thought it a dubiofossil. Others have suggested that this form was a Protochordate

Protodipleurosoma asymmetrica (holotype)
Occurrence: ?
Locality: Karanyurt, Askyn River, Bashkeria, southern Urals
Rock Unit: Ashynskaya Sequence, Zeganskaya Series
Description: Disc-shaped impression, with central disc clearly defined by a a depression with peculiar radial imprints (probably due to deformation), lacking any concentric zonation; central disc occupies about half of the surface area of the structure, with raised central part; peripheral part of disc defined as a raised band with faint radial furrows; diameter of disc, 42 mm; diameter of central disc, 20 mm; diameter of cental raised boss, 4 mm
Reference: Bekker, 1996
Type Specimen: Geological Museum Novosibersk, TSGM 1/11406 (316/18)
Classification: Perhaps holdfast of a frond-like organism or Phylum: Cnidaria

Protodipleurosoma paula (PIN 673-34)
Occurrence: Rare and poorly preserved material
Locality: Ryauzyak River, South Urals, Russia
Rock Unit: Basa Formation, Asha Series
Description: Small, well defined, rounded impression with distinct outer ring around periphery. Inner disc divided by central groove into nearly equal parts. On one side, central groove cuts outer ring, which it slightly deforms at intersection. Small grooves radiating from central groove are flattened and do not reach outer ring
Reference: Becker, 1977; Fedonkin, 1980, 1985
Type Specimen: CGM 1/11406 sample 176/86
Classification: Phylum: Cnidaria; Class: Cyclozoa; Family: Uncertain

"Protodipleurosoma rugulosa" (holotype; See Aspidella terranovica)
Occurrence: ?
Locality: Zimnii Bereg, White Sea, Russia; Dniester River Basin, Podolia, Ukraine
Rock Unit: Ust'-Pinega Formation, Valdai Group (Verkovka Formation of Grazhdankin, 2004); Mogilev Formation
Other Names: Even though *P. wardi* has been combined (synonymized) with *Ediacaria flindersi* (= *Aspidella terranovica*), *P. rugulosum* may yet be recognized as independent of *P. wardi*. Material from Russia is much better preserved and more abundant than that known from Australia, and may be unique (Fedonkin, 1985), but Gehling *et al.* have assigned this form to *Aspidella terranovica*. Not all researchers agree with this synonymy (Serezhnikova, *pers. com.*, 2006)

Protechiurus edmondsi

Protodipleurosoma asymmetrica

Protodipleurosoma paula

Protodipleurosoma rugulosa

Description: Rounded impression with broad outer ring covered by narrow concentric grooves; middle of disc with ovoid, irregularly compressed central area divided into lobes, irregular radial grooves. Primary grooves deep. Groove in middle section of form divides central disc into two unequal elongated lobes. Minor furrows and ridges, radiating from medial groove thin toward the outside of this central disc. None of this topography affects peripheral zone. Peripheral zone expanded in large forms and very reduced in smaller forms. On holotype, semicircular processes occur around outside of body. Diameter ranges from 15 to 80 mm; diameter of central disc from 15 to 60 mm. Main difference from *P. wardi* is that *P. rugulosum* has larger number of ridges, is larger and possesses lateral processes in largest specimens.
Reference: Fedonkin, 1980; Gehling *et al.*, 2000
Type Specimen: Geological Institute, Moscow, GIN 4482/21

"Protodipleurosoma wardi"
(holotype)
Occurrence: Rare and poorly preserved material
Locality: Flinders Ranges, South Australia
Rock Unit: Rawnsley Quartzite, Wilpena Group
Other Names: *Aspidella terranovica* and may be the same as *Ediacaria flindersi* and thus a junior synonym. Not all researchers would agree with this synonymy (Serezhnikova, *pers. com*, 2006)
Description: A circular impression, the outer ring concentrically striated and wide. Middle area large and with lobate feature in very center. Irregularly developed radial grooves that branch.
Reference: Sprigg, 1949; Gehling *et al.*, 2000
Type Specimen: South Australian Museum SAM T36-2023

Protodipleurosoma wardi

Protoniobea wadea (holotype)
Occurrence: ?
Locality: Mt John Osmond Range, Kimberleys, Western Australia
Rock Unit: "Lower Cambrian flags" thought to be Neoproterozoic, but needs confirmation
Description: Circular impression, 4.1 mm in diameter with a few prominent annular undulations. Nodular structures centrally arranged in a polygonal pattern, with central depressed zone. The platform with the nodular structures is slightly wider than the central area and surrounded by a deep groove. Annular ridges present, separated by a groove; possibly some radial canals
Reference: Sprigg, 1949
Type Specimen: Commonwealth Palaeontology Collection, Canberra, CPC 102
Classification: Possibly a cnidarian

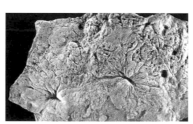

Protoniobea wadea

Pseudorhizostomites howchini
Occurrence: Very common
Locality: Flinders Ranges, South Australia; Zimnii Bereg, White Sea, Russia; Dniester River Basin, Podolia, Ukraine; Gornaya Baskkiria, Ryauzyak Basin
Rock Unit: Ediacara Member, Rawnsley Quartzite, Wilpena Group; Ust'-Pinega Formation, Valdai Group (Verkovka and Yorga formations, Grazhdankin 2004); Mogilev Formation; Basa Formation, Asjha Series (Bashkiria, southern Urals)
Other Names: *Pseudorhopilema chapmani*; possibly *Wigwamiella enigmatica*
Description: Grooves radiating from center, irregularly branching and thinning towards outside. No distinct peripheral boundary present. Central part often strongly depressed (negative hyporelief). Diameter of radial structures can vary significantly – from 9 to 60 mm
Reference: Sprigg, 1949; Glaessner & Wade, 1966; Wade, 1968
Type Specimen: South Australian Museum, SAM T73-2034
Classification: Pseudofossil. Gas escape structure or perhaps structure

Pseudorhizostomites howchini

created where a holdfast has been pulled by currents or some other force from the sediments that enclosed it

"Pseudorhopilema chapmani"
(holotype; See Pseudorhizostomites howchini)
Locality: Ediacara Hills, Flinders Ranges, South Australia
Rock Unit: Ediacara Member, Rawnsley Quartzite, Wilpena Group
Other Names: *Pseudorhizomites howchini*
Reference: Sprigg, 1949; Glaessner & Wade, 1966
Type Specimen: South Australian Museum SAM T74-2036

Pseudovendia charnwoodensis
(holotype)
Occurrence: Rare
Locality: The Outwoods, Charnwood Forest, near Loughborough, Leicestershire,UK; Suz'ma River, White Sea, Russia
Rock Unit: Woodhouse Beds, Maplewell Series, Charnian Supergroup; Ust'-Pinega Formation, Valdai Group (Verkovka Formation, Grazhdankin, 2004)
Description: Medium sized, @25 mm long and 20 mm wide, with largest paired lobes 6 mm long and 3 mm wide, roughly triangular-shaped lobate impression. Anterior end not well preserved; three distinct pairs of lobes present, possibly a fourth pair located anteriorly, but specimen too poorly preserved for certainty; lobes bluntly rounded at ends, straight and slightly inclined posteriorly; lobes join along indistinct axis; with one large, ovate and rounded lobe at the posterior end, about 7 mm long and 12 mm wide
Reference: Boynton & Ford, 1979; Fedonkin 1985
Type Specimen: Leicestershire Museum, UK, No. 6456/1978
Classification: Phylum uncertain, but in the past has been placed in the Phylum Proarticulata based on its "symmetry of gliding reflection" where segments meet along the midline in a zigzag fashion. Class: Vendiomorpha; Family: Vendomiidae; possibly a dickinsoniomorph; see discussion for *Dickinsonia costata*; thought by some to be an arthropod (Boynton & Ford, 1979)

"Pteridinium carolinaense"
(holotype; See Pteridinium simplex)
Occurrence: Rare
Locality: North Carolina
Rock Unit: McManus Formation, Albemarle Group
Other names: *Pteridinium simplex*
Reference: St Jean, 1973; Gibson *et al.*, 1984
Type Specimen: University of North Carolina, Department of Geography and Earth Sciences (Charlotte) 3602

"Pteridinium latum" = Inkryolovia lata
Occurrence: ?
Locality: Suz'ma, White Sea, Russia
Rock Unit: Ust'-Pinega Formation, Valdai Group (Verkovka Formation, Grazhdankin, 2004)
Other Names: *Pteridinium nenoxa, Inkrylovia lata, Podolimirus mirus, Valdaina plumosa*
Reference: Palij, Posti & Fedonkin, 1979
Type Specimen: Same specimen as *Inkrylovia lata*

"Pteridinium nenoxa" (PIN 3992/400; See Pteridinium simplex)
Occurrence: Common in Russia
Locality: Suz'ma and Zimnii Coast, White Sea, Russia; Flinders Ranges, South Australia; and other places
Rock Unit: Ust'-Pinega Formation, Valdai Group (Verkhovka, Zimnii Gory formations of Grazhdankin, 2004); Ediacara Member, Rawnsley Quartzite, Wilpena Group
Other Names: *Pteridinum simplex*
Description: Leaf-shaped, elongated, bilaterally symmetric segmented organism with fine intersegmental ribs that are curved in one direction and

Pseudorhopilema chapmani

Pseudovendia charnwoodensis

Pteridinium carolinaense

Pteridinium nenoxa (latex peel)

converge towards their lateral ends. Segments divided by walls. Entire body segmented with segments not opposite but alternating. The Russian and Ukranian forms very similar to those in Australia but differ from the Namibian forms in the morphology of the segments and their number, absence of smooth zone and with much less complex morphology of mid-zone **Reference:** Keller *et al.*, 1974; Fedonkin, 1981; Gibson *et al.*, 1984; Grazhdankin & Seilacher, 2002. **Type Specimen:** Paleontological Institute, Moscow, PIN 3992/1a

Pteridinium simplex (unnumbered, Namibian Geological Survey specimen)

Occurrence: Common in Namibia **Locality:** Namibia; Flinders Ranges, South Australia; White Sea, Russia; ?Mackenzie Mountains, Northwest Territories, Canada (and see above) **Rock Unit:** Dabis Formation, Nama Group; Ediacara Member, Rawnsley Quartzite, Wilpena Group; Ust'Pinega Formation; Blue Flower Formation, Windermere Supergroup **Other Names:** *Inkrylovia lata, Pteridinium carolinaense, cf. Podolimirus mirus, Pteridinium lata, Pteridinium nenoxa* **Description:** Organism consisting of three walls joined along axis by alternating insertion of tubular chambers that make up each wall. Ends possibly open, as tubes are often partly sand filled in preservation **Reference:** Gürich, 1930; Narbonne, 2005 **Type Specimen:** ? **Classification:** Phylum: Petalonamae; Family: Pteridiniidae; Narbonne has allied with dickinsoniomorphs, but this questioned by Gehling *et al.* (2005) due to the very different manner of its preservation

Pteridinium simplex

Ramellina pennata (holotype)

Occurrence: Rare **Locality:** Zimnii Bereg, White Sea, Russia **Rock Unit:** Ust'-Pinega Formation, Valdai Group (Verkovka Formation of Grazhdankin, 2004) **Description:** Small bipinnate, leaf-shaped forms preserved in positive hyporelief with midline ridge and from this branch secondary structures. Structures meet the mid-line in alternating arrangement, not opposite. One end of organism rounded, the other pointed. Length of axis, 20-30 mm; width 1-2 mm; length of secondary branches, mid-organism 4.5-6 mm, width 1-1.5 mm. About 20 branches can be observed **Reference:** Fedonkin, 1980, 1985 **Type Specimen:** Geological Institute, Moscow, GIN 4482/47 **Classification:** Phylum: Petalonamae; Family: Uncertain

Ramellina pennata

"Rangea arborea" (holotype; same specimen as Arborea arborea; See Charniodiscus arboreus)

Locality: Flinders Ranges, South Australia **Rock Unit:** Ediacara Member, Rawnsley Quartzite, Wilpena Group **Other Names:** *Charniodiscus arboreus, Charniodiscus concentricus;* same specimen as *Arborea arborea* **Reference:** Glaessner, 1959; Glaessner & Daily, 1959; Jenkins & Gehling, 1978 **Type Specimen:** South Australian Museum SAM P12891

"Rangea brevior"

(*See Rangea sneiderhoehni*) **Locality:** Kuibis, Namibia **Rock Unit:** Dabis Formation, Nama Group **Other Names:** *Rangea sneiderhoehni* **Reference:** Gürich, 1930, 1933; Jenkins, 1985; Pickford, 1995 **Type Specimen:** ?

Rangea arborea

"Rangea grandis" (holotype; See Charnia masoni)

Occurrence: Rare **Locality:** Ediacara Hills, Flinders Ranges, South Australia **Rock Unit:** Rawnsley Quartzite, Wilpena Group **Other Names:** Same specimen as *Glaessneria grandis* **Description:** Probably referable to *Charnia masoni* **Reference:** Glaessner & Wade, 1966; Germs, 1972, 1973; Jenkins & Gehling, 1978 **Type Specimen:** South Australian Museum SAM P12897

"Rangea longa" (holotype; See Charniodiscus longus, same specimen as Glaessnerina longa)

Locality: Flinders Ranges, South Australia **Rock Unit:** Ediacara Member, Rawnsley Quartzite, Wilpena Group **Other names:** *Charniodiscus longus, Glaessnerina longa* **Reference:** Glaessner & Wade, 1966 **Type Specimen:** Same specimen as *Charniodiscus longus*

Rangea schneiderhoehni

Occurrence: Not common **Locality:** Farm Kuibis, Aus area, Namibia, southern Africa **Rock Unit:** Dabis Formation, Nama Group? **Other Names:** *Rangea brevior*, and may have some relationships with *Bomakellia* and *Mialsemia* **Description:** *Rangea* has bipolar growth and exhibits fractal quilting and individuals have "leaf-like bodies with primary and secondary branches, or vein-like structures. These are now interpreted as being part of a network of pneu morphology. One specimen is 6.5 cm wide and 7.5 cm long. Primary branches form angle of 40-50 degrees with the somewhat obscure median line. Secondary branches bilateral and form angles of 45-55 degrees with primary branches in plane of 'leaf.' Spacing between primary furrows is 4-5 mm, and that between secondary furrows is 0.9-1.1 mm. Near axis there are commIsure ribs between prominent primary ones. Inner extremities of prImary branches slope and taper towards median line and then show secondary branches in one direction." (Pickford, 1995) **Reference:** Gürich, 1930; Germs, 1973; Pickford, 1995; Grazhdankin & Seilacher, 2005, Narbonne, 2005 **Type Specimen:** Hamburg University HU 179 (2 specimens in Hamburg, 21 in the Pflug collection at Justis Liebig University, Giessen, Germany) **Classification:** Phylum: Petalonamae; Family: Rangeidae. Dzik (2002) has suggested *Rangea* could be a ctenophore; Narbonne has grouped this with other taxa from Newfoundland with fractal architecture, the rangeomorphs

"Rangea siberica" (holotype; same specimen as Charnia siberica)

Locality: Olenyok Uplift, Siberia **Rock Unit:** Khatysput Formation **Other Names:** *Charnia siberica, Charnia masoni* **Reference:** Sokolov, 1972 **Type Specimen:** Same specimen as *Charnia siberica*

Rugoconites enigmaticus

Occurrence: Moderately common **Locality:** Flinders Ranges, South Australia **Rock Unit:** Ediacara Member, Rawnsley Quartzite, Wilpena Group **Other Names:** *Lorenzinites rarus* **Description:** Conical organism with small central depressed area and sets of coarse, radiating and branching ridges that bifurcate towards outer margin. In larger specimens, branching order obscured, except for final division near margin, while in smaller, marginal branches and grooves are not always obvious. In some specimens, three bosses at apex and parts of annulus that may represent marginal flange. Based on nature of collapsed

Rangea grandis

Rangea longa

Rangea schneiderhoehni (cast)

Rangea siberica

depression preserved in sandstone bed above some specimens, *R. enigmaticus* probably had a height about equal to its radius
Reference: Glaessner & Wade, 1966; Jenkins, 1992
Type Specimen: South Australian Museum SAM P13781
Classification: *Rugoconites* was originally thought to be the medusa of a jellyfish, but it does not easily fit into any known phylum. Dolf Seilacher suggested that it, like *Tribrachidium*, had a sponge grade of organization

Rugoconites tenuirugosus
(holotype)
Occurrence: Uncommon
Locality: Flinders Ranges, South Australia
Rock Unit: Ediacara Member, Rawnsley Quartzite, Wilpena Group
Description: Conical organism with branching ridges arising from central area and radiating and bifurcating towards margin. At margin, main branches end in numerous sets of fine furrows and ridges. This, together with initial width of ridges and more radial placement of them, are major distinctions between this form and *R. enigmaticus*. Known specimens of *R. tenuirugosus* are generally larger and shallower in relief than *R. enigmaticus*, but, larger specimens of *R. enigmaticus* also tend to be lower domed when compared to smaller specimens
Reference: Wade, 1972; Jenkins, 1992
Type Specimen: South Australian Museum SAM P17461
Classification: *Rugoconites* was originally thought to be the medusa of a jellyfish, but it does not easily fit into any known phylum. Dolf Seilacher suggested that it, like *Tribrachidium*, had a sponge grade of organization

Sekwia excentrica
Occurrence: ?
Locality: Mackenzie Mountains, Northwest Territories and Wernecke Mountains, Yukon Territory, Canada; Nainital District, Uttar Pradesh, India
Rock Unit: Blueflower Formation, Windermere Group; Vampire Formation; Krol Formation
Description: Discoidal fossil preserved as a subcircular depression about 2.2 cm across and 1 mm deep, with faint eccentric markings and indistinct annular depression at margin
Reference: Hofmann, 1981; Hofmann, et al., 1983
Type Specimen: Geological Survey of Canada GSC66173a
Classification: ?Phylum: Petalonamae

"Sekwia kaptarenkoe" (holotype;
See *Nemiana simplex*)
Occurrence: Abundant
Locality: Vinozh Village, Ladova River, a tributary of the Dniester River, Ukraine
Rock Unit: Mogilev Formation
Other Names: Considered by Fedonkin conspecific with *Nemiana simplex*
Description: Small, discoidal form preserved in positive hyporelief, body cylindrical in shape, with large, hemispherical central knob; outer marginal zone flattened; very often occurs in groups
Reference: Gureev, 1987
Type Specimen: Geological Museum, Institute of Geological Science, Kiev No. 2127/11

Shepshedia palmata (holotype)
Occurrence: Rare
Locality: Ives Head, Shepshed, Charnwood Forest, Leicestershire, UK
Rock Unit: Lub Cloud Greywacke Member, Ives Head Formation, Blackbrook Group
Description: "The palmate impression measures 120 x 80 mm and consists of three main branches which dichotomise towards the margin yielding eleven terminations about 1 mm wide. There is a gap at the margin which suggests the former presence of a stem." Boynton & Ford, 1995
Reference: Boynton & Ford, 1995
Type Specimen: Holotype *in situ* on Ives Head, Shepshed, Charnwood

Rugoconites enigmaticus

Rugoconites tenuirugosus

Sekwia excentrica

Sekwia kaptarenkoe

Shepshedia palmata

Forest Leicestershire Museum, UK, SK477 170. Casts in Leicester University Geology Department, Collections Accession No. 115573 and in Leicestershire Museum Geology Collection, UK, Accession No. G34/1994
Classification: Phylum: ?Cnidaria;

Skinnera brooksi (holotype)
Occurrence: Common in the Northern Territory but uncommon in South Australia
Locality: Mt Skinner, Northern Territory, Australia; Flinders Ranges, South Australia
Rock Unit: Central Mount Stuart Formation; Ediacara Member, Rawnsley Quartzite, Wilpena Group
Description: Circular, disc-shaped body with three pouch-shaped depressions around the apex, surrounded by secondary depressions, which splay outward and join the margin
Reference: Wade, 1969
Type Specimen: South Australian Museum SAM F16474a
Classification: Phylum: Cnidaria; Class: Trilobozoa; Family: Uncertain but perhaps related to *Tribrachidium*

Solza margarita (holotype)
Occurrence: Not common
Locality: Solza River, White Sea, Russia
Rock Unit: Ust'-Pinega Formation, Valdai Group (Verkovka Formation, Grazhdankin, 2004)
Description: Body low, conical with flattened margin, egg-shaped. Dorsal surface with system of canals, which bifurcate towards periphery once or twice, anastomosing centrally, opening along body margin; where canals meet; medially body surface covered with a series of pores; at present function of such a body plan uncertain
Reference: Ivantsov, Malakhovskaya & Serezhnikova., 2004
Type Specimen: Paleontological Institute PIN 4853/60
Classification: Uncertain, but might be a "shell" of a form close to *Kimberella*

"Spriggia annulata" (holotype;
See *Aspidella terranovica*)
Occurrence: ?
Locality: Flinders Ranges, South Australia
Rock Unit: Ediacara Member, Rawnsley Quartzite, Wilpena Group
Other Names: *Aspidella terranovica*, *Madigania annulata*, (same type specimen as *Madigania annulata*); *Cyclomedusa davidi*
Reference: Sprigg, 1949; Jenkins, 1984; Sun, 1986; Gehling et al., 2000
Type Specimen: South Australian Museum SAM T30-2031

"Spriggia wadea" (holotype; See
Aspidella terranovica)
Occurrence: ?
Locality: Brachina Gorge, Flinders Ranges, South Australia; Wernecke Mountains, NW Canada
Rock Unit: Rawnsley Quartzite, Wilpena Group; Wendermere Supergroup
Other Names: *Aspidella terranovica*, *Cyclomedusa*?
Reference: Sun, 1986; Gehling et al., 2000
Type Specimen: South Australian Museum SAM P40325

"Spriggina" borealis (holotype)
Occurrence: Rare
Locality: Suz'ma, White Sea, Russia
Rock Unit: Ust'-Pinega Formation, Valdai Group (Verkovka Formation, Grazhdankin, 2004)
Other Names: Similar to *Spriggina* but probably not in this genus
Description: Horseshoe-shaped head region, its width slightly greater than that of segmented body in its broadest middle section. Largest segments are both broadest and longest. Shallow median groove present. Structure of head and type of segmentation similar to that seen in *Spriggina floundersi* but *S. borealis* differs in having many fewer segments, although it has relatively

Skinnera brooksi

Solza margarita (latex peel)

Spriggia annulata

Spriggia wadea

larger body. Shape of individual segments also differs. Length of head region is 5.5 mm, width of segmented part of body up to 20 mm; maximum length of each segment, 2.5 mm, width of mid-groove up to 1.5 mm; visible number of segments, 19
Reference: Keller & Fedonkin, 1977; Palij, Posti & Fedonkin, 1979
Type Specimen: Geological Institute GIN 4469/110 – specimen lost
Classification: Phylum: Arthropoda; Class: Paratrilobita; Family: Sprigginidae

Spriggina floundersi (SAM 29801)
Occurrence: Uncommon
Locality: Flinders Ranges, South Australia
Rock Unit: Ediacara Member, Rawnsley Quartzite, Wilpena Group
Other Names: Perhaps *Marywadea* Glaessner, 1976) and *Praecambridium* (Glaessner & Wade, 1966) are junior synonyms of *Spriggina* (Birket-Smith, 1981)
Description: A segmented, elongate form with head distinct from rest of body and composed of regular angular segmental elements that meet along midline groove. Head shield U-shaped, with tines that sweep back to first segmental ends represented by prominent bulge in centre of head that is slightly asymmetric, likely due to compaction of high point on cephalon (head)
Reference: Glaessner, 1958
Type Specimen: South Australian Museum SAM P18887
Classification: Phylum: Arthropoda; Family: Sprigginidae; Ivantsov (*in press*) suggests placement in Phylum: Proarticulata; Class: Cephalozoa; Family: Sprigginidae

"Spriggina ovata" (holotype; See Marywadea ovata)
Occurrence: Rare
Locality: Flinders Ranges, South Australia
Rock Unit: Ediacara Member, Rawnsley Quartzite, Wilpena Group
Other Names: *Marywadea ovata*
Description: Elongate, oval, segmented form with head distinct from the rest of body and composed of regular angular segmental elements that meet along midline groove. Angular segments and head shield resemble those of an arthropod. Head shield U-shaped, with segments (tines) that sweep back to first segmental ends represented by prominent bulge in center of head that is slightly asymmetric and tends to be broader than *S. floundersi*, its body being more similar in shape to that of *Marywadea ovata*
Reference: Glaessner & Wade, 1966
Type Specimen: South Australian Museum SAM P13754
Classification: Phylum: Arthropoda; Family: Sprigginidae. Although originally thought to be related to polychaete worms, with further study, *S. ovata* seems more similar to some kind of arthropod. Ivantsov (2004) suggested it might belong in the Phylum: Proarticulata

Staurinidia crucicula
Occurrence: Rare
Locality: Zimnii Bereg, White Sea, Russia
Rock Unit: Ust'-Pinega Formation, Valdai Group (Zimnii Gory Formation of Grazhdankin, 2004)
Description: Small discoid form with four radial canals, originating from small cavity in center of disc. Ends of canals towards periphery are swollen. Sometimes there are tentacle-like structures around the outside, but they are rarely preserved. These forms leave rather deep impressions in clays, suggesting that they had significant relief. Margins of disc very thin with no indication of encircling ridge or canal. Diameter of disc ranges from 6 to 10 mm, minimum width of radial canals up to 1 mm and maximum up to 3 mm; length of marginal tentacles 4 to 5 mm
Reference: Fedonkin, 1985
Type Specimen: Paleontololgical

Spriggina" borealis

Spriggina floundersi

1cm

Spriggina ovata

Staurinidia crucicula

Institute, Moscow, PIN 3993/392a
Classification: Phylum: Cnidaria; Class: Scyphozoa; Family: Uncertain

Suvorovella aldanica (holotype)
Occurrence: ?
Locality: Farm Aar, Aus area, Namibia, southern Africa; Maya River (Russia)
Rock Unit: Dabis Formation, Nama Group; Yudoma Formation
Description: Structure saucer-shaped, with diameter up to 30 mm; surface covered with small, raised rhomboidal shapes that lie in intersecting, curved rows
Reference: Vologdin & Maslov, 1960; Pflug, 1966; Sokolov, 1973; Glaessner, 1979; Runnegar, 1992
Type Specimen: Paleontological Institute, Moscow, PINAN SSSR 1766 (Obr.No. 35-36a)
Classification: Of uncertain affinities

Suzmites tenuis (holotype)
Occurrence: Rare
Locality: Suz'ma, White Sea, Russia
Rock Unit: Ust'-Pinega Formation, Valdai Group (Verkovka Formation, Grazhdankin, 2004)
Description: Succession of concentric ridges parallel to one another with size increasing from one end to the other, presumed to be from center to periphery; distance between paired ridges also increases from center to periphery; total size of structure quite small, ranging up to only 7 mm; central area contains series of small pustules; form is conical with angle at apex from 30 to 45 degrees
Reference: Fedonkin, 1981
Type Specimen: Geological Institute, Moscow, GIN 4464/181
Classification: Ichnofossil (passive), the imprint of the outside of a conical organism left in the substrate while the animal was alive. Conical structure was attached, and its movement left the "swing" mark

Suzmites volutatus (holotype)
Occurrence: Rare
Locality: Suz'ma, White Sea, Russia
Rock Unit: Ust'-Pinega Formation, Valdai Group (Verkovka Formation of Grazhdankin, 2004)
Other Names: Very likely a drag mark of *Pteridinium* (Fedonkin, 1981)
Description: Succession of parallel bands, bound by narrow ridges of equal width about 1 mm wide; no lateral margin of structure is apparent; width of the bands gradually decreasing from 6 mm on one side to 4 mm on the other; in cross section, perpendicular to bedding plane, imprints of *Pteridinium* are observed; the size of the bands in *Suzmites* corresponds to the size of the segments in *Pteridinium* – thus it may well be this organism is the source of this pattern
Reference: Fedonkin, 1981
Type Specimen: Geological Institute, Moscow, GIN 4310/184
Classification: Ichnofossil (passive) of *Pteridinium*

Swartpuntia germsi (Geological Survey Namibia specimen)
Occurrence: Uncommon
Locality: Namibia, White Sea, Russia
Rock Unit: Nama Group; Valdai Group
Description: Ovate, leaf-shaped frond with at least three petaloids or subdivisions that are attached longitudinally to transversely segmented central stalk. Each petaloid forms quilted sheet of 2-3 mm wide tubular segments. Length of tubular segments and their angle of branching from central stalk decreases distally. Bilaterally symmetric and quite large, with lengths reaching 180 mm and width 140 mm. Outer margin serrate
Reference: Narbonne *et al.*, 1997; Narbonne, 2005
Type Specimen: Geological Survey of Namibia GSN F238-H
Classification: Phylum: Petalonamae; Family: Charniidae; but this frond-like form may have been a colonial animal or even a large alga (Gehling, *pers. obs.*, 2005); Narbonne has allied with the dickinsoniomorphs, but this

Suvorovella aldanica

Suzmites tenuis

Suzmites volutatus

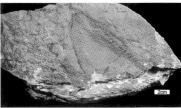

Swartpuntia germsi

questioned by Gehling *et al.* (2005) due to the very different nature of preservation

Tamga hamulifera (holotype)
Occurrence: ?
Locality: Zimnii Bereg, White Sea, Russia
Rock Unit: Ust'-Pinega Formation, Valdai Group (Yorga Formation of Grazhdankin, 2004)
Description: Small, egg-shaped organism; body has flattened periphery and convex upward central region; central region made up of seven, hook-like, posteriorly curved isomeres, left and right isomeres meet along midline in alternating manner (gliding reflection); size of isomere increases from narrow end of fossil to the wider
Reference: Ivantsov, *in press*
Type Specimen: Paleontological Institute, Moscow, PIN 3993/5508
Classification: Uncertain

"Tateana inflata" (SAMT24; 2018; See *Aspidella terranovica*)
Occurrence: ?
Locality: Flinders Ranges, South Australia
Rock Unit: Ediacara Member, Rawnsley Quartzite, Wilpena Group
Other Names: *Aspidella terranovica*
Reference: Sprigg, 1949; Jenkins, 1984; Sun, 1986; Gehling *et al.*, 2000
Type Specimen: South Australian Museum SAM T-11

Temnoxa molluscula (holotype)
Occurrence: Rare
Locality: Solza, White Sea, Russia
Rock Unit: Ust'-Pinega Formation, Valdai Group (Verkovka and Yorga formations of Grazhdankin, 2004)
Description: Body elongate oval, clearly divided into two parts, a wide crescent-shaped cephalic region and a narrow, oval, unsegmented body. Deep depression extends across cephalic region parallel to anterior margin. Body with nearly flat lateral margins, which could represent maximum extent of retractable foot. Organism strongly convex dorsally over more than half of body, but not extending to very end of trunk. Similarities of this form to *Parvancorina*, but differences lie in the fact that body of *Temnoxa* divided into two parts. In *Temnoxa* "cephalic" region lacks any dorsally convex structure, which is restricted to the "trunk"
Reference: Ivantsov, Malakhovskaya & Serezhnikova, 2004
Type Specimen: Paleontological Institute PIN 4852/104
Classification: Uncertain, though there are hints that it could be related to the Phylum: Mollusca or even Arthropoda

Thectardis avalonensis (field specimen)
Occurrence: Common
Locality: Avalon Peninsula, Newfoundland
Rock Unit: Drook Formation and Mistaken Point Formation, Conception Group
Description: A cm-scale imprint in the shape of a triangle, with prominent raised margin and featureless interior or exhibits faint transverse markings. Holotype elongate triangular-shaped fossil 90 mm long and 30 mm wide at triangular base, with prominent 5-7 mm-wide raised margin that tapers toward apex
Reference: Clapham, Narbonne, Gehling, Greentree & Anderson, 2004
Type Specimen: Royal Ontario Museum, Toronto, ROM 38632
Classification: Uncertain

Tirasiana cocarda (holotype)
Occurrence: Rare
Locality: Perm, Shirokovskoe Reservoir, Nyar Village, Ural Mountains, Russia
Rock Unit: Cherny Kamen Formation, Sylvitsa Series
Other Names: Would be worth comparing to *Aspidella terranovica*
Description: Flattened form with concentric ridges and tubercle in center. Overall shape an ellipsoid with

Tamga hamulifera

Tateana inflata

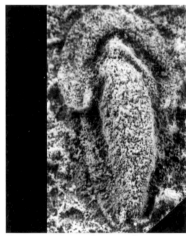
Temnoxa molluscula (latex peel)

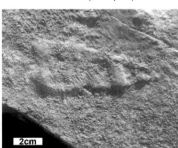
Thectardis avalonensis

Tirasiana cocarda

five annular ridges. First two outer ridges quite distinct but not over entire circumference. Third ridge from outside exhibits greatest relief. Distinct from other species in having stronger relief – more convex, and more intricate ornamentation, five rings and five longitudinal ridges. Differs from most species of *Cyclomedusa* in having hardly any radial ribs; smaller than *C. annulata*
Reference: Becker, 1985
Type Specimen: Central Geological Museum, St Petersburg, CGM 1/11406 and 2/1972
Classification: Phylum: ?Cnidaria or attachment disc for frond-like organism; Phylum: Petalonamae

"Tirasiana concentralis" (CGM 232/1972; See *Aspidella terranovica*)
Occurrence: ?
Locality: Perm district, Shirokovskoe Reservoir, Nyar Village, Ural Mountains, Russia
Rock Unit: Cherny Kamen Formation, Sylvitsa Series
Other Names: *Aspidella terranovica*, synonymy (Serezhnikova, *pers. com.*
Description: Rounded organism preserved on bedding surface with concentric ornamentation and tubercle at center. Four annular ridges with decreasing diameter towards center and meet central tubercle. Impression appears to be flattened cone. Moving from middle to outside, second and fourth ridges have most relief. Diameter ranges from 20 to 50 mm
Reference: Becker, 1985; Gehling *et al.*, 2000
Type Specimen: Central Geological Museum, St Petersburg, CGM 1/11406, sample 232/72

"Tirasiana coniformis" (holotype); See *Aspidella terranovica*)
Occurrence: Relatively common
Locality: Moldavia; Zimnii Coast, White Sea, Russia
Rock Unit: Bernashev Beds, Yarychnev Formation; Ust'-Pinega Formation, Valdai Group
Other Names: *Aspidella terranovica*, but other researchers not in agreement with this synonymy (Serezhnikova, *pers. com.*)
Description: Cast of sedentary organism preserved in positive hyporelief, which has conical shape with concentric, step-wise construction of surface; with rounded knob in middle, appears constructed of three discs of decreasing diameter – smallest with central knob; largest disc diameter, 35 mm; second, 25 mm; smallest, 20 mm maximum; central knob 6 mm, vertical height up to 11 mm
Reference: Palij, 1976; Gehling *et al.*, 2000
Type Specimen: Kiev State University KGU 1831

"Tirasiana disciformis" (holotype; See *Aspidella terranovica*)
Occurrence: Relatively common, but rare in the Urals
Locality: Mogilev-Podolski region near villages of Ataki and Serebiya, Dniester River, Moldavia; Zimnii Bereg, White Sea, Russia; Olenek Uplift, Siberia; Shirokovskoe Reservoir, Nyar Village, central Urals, Russia
Rock Unit: Bernashevka Beds, Yaryshev Formation; Ust'-Pinega Formation; Khatyspyt Formation; Bernashev Beds, Mogilev Formation; Cherny Kamen Formation, Sylvitsa Series
Other Names: *Aspidella terranovica*; some forms assigned to this taxon may be *Nemiana simplex*
Description: Impressions convex to conical with series of concentric rings forming surface ornamentation – shelves and ridges. Distinct tubercle present in center of disc. Sometimes individual specimens found and at other times they occur in concentrations. Fossils medium-sized (10 to 27 mm in diameter), discs preserved on lower part of bedding planes. May well be a holdfast of an organism such as *Charnia* or some other similar form. Differs from

Tirasiana concentralis

Tirasiana coniformis

Tirasiana disciformis

other species in genus in its simple structure. Unlike *Ediacaria flindersi*, it is characterized by the total absence of radial ribbing and gigantic forms
Reference: Palij, 1976; Fedonkin, 1981
Type Specimen: Kiev State University, KGU No. 1826

Tribrachidium heraldicum
(holotype)
Occurrence: Moderately common
Locality: Flinders Ranges, South Australia; Dniester River Basin, Podolia, Ukraine; Solza and Suz'ma rivers, White Sea of Russia (Zimnii Bereg) and Suz'ma)
Rock Unit: Ediacara Member, Rawnsley Quartzite, Wilpena Group; Mogilev Formation; Ust'-Pinega Formation, Valdai Group (Verkhovka, Zimnii Gory and Yorga formations of Grazhdankin, 2004)
Description: Disc with three bent, fringed arms. Small specimens and those that have not been compressed almost conical in shape. Usually preserved with upper surface of fossil covered with three sets of anastomosing ribs or tubes originating on inside arc of bent arms within body. Three sickle-shaped areas most resilient organs. Arms originate in triple junction in center of disc, and extend out radially before elbow bend where they eventually lie parallel to raised edge of disc. Arms taper and have sharp top edges, each arm has boss or bulla attached at "elbow. " Rarer specimens show sets of long, equal length bristles that form straight or sweeping curves from apex of disc and arms to outer margin. In type specimen, fine bifurcating ridges apparent on outer edges of arms due to the coarseness of preserving sand. No distinct opening or mouth near center of organ is discernable. Size of *Tribrachidium* ranges from diameter of 0.3 to 5 cm
Reference: Glaessner, 1959; Glaessner & Wade, 1966
Type Specimen: South Australian Museum SAM P12898
Classification: The unusual design of this animal has no analogue in modern groups. The true relationship of *Tribrachidium* is still unknown, but it has been compared by some to edrioasteroids, which have a three-fold symmetry as well as lophophorates, and to sponges and even to some of the small triradiate cones (*e.g. Anabarites*) that are abundant in the Cambrian, a part of the Small Shelly Fauna (SSF) (Narbonne, 2005)

Triforillonia costellae *(holotype)*
Occurrence: Uncommon
Locality: Avalon Peninsula, Newfoundland
Rock Unit: Fermuese Formation, St John's Group
Description: Preserved as a shallow cast, in positive relief on the base of a sedimentary bed. Three lobed organism with central rosette and marginal rim. Radial ribs within each lobe may be caused by compaction of the sediments. Preservation as flat cast suggest that *Triforillonia* was less resilient than other triradial forms such as *Rugoconites*, *Albumares* and *Anfesta*
Reference: Gehling *et al.*, 2000
Type Specimen: Geological Survey of Canada GSC116860
Classification: Triradial disc or cone of unknown affinity

Vaizitsinia sophia *(holotype)*
Occurrence: Rare (only two specimens)
Locality: Vaizitsa Borehole, Zimnii Bereg, White Sea, Russia
Rock Unit: Ust'-Pinega Formation, Valdai Group (Verkovka Formation, Grazhdankin 2004)
Description: Small, pinnate colonies with some similarity in gross form to modern sea pens (Pennatularia), to which such forms as *Charniodiscus* have been related by some researchers, though not all would agree. *Vaizitsinia* has leaf-shaped colony with elongate, egg-shaped basal part, which could have served as an attachment in mud. Shape of this differs significantly from

Tribrachidium heraldicum

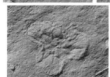

Triforillonia costellae

Vaizitsinia sophia

that of *Charniodiscus*. From median part of organism there are secondary branches which are longest in central part of organism, and then size decreases towards what was probably top. Distal ends of branches rounded. Outside edge of specimen distinct and preserved as broad, flat ridge. Observable length of organism, 70 mm; maximum width, up to 37 mm; greatest thickness of lateral processes up to 3.5 mm; length of lateral processes up to 15 mm; length of midline axis of colony 18 mm
Reference: Sokolov & Fedonkin, 1984; Fedonkin, 1985
Type Specimen: Paleontological Institute, Moscow, PIN 3993/1305
Classification: Phylum: Petalonamae

Valdainia plumosa *(holotype)*
Occurrence: Rare
Locality: Novodnestrovskaya Hydroelectric Power Station, Dniester River, Ukraine
Rock Unit: Mogilev Formation
Other Names: Similarities to *Pteridinium simplex* and others in this genus, *Podolimirus mirus, Inkrylovia lata*
Description: Bipinnate, segmented organism with a short and broad body. Midline forms zigzag pattern. Segments of organism smooth and gently curved, distinctly alternating where they meet. Segments divided by pair of grooves that have long, narrow ridge in between. Narrowing of segments occurs from proximal to distal. Largest segments form nearly right angle with main axis but near area of growth, smaller segments reduces and this angle to between 30 to 40 degrees. Preserved, but not entire length, reaches 70 mm; maximum width 82 mm; width of segments range from 4 to 12 mm; width of intersegmental ridges range from 1-3.5 mm
Reference: Fedonkin, 1983; Velikanov, Aseeva & Fedonkin, 1983
Type Specimen: Paleontological Institute, Moscow, PIN 3994/276
Classification: Phylum Uncertain, but in the past has been placed in the Phylum Proarticulata based on its "symmetry of gliding reflection" where segments meet along the midline in a zigzag fashion. Class: Vendomorpha; Family: Vendomiidae but in past classifications was placed in the Petalonamae; see discussion for *Dickinsonia costata*

Vaveliksla vana *(holotype)*
Occurrence: Rare in Russia; uncommon in South Australia
Locality: Zimnii Bereg, White Sea, Russia (near Zlmnegorsk Lighthouse); Chace Range, Flinders Ranges, South Australia
Rock Unit: Ust'-Pinega Formation, Valdai Group (Yorga Formation, Grazhdankin 2004); Ediacara Member, Rawnsley Quartzite, Wilpena Group
Preservation Style: Fossils were found on the basal surface of thin layers of fine-grained sandstones, represented by positive mold
Description: Small, sac-like organisms with radial symmetry; top and bottom of organism with different morphologies. Body divided into two different parts, one being rather convex and massive attachment disc weakly connected to capsule-like body. Capsular body sac elongated to along central axis, and incised by a number of longitudinal, subparallel furrows, prominent in middle of body and ends. Small, star-like depressions present over surface of body. Disc ranges from 7 to 15 mm in diameter. Body walls relatively thin and likely perforated, as indicated by presence of numerous small pits. Based on mode of preservation, organism appears to be attached, sessile form. Differs from *V. velikanovi* in its proportions and smaller diameter, as well as having more convex surface of attachment disc
Reference: Ivantsov *et al.*, 2004
Type Specimen: Paleontological Institute, Moscow, PIN 3993/5217-1
Classification: Phylum: Possibly Porifera

Valdainia plumosa

Vaveliksia vana

Vaveliksia velikanovi (holotype)

Occurrence: Not common
Locality: Novodnestrovskaya Hydroelectric Power Station, Dniester River and near Ozarintsy Village, Podilia, Ukraine
Rock Unit: Lomozov Beds, Mogilev Formation (near power station) and Bernashevka Beds, Yaryshev Formation (Ozarintsy Village)
Description: Sac-like organism joined to a disc by short stalk. End opposite disc bears crown of short tentacle-like structures surrounding what may be mouth or oral aperture. Surface of sac smooth in parts or hosting some tubercles in other parts, possibly dependent on whether inner or outer surface of sacs is impressed in sediment. Length from 30 to 80 mm; maximum width of sacs, 35 mm; diameter of discoid attachment from 8 to 20 mm
Reference: Fedonkin, 1983
Type Specimen: Paleontological Institute, Moscow, PIN 3994/581
Classification: Uncertain, but possibly poriferan level of complexity

Velancorina martina (holotype)

Occurrence: Rare
Locality: Farm Aar and Schakalskuppe, Namibia
Rock Unit: ?Kuibis Formation, Nama Group
Description: Body roughly oval or elliptical and bipolar, with distinct furrow down long axis of oval. Furrows present that lead laterally from central groove, indicating presence of serial pneu structure. Fossil not well preserved. Fossil described by Pflug, about 24 mm long and has different structures at either end (Pickford, 1995)
Reference: Pflug, 1966
Type Specimen: Geological Survey of Namibia GSN F321
Classification: Phylum: Petalonamae

Vendella haelenicae (holotype)

Occurrence: Rare
Locality: Mogilev-Podolski, Derlo River, a tributary of the Dniester River, Ukraine
Rock Unit: Yaryshev Formation
Other Names: Possibly Nemiana
Description: Discoidal forms with high relief preserved on soles of beds as flattened casts with small central depression; diameter of casts ranges from 12 to 25 mm
Reference: Gureev, 1987
Type Specimen: Geological Museum, Institute of Geological Sciences, Kiev No. 2127/49
Classification: Class: Cyclozoa; Family: Vendellidae

"Vendella larini" (holotype); See Aspidella terranovica)

Occurrence: Rare, only three specimens known
Locality: Sokolets, Ushitsa River, a tributary of the Dniester River, Ukraine
Rock Unit: Kanilov Formation
Other Nmes: Aspidella terranovica
Description: Discoidal form preserved in positive hyporelief, with concentric ridges, which encircle central depression; surface of ridges bears regularly spaced transverse furrows; diameter of disc varies from 4 to 8 mm. and width of ridge is 1 to 2 mm
Reference: Gureev, 1987
Type Specimen: Geological Museum, Institute of Geological Sciences, Kiev No. 2089/27

Vendella sokolovi (from plate 38, Gureev, 1985)

Occurrence: Not common
Locality: Ladova and Nemia rivers, tributaries of the Dniester River, Ukraine
Rock Unit: Mogilev Formation
Other Names: Originally described as Medusinites sokolovi, but renamed as a new genus
Description: Middle-sized casts preserved on sole of beds in positive hyporelief, so cast resembles non-sculptured, broad ring (doughnut-shaped) with flattened central area; diameter of whole organism 15 to 35 mm; diameter of central depression up

Vaveliksia velikanovi

Velancorina martina

Vendella haelenicae

Vendella larini

to 5 mm with ratio between diameter of whole width and width of central depression 5/1
Reference: Gureev, 1987
Type Specimen: Geological Museum, Institute of Geological Sciences, Kiev IGN ANNo.2069/27
Classification: Phylum Uncertain, but in the past has been placed in the Phylum Proarticulata based on its "symmetry of gliding reflection" where segments meet along the midline in a zigzag fashion. Class: Vendiamorpha; see discussion for *Dickinsonia costata*

"Vendia" janae (holotype; same specimen as Paravendia janae; See Paravendia janae)

Occurrence: Relatively rare
Locality: Zimnii Bereg, Winter Coast, Russia
Rock Unit: Ust'-Pinega Formation, Valdai Group (Yorga Formation, Grazhdankin, 2004)
Other Names: See *Paravendia janae*
Reference: Ivantsov, 2001; Ivantsov, 2004 (named *Paravendia*)
Type Specimen: Same specimen as *Paravendia janae*, PIN 3993/5070

Vendia rachiata (holotype)

Occurrence: Relatively rare
Locality: Solza, White Sea, Russia
Rock Unit: Ust'-Pinega Formation, Valdai Group (Verkovka Formation of Grazhdankin, 2004)
Description: Organism small and has elongated oval-shaped body that is completely segmented; segments arranged alternately in two rows and have rounded edges; flexed posteriorly, with each row containing five or fewer segments. Larger segments cover smaller ones externally and partially overlap them, but posterior ends of all segments remain free. Size of segments decreases posteriorly. What appears to be digestive-distributive system consists of a simple axial tube and short non-branching lateral extensions located along boundaries between segments. Except for first segment pair, all have lateral appendages; number of isomere pairs is five or less; isomeres have rounded ends; lateral outgrowth of digestive system short
Reference: Ivantsov, 2004
Type Specimen: Paleontological Institute, Moscow, PIN 4853/63
Classification: Phylum uncertain, but in the past has been placed in the Phylum Proarticulata based on its "symmetry of gliding reflection" where segments meet along the midline in a zigzag fashion. Class: Vendomorpha; Family: Vendomiidae; see discussion for *Dickinsonia costata*

Vendia sokolovi (holotype)

Occurrence: One specimen only
Locality: Yarensk Borehole at depth 1552 m, Russia
Rock Unit: Redkino Horizon, Ust'-Pinega Formation, Valdai Group
Description: Small bilateral animal with proarticulatan body plan; body slender oval shape with smooth margin; body subdivided into isomeres flexed posteriorly; tubular distributive system (perhaps digestive) consists of medial channel with relatively long lateral outgrowths, one per each isomere; seven pairs of isomeres present that seem to taper distally
Reference: Keller, 1969
Type Specimen: Paleontological Institute, Moscow, PIN3593/1
Classification: Phylum uncertain, but in the past has been placed in the Phylum Proarticulata based on its "symmetry of gliding reflection" where segments meet along the midline in a zigzag fashion. Class: Vendomorpha; Family: Vendomiidae; the relationships of this form are most unclear. Some have suggested it is similar to trilobites. Some palaeontologists have suggested it was a medusoid form and others that it was some primitive coelenterate

"Vendia janae" (latex peel)

Vendia rachiata

Vendia sokolovi (latex peell)

Vendella sokolovi

290

Vendoconularia triradiata
(holotype)
Occurrence: Extremely rare
Locality: Yarnema, Onega River, White Sea, Russia
Rock Unit: Ust'-Pinega Formation, Valdai Group (Verkovka Formation of Grazhdankin, 2004)
Description: Conical form with six faces and very characteristic pattern of rod arrangement demonstrating two series of major and secondary rods, both having short spines directed towards aperture
Reference: Ivantsov & Fedonkin, 2002
Type Specimen: Paleontological Institute, Moscow, PIN 4564/1025
Classification: Phylum: Cnidaria

Vendoconularia triradiata

Vendomia menneri *(holotype)*
Occurrence: Extremely rare
Locality: Suz'ma, White Sea, Russia
Rock Unit: Ust'-Pinega Formation, Valdai Group (Verkovka Formation, Grazhdankin 2004)
Other Names: *Dickinsonia* sp.
Description: Organism with small ovoid body of which more than one-third occupied by a broad, semicircular "head" region with two symmetrical depressions, which could be the remains of eyes? Median ridge originates in this area; paired, elongate segments diverge from this median ridge and become smaller towards narrower end of body. Length, 4 mm; width, 3 mm; width of median ridge, 0.2 mm; number of segments, six
Reference: Keller & Fedonkin, 1977; Ivantsov, *in press*
Type Specimen: Geological Institute GIN 4464/57; Paleontological Institute, Moscow, PIN3992/57
Classification: Uncertain

Vendomia menneri

Ventogyrus chistyakovi *(PIN 4564/1009)*
Occurrence: Moderate abundance at type locality
Locality: Yarnema, Onega River, White Sea, Russia
Rock Unit: Ust'-Pinega Formation, Valdai Group (Verkovka Formation of Grazhdankin, 2004). Some researchers have suggested it belongs to the uppermost part of the Vendian complex (Erga or Padun Formation)
Other Names: Cf *Arborea* sp. (Christyakov *et al.*, 1984); not all researchers agree
Description: Internal sand casts, usually with egg-shape, bipolar form; three-fold, radially situated modules, each with a large basal chamber and two meridional rows of smaller chambers decreasing in size towards apical end. Three longitudinal and numerous transverse soft septa subdivide internal space into chambers; each small chamber with some triangular minor septa; on each side of each major septa are complex, identical systems of branching channels coursing along the axial rod of this structure with alternating lateral branches, which bifurcate up to five times before they reach outer surface; internal space of each module subdivided into two rows of chambers by longitudinal septum and transverse septa originating from the latter; modules have a single large, non-paired chamber at wide end; internal sides of longitudinal septa join edges of axial pyramid; largest septa of the first order extends from axial pyramid to outer side of module, defines chambers of first order alternating relative to each other by about half of their length; transverse septa of second order are about four times shorter than first order septa, and so only partially subdivide space of first order chamber; septae of the third and fourth orders are even shorter and divide space of corresponding chambers of the second and next smaller chambers; each face of axial pyramid and both surfaces of longitudinal septum bear imprints of regularly branching channels of distribution system, which consists of the major branches; each branch has axial channel coursing along face of axial pyramid, with lateral channels spreading over surfaces of two adjacent

Ventogyrus chistyakovi

longitudinal septa; each lateral channel bifurcates a few times towards the periphery; inside each chamber of first order in any one module, only one lateral channel with all its branches present; ends of channels intersect outer surface of module, its longitudinal septum and transverse septa so that every transverse septum reflects a specific branch end of the channel system
Reference: Ivantsov & Grazhdankin, 1997; Fedonkin & Ivantsov, *in press*
Type Specimen: Paleontological Institute, Moscow, PIN 4564/25
Classification: Phylum: Petalonamae; Family: ?Pteridiniidae or perhaps a cnidarian

Veprina undosa *(holotype)*
Occurrence: Rare, only one well-preserved specimen and a few fragments
Locality: Zimnii Bereg, White Sea, Russia
Rock Unit: Ust'-Pinega Formation, Valdai Group (Verkovka Formation of Grazhdankin, 2004)
Description: Large, umbrella-like form clearly divided into two broad zones, separated by depression. Outer zone has sharp relief and regular radial ribs, separated by relatively narrow furrows. Inner zone, similar in width to outer zone, has less regular ribbing and extends around an elongated depression in center of this form. Elongated central depression imparts bilateral character to body. Diameter approximately 60 mm; width of outer ribbed zone 10-15 mm; width of the inner ribbed zone 6-11 mm; width of central depression, 2-4 mm; width of the ribs in outer zone 1.5-3 mm; width of the fine ridges, 0.3-0.5 mm and length up to 15 mm. As noted in *Treatise of Invertebrate Paleontology*, *Veprina* has some similarities to *Peytoia* from the mid-Cambrian Burgess Shale Fauna of Canada
Reference: Fedonkin, 1980
Type Specimen: Geological Institute, Moscow, GIN 4482/29
Classification: Phylum: Cnidaria; Class: Cyclozoa; Family: Uncertain

Veprina undosa

Vladimissa missarzhevskii *(holotype)*
Occurrence: Rare
Locality: Zimnii Bereg, White Sea, Russia (5 km north of Zimnegorsky Lighthouse)
Rock Unit: Ust'-Pinega Formation, Valdai Group (Yorga Formation of Grazhdankin, 2004)
Description: Flat, leaf-shaped organism broad in middle with acutely pointed anterior and posterior ends. Margin finely scalloped, and two rows of small, longitudinal tubercles present. Body seems to have considerable thickness (several mm). Body surface smooth. Along mid-body are two rows of similar sized, rounded tubercles, closely spaced. Each row oriented longitudinally and contains four tubercles. No mouth discerned. Length about 45 mm, width, 32 mm; width of tubercles, 1.5 mm with maximum spacing of 2 mm
Reference: Fedonkin, 1985
Type Specimen: Paleontological Institute, Moscow, PIN 3993/204
Classification: Thought to be related to worms in the Platyhelminthes, similar in form to living rhabdocoelan turbellarians

Wadea gen. nov.
(See Rugoconites tenuirugosus)

Wigwamiella enigmatica *(holotype)*
Occurrence: Rare
Locality: Mount Scott Range, South Australia
Rock Unit: Ediacara Member, Rawnsley Quartzite, Wilpena Group
Other Names: Possibly *Pseudorhyzostomites*, an escape structure
Description: Circular to oval in shape, with defined outer rim and furrows

Vladimissa missarzhevskii

Wigwamiella enigmatica

radiating from a center. Furrows irregularly branch and thin towards rim where series of smaller, finer furrows tend to parallel main ones. Central part often strongly depressed (negative hyporelief)
Reference: Runnegar, 1991
Type Specimen: South Australian Museum SAM P27978
Classification: Phylum: Cnidaria or Unknown – could be pull away structure when attachment bulb dislodged from the sea bottom

Windermeria aitkeni *(holotype)*
Occurrence: Rare, single specimen
Locality: Sekwi Brook North, Mackenzie Mountains, Northwestern Canada
Rock Unit: Upper Blueflower Formation, Windermere Group
Description: Small (16.4 mm x 7.9 mm), bipolar, parallel-sided, segmented ovoid with semicircular terminations. Eight nearly equal-sized segments arranged subtransverse to medial furrow in opposite arrangement *(i.e.* symmetry of "gliding reflection" – that is, segments on one side of the midline do not match up, they are offset)
Reference: Narbonne, 1994
Type Specimen: Geological Survey of Canada, GSC 102374
Classification: Primitive bilaterian; Family Dickinsonidae

Yarnemia acidiformis *(holotype)*
Occurrence: ?
Locality: Yarnema, Onega Peninsula, White Sea, Russia
Rock Unit: Ust'-Pinega Formation, Valdai Group (Verkovka Formation of Grazhdankin, 2004)
Description: Organism with sac-like, morphologically changeable shape. Integument quite thick. May have two apertures that open to environment. Preserved in much the same fashion as *Ventogyrus*
Reference: Chistyakov *et al.*, 1984
Type Specimen: Paleontological Institute, Moscow, PIN 4564/2 (1/12174) (possibly lost)
Classification: Nessov suggested a relationship to the ascidians, or tunicates, and thus assignment to the Phylum: Chordata. Illustrations can be accessed at: www.vend.paleo.ru/pub/Chistyakov_et_al_1984.pdf

Yelovichnus gracilis *(holotype)*
Occurrence: ?
Locality: Zimnii Bereg, White Sea, Russia
Rock Unit: Ust'-Pinega Formation, Valdai Group (Verkovka Formation of Grazhdankin, 2004)
Description: Fossil in positive hyporelief, forming broad meandering curves, individual traces not regular but lay in dense concentrations, contacting each other
Reference: Fedonkin, 1985; Narbonne, 2005, Droser *et al.*, 2005
Type Specimen: Paleontological Institute, Moscow, PIN 3993/1309.
Classification: Possibly a colonial alga or protist, though previously described as a trace fossil

Yorgia waggoneri *(holotype)*
Occurrence: Relatively common in Russia; rare in South Australia
Locality: Zimnii Bereg, White Sea, Russia; Flinders Ranges, South Australia
Rock Unit: Ust'-Pinega Formation, Valdai Group (Verkhovka, Zimnii Gory and Yorga formations of Grazhdankin 2004); Ediacara Member, Rawnsley ,Quartzite, Wilpena Group
Description: Large bilateral animal with proarticulate body plan; shape of body round in top view; head area unsegmented – well expressed in all stages of development; isomeres wide, elongate, with ends that taper distally, flexed posteriorly; isomeres number 40 or less; head area contains fine branching channels; probable digestive distribution system consists of an axial channel with one outgrowth per isomere; resting or feeding tracks of *Yorgia* reflects the ventral (underside)

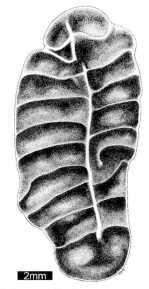
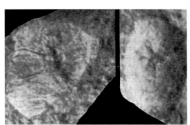
Windermeria aitkeni

Yarnemia acidiformis

Yelovichnus gracilis

Yorgia waggoneri

structure of the isomeres but axial lobe not visible
Reference: Ivantsov, 1999
Type Specimen: Paleontological Institute, Moscow, PIN 3993/5024
Classification: Bilaterian but Phylum Uncertain; in the past has been placed in the Phylum Proarticulata based on its "symmetry of gliding reflection" where segments meet along the midline in a zigzag fashion. Class: Cephalozoa; Family: Yorgiidae

Zolotytsia biserialis *(holotype)*
Occurrence: Rare
Locality: Zimnii Bereg, White Sea, Russia; Dniester River Basin, Podolia, Ukraine; Nainital district, Uttar Pradesh, India
Rock Unit: Ust'-Pinega Formation, Valdai Group (Verkovka Formation of Grazhdankin, 2004); Bernashevka Beds, Mogilev Formation; Krol Formation
Description: Two distinct rows of oval shaped bodies on either side of mid-line groove. Each row divided by deep groove, narrower than median groove. Bodies wide at one end and narrower at other. Oval bodies not symmetric, possibly due to postmortem deformation
Reference: Fedonkin, 1981
Type Specimen: Paleontological Institute, Moscow, PIN 3993/343
Classification: Phylum: Petalonamae; Family: Uncertain

Zolotytsia biserialis

BIBLIOGRAPHY

Aceñolaza, F. G. and Durand, F. R., 1984. The trace fossil *Oldhamia*. Its interpretation and occurrence in the Lower Cambrian of Argentina. *Neues Jahrbuch Geologie und Palaeontologie* 12: 728-740.

Aceñolaza, F. G. and Durand, F. R., 1986. Upper Precambrian-Lower Cambrian biota from the northwest of Argentina. *Geological Magazine* 123 (4): 367-375.

Aceñolaza, G. F., 2004. Precambrian-Cambrian ichnofossils, an enigmatic "annelid tube" and microbial activity in the Puncoviscana Formation (La Higuera, Tucuman Province, NW Argentina). *Geobios* 37: 127-133.

Aceñolaza, G., Fedonkin, M., Aceñolaza, F. and Vickers-Rich, P., 2005. The Ediacaran/Early Cambrian transition in northwest Argentina: New paleontological evidence along the proto-margin of Gondwana. 2nd International Symposium on Neoproterozoic-Early Paleozoic Events in Southwestern Gondwana. Extended Abstracts. *Geological Survey of Namibia*, Windhoek: 2-4.

Aceñolaza, G. F. and Tortello, M. F., 2003. El Alisal: A new locality with trace fossils of the Puncoviscana Formation (late Precambrian-early Cambrian) in Salta Province, Argentina. *Geologica Acta* 1 (1): 95-102.

Adams, C. and Shapiro, J., 2001. The shape of the universe: Ten possibilities. *American Scientist* 89: 443-453.

Adams, F., 2002. *Origins of Existence. How Life Emerged in the Universe.* London: The Free Press.

Ainsworth, C., 2003. The facts of life. *New Scientist*, 31 May: 28-31.

Aitken, J. D., 1987. First appearance of trace fossils and small shelly fossils in Mackenzie Mountains, northwest Canada, in relation to the highest glacial deposits. Abstract. *New York State Museum Bulletin* 463: 8.

Aitken, J. D., 1988. Giant "algal" reefs, Middle/Upper Proterozoic Little Dal Group (> 780, < 1200 Ma), Mackenzie Mountains, N.W.T., Canada. In: *Reefs, Canada and Adjacent Areas.* Edited by H. H. J. Geldsetzer, N. P. James and G. E. Tebbutt. *Canadian Society of Peroleum Geogists, Memoir* 13: 13–23.

Aitken, J. D., 1989. A "sea level" curve for the terminal Proterozoic, Mackenzie Mountains, N.W.T. – progress report. *Program with Abstracts, Geological Association of Canada* 14: A99.

Aitken, J. D., 1989. Uppermost Proterozoic formations in Central Mackenzie Mountains, Northwest Territories. *Geological Survey of Canada Bulletin* 368: 1-26.

Aitken, J. D. and Narbonne, G. M., 1989. Two occurrences of Precambrian thrombolites from the Mackenzie Mountains, northwestern Canada. *Palaios* 4: 384-388.

Albarede, F., 2003. *Geochemistry. An Introduction.* Cambridge: Cambridge University Press.

Allen, P. A. and Hoffman, P. F., 2005. Extreme winds and waves in the aftermath of a Neoproterozoic glaciation. *Nature* 433: 123-127.

Alpher, R. A., Bethe, H. and Gamow, G., 1948. The origin of chemical elements. *Physics Review* 73: 803.

Altermann, W. and Corcoran, P. L., eds., 2002. Precambrian Sedimentary Environments. A Modern Approach to Ancient Depositional Systems. *Special Publication of the International Association of Sedimentologists*, 33. Oxford: Blackwell Science.

Alves, R., Chaleil, R. A. G. and Sternberg, M. J. E., 2002. Evolution of enzymes in metabolism: A network perspective. *Journal of Molecular Biology* 320 (4): 751-770.

Amelin, Y., 2005. A tale of early Earth told in zircons. *Science* 310: 1914-1915.

Amthor, J. E., Grotzinger, J. P., Schröder, S., Bowring, S. A., Ramezani, J., Martin, M. W. and Matter, A., 2003. Extinction of *Cloudina* and *Namacalathus* at the Precambrian-Cambrian boundary in Oman. *Geology* 31: 431-434.

Anbar, A. D. and Knoll, A. H., 2002. Proterozoic ocean chemistry and evolution: A bioinorganic bridge. *Science* 297: 1137-1142.

Anderson, M. M., 1976. Fossil Metazoa of the Late-Precambrian Avalon Fauna, southeastern Newfoundland. *Geological Society of America*, Abstracts with Programs 8: 754.

Anderson, M. M., 1978. Ediacaran fauna. In: *Yearbook of Science and Technology.* Edited by D. N. Lapedes. New York: McGraw-Hill Book Co.: 146-149.

Anderson, M. M. and Conway Morris, S., 1982. A review, with descriptions of four unusual forms, of the soft-bodied fauna of the Conception and St. John's Groups (late Precambrian), Avalon Peninsula, Newfoundland. *Proceedings of the 3rd Northern American Paleontological Convention* 1: 1-8.

Anderson, M. M. and Misra, S. B., 1968. Fossil found in the Precambrian Conception Group in Southeastern Newfoundland. *Nature* 220: 680-681.

Anderson, M. M. and Misra, S. B., 1969. Criteria for recognizing Pre-Cambrian fossils (Reply to R. Goldring). *Nature* 223: 1076.

Antcliffe, J. B. and Brasier, M. D., 2007. *Charnia* and sea pens are poles apart. *Journal of the Geological Society, London* 164: 49-51.

Arthur, W., 1997. *The Origin of Animal Body Plans.* Cambridge: Cambridge University Press.

Asimov, I., 1965. *Life and Energy.* New York: Bantam Pathfinder Books.

Aseeva, E. A., 1983. Stratigraphic significance of the Late Precambrian microfossils of the south-western part of the East European Platform. In: *Stratigraphy and Formations of the Part of the Precambrian of the Ukraine.* Edited by V.A. Ryabenko. Kiev: Naukova Dumka: 148-156. (in Russian)

Atkins, P., 1992. *Creation Revisited. The Origin of Space, Time and the Universe.* London: Penguin.

Atreya, S. K., 2007. The my7stery of methane on Mars and Titan. *Scientific American* 296 (5): 24-33.

Aubay, S., 2000. *Origins of Life on the Earth and in the Cosmos.* 2nd Edition. San Diego: Academic Press.

Awramik, S. M. and Buchheim, H. P., 2001. Late Archaean lacustrine carbonates, stromatolites, and transgressions. In: *International Archaean Symposium Extended Abstracts.* Edited by K.F. Cassidy, J. M. Dunphy and M. J. Van Kranendonk. *AGSO – Geoscience Australia Record* 2001/37: 222-223.

Ayala, F. J., 1997. Vagaries of the molecular clock. *Proceedings of the National Academy of Sciences* 94: 7776-7783.

Ayala, F. J., Rzhetsky, A. and Ayala, F. J., 1998. Origin of the metazoan phyla: Molecular clocks confirm paleontological estimates. *Proceedings of the National Academy of Sciences of the United States of America* 95: 606-611.

Babcock, L. E., ed., 2005. Interpretation of Biological and Environmental Changes Across the Neoproterozoic-Cambrian Boundary. *Palaeogeography, Palaeoclimatology, Palaeoecology*, 220, Issues 1-2: 1-225.

Babcock, L. E., Grunow, A. M., Sadowski, G. R. and Leslie, S. A., 2005. *Corumbella*, an Ediacaran-grade organism from the Late Neoproterozoic of Brazil. *Palaeogeography, Palaeoclimatology, Palaeoecology* 220: 7-18.

Bada, J. L. and Lazcano, A., 2002. Some like it hot, but not the first biomolecules. *Science*, 296 (5575): 1982-1983.

Bailey, J. V., Joye, S. B., Kalanetra, K. M., Flood, B. E. and Corsetti, F. A., 2007. Evidence of giant sulphur bacteria in Neoproterozoic phosphorites. Nature, 445: 198-201

Baldauf, S. L., 2003. Phylogeny for the faint of heart: A tutorial. *Trends Genet.* 19 (6): 345-351.

Baldauf, S. L., 2003. The deep roots of eukaryotes. *Science* 300 (5626): 1703-1706.

Baldauf, S. L., Roger, A. J., Wenk-Siefert, I., *et al.,* 2000. A kingdom-level phylogeny of eukaryotes based on combined protein data. *Science* 290 (5493): 972-977.

Baltscheffsky, H. and Baltscheffsky, M., 1994. Molecular origin and evolution of early biological energy conservation. In: *Early Life on Earth.* Edited by S. Bengston. New York: Columbia University Press: 81-90.

Barbieri, M., 2003. *The Organic Codes. An Introduction to Semantic Biology.* Cambridge: Cambridge University Press.

Barghoorn, E. S. and Schopf, J. W., 1965. Microorganisms from the Late Precambrian of Central Australia. *Science* 150: 337-339.

Barghoorn, E. S. and Tyler, S. A., 1965. Microorganisms from the Gunflint Chert. *Science* 147: 563-577.

Barrow, J. D. and Silk, J., 1983. *The Left Hand of Creation. The Origin and Evolution of the Expanding Universe.* London: Heinemann.

Bartley, J. K., Pope, M., Knoll, A. H., Semikhatov, M. A. and Petrov, P. Yu., 1998. A Vendian-Cambrian boundary succession from the northwestern margin of the Siberian Platform: Stratigraphy, palaeontology, chemostratigraphy and correlation. *Geological Magazine* 135 (4): 473-494.

Barton, N. H. and Charlesworth, B., 1998. Why sex and recombination? *Science* 281: 1986-1990.

Bartusiak, M., 2005. Beyond the Big Bang. Einstein's evolving Universe. *National Geographic* 207 (5): 110-121.

Basslera, R. S., 1941. A supposes jellyfish from the Precambrian of The Grand Canyon. *Proceedings of the U.S. National Museum* 89:519-522.

Batten, K., Narbonne, G. M. and James, N. P., 2004. Early Neoproterozoic (Tonian) ramp carbonates and microbial buildups, Little Dal Group, Northwest Territories, Canada. *Precambrian Research* 133: 249-269

Baudet, D., Aitken, J. D. and Vanguestaine, M., 1989. Palynology of uppermost Proterozoic and lowermost Cambrian formations, central Mackenzie Mountains, northwestern Canada. *Canadian Journal of Earth Sciences,* 26: 129-148.

Beaumont, V. and Robert, F., 1999. Nitrogen isotope ratios of kerogens in Precambrian cherts: A record of the evolution of atmosphere chemistry? *Precambrian Research* 96 (1-2): 63-82.

Becker, Yu. R. 1977. First paleontological finds in the Riphean of the Urals. *Izvestiya Akademii Nauk SSSR, Ser. Geol.*, 3: 90-100. (in Russian)

Becker, Yu. R., 1985. Metazoa iz venda Urala. In: *Vendskaya Sistema 1: Istoriko-geologicheskoe i paleontologicheskoe obosnovanie.* Edited by B. S. Sokolov and A. B. Iwanowski. Moscow: Nauka: 107-112.

Becker, Yu. R., 1990. Vendian Metazoa from the Urals. In: *The Vendian System.* Vol. 1. *Paleontology.* Edited by B. S. Sokolov and A. B. Iwanowski. Berlin: Springer-Verlag: 121-131.

Becker, Yu. R., 1992. Oldest Ediacara biota of the Urals. *Izvestiya Akademii Nauk SSSR*, Ser. Geol., No. 6, c. 16-24. (in Russian).

Becker, Yu. R., 1992. Drevnejshaja ediakarskaja biota urala. *Izvestija Akademij Nauk, Serija Geologicheskaja* 1992 (6): 16-24.}

Becker, Yu. R. and Kishka, N. V., 1989. Discovery of the Ediacara biota in Southern Urals. *Transactions of the All-Union Paleontological Society, Session 33: Theoretical and Applied Aspects of Modern Paleontology.* Leningrad: Nauka: 109-120. (in Russian).

Becker, Yu. R. and Kishka, N. V. 1989. Otkrytie ediakarskoj bioty na juzhnom urale. In *Teoreticheskie i prikladnye aspekty sovremennoj paleontologii.* Edited by T. N. Bogdanova and L. I. Khozatskij. *Trudy XXXIII Sessii Vsesojuznogo paleontologicheskogo Obshchestva.* Leningrad: Nauka: 109-120.

Bekker Yu. R., 1996. Discovery of the Ediacaran biota at the top of the Vendian of South Urals. *Regionalnaya Geologia i Metallogenia.* № 5. P. 111-135/ (in Russian).

Bekker, A., Kaufman, A. J., Karhu, J. A., Beukes, N. J., Swart, Q. D., Coetzee, L. L. and Eriksson, K. A., 2001. Chemostratigraphy of the Paleoproterozoic Duitschland Formation, South Africa: Implications for coupled climate change and carbon cycling. *American Journal of Science* 301: 261-285.

Bendrick, R. and Pfluger, F., 2001. A paleoecological analysis of the Ediacaran Mistaken Point Fauna, E-Horizon, SW Newfoundland. Abstracts. *Geological Association of Canada Annual Meeting,* St John's, Newfoundland.

Bengtson, S., ed., 1993. *Early Life on Earth.* Nobel Symposium No. 84. New York: Columbia University Press.

Bengtson, S., 2002. Origins and early evolution of predation. *Paleontological Society Paper* 8: 289-317.

Bengtson, S. and Budd, G., 2004. Comment on "Small Bilaterian Fossils from 40 to 55 Million Years Before the Cambrian." *Science* 306: 1290a-1291a.

Bengtson, S. and Conway Morris, S., 1992. Early radiation of biomineralizing phyla. In: *Origin and Early Evolution of the Metazoa* Edited by J. Lipps and P. W. Signor. New York: Plenum Press.

Bengtson, S., Conway Morris, S., Cooper, B. J., Jell, P. A. and Runnegar, B. N., 1990. Early Cambrian fossils from South Australia. *Association of Australasian Palaeontologists Memoir* 9: 1-364.

Bengtson, S. and Yue, Z., 1992. Predatorial borings in late Precambrian mineralized exoskeletons. *Science* 257: 367-369.

Bennett, J., Shostak, S. and Jakosky, B., 2003. *Life in the Universe.* San Francisco: Addison Wesley.

Benus, A. P., 1988. Sedimentological context of a deep-water Ediacaran fauna (Mistaken Point, Avalon Zone, eastern Newfoundland). In: *Trace Fossils, Small Shelly Fossils and the Precambrian-Cambrian Boundary.* Edited by E. Landing, G. M. Narbonne and P. M. Myrow. *New York State Museum and Geological Survey Bulletin* 463: 8-9.

Benz, W., Kallenbach, R. and Lugmair, G. W., 2000. *From Dust to Terrestrial Planets.* Dordrecht: Kluwer.

Berry, W. B. N., 1978. Progressive ventilation of the oceans – an explanation for the distribution of the Lower Paleozoic black shales. *American Journal of Science* 285: 257-275.

Beukes, N. J., Dorland, H., Gutzmer, J., Nedach, M. and Ohmoto, H., 2002. Tropical laterites, life on land and the history of atmosphereic oxygen in the Paleoproterozoic. *Geology* 30: 491-494.

Billings, E., 1872. On some fossils from the primordial rocks of Newfoundland. *Canadian Naturalist and Quarterly Journal of Science* n.s. 6: 465-479.

Birket - Smith, S. J. R., 1981. A reconstruction of the PreCambrian *Spriggina. Abteilung für Anatomie und Ontogenie der Tiere* 105, 237-258.

Bjerrum, C. J. and Canfield, D. E., 2002. Ocean productivity before about 1.9 Gyr ago limited by phosphorus adsorption onto iron oxides. *Nature* 417: 159-162.

Blair, J. E. and Hedges, S. B., 2005. Molecular clocks do not support the Cambrian Explosion. *Molecular Biology and Evolution* 22 (3): 387-390.

Bland, B. H., 1984. *Arumberia* Glaessner & Walter, a review of its potential for correlation in the region of the Precambrian-Cambrian boundary. *Geological Magazine* 121: 625-633.

Blankenship, R. E. and Hartman, H., 1998. The origin and evolution of oxygenic photosynthesis. *Trends in Biochemical Sciences* 23: 94-97.

Bodiselitsch, B., Koeberl, C., Master, S. and Reimold, W. U., 2005. Estimating duration and intensity of Neoproterozoic Snowball glaciations from Ir anomalies. *Science* 308: 239-242.

Bonner, J. T., 1988. *The Evolution of Complexity.* Princeton: Princeton University Press.

Bonner, W. N. and Walton, D. W. H., 1985. *Antarctica. Key Environments.* Oxford: Pergamon Press.

Born, M., 1935. *The Restless Universe.* London: Blackie & Son.

Bottjer, D. J., 2005. Geobiology and the fossil record: Eukaryotes, microbes, and their indicators. In: *Geobiology: Objectives, Concepts, Perspectives.* Edited by N. Noffke. Philadelphia: Elsevier: 5-21.

Bottjer, D. J., 2005. The early evolution of animals. *Scientific American* 239 (2): 30-35.

Bowring, S. A. and Condon, D., 2006 (*in prep.*). Sequencing the Neoproterozoic: The importance of high-precision geochronology.

Bowring, S. A. and Schmitz, M. D., 2003. High-precision U-Pb zircon geochronology and the stratigraphic record. *Reviews in Mineralogy and Geochemistry* 53: 305-326.

Boynton, H. E., 1978. Fossils from the Precambrian of Charnwood Forest, Leicestershire. *Mercian Geologist* 6 (4): 291-296.

Boynton, H. E. and Ford, T. D., 1979. *Pseudovendia charnwoodensis* – a new Precambrian arthropod from Charnwood Forest, Leicestershire. *Mercian Geologist* 7 (2): 175-177.

Boynton, H. E. and Ford, T. D., 1995. Ediacaran fossils from the Precambrian (Charnian Supergroup) of Charnwood Forest, Leicestershire, England. *Mercian Geologist* 13: 165-182.

Brack, A., ed., 1998. *The Molecular Origins of Life.* Cambridge: Cambridge University Press.

Bragg, W., 1948. *Concerning the Nature of Things.* New York: Dover.

Brain, C. K., Prave, A. R., Fallick, A. E. and Hoffmann, K.-H., 2004. Progress in a search for fossils of ancestral invertebrates in Proterozoic limestones of the Otavi Group in Namibia. In: *Geoscience Africa 2004. Abstract Volume.* Edited by L. D. Ashwal. Johannesburg: University of Witwatersrand: 85.

Brandt, S. and Dahmen, H. D., 1995. *The Picture Book of Quantum Mechanics.* New York: Springer-Verlag.

Brasier, M. and Antcliffe, J., 2004. Decoding the Ediacaran enigma. *Science* 305: 1115-1117.

Brasier, M. D., Green, O. R., Jephcoat, A. P., Kleppe, A. K. and Van, N. V., 2002. Questioning the evidence for Earth's oldest fossils. *Nature* 416: 76-81.

Brasier, M. D. and Lindsay, J. F., 1998. A billion years of environmental stability and the emergence of eukaryotes: New data from northern Australia. *Geology* 26: 555-559.

Brasier, M. D. and Lindsay, J. F., 2001. Did supercontinental amalgamation trigger the "Cambrian Explosion"? In: *The Ecology of the Cambrian Radiation.* Edited by A. Yu. Zhuravlev and R. Riding. New York: Columbia University Press: 69-89.

Brasier, M. D. and McIlroy, D., 1998. *Neonereites uniserialis* from c. 600 Ma year old rocks in western Scotland and the emergence of animals. *Journal of the Geological Society, London* 155: 5-12.

Brennan, S. T., Lowenstein, T. K. & Horita, J., 2004. Seawater chemistry and the advent of biocalcification. *Geology* 32 (6): 473-476.

Briggs, D. E. G., ed., 2005. *Evolving Form and Function: Fossils and Development.* Special Publication of the Peabody Museum of Natural History, New Haven: Yale University.

Briggs, D. E. G. and Crowthers, P. R., eds., 2001. *Palaeobiology II.* London: Blackwell Science.

Brocks, J. J., Buick, R., Logan, G. A. and Summons, R. E., 2003. Composition and syngenecity of molecular fossils from the 2.78 to 2.45 billion-year-old Mount Bruce ?Supergroup, Pilbara Craton, Western Australia. *Geochimica et Cosmochimica Acta* 67 (22): 4289-4319.

Brocks, J. J., Logan, G. A., Buick, R. and Summons, R. E., 1999. Archean molecular fossils and the early rise of eukaryotes. *Science* 285: 1033-1036.

Brocks, J. J., *et al.*, 1999. Archean molecular fossils and the early rise of eukaryotes. *Science* 285: 1033-1036.

Brocks, J. J., Love, G. D., Summons, R. E., Knoll, A. H., Logan, G. A. & Bowden, S. A., 2005. Biomarker evidence for green and purple sulphur bacteria in a stratified Palaeoproterozoic sea. *Nature* 437 (6): 866-870.

Bromham, L., Rambaut, A., Fortey, R., Cooper, A. and Penny, D., 1998. Testing the Cambrian Explosion hypothesis by using a molecular dating technique. *Proceedings of the National Academy of Sciences of the United States of America* 95: 12386-12389.

Bromley, R. G., 1990. *Trace Fossils. Biology and Taphonomy.* London: Unwin Hyman.

Bronowski, J., 1973. *The Ascent of Man.* Boston: Little, Brown and Company.

Brown, B. and Morgan, L., 1989. *The Miracle Planet.* Sydney: Child & Associates.

Brown, J. R. and Doolittle, W.F., 1997. Archaea and the prokaryote-to-eukaryote transition. *Microbiology Molecular Biology Review* 61: 456.

Bryant, C., ed., 1991. *Metazoan Life Without Oxygen.* London: Chapman and Hall.

Buatois, L. A. and Mangano, M. G., 2003. Early colonization of the deep sea: Ichnologic evidence of deep-marine benthic ecology from the Early Cambrian of northwest Argentina. *Palaios* 18: 572-581.

Buatois, L. A. and Mangano, G., 2004. Terminal Proterozoic-Early Cambrian ecosystems: Ichnology of the Puncoviscana Formation, northwest Argentina. *Fossils and Strata* 51: 1-16.

Buatois, L. A., Mangano, M. G., Aceñolaza, F. G. and Esteban, S. B., 2000. The Puncoviscana ichnofauna of northwest Argentina: A glimpse into the ecology of the Precambrian-Cambrian transition. In: *Cambrian from the Southern Edge.* Edited by G. F. Aceñolaza and S. Peralta: INSUGEO, *Miscelanea*: 82-84.

Budd, G. E. and Jensen, S., 2000. A critical reappraisal of the fossil record of the bilaterian phyla. *Biological Review* 75: 253-295.

Budin, K. and Philippe, H., 1998. New insights into the phylogeny of eukaryotes based on ciliate Hsp70 sequences. *Molecular Biology and Evolution* 15: 943-956.

Budyko, A., 1969. The effect of solar radiation variations on the climate of the Earth. *Tellus* 21: 611-619.

Buick, R., 1992. The antiquity of oxygenic photosynthesis evidence – from stromatolites in sulfate-deficient Archean lakes. *Science* 255: 74-77.

Buick, R., 2001. Life in the Archean. In: *Palaeobiology II.* Edited by D. E. G. Briggs and P. R. Crowther. London: Blackwell Scientific Publications: 13-21.

Buick, R., Thornett, J. R., McNaughton, N. J., Smith, J. B., Barley, M. E. and Savage, M., 1995. Record of emergent continental crust ~3.5 billion years ago in the Pilbara craton of Australia. *Nature* 375: 574-577.

Buss, L.W., 1987. *The Evolution of Individuality.* Princeton: Princeton University Press.

Buss, L. W. and Seilacher, A., 1994. The Phylum Vendobionta: A sister group of the Eumetazoa? *Paleobiology* 20: 1-4.

Butterfield, N. J., 2000. *Bangiomorpha pubescens*, n. gen. n. sp.: Implications for the evolution of sex, multicellularity, and the Mesoproterozoic/Neoproterozoic radiation of eukaryotes. *Paleobiology* 263: 386-404.

Butterfield, N. J., 2004. A vaucheriacean alga from the middle Neoproterozoic of Spitzbergen: Implications for the evolution of Proterozoic eukaryotes and the Cambrian explosion. *Paleobiology* 30: 231-232.

Butterfield, N. J., 2005. Probable Proterozoic fungi. *Paleobiology* 31 (1): 165-182.

Butterfield, N. J., Knoll, A. H. and Swett, K., 1994. Paleobiology of the Neoproterozoic Svanbergfjellet formation, Spitsbergen. *Fossils and Strata* 34: 1-84.

Butterfield, N. J. and Rainbird, R. H., 1998. Diverse organic-walled fossils, including "possible dinoflagellates," from the early Neoproterozoic of arctic Canada. *Geology* 26 (11): 963-966.

Cairncross, B., Beukes, N. J. and Gutzmer, J., 1997. *The Manganese Adventure. The South African Manganese Fields.* Johannesburg: The Associated Ore & Metal Corporation.

Cairns-Smith, A. G., 1982. *Genetic Takeover and the Mineral Origins of Life.* Cambridge: Cambridge University Press.

Calver, C. R., Black, L. P., Everard, J. L. and Seymour, D. B., 2004. U-Pb zircon age constraints on late Neoproterozoic glaciation in Tasmania. *Geology* 32: 893-896.

Calver, C. R. and Lindsay, J. F., 1998. Ediacarian sequence and isotope stratigraphy of the Officer Basin, South Australia. *Australian Journal of Earth Sciences* 45: 513-532.

Campins, H., Swindle, T. D. and Kring, D. A., 2004. Evaluating comets as a source of Earth`s water. In: Origins. Genesis, Evolution and Diversity of Life. Edited by J. Seckbach. Dordrecht: Kluwer Academic Publisher: 571-579.

Canfield, D. E., 1998. A new model for Proterozoic ocean chemistry. *Nature*, 396: 450-453.

Canfield, D. E., 2005. The early history of atmospheric oxygen; Homage to Robert M. Garrels. *Annual Review of Earth and Planetary Science* 33: 1-36.

Canfield, D. E., Habicht, K. S. and Thamdrup, B., 2000. The Archean sulfur cycle and the early history of atmospheric oxygen. *Science* 288 (5466): 658-661.

Canfield, D. E., Poulton, S. W. and Narbonne, G. M., 2007. Late-Neoproterozoic deep-ocean oxygenation and the Rise of Animal Life. *Science* 315: 92-95.

Canfield D. E. and Teske, A., 1996. Late Proterozoic rise in atmospheric oxygen concentration inferred from phylogenetic and sulphur-isotope studies. *Nature* 382 (6587): 127-132.

Canup, R. M. and Asphaug, E., 2001. Origin of the Moon in a giant impact near the end of the Earth's formation. *Nature* 412: 708-712.

Carr, M. H., *et al.,* 1981. *The Surface of Mars.* New Haven: Yale University Press.

Carr, M. H., *et al.,* 1998. Evidence for a subsurface ocean on Europa. *Nature* 391: 363.

Carroll, S. B., 2001. Chance and necessity: The evolution of morphological complexity and diversity. *Nature* 409: 1102-1109.

Carroll, S. B., 2005. *Endless Forms Most Beautiful. The New Science of Evo Devo and the Making of the Animal Kingdom.* New York: W. W. Norton & Co.

Carroll, S. B., Crenier, J. K. amd Weatherbee, S. D., 2001. *From DNA to Diversity: Molecular Genetics and the Evolution of Animal Design.* London: Backwell Science.

Cas, R. A. F., Beresford, S. and Appel, P., 2002. The oldest volcanics and sediments from the 3.7 – 3.8 GA Isua Greenstone Belt, Greenland: Implications for the earliest known palaeoenvironments on Earth. *Abstracts. 17th Australian Geological Convention*, Adelaide.

Cassidy, K. F., Dunphy, J. M. and Van Kranendonk, M. J., eds., 2001. *International Archaean Symposium. Extended Abstracts.* Canberra: AGSO – *Geoscience Australia Record* 2001/37.

Casti, J. L. and Karlquist, A., eds., 1994. *Cooperation and Conflict in General Evolutionary Processes.* New York: Wiley-Interscience.

Cattermole, P., 1994. *Venus. The Geological Story.* Baltimore : Johns Hopkins University Press.

Chang, S., 1988. Planetary environments and the conditions of life. *Philosophical Transactions of the Royal Society (London A)* 325: 601-610.

Chang, S., 1992. *The Planetary Setting of Pre-biotic Evolution.* Space Science Division, NASA Ames Research Centre, Moffett Field, California.

Chen, J.-Y., Bottjer, D. J., Davidson, E. H., Donrbos, S. Q., Gao X., Yang, Y - H., Li, C.-W, Li, G., Wang, X-Q., Xian, D – C., Wu, H – J., Hwu, Y – K.and Tafforeau, P, 2006. Phosphatized polar lobe-forming embryos from Precambrian of southwest China. *Science* 312: 1644-1646.

Chen, J.-Y., Bottjer, D. J., Oliveri, P., Donrbos, S. Q., Gao, F., Ruffins, S., Chi, H., Li, C.-W. and Davidson, E. H., 2004. Small bilaterian fossils from 40-55 million years before the Cambrian. *Science* 305: 218-222.

Chen, J.-Y., Cheng, Y.-N. and Iten, H. V., 1997. *The Cambrian Explosion and the Fossil Record.* National Museum of Natural Science. *Bulletin of the National Museum of Natural Science*, 10. Taiwan: Taichung.

Chen, J.-Y., Huang, D.-Y. and Li, C.-W., 1999. An early Cambrian craniate-like chordate. *Nature*, 402: 518-522.

Chen, C., Oliveri, P., Gao, F., Dornbos, S,Q., Li, C.-W., Bottjer, D. J., and Davidson, E.H., 2002. Precambrian animal life: Probable developmental and adult cnidarian forms from southwest China. *Developmental Biology* 248: 182-196.

Chen, C., Oliveri, P., Li, C.-W., Zhou, G.-Q., Gao, F., Hagadorn, J. W., Peterson, K. J. and Davidson, E. H., 2000. Precambrian animal diversity: Putative phosphatised embryos from the Doushantuo Formation of China. *Proceedings of the National Academy of Sciences USA* 97: 4457-4462.

Chen, J.-Y., Waloszek, D. and Maas, A., 2004. A new 'great-appendage' arthropod from the Lower Cambrian of China and homology of chelicerate chelicerae and raptorial antero-ventral appendages. *Lethaia* 37: 3-20.

Chistyakov, V. G., Kalmykova, N. A., Nesov, L. A. and Suslov, G. A., 1984. On the presence of the Vendian deposits in the middle stream of Onega River and on probable existence of tunicates (Tunicata: Chordata) in Precambrian (O nalichii vendskikh otlozhenij v srednem techenii r. Onegi i vozmozhnom sushchestvovanii obolochikov (Tunicata: Chordata) v dokembrii). *Vestnik Leningradskogo Gosudarstvennogo Universiteta* No. 6: 11-19. (in Russian)

Christianson, G. E., 1995. *Edwin Hubble. Mariner of the Nebulae.* New York: Farrar, Straus and Giroux.

Christie-Blick, N., Sohl, L. E. and Kennedy, M. J., 1999. Considering a Neoproterozoic Snowball Earth. *Science* 284: 1087-1088.

Chumakov, N. M., 1997. Warm biosphere. *Priroda* 5: 1-4.

Chumakov, N. M., 2001. General trend in climatic change on Earth during past 3 billion years. *Doklady Akademii Nauk* 381: 1-4. (in Russian and English)

Chyba, C. F., 2005. Rethinking Earth's early atmosphere. *Science* 308: 962-963.

Chyba, C. F., Thomas, P. J., Brookshaw, L. and Sagan, C., 1990. Comet delivery of organic molecules to the early Earth. *Science* 249: 366-373.

Clapham, M. E. and Narbonne, G. M., 2002. Ediacaran epifaunal tiering. *Geology* 30: 627-630.

Clapham, M. E., Narbonne, G. M. and Gehling, J. G., 2003. Paleoecology of the oldest known animal communities: Ediacaran assemblages at Mistaken Point, Newfoundland. *Paleobiology* 29 (4): 527-544.

Clapham, M. E., Narbonne, G. M., Gehling, J. G., Greentree, C. and Anderson, M. M., 2004. *Thectardis avalonensis*: a new Ediacaran fossil from the Mistaken Point Biota, Newfoundland. *Journal of Paleontology* 78 (6): 1031-1036.

Clarke, A., 1983. Life in cold water: The physiological ecology of polar marine ectotherms. *Theoretical Population Biology* 16: 267-282.

Clarke, A., 1993. Temperature and extinction in the sea: A physiologist's view. *Paleobiology* 19 (4): 499-518.

Close, F. E., 1990. *End: Cosmic Catastrophe and the Fate of the Universe*. London: Penguin.

Cloud, P. E., 1968. Atmospheric and hydrospheric evolutin on the primitive Earth. *Science* 160: 729-736.

Cloud, P., 1976. Major features of crustal evolution. A. L. du Toit Memorial Lecture Series 14. *Transactions of the Geological Society of South Africa*, 79: 1-33.

Cloud, P. and Glaessner, M. F., 1982. The Ediacarian Period and System: Metazoa inherit the Earth. *Science* 217: 783-792.

Coleman, J. A., 1963. *Modern Theories of the Universe*. New York: The New American Library.

Collerson, K. D. and Kamber, B., 1999. Evolution of the continents and the atmosphere inferred from Th-U-Nb systematics of the depleted mantle. *Science* 283: 1519-1522.

Collins, D. H., Briggs, D. E. G. and Conway Morris, S., 1983. New Burgess Shale fossil sites reveal Middle Cambrian faunal complex. *Science* 222: 163-167.

Condie, K. C., 1993. *Plate Tectonics and Crustal Evolution*. New York: Pergamon.

Condie, K. C., 2001. *Mantle Plumes and Their Record in Earth History*. Cambridge: Cambridge University Press.

Condie, K. C., 2001. Rodinia and continental growth. *Gondwana Research* 4 (2): 154-155.

Condie, K. C., 2003. Supercontinents, superplumes and continental growth: The Neoproterozoic record. In: *Proterozoic East Gondwana: Supercontinent Assembly and Breakup*. Edited by M. Yoshida *et al.*. London: The Geological Society: 1-33.

Condon, D. and Bowring, S., 2005. A high-resolution temporal framework for the Ediacaran Period. The Palaeontological Association, 49th Annual Meeting, *Abstracts Newsletter*, Oxford: 10.

Condon, D., Zhu, M., Bowring, S., Wang, W., Yang, A. and Jin, Y., 2005. U-Pb ages from the Neoproterozoic Doushantuo Formation, China. *Science* 308: 95-98.

Conway Morris, S., 1990. Late Precambrian and Cambrian soft-bodied faunas. *Annual Review of Earth and Planetary Science* 18: 101-122.

Conway Morris, S., 1993. Ediacaran-like fossils in Cambrian Burgess shale-type faunas of North America. *Palaeontology* 36: 593-635.

Conway Morris, S., 1998. Southeastern Newfoundland and adjacent areas (Avalon Zone). In: *The Precambrian-Cambrian Boundary*. Edited by J. W. Cowrie and M. D. Brasier. Oxford: Clarendon Press: 7-39.

Conway Morris, S., 1998. *The Crucible of Creation. The Burgess Shale and the Rise of Animals*. Oxford: Oxford University Press: 1-242.

Conway Morris, S. and Grazhdankin, D., 2005. Enigmatic worm-like organisms from the Upper Devonian of New York: An apparent example of Ediacaran-like preservation. *Palaeontology* 48 (2): 395-410.

Conway Morris, S., Mattes, B. W. and Cheng Meng, 1990. The early skeletal organism *Cloudina*: New occurrences from Oman and possibly China. *American Journal of Science* 290A: 245-260.

Cook, P. J. and Shergold, J. H., 1984. Phosphorus, phosphorites and skeletal evolution at the Precambrian-Cambrian boundary. *Nature*, 308: 231-236.

Cooper, A. and Fortey, R., 1998. Evolutionary explosions and the phylogenetic fuse. *TREE* 13: 151-156.

Cooper, G. M. and Hausman, R. E., 2004. *The Cell. A Molecular Approach*. Washington: ASM Press.

Cope, J. C. W., 1977. An Ediacara-type fauna from South Wales. *Nature* 268: 624.

Cope, J. C. W., 1982. Precambrian fossils of the Carmarthen area, Dyfed. *Nature in Wales, New Series* 1 (2): 11-16.

Cope, J. C. W. and Bevins, R. E. 1993. The stratigraphy and setting of the Precambrian rocks of the Llangynog Inlier, Dyfed, South Wales. *Geological Magazine* 130: 101-111.

Cordani, U. G., D'Agrella-Filho, M. S., Brito-Neves, B. B. and Trindade, R. I. F., 2003. Tearing up Rodinia: The Neoproterozoic palaeogeography of South American cratonic fragments. *Terra Nova* 15 (5): 350-359.

Cordani, U. G., Thomaz-Filho, A., Brito-Neve, B. B. and Kawashita, K., 1985. On the applicability of the Rb-Sr method to argillaceous sedimentary rocks: Some examples from Precambrian sequences of Brazil. *Journal of Geology* 47: 253-280.

Cotter, K. L, 1999. Microfossils from Neoproterozoic Supersequence 1 of the Officer Basin, Western Australia. *Alcheringa* 23: 63-86.

Cowen, R., 2004. History of Life. Oxford: Blackwell Publishing.

Couper, H. and Pelham, D., 1985. *Universe. A Three Dimensional Study*. London: Century Publishing Co.

Crimes, T. P., 1992. Changes in the trace fossil biota across the Proterozoic-Phanerozoic boundary. *Journal of the Geological Society of London*: 149: 637-646.

Crimes, T. P., 2001. Evolution of the deep-water benthic community. In: *The Ecology of the Cambrian Radiation*. Edited by Zhuravlev, A. Y and Riding, R. New York: Columbia University Press: 275-290.

Crimes, T. P. and Anderson, M. M., 1985. Trace fossils from Late Precambrian-Early Cambrian strata of southeastern Newfoundland (Canada): Temporal and environmental implication. *Journal of Paleontology* 59 (2): 310-343.

Crimes, T. P. and Droser, M. L., 1992. Trace fossils and bioturbation: The other fossil record. *Annual Reviews of Ecology and Systematics* 23: 339-360.

Crimes, T. P. and Fedonkin M. A., 1994. Evolution and dispersal of deep sea traces. *Palaios* 9 (1): 74-83.

Crimes, T. P. and Fedonkin, M. A., 1996. Biotic changes in platform communities across the Precambrian-Phanerozoic boundary. *Revista Italiana di Paleontologia e Stratigrafia* 102 (3): 317-331.

Crimes, T. P. and Germs, G. J. B., 1982. Trace fossils from the Nama Group (Precambrian-Cambrian of Southwest Africa [Namibia]). *Journal of Paleontology* 56 (4): 890-907.

Crimes, T. P. and Harper, J. C., eds., 1977. Trace fossils 2. *Proceedings of an International Symposium held in Sydney, Australia, 23-24 August 1976 as part of the 25th International Geological Congress*. Liverpool: Seel House Press.

Crowe, M. J., 1986. *The Extraterrestrial Life Debate 1750-1900. The Idea of a Plurality of Worlds from Kant to Lowell*. Cambridge: Cambridge University Press.

Crowley, J. L., Myers, J. S., Sylvester, P. J. and Cox, R. A., 2005. Detrital zircon from the Jack Hills and Mount Narryer, Western Australia: Evidence for diverse 4.0 Ga source rocks. *Journal of Geology* 113: 239-263.

Cruse, T., Harris, L. B., and Rasmussen,B., 1993. The discovery of Ediacaran trace and body fossils in the Stirling Range Formation, Western Australia: Implications for sedimentation and reformation during the 'Pan-African orogenic cycle.' *Australian Journal of Earth Science* 40 (3): 293-296.

Daily, B., 1956. The Cambrian in South Australia. In: *El Sistema Cambrico, su Paleogeografia y el Problema de su Basee*. Edited by J. Rodgers. Mexico: *Report of the 20th International Geological Congress 1956* (2): 91-147.

Daily, B., 1973. Discovery and significance of basal Cambrian Uratanna Formation, Mt Scott Range, Flinders Ranges, South Australia. *Search* 4: 202-205.

Dalrymple, R. W. and Narbonne, G. M., 1996. Continental slope sedimentation in the Sheepbed Formation (Neoproterozoic, Windermere Supergroup), Mackenzie Mountains, N. W. T. *Canadian Journal of Earth Sciences* 33: 848-862.

Dalziel, I. W. D., 1997. Neoproterozoic-Paleozoic geography and tectonics: Review, hypothesis, environmental speculation. *Geological Society of America Bulletin* 109: 16-42.

Dalziel, I. W. D., Mosher, S. and Gahagan, L. M., 2000 Laurentia-Kalahari collision and the assembly of Rodinia. *Journal of Geology* 108: 499-513.

Darling, D., 2001. *Life Everywhere: The Maverick Science of Astrobiology*. New York: Basic Books.

Darwin, C., 1859. *On the Origin of Species*. London: John Murray.

Davidson, E. H., Peterson, K. J. and Cameron, R. A., 1995. Origin of bilaterian body plans: Evolution of developmental regulatory mechanisms. *Science*, 270: 1319-1325.

Davies, P.C.W., 1994. *The Last Three Minutes. Conjectures about the Ultimate Fate of the Universe*. New York: Basic Books (A Division of HarperCollins).

Davies, P. C. W., 1998. *The Fifth Miracle: The Search for the Origin of Life*. New York: Simon & Schuster.

Davies, P. C. W., 2004. When time began. *New Scientist Supplement*. State of the Universe, 9 October: 4-7.

Debrenne, F. and Nard, G., 1981. Méduse et traces fossiles supposées péecambrennes ans la formation de San Vito, Sarrabus, Sud-Est de la Sandaigne. *Bulletin de la Société Géologique de France* 23: 23-31.

Dec, T., O'Brien, S. J. and Knight, I., 1992. Late Precambrian volcaniclastic deposits of the Avalonian Eastport basin (Newfoundland Appalachians): Petrofacies, detrital clinopyroxene geochemistry and palaeotectonic implications. *Precambrian Research* 59: 243-262.

de Duve, C., 1984. *A Guided Tour of the Living Cell*. New York: Freeman.

de Duve, C., 1991. *Blueprint for a Cell*. Burlington: Paterson.

de Duve, C., 1995. *Vital Dust*. New York: Basic Books.

de Duve, C., 1996. The birth of complex cells. *Scientific American* 274: 38-45.

De Laeter, J.R. and Trendall, A. F., 2002. The oldest rocks: The Western Australian connection. *Journal of the Royal Society of Western Australia* 85: 153-160.

de Pater, I. and Lissaur, J. J., 2001. *Planetary Sciences*. Cambridge: Cambridge University Press.

Delsemme, A. H., 2002. An argument for the cometary origin of the biosphere. *American Scientist* 89: 432-442.

Deppenmeier, U., 2002. The unique biochemistry of methanogenesis. *Progress Nucleic Acid Research in Molecular Biology* 71: 223-283.

DeRobertis, E. M., 1995. Homeotic genes and the evolution of body plans. In: *Evolution and the Molecular Revolution*. Edited by C. R. Marshall and J. W. Schopf. Boston: Jones & Bartlett Publishers: 109-124.

Derry, L. A., 2006. Fungi, weathering and the emergence of animals. *Science* 311: 1386-1387.

Des Marais, D., 2000. When did photosynthesis emerge on Earth? *Science* 289: 1703.

Des Marais, D. J., Strauss, H., Summons, R. E. and Hayes, J. M., 1992. Carbon isotope evidence for the stepwise oxidation of the Proterozoic environment. *Nature* 359: 605-609.

De Wit, M., Roering, C., Hart, R. J., Armstrong, R. A., de Ronde, C. E. J., Green, R. W. E., Tredoux, M., Peberdy, E. and Hart, R., 1992. Formation of an Archaean continent. *Nature* 357: 553-562.

Diamond, J., 2004. *Collapse*. New York: Viking.

Di Giulio, M., 2003. The universal ancestor was a thermophile or a hyperthermophile: Test and further evidence. *Journal of Theoretical Biology* 221 (3): 425-436.

Dimroth, E. and Kimberkey, M. M., 1976. Precambrian atmospheric oxygen; evidence in the sedimentary distributions of carbon, sulfur, uranium, and iron. *Canadian Journal of Earth Sciences* 13: 1161-1185.

Ding, L., Li, Y., Hu, X., Xiao, Y., Su, C. and Huang, J., 1996. *Sinian Miaohe Biota*. Beijing: Geological Publishing House.

Domenico, P. A. and Schwartz, F. W., 1998. *Physical and Chemical Hydrology*. 2nd ed. New York: John Wiley & Sons.

Donoghue, P. C. J., 2007. Embryonic identity crisis. *Nature* 445: 155-156.

Downie, C., Evitt, W. R. and Sarjeant, W. A. S., 1963. Dinoflagellates, hystrichospheres, and the classification of the acritarchs. *Stanford University Publications, Geological Sciences* 7: 1–16.

Dozy, J. J., 1984. A late Precambrian Ediacara-type fossil from Calica (NW Spain). *Geologie en Mijnbouw* 63: 71-74.

Droser, M. L., Gehling, J. G. and Jensen, S., 1999. When the worm turned: Concordance of Early Cambrian ichnofabric and trace fossil record in siliciclastics of South Australia. *Geology* 27: 625-628.

Droser, M. L., Gehling, J. G. and Jensen, S., 2005. The palaeoecology of the Ediacara biota: Evidence from successively excavated beds, South Australia. *The Palaeontological Association, 49th Annual Meeting, Abstracts Volume, Newsletter* 60: 11.

Droser, M. L., Jensen, S. and Gehling, J. G., 2002. Trace fossils and substrates of the terminal Proterozoic-Cambrian transition: Implications for the record of early bilaterians and sediment mixing. *PNAS* 99 (20): 12572-12576.

Droser, M. L., Jensen, S., Gehling, J. G., Myrow, P. M. and Narbonne, G. M., 2002. Lowermost Cambrian ichnofabrics from the Chapel Island Formation, Newfoundland: Implications for Cambrian substrates. *Palaios* 17: 3-15.

Dunbar M.J., ed., 1977. *Polar Oceans*. Calgary: Arctic Institute of North America.

Durham, J. W., 1978. The probable metazoan biota of the Precambrian as indicated by the subsequent record. *Annual Review of Earth and Planetary Sciences* 6: 21-42.

Dymek, R. F. and Klein, C., 1988. Chemistry, petrology and origin of banded iron-formation lithologies from the 3800 Ma Isua supracrustal belt, West Greenland. *Precambrian Research* 39: 247-302.

Dyson, F. J., 1999. *Origins of Life*. Cambridge: Cambridge University Press.

Dyson, I. A., 1985. Frond-like fossils from the base of the late Precambrian Wilpena Group, South Australia. *Nature* 318: 283-285.

Dyson, I. A. and von der Borch, C. C. 1994. Sequence stratigraphy of an incised-valley fill: The Neoproterozoic Seacliff Sandstone, Adelaide Geosyncline, South Australia. *SEPM Special Publication* 51: 209-222.

Dzik, J., 1999. Organic membranous skeleton of Precambrian metazoans from Namibia. *Geology* 27: 519-522.

Dzik, J., 2000. The origin of the mineral skeleton in chordates. *Evolutionary Biology* 31: 105-154.

Dzik, J., 2002. Early diversification of organisms in the fossil record. In *Fundamentals of Life*. Edited by G. Palyi, C. Zucchi and L. Caglioti. Paris: Elsevier Science S.A.: 219-248.

Dzik, J., 2002. Possible Ctenophoran affinities of the Precambrian "sea-pen" *Rangea*. *Journal of Morphology* 252: 315-334.

Dzik, J., 2003. Anatomical information content in the Ediacaran fossils and their possible zoological affinities. *Integrative Comparative Biology* 43: 114-126.

Dzik, J., 2004. Anatomy and relationships of the Early Cambrian worm *Myoscolex*. *Zoologica Scripta* 33: 57-69.

Dzik, J., 2004. The Verdun Syndrome (Abstract). *Symposium IGCP493. The Rise and Fall of the Vendian Biota*. Edited by P. Komarower and P; Vickers-Rich. Prato: Monash University: 15-19.

Dzik, J. and Ivantsov, A. Yu., 1999. An asymmetric segmented organism from the Vendian of Russia and the status of the Dipleurozoa. *Historical Biology* 13: 255-268.

Dzik, J. and Ivantsov, A. Yu., 2002. Internal anatomy of a new Precambrian dickinsoniid dipleurozoan from northern Russia. *Neues Jahrbuch Geologie und Palaontologie Mh.* 7: 385-396.

Edgell, H. S., 1964. Precambrian fossils from the Hamersley Range, Western Australia, and their use in stratigraphic correlation. *Journal of the Geological Society of Australia* 11(2): 235-259.

Edgeworth David, T. W. and Tillyard, R. J., 1936. *Memoir on Fossils of the Late Pre-Cambrian (Newer Proterozoic) from the Adelaide Series, South Australia*. Sydney: Angus & Robertson Ltd. and the Royal Society of New South Wales.

Eerola, T., 2006. Neoproterozoic glaciations. Research on southern Brazil. *Geologi* 58: 164-174. (in Finnish with English summary) (www.geologin-nseura.fi/geologi-lehti/5-2006/Brasilia.pdf)

Eeola, T., 2007. No tillites, no cap carbonates? What's wrong with the "Snowball Earth's" southernmost Brazil? *Geologi* 59: 39-47. (in Finnish, with English summary)

Eernisse, D. J. and Peterson, K. J., 2004. The history of animals. In *Assembling the Tree of Life*. Edited by J. Cracraft and M. J..Donoghue. Oxford: Oxford University Press: 197-208.

Ehrenfreund, P. and Charnley, A., 2000. Organic molecules in the interstellar medium, comets, and meteorites: A voyage from dark clouds to the early Earth. *Annual Review of Astronomy and Astrophysics* 38: 427.

Eiseley, L., 1969. *The Unexpected Universe*. New York : Harcourt, Brace & World.

Elliott, T., Zindler, A. and Bourdon, B., 1999. Exploring the kappa conundrum: The role of recycling in the lead isotope evolution of the mantle. *Earth and Planet Science Letters* 169 (1-2): 129-145.

Ellis, G. F. R., 2003. Cosmology: The shape of the universe. *Nature* 425: 566-567.

Ellis, R., 2001. *Aquagenesis. The Origin and Evolution of Life in the Sea*. Hammondsworth: Penguin.

Emiliani, C., 1988**.** *The Scientific Companion. Exploring the Physical World with Facts, Figures, and Formulas*. Wiley Science Editions. New York: John Wiley & Sons.

Emsley, J., 1998. *The Elements*. 3rd ed. Oxford: Clarendon Press.

Endo, K., 2006. Origins of skeletal biomineralization; Molecular evidence. Abstract volume, IGCP 493 Symposium, Kyoto University, Japan, 27-31 January 2006. Edited by P. Vickers-Rich and T. Ohno. Kyoto: Kyoto University: 9.

Eriksson, P. G., Altermann, W., Nelson, D. R., Mueller, W. U. and Catuneanu, eds., 2004. *The Precambrian Earth: Tempos and Events. Developments in Precambrian Geology* 12. New York: Elsevier.

Erwin, D. H., 1999. The origin of body plans. *American Zoologist* 39: 617-629.

Erwin, D. H., 2005. Development, ecology, and environment in the Cambrian metazoan radiation. *Proceedings of the California Academy of Sciences*, 56, *Supplement* 1 (3): 24-31.

Erwin, D. H., 2005. Invention and innovation in the early evolution of animals. The Palaeontological Association, *49th Annual Meeting, Abstracts Newsletter* Oxford: 11.

Erwin, D. H. and Davidson, E. H., 2002. The last common bilaterian ancestor. *Development* 129 (13): 3021-3032.

Erwin, D. H., Valentine, J. and Jablonski, D., 1997. The origin of animal body plans. *American Scientist* 85: 126-137.

Etienne, J. L., Allen, P., Le Guerroue, E. and Riev, R., 2006. Snowball Earth. *Swiss National Foundation. Abstract Volume*, Monteverita, 16-21 July 2006: 128 pp. (http://www.igcp512.com and http://wwww.snowballearth.org/)

Evans, D. A., 1998. *I. Neoproterozoic-Paleozoic supercontinental tectonics and true Polar Wander. II. Temporal and spatial distributions of Proterozoic glaciations*. Ph.D. Thesis, California Institute of Technology.

Evans, D.A.D., 2003. A fundamental Precambrian-Phanerozoic shift in earth's glacial style. *Tectonophysics* 375: 353-385.

Evans, D. A. D., 2006. Proterozoic low orbital obliquity and axial-dipolar geomagnetic field from evaportite palaeolatitudes. *Nature* 444 (2): 51-55.

Evans, K. V., Aleinikoff, J. H., Obradovich, J. D. and Fanning, C. M., 2000. SHRIMP U-Pb geochronology of volcanic rocks, Belt Supergroup, western Montana: Evidence for rapid deposition of sedimentary strata. *Canadian Journal of Earth Sciences* 37: 1287-1300.

Evitt, W. R., 1963. A discussion and proposals concerning fossil dinoflagellates, hystricospheres, and acritarchs. *Proceedings of the National Academy of Sciences* 49: 158-164.

Eyles, N. and Eyles, C. H., 1989. Glacially-influenced deep-marine sedimentation of the Late Precambrian Gaskers Formation, Newfoundland, Canada. *Sedimentology* 36: 601-620.

Farmer, J., Vidal, G., Moczydlowska, M., Strauss, H., Ahlberg, P. and Siedlecka, A., 1992. Ediacaran fossils from the Innerelv Member (late Proterozoic) of the Tanafjorden area, northeastern Finnmark. *Geological Magazine* 129 (2): 185-195.

Farquhar, J., *et al.*, 2000. Atmospheric influence of Earth's earliest sulfur cycle. *Science* 289: 756-758.

Farquhar, J. and Wing, B. A., 2003. Multiple sulfur isotopes and the evolution of the atmosphere. *Earth and Planetary Science Letters* 213 (1-2): 1-13.

Fedo, C. M. and Whitehouse, M. J., 2002. Metasomatic origin of Quartz-Pyroxene rock, Akilia, Greenland, and implications for Earth's earliest life. *Science* 296: 1448-1452.

Fedonkin, M.A., 1977. Precambrian-Cambrian ichnocoenoses of East-European platform. In: *Trace Fossils 2*. Edited by T.P. Crimes and J.C. Harper. *Geological Journal Special Issue* 9. Liverpool: See House Press: 183-194.

Fedonkin M.A. 1978. New locality of non-skeletal Metazoa in the Vendian of Zimnii Shore. *Doklady Akademii Nauk SSSR* 239 (6): 1423-1426. (in Russian)

Fedonkin, M.A., 1978. Oldest trace fossils and the ways of behavioral evolution of mud eaters. *Paleontologicheskii Zhurnal* 2: 106-112. (in Russian)

Fedonkin, M. A., 1980. Fossil traces of Precambrian Metazoa. *Izvestia Akademeyi Nauk, SSSR Series Geology* 1: 39-46. (in Russian)

Fedonkin, M. A., 1981. White Sea Biota of Vendian. Precambrian Non-Skeletal Fauna of the Russian Platform North. *Transactions of the Geological Institute, Academy of Sciences of the USSR* 342: 1-100.

Fedonkin M. A., 1982. Precambrian non-skeletal fauna and the earliest stages of metazoan evolution. Third North American Paleontological Convention, Abstracts of papers. *Journal of Paleontology* 56 (2), Supplement 9.

Fedonkin M. A. 1982. Precambrian soft-bodied fauna and the earliest radiation of invertebrates. *Third North American Paleontological Convention Proceedings* 1: 165-167.

Fedonkin, M. A., 1983. Organic World of the Vendian. *Transactions of the Institute of Scientific and Technical Information. Itogi nauki I techniki: Stratigraphy, Paleontology. VINITI* 12: 1-128. (in Russian)

Fedonkin, M. A., 1983. Promorphology of the Vendian Radialia as a key to understanding the early evolution of Coelenterata. *Fourth International Symposium on Fossil Cnidaria. Abstracts.* 7-12 August 1983, Washington, D.C.: 6.

Fedonkin, M. A., 1984. Promorphology of Vendian Radialia. In: *Stratigraphy and Paleontology of the Most Ancient Phanerozoic.* Edited by A. B. Ivanovsky and A. B. Ivanov. Moscow: Nauka: 30-58.

Fedonkin, M. A., 1985. Paleoichnology of Vendian Metazoa. In: *The Vendian System: Historic-Geological and Paleontological Basis* Vol. 1. Edited by B. S. Sokolov and M. A. Ivanovskiy. Moscow: Nauka: 112-116 (in Russian)

Fedonkin, M. A., 1985. Precambrian metazoans: The problems of preservation, systematics and evolution. *Philosophical Transactions of the Royal Society of London* B311: 27-45.

Fedonkin, M. A., 1986. Precambrian problematic animals: Their body plan and phylogeny. In: *Problematic Fossil Taxa.* Edited by A. Hoffman and M.H. Nitecki. Oxford: Oxford University Press: 59-67.

Fedonkin, M. A., 1987. Fossil traces. *Priroda*, 864 (8): 82-94. (in Russian)

Fedonkin, M. A., 1990. Non-skeletal fauna of the Vendian: Promorphological analysis. In: *The Vendian System. 1. Paleontology.* Edited by B.S. Sokolov and A.B. Iwanowski. New York: Springer-Verlag: 7-120, 132-137.

Fedonkin, M. A., 1990. Precambrian metazoans. In: *Palaeobiology. A Synthesis.* Edited by D.E.G. Briggs and P.R. Crowther. New York: Blackwell Scientific: 17-24.

Fedonkin, M. A., 1992. Neoproterozoic ecosystem restructuring: From net to pyramid. In: *Fifth International Conference on Global Bioevents.* Abstract volume. Goettingen: 33-34.

Fedonkin, M. A., 1992. Vendian faunas and the early evolution of Metazoa. In: *Origin and Early Evolution of the Metazoa.* Edited by J. H. Lipps and P. W. Signor. New York: Plenum Press: 87-129.

Fedonkin, M. A., 1994. Vendian body fossils and trace fossils. In: *Early Life on Earth (Nobel Symposium 84).* Edited by S. Bengtson. New York: Columbia University Press: 370-388.

Fedonkin, M. A., 1996. Cold-water cradle of animal life. *Paleontologicheskii Zhurnal* (English version) 30 (6): 669-673.

Fedonkin, M. A., 1996. Geobiological trends and events in the Precambrian biosphere. In *Global Events and Event Stratigraphy in the Phanerozoic: Results of the International Interdisciplinary Cooperation in the IGCP.* Edited by O. H. Walliser. New York: Springer: 89-112.

Fedonkin, M. A., 1996. Precambrian fossil record: New insight of life. In: *Systematic Biology as an Historical Science.* Edited by M. Ghiselin and G. Pinna. *Memorie della Societa Italiana di Scienze Naturali e del Museo Civico di Storia Naturale di Milano,* XXVII Fascicolo 1: 41-48.

Fedonkin, M. A., 1996. The oldest fossil animals in ecological perspective. In: *New Perspectives on the History of Life.* Edited by M. Ghiselin and G. Pinna. *Memoirs of the California Academy of Sciences* 20: 31-45.

Fedonkin, M. A., 1998. Metameric features in the Vendian metazoans. *Italian Journal of Zoology* 65: 11-17.

Fedonkin, M. A., 2000. Cold dawn of animal life. *Priroda* 9: 3-11.

Fedonkin M. A., 2001. Glimpse into 600 million years ago. *Science in Russia* 6: 4-15. (in English and Russian)

Fedonkin M. A., 2001. Proterozoic fossil record and origin of Metazoa. *The 17th International Symposium in Conjunction with Award of the International Prize for Biology. The Origin and Early Evolution of Metazoa.* 5-6 December 2001, Kyoto University, Kyoto, Program and Abstracts: 5-7.

Fedonkin, M. A., 2002. *Andiva ivantsovi* gen. et sp. n. and related carapace-bearing Ediacaran fossils from the Vendian of the Winter Coast, White Sea, Russia. *Italian Journal of Zoology* 69: 175-181.

Fedonkin, M. A., 2003. The origin of the Metazoa in the light of the Proterozoic fossil record. *Paleontological Research* 7 (1): 9-41.

Fedonkin, M. A., *et al.*, 1983. *Upper Precambrian and Cambrian Palaeontology of the East-European Platform. Contribution of the Soviet-Polish Working Group on the Precambrian-Cambrian Boundary Problem.* Warzawa: Wydawnictwa Geologiczne.

Fedonkin, M. A. and Ivantsov, A. Yu., 2001. Faunal succession in the Vendian (Terminal Proterozoic) deposits of the White Sea Region, north of the Russian Platform. *North American Paleontological Convention, Berkeley, California.* Abstract: 50.

Fedonkin, M. A. and Runnegar, B. N., 1992. Proterozoic metazoan trace fossils. In: *The Proterozoic Biosphere – A Multidisciplinary Study.* Edited by J. W. Schopf and C. Klein. New York: Cambridge University Press: 389-395.

Fedonkin, M. A., Simonetta, A. and Ivantsov. A., 2004. New data on *Kimberella*, the Vendian mollusc-like organism (White Sea Region, Russia): Paleoecological and evolutionary Implications. *Symposium IGCP493. The Rise and Fall of the Vendian Biota.* Edited by P. Komarower and P. Vickers-Rich. Prato: Monash University: 23-25.

Fedonkin, M. F., Vickers-Rich, P. and Swalla, B., in prep. The world's oldest tunicate? A new taxon from the Vendian sequence of the White Sea, Russia with affinities to the Namibian *Ausia*.

Fedonkin, M. A. and Waggoner, B. M., 1996. The Vendian fossil *Kimberella*: The oldest mollusc known. *Geological Society of America Abstracts with Programs* 28 (7): 53.

Fedonkin, M. A. and Waggoner, B. M., 1997. The Late Precambrian fossil *Kimberella* is a mollusc-like bilaterian organism. *Nature* 388: 868-871.

Fedonkin, M. and Yochelson, E. L., 2002. Middle Proterozoic (1.5 Ga) *Horodyskia moniliformis* Yochelson and Fedonkin, the oldest known tissue-grade colonial eucaryote. *Smithsonian Contributions to Paleobiology* 94: 1-29.

Fedonkin, M. A., Yochelson, E. L. and Horodyski, R. J., 1994. Ancient Metazoa (A search for ancient Metazoa: The Appekunny Formation, Glacier National Park, Montana). *National Geographic Research and Exploration* 10 (2): 200-223.

Fenchel, T., 2002. *Origin and Early Evolution of Life.* Oxford: Oxford University Press.

Fenchel, T. and Finlay, B. J., 1992. Production of methane and hydrogen by anerobic ciliates containing symbiotic methanogens. *Archives of Microbiology* 157: 457-480.

Fenchel, T. and Finlay, B. J., 1994. The evolution of life without oxygen. *American Scientist* 82: 22-29.

Feng, L. R., Donaldson, J. A. and Holland, H. D., 2000. Alteration rinds on glacial diamictite clasts in the Gowganda formation: Possible indicators of low atmospheric oxygen ca. 2.3 Ga *International Geological Review* 42 (8): 684-690.

Ferris, T., 1997. *The Whole Shebang. A State-of-the-Universe(s) Report.* London: Weidenfeld and Nicolson.

Feynman, R. P., 1985. *QED. The Strange Theory of Light and Matter.* Princeton: Princeton University Press.

Finlay, B. J. and Fenchel, T., 1993. Methanogens and other bacteria as symbionts of free-living anaerobic ciliates. *Symbiosis* 14: 375-390.

Fisk, M. R., Giovannoni, S.J. and Thorseth, I. H., 1998. Alteration of oceanic volcanic glass: Textural evidence of microbial activity. *Science* 281 (5379): 978-980.

Ford, T. D., 1958. Precambrian fossils from Charnwood Forest. *Proceedings of the Yorkshire Geological Society* 31: 211-217.

Ford, T. D., 1962. The oldest fossils. *New Scientist* 15: 191-194.

Ford, T. D., 1963. The Pre-Cambrian fossils of Charnwood Forest. *Transactions of the Leicester Literary and Philosophical Society* 57: 57-62.

Ford, T. D., 1975. The place of Charnwood Forest in the history of geological science. *Transactions of the Leicester Literary and Philosophical Society* 68: 23-31.

Ford, T. D., 1979. Precambrian fossils and the origin of the Phanerozoic phyla. In: *The Origin of Major Invertebrate Groups.* Edited by M. R. House. *Systematics Association Special Volume* 12. London: Academic Press: 7-21.

Ford, T. D., 1980. The Ediacarian fossils of Charnwood Forest, Leicestershire. *Proceedings of the Geologists' Association* 91: 81-83.

Ford, T. D. and Breed, W. J., 1973. The problematical Precambrian fossil *Chuaria. Palaeontology* 16: 535-550.

Fortey, R., 1997. *Life. A Natural History of the First Four Billion Years of Life on Earth.* New York: Vintage Books.

Fortey, R., 2001. The Cambrian explosion exploded. *Science* 293: 438-439.

Fowler, C. M. R., Ebinger, C. J. and Hawkesworth, C. J., eds., 2002. *The Early Earth: Physical, Chemical and Biological Development.* Geological Society Special Publication 199. London: Geological Society.

Foyn, S. and Glaessner, M. F., 1979. *Platysolenites*, other animal fossils, and the Precambrian-Cambrian transition in Norway. *Norsk Geologiska Tidsskrift* 59: 25-46.

Fraser, J. T., 1982. *The Genesis and Evolution of Time. A Critique of Interpretation in Physics.* Amherst: University of Massachusetts Press.

Frausto da Silva, J. J. R. and Williams, R. J. P., 1991. *The Biological Chemistry of the Elements. The Inorganic Chemistry of Life.* Oxford: Clarendon Press.

Frey, R. W., 1975. *The Study of Trace Fossils. A Synthesis of Principles, Problems, and Procedure in Ichnology.* New York, Heidelberg and Berlin: Springer-Verlag.

Friedmann, E. I., ed., 1993. *Antarctic Microbiology.* New York: Wiley-Liss.

Fritz, W. H., 1980. International Precambrian-Cambrian Boundary Working Group's 1979 field study to Mackenzie mountains, Northwest Territories, Canada. *Geological Survey of Canada Paper* 80-1A: 41-45.

Fuhrman, J. A. and Capone, D. G., 2001. Nifty nanoplankton. *Nature* 412: 593-594.

Fuller, M. and Jenkins, J. F., 1995. *Arrowipora fromensis* a new genus and species of tabulate-like coral from the Early Cambrian Moorowie Formation, Flinders Ranges, South Australia. *Transactions of the Royal Society of South Australia* 119 (2): 75-82.

Galimov, E. M., 2001. *Fenomen zhizni: Mezhdu ravnovesiem I nelineinost'yu. Proiskhozhdenie I pritsipy evolyutsii (Phenomenon of Life: Between Equilibrium and Nonlinearity: Origin and Principles of Evolution).* Moscow: Editoria URSS.

Galimov, E. M., 2005. Redox evolution of the Earth caused by a multi-stage formation of its core. *Earth and Planetary Science Letters* 233: 263-276.

Gamow, G., 1952. *The Creation of the Universe.* New York: A Mentor Book, The New American Library.

Gamow, G., 1963. *A Star Called the Sun.* New York: Bantam Pathfinder Editions.

Garcia-Ruiz, J. M., Hyde, S. T., Carnerup, A. M., Christy, A. G., van Kranendonk, M. J. and Welham, N. J., 2003. Self-assembled silica-carbonate structures and detection of ancient microfossils. *Science* 302: 1194-1197.

Gardiner, S. and Hiscott, R. N., 1988. Deep-water facies and depositional setting of the Lower Conception Group (Hadrynian), southern Avalon Peninsula, Newfoundland. *Canadian Journal of Earth Sciences* 25: 1579-1594.

Gardiner, S. L. and Jones, M. L., 1993. *Microscopic Anatomy of Invertebrates.* 12. New York: Wiley-Liss Inc.

Garrett, R., and Klenk, H. P. (ed.), 2006. *Archaea.* Oxford: Blackwell Publishing.

Gauthier-Lafaye, F. and Weber, F., 2003, Natural nuclear fission reactors: Time constraints for occurrence, and their relation to uranium and manganese deposits and to the evolution of the atmosphere. *Precambrian Research* 120: 81-100.

Gee, D. G., Beliakova, L., Pease, V., *et al.,* 2000. New, single zircon (Pb-evaporation) ages from Vendian intrusions in the basement beneath the Pechora Basin, northeastern Baltica. *Polarforschung* 68: 161-170.

Gee, D. G. and Pease, V., eds, 2004. *The Neoproterozoic Timanide Orogen of Eastern Baltica.* London: *Geological Society of London Memoir* 30.

Gehling, J. G. 1982. The sedimentology and stratigraphy of the late Precambrian Pound Subgroup, central Flinders Ranges, South Australia, University of Adelaide, M.Sc. Thesis (unpublished).

Gehling, J. G., 1987. Earliest known echinoderm – a new Ediacaran fossil from the Pound Subgroup of South Australia. *Alcheringa* 11: 337-345.

Gehling, J. G., 1988. A cnidarian of actinian-grade from the Ediacaran Pound Subgroup, South Australia. *Alcheringa* 12: 299-314.

Gehling, J.G., 1991. The case for Ediacaran fossil roots to the Metazoan tree. In: *The World of Martin Glaessner.* Edited by B. P. Radhakrishna. *Geological Society of India Memoir* 20: 181-224.

Gehling, J. G., 1996. Taphonomy of the Terminal Proterozoic Ediacara biota, South Australia. Unpublished Ph.D. Thesis, U.C.L.A.

Gehling, J. G., 1999. Microbial mats in terminal Proterozoic siliciclastics: Ediacaran death masks. *Palaios* 14: 40-57.

Gehling, J. G., 2000. Environmental interpretation and a sequence stratigraphic framework for the terminal Proterozoic Ediacara Member within the Rawnsley Quartzite, South Australia. *Precambrian Research* 100: 65-95.

Gehling, J.G., 2000. Sequence stratigraphic context of the Ediacara Member, Rawnsley Quartzite, South Australia: A taphonomic window into the Neoproterozoic biosphere. *Precambrian Research,* 100: 65-95.

Gehling, J. G., Droser, M. L., Jensen, S. and Runnegar, B. N., 2005. *Ediacara Organisms: Relating Form to Function.* In: *Evolving Form and Function; Fossils and Development.* Edited by D. E. G. Briggs. *Proceedings of a Symposium honouring Adolph Seilacher for his contributions to paleontology in celebration of his 80th birthday.* New Haven: Peabody Museum of Natural History, Yale University: 1-25.

Gehling, J. G., Jensen, S., Droser, M. L., Myrow, P. M. and Narbonne, G. M., 2001. Burrowing below the basal Cambrian GSSP, Fortune Head, Newfoundland. *Geological Magazine* 138: 213-218.

Gehling, J. G., Narbonne. G. M. and Anderson, M. M., 2000. The first named Ediacaran body fossil: *Aspidella terranovica* Billings 1872. *Palaeontology* 43: 427-456.

Gehling, J. G. and Rigby, J. K., 1996. Long-expected sponges from the Neoproterozoic Ediacara Fauna, Pound Subgroup, South Australia. *Journal of Paleontology* 70: 185-195.

Gehling, J. G., Runnegar, B. and Seilacher, A. 1996. Rasping markings of large metazoan grazers, terminal Neoproterozoic of Australia nad Cambrian of Saudi Arabia. *SEPM Meeting, Abstracts*

Gerasimenko, L. M., Hoover, R. B., Rozanov. A. Yu., Zhegallo, E. A. and Zhmur, S. I., 1999. Bacterial paleontology and studies of carbonaceous chondrites. *Paleontological Journal* 33 (4): 439-459.

Gerdes, G., Kelnke, T. and Noffke, N., 2000. Microbial signatures in peritidal siliciclastic sediments: A catalogue. *Sedimentology* 47: 279-308.

Germs, G. J. B., 1968. Discovery of a new fossil in the Nama System, South West Africa. *Nature* 219: 53-54.

Germs, G. J. B., 1972. New shelly fossils from Nama Group, South West Africa. *American Journal of Science* 272: 752-761.

Germs, G. J. B., 1972. The stratigraphy and paleontology of the lower Nama Group, South West Africa. *Bulletin of the Precambrian Research Unit, University of Cape Town* 12.

Germs, G. J. B., 1972. Trace fossils from the Nama Group, South-West Africa. *Journal of Paleontology,* 46 (6): 864-870.

Germs, G. J. B., 1973. A reinterpretation of *Rangea schneiderhoehni* and the discovery of a related new fossil from the Nama Group, South West Africa. *Lethaia* 6: 1-10.

Germs, G. J. B., 1973. Geochemistry of some carbonate members of the Nama Group, south West Africa. *South African Journal of Science* 69: 14-17.

Germs, G. J. B., 1973. Possible sprigginid worm and a new trace fossil from the Nama Group, South West Africa. *Geology* 1973: 69-70.

Germs, G. J. B., 1974. The Nama Group in South West Africa and its relationship to the Pan-African Geosyncline. *Journal of Geology* 82: 301-317.

Germs, G. J. B., 1983. Implications of a sedimentary facies and deposition environmental analysis of the Nama Group in South West Africa/Namibia. *Special Publication of the Geological Society of South Africa* 11 (1983): 89-114.

Germs, G. J. B., 1995. The Neoproterozoic of southwestern Africa, with emphasis on platform stratigraphy and paleontology. In: *Neoproterozoic Stratigraphy and Earth History.* Edited by A. H. Knoll and M. Walter. *Precambrian Research* 73: 137-151.

Germs, G. J. B., Knoll, A. H. and Vidal, G., 1986. Latest Proterozoic microfossils from the Nama Group, Namibia (South West Africa). *Precambrian Research* 32 (1986): 45-62.

Gibbons, W., 1989. Trace fossils from late Precambrian Carolina Slate Belt, south-central Carolina. *Journal of Paleontology* 63: 1-10.

Gibbons, W. and Horak, J. M., 1996. The evolution of the Neoproterozoic Avalonian subduction system: Evidence from the British Isles. In: *Avalonian and Related Peri-Gondwanan Terranes of the Circum-North Atlantic*. Edited by R. D. Nance and M. D. Thompson. *Geological Society of America, Special Paper* 304: 269-280.

Gibson, G. G., Teeter, S. A. and Fedonkin, M. A. 1984. Ediacaran fossils from the Carolina slate belt, Stanley County, North Carolina. *Geology* 12 (7): 387-390.

Gilmore, I. and Sephton, M. A., 2003. *An Introduction to Astrobiology*. Cambridge: The Open University and Cambridge University Press.

Gilmore, R., 1995. *Alice in Quantumland*. New York: Copernicus (Springer-Verlag).

Giribet, G., 2003. Molecules, development and fossils in the study of metazoan evolution; Articulata vs. Ecdysozoa revisited. *Zoology* 106: 303-326.

Glaessner, M. F., 1958. New fossils from the base of the Cambrian in South Australia. *Transactions of the Royal Society of South Australia*, 81: 185-188.

Glaessner, M. F., 1959. Precambrian Coelenterata from Australia, Africa and England. *Nature* 183: 1472-1473.

Glaessner, M. F., 1961. Precambrian Animals. *Scientific American* 204: 72-78.

Glaessner, M. F., 1963. Zur Kenntnis der Nama-Fossilien Sudwest-Afrikas. *Annales Naturhistorie Museum Wien* 66: 13-120.

Glaessner, M. F., 1969. Trace fossils from the Precambrian and basal Cambrian. *Lethaia* 2: 368-393.

Glaessner, M. F., 1977. Re-Examination of *Archaeichnium*, a fossil from the Nama Group. *Annals of the South African Museum* 74 (13): 335-342.

Glaessner, M. F., 1979. An echiurid worm from the late Precambrian. *Lethaia* 12 (2): 121-124.

Glaessner, M. F., 1979. Precambrian. In: *Treatise on Invertebrate Paleontology. Part A*. Edited by R. A. Robison and C. Teichert. Geological Society of America: 79-118.

Glaessner, M. F., 1980. *Parvancorina* – an arthropod from the Late Precambrian (Ediacaran) of South Australia. *Annals Naturhistorie Muse. Wien* 83: 83-90.

Glaessner, M. F., 1984. *The Dawn of Animal Life — A Biohistorical Study*. Cambridge: Cambridge University Press.

Glaessner, M. F. and Daily, B., 1959. The geology and late Precambrian fauna of the Ediacara Fossil Reserve. *Records of the South Australian Museum* 13: 369-401.

Glaessner, M. F. and Wade, M., 1966. The late Precambrian fossils from Ediacara, South Australia. *Palaeontology* 9: 599-628.

Glaessner, M. F. and Wade, M., 1971. *Precambridium* - a primitive arthropod. *Lethaia* 4: 71-77.

Glaessner, M. F. and Walter, M. R., 1975. New Precambrian fossils from the Arumbera Sandstone, Northern Territory, Australia. *Alcheringa* 1 (1): 59-69.

Glendenning, N. K., 2007. *Our Place in the Universe*. London: Imperial College Press.

Glenner, H., Hansen, A. J. Sorensen, M. V., Ronquist, F., Huelsenbeck, J. P. and Willerslev, E., 2004. Bayesian inference of the Metazoan phylogeny: A combined molecular and morphological approach. *Current Biology* 14: 1644-1649.

Gnilovskaya, M. B., 1990. Vendian Actinomycetes and organisms of uncertain systematic postion. In: *The Vendian System. Vol. 1. Paleontology*. Edited by B. S. Sokolov and A. B. Iwanowski. Berlin and New York: Springer-Verlag: 148-153.

Gnilovskaya, M. B., 1990. Vendotaenids – Vendian Metaphytes. In: *The Vendian System. Vol. 1. Paleontology*. Edited by B. S. Sokolov and A. B. Iwanowski. Berlin and New York: Springer-Verlag: 138-147.

Gnilovskaya, M. B., 1996. New saarinids from the Vendian of Russian Platform. *Doklady Akademii Nauk* 348: 89-93. (in Russian and English)

Gnilovskaya, M. B., 1998. The oldest annelidomorphs from the Upper Riphean of Timan. *Doklady Akademii Nauk* 359: 369-372.

Gnilovskaya, M. B., Becker, Yu. R., Weiss, A. F., Olovyanishnikov, Vs. G. and Raaben, M. E., 2000. Pre-Ediacaran fauna of Timan (Upper Riphean annelidomorphs). *Stratigraphy and Geological Correlation* 8 (4): 11-39.

Gnilovskaya, M. B., Istchenko, A. A., Kolesnikov, C. M., Korenchuk, L. V. and Udaltsov, A. P., 1988. *Vendotaenids of the East European Platform*. Nauka, Leningrad. (in Russian)

Gold, T., 1987. *Power from the Earth. Deep Earth Gas – Energy for the Future*. Melbourne: J.M. Dent & Sons.

Gold, T., 1992. *The Deep Hot Biosphere*. New York: Copernicus.

Goldring, R., 1969. Criteria for recognizing Pre-Cambrian fossils. *Nature* 223: 1076.

Goldring, R. and Curnow, C. N., 1967. The stratigraphy and facies of the Late Precambrian at Ediacara, South Australia. *Journal of the Geological Society of Australia* 14: 195-214.

Gonick, L. and Huffman, A., 1991. *The Cartoon Guide to Physics*. New York: Harper Perennial (A Division of HarperCollins).

Goodenough, U., 1998. *The Sacred Depths of Nature*. Oxford: Oxford University Press.

Gostin, V. A., Haines, P. W., Jenkins, R. J. F., Compston, W. and Williams, G. E., 1986. Impact ejecta horizon within late Precambrian shales, Adelaide Geosyncline, South Australia. *Science* 233: 198-200.

Gostin, V. A., Keays, R. R. and Wallace, M. W., 1989. Iridium anomaly from the Acraman impact ejecta horizon: Impacts can produce sedimentary iridium peaks. *Nature* 340 (6234): 542-544.

Gould, S. J., 1989. *Wonderful Life. The Burgess Shale and the Nature of History*. New York and London: W. W. Norton & Company.

Gould, S. J., 1998. On embryos and ancestors. *Natural History* 107 (6): 20-22, 58-65.

Gradstein, F. M., Ogg, J. G., Smith, A. G., Bleeker, W. and Lourens, L. J., 2004. A New Geologic Time Scale with reference to Precambrian and Neogene. *Episodes* 27 (2): 83-100.

Grant, S. W. F., 1990. Shell structure and distribution of *Cloudina*, a potential index fossil for the terminal Proterozoic. *American Journal of Science* 290A: 261-294.

Grant, S. W. F., Knoll, A. H. and Germs, G. J. B., 1991. Probable calcified metaphytes in the latest Proterozoic Nama Group, Namibia: Origin, diagenesis and implications. *Journal of Paleontology* 65 (11): 1-18.

Gravenor, C. P., 1980. Heavy minerals and sedimentological studies on the glaciogenic late Precambrian Gaskiers Formation of Newfoundland. *Canadian Journal of Earth Sciences* 17: 1331-1341.

Gravestock, D. I., 1984. Archaeocyatha from lower parts of the Cambrian carbonate sequence in South Australia. *Association of Australasian Palaeontologists Memoir* 2: 1-139.

Gravestock, D. I., Morton, J. G. G. and Zang, W.-L., 1997. Biostratigraphy and correlation. In: *Petroleum Geology of South Australia. Volume 3: Officer Basin. South Australia*. Edited by J. G. G. Morton and J. F. Drexel. *Department of Mines and Energy Resources Report Book* 97/19: 87-97.

Grazhdankin, D. V., 2000. The Ediacaran genus *Inaria*: A taphonomic/morphodynamic analysis. *Neues Jahrbuch fur Geologie und Palaontologie, Abhandlungen* 216: 1-34.

Grazhdankin D. V., 2003. Structure and depositional environment of the Vendian Complex in the Southeastern White Sea area. *Stratigraphy and Geological Correlation* 11 (4): 313-331.

Grazhdankin, D., 2004. Late Neoproterozoic sedimentation in the Timan foreland. In: *The Neoproterozoic Timanide Orogen of Eastern Baltica*. Edited by D. G. Gee and V. Pease. *Geological Society of London, Memoir* 30: 37-46.

Grazhdankin, D., 2004. Patterns of distribution in the Ediacaran biotas: Facies versus biogeography and evolution. *Paleobiology* 30 (2): 203-221.

Grazhdankin, D. V. and Bronnikov, A. A., 1997. New locality of Late Vendian remains of jellyfish-type organisms in the Onega Peninsula. *Doklody Rossia Akademie Nauk* 357 (6): 792-796.

Grazhdankin, D. V. and Ivantsov, A. Yu., 1996. Reconstruction of biotopes of ancient metazoa of the late Vendian White Sea biota. *Paleontological Journal* 30 (6): 674-678.

Grazhdankin, D. V., Maslov, A. V., Mustill, T. M. R. and Krupenin, M. T., 2005. The Ediacaran White Sea Biota in the central Urals. *Doklady Earth Sciences* 401A (3): 382-385.

Grazhdankin, D. V., Podkovyrov, V. N. and Maslov, A. V., 2005. Paleoclimatic environments of the formation of Upper Vendian rocks on the Belomorian-Kuloi Plateau, southeastern White Sea Region. *Lithology and Mineral Resources* 40 (3): 232-244.

Grazhdankin, D. and Seilacher, A., 2002. Underground Vendobionta from Namibia. *Palaeontology* 45 (1): 57-78.

Grazhdankin, D. and Seilacher, A., 2005. A re-examination of the Nama-type Vendian organism Rangea schneiderhoehni. *Geological Magazine* 142 (4): 1-12.

Green, B. R. and Gantt, E., 2000. Is photosynthesis really derived from purple bacteria? *Journal Phycology* 36: 983-985.

Greenberg, R., 2005. *Europa. The Ocean Moon: Search for an Alien Biosphere*. Berlin: Springer-Verlag.

Greene, B., 1999. *The Elegant Universe: Superstrings, Hidden Dimensions, and the Quest for the Ultimate Theory*. New York: Norton.

Grey, K. 1998. Ediacarian acritarch biozonation in Australia. In: *Inaugural Sprigg Symposium — The Ediacaran Revolution*. 24 June 1998. Compiled by J.G. Gehling. *Geological Society of Australia Abstracts* 51: 22-23.

Grey, K., 2002. Toward Neoproterozoic biozonation in Australia. *First International Palaeontological Congress. Abstracts* 68: 70.

Grey, K., 2005. A baptism of ice and fire. *Australasian Science* 26 (3): 26-28.

Grey, K., 2005. Ediacaran palynology of Australia. *Association of Australasian Palaeontologists Memoir* 31: 1-439.

Grey, K. and Cotter, K. L. 1996. Palynology in the search for Proterozoic hydrocarbons. *Western Australia Geological Survey Annual Review for 1995–96*: 70–80.

Grey, K., Hocking, R. M., Stevens, M. K., Bagas, L., Carlsen, G. M., Irimies, F., Pirajno, F., Haines, P. W. and Apak, S. N., 2005. Lithostratigraphic nomenclature of the Officer Basin and correlative parts of the Paterson Orogen, Western Australia. *Geological Survey of Western Australia* Report 93.

Grey, K., Walter, M. F. and Calver, C. R., 2003. Neoproterozoic biotic diversification: Snowball Earth or aftermath of the Acraman impact? *Geology*, 31 (5): 459-462.

Grey, K. and Williams, I. R., 1987. Possible megascopic algae from the Middle Proterozoic Manganese Group, Bangemall Basin, Western Australia. In: *Abstracts, 4th International Symposium on Fossil Algae*. Edited by S. Beadle. *Friends of Algae Newsletter* 8: 36-37.

Grey, K. and Williams, I. R., 1990. Problematic bedding-plane markings from the Middle Proterozoic Manganese Group, Bangemall Basin, Western Australia. *Precambrian Research* 46: 307-327.

Grey, K., Williams, I. R., Martin, D. McB.., Fedonkin, M. A., Gehling, J. G., Runnegar, B. N. and Yochelson, E. L., 2002. New occurrences of "strings of beads" in the Bangemall Supergroup, a potential biostratigraphic marker horizon. *Western Australian Geological Survey Annual Review* 2000-2001: 69-73.

Gross, G. A., 1983. Tectonic systems and the deposition of iron-formations. *Precambrian Research* 20: 171-187.

Grotzinger, J. P., Bowring, S. A., Saylor, B. Z. and Kaufman, A. J., 1995. Biostratigraphic and geochronologic constraints on early animal evolution. *Science* 270: 598-604.

Grotzinger, J. P. and Rothman, D. H., 1996. An abiotic model for stromatolite morphogenesis. *Nature* 383: 423-425.

Grotzinger, J. P., Watters, W. A. and Knoll, A. H., 2000. Calcified metazoans in thrombolite-stromatolite reefs of the terminal Proterozoic Nama Group, Namibia. *Paleobiology* 26 (3): 334-359.

Gupta, R. S., 1998. Protein phylogenies and signature sequences: A reappraisal of evolutionary relationships among archaebacteria, eubacteria, and eukaryotes. *Microbiology and Molecular Biology Reviews* 62: 1435-1491.

Gureev, Y.A., 1987, Morphological analysis and systematics of Vendiata. *Akademiya Nauk Ukrainian SSR, Institute of Geological Science Preprint:* 87-15, 1-54. (in Russian)

Gureev, Y.A., Chikanov, V. A. and Ivanchenko, V. A., 1985. Besskechetial Faonavo Hoshenyach. *Baitilskoi I Bereshchovskoi Seriv Pochokii, Dokhlady – AN UCCP (USSR)*, B (6): 10-13. (in Russian)

Gürich, G., 1920. Die altesten Fossilien Sudafrikas. *Zeitschrift Praktik Geologie* 37: 85.

Gürich, G., 1930. Die bislang altesten Spuren von Organismen in Sudafrick. *International Geological Congress of South Africa* 1929, 15: 670-680.

Gürich, G., 1930. Uber den Kuibisquarzit in Sudwest-afrika. *Zeitschrift Deutsch Geologie Gesellschaft* 82: 637.

Gürich, G., 1933. Die Kuibis-Fossilien der Nama Formation von Sudwestafrika. *Palaeontologische Zeitschrift* 15: 137-154.

Guth, A., 1997. *The Inflationary Universe: The Quest for a New Theory of Cosmic Origins*. Reading: Addison-Wesley.

Hagadorn, J. W. and Bottjer, D. J., 1997. Wrinkle structures: Microbially mediated sedimentary structure in siliciclastic settings at the Proterozoic-Phanerozoic transition. *Geology* 25: 1047-1050.

Hagadorn, J. W. and Waggoner, B., 2000. Ediacaran fossils from the Southwestern Great Basin, United States. *Journal of Paleontology* 74 (2): 349-359.

Hagadorn, J. W., Xiao, S., Donoghue, P. C. J., Bengston, S., Gostling, N. J., Pawlowska, M., Raff, E. C., Raff, R. A., Turner, F. R., Chonguyu, Y., Zhou, C., Uyan, X., Mc Feely, M. B., Stampanoni, M. and K. H. Nealson, 2006. Cellular and subcellular structure of Neoproterozoic animal embryos. *Science* 314 (5797): 291-294.

Hahn, G., Hahn, R., Leonardos, O. H., Pflug, H. D. and Walde, D. H. G., 1982. Koperlich erhaltene Scyphozoen-Reste aus dem Jungprakambrium Brasiliens. *Geologie und Palaeontologie* 16: 1-18.

Hahn, G. and Pflug, H. D., 1980. Ein neuer Medusen-fund aus dem Jung-Prakambrium von Zentral-Iran. *Senckenbergiana Lethaea* 60: 449-461.

Hahn, G. and Pflug, H. D., 1985. Die Cloudinidae n. fam., Kalk-Rohren aus dem Vendium und Unter-Kambrium. *Senckenbergiana Lethaea* 65: 413-431.

Hahn, G. and Pflug, H. D., 1985. Polypenartige Organismen aus dem Jung-Prakambrium (Nama-Gruppe) von Namibia. *Geologica et Palaeontologica* 19: 1-13.

Hahn, G. and Pflug, H. D., 1988. Zweischalige organismen aus dem Jung-Prakambrium (Vendium) von Namibia (SW-Afrika). *Geologica Palaeontologica* 22: 1-19.

Haines, P. W., 1990. A late Proterozoic storm-dominated carbonate shelf sequence: The Wonoka Formation in the central and southern Flinders Ranges, South Australia. In: *The Evolution of a Late Precambrian – Early Palaeozoic Rift Complex: The Adelaide Geosyncline*. Edited by J.B. Jago and P.S. Moore. *Geological Society of Australia Special Publication* 16: 215-229.

Haines, P. W., 2000. Problematic fossils in the late Neoproterozoic Wonoka Formation, South Australia. *Precambrian Research* 100: 97-108.

Haldane, J. B. S., 1929. The origin of life. Appendix. In: *The Origin of Life*. Edited by J.D. Bernal. London: Weidenfeld and Nicolson: 242-249.

Hale, C. J., 1987. Paleomagnetic data suggest link between the Archean-Proterozoic boundary and inner-core nucleation. *Nature,* 329 (6236): 233-236.

Halliday, A. N., 2001. In the beginning… *Nature* 409: 144-145.

Halverson, G. P., Hoffman, P. F. and Schrag, D. P., 2002. A major perturbation of the carbon cycle before the Ghaub glaciation (Neoproterozoic) in Namibia: Prelude to Snowball Earth. *Geochemistry, Geophysics, Geosystems* 3 (6): 1036.

Halverson, G. P., Hoffman, P. F., Schrag, D. P., Maloof, A. C. and Rice, A. H. N., 2005. Toward a Neoproterozoic composite carbon-isotope record. *Geological Society of America Bulletin* 117 (9/10): 1181-1207.

Hamblin, K. and Christiansen, E., 1998. *The Earth's Dynamic Systems*. New York: Macmillan Publishing Co.

Han, T. M. and Runnegar, B., 1992. Megascopic eukaryotic algae from the 2.1-billion-year-old Negaunee Iron-Formation, Michigan. *Science* 257: 32-235.

Häntzschel, W., 1962. Trace fossils and problematica. In: *Treatise on Invertebrate Paleontology*. Edited by R. C. Moore. Geological Society of America and University of Kansas Press. Vol. W: 177-245.

Harland, W. B., Armstrong, R. L., Cox, A. V., Craig, L. E., Smith, A. G. and Smit, D. G., 1989. *A Geologic Time Scale*. Cambridge: Cambridge University Press.

Harland, W. B., Cox, A. V., Llewellyn, P. G., Picton, C. A. G., Smith, A. G. and Walters, R., 1982. *A Geologic Timescale*. Cambridge: Cambridge University Press.

Harold, F.M., 2001. *The Way of the Cell – Molecules, Organisms and the Order of Life*. Oxford: Oxford University Press.

Harrison, E., 1987. *Darkness at Night. A Riddle of the Universe*. Harvard University Press.

Harrison, T. M., Blicert-Toft, J., Muller, Albarede, F., Holden, P. and Mojzsis, S. J., 2005. Heterogeneous hadean hafnium: Evidence of continental crust at 4.4 to 4.5 Ga. *Science* 310: 1947-1950.

Hartz, E. H. and Tosvik, T. H., 2002. Baltica upside down: A new plate tectonic model for Rodinia and the Iapetus Ocean. *Geology* 30: 255-258.

Hawking, S. W., 1988. *A Brief History of Time. From the Big Bang to Black Holes*. New York: Bantam.

Heckman, D. C., *et al.*, 2001. Molecular evidence of early colonization of land by fungi and plants. *Science* 293: 1129-1133.

Hedges, S. B., Blair, J. E., Venturi, M. L. and Shoe, J. L., 2004. A molecular timescale of eukaryote evolution and the rise of compex multicellular life. *BMC Evolutionary Biology* 4 (2): 1471-1421.

Hedges, S. B. and Kumar, S., 2003. Genomic clocks and evolutionary timescales. *Trends in Genetics* 19 (4): 200-206.

Heisenberg, W., 1971. *Physics and Beyond*. New York: Harper & Row.

Hengeveld, R. and Fedonkin, M. A., 2004. Causes and consequences of eukaryotization through mutualistic endosymbiosis and compartmentalization. *Acta Biotheoretica* 52: 105-154.

Herman, T. N., 1990. *Organic World, a Billion Years Ago*. Leningrad: Nauka.

Hermann, T. N. & Podkovyrov, V. N., 2006. Fungal remains from the Late Riphean. *Paleontological Journal* 40 (2): 207-214.

Hill, A. C., Cotter, K. L. and Grey, K., 2000. Mid-Neoproterozoic biostratigraphy and isotope stratigraphy in Australia. *Precambrian Research* 100: 283–300.

Hill, A. C., Grey, K., Gostin, V. A. and Webster, L. J., 2004. New records of Late Neoproterozoic Acraman ejecta in the Officer Basin. *Australian Journal of Earth Sciences* 51: 47–51.

Hill, A. C. and Walter, M. R., 2000. Mid-Neoproterozoic (~830–750 Ma) isotope stratigraphy of Australia and global correlation. *Precambrian Research* 100: 181–211.

Hill, E. and Bonney, T. G., 1877. The Precarboniferous rocks of Charnwood Forest. *Quarterly Journal of the Geological London* 33: 754-789.

Hir, G. L., *et al.*, 2007. Investigating plausible mechanisms to trigger a deglaciation from a hard Snowball earth. *Comptes Rendus Gescience* 339 (3-4): 274-287.

Hoffman, P. F., Kaufman, A. J., Halverson, G. P. and Schrag, D. P., 1998. A Neoproterozoic Snowball Earth. *Science* 281: 1342-1346.

Hoffman, P. F. and Schrag, D. P., 2000. Snowball Earth. *Scientific American* 282 (1): 50-57.

Hoffman P. F. and Schrag D. P., 2002. The Snowball Earth hypothesis: Testing the limits of global change. *Terra Nova* 14 (3): 129-155.

Hofmann, H. J., 1967. Precambrian fossils (?) near Elliot Lake, Ontario. *Science* 156: 500-504.

Hofmann, H. J., 1971. Precambrian fossils, pseudofossils, and problematica in Canada. *Bulletin of the Geological Survey of Canada,* 189: 1-146.

Hofmann, H. J., 1981. First record of a Late Proterozoic faunal assemblage in the North American Cordillera. *Lethaia* 14: 303-310.

Hofmann, H. J., 1992. Megascopic dubiofossils. In: *The Proterozoic Biosphere – a Multidisciplinary Study*. Edited by J. W. Schopf and C. Klein. New York: Cambridge University Press: 413-419.

Hofmann, H.J., 1985. The mid-Proterozoic Little Dal macrobiota, Mackenzie Mountains, northwest Canada. *Palaeontology* 28: 331–354.

Hofmann, H. J., 1994. Proterozoic carbonaceous compression ("metaphytes" and "worms"). In: *Early Life on Earth*. Edited by S. Bengtson. *Nobel Symposium No. 84*. New York: Columbia University Press: 342-357.

Hofmann, H. J., 1998. Synopsis of Precambrian fossil occurrences in North America. In: *Geology of the Precambrian Superior and Grenvill Provinces and Precambrian Fossils in North America*. Edited by Lucas, S. B. and St-Onge, M. R. *Geological Survey of Canada, Geology of Canada* 7: 271-376.

Hofmann, H. J., 2001. Ediacaran enigmas, and puzzles from earlier times. *Abstract Volume of the Geological Association of Canada Abstract Volume,* 26: 64-65.

Hofmann, H. J., 2005. ?*Hiemalora* and other Ediacaran fossils of northeastern Newfoundland, and correlations within Avalonia. *Abstract Volume of the Geological Society of America Annual Meeting*, Salt Lake City 37 (7): 485.

Hofmann, H.J. and Aitken, J.D., 1979. Precambrian biota from the Little Dal Group, Mackenzie Mountains, northwestern Canada. *Canadian Journal of Earth Sciences* 16: 150–166.

Hofmann, H. J. and Chen, J., 1981. Carbonaceous megafossils from the Precambrian (1800 Ma) near Jixian, northern China. *Canadian Journal of Earth Sciences* 18: 443-447.

Hofmann, H. J., Fritz, W. H. and Narbonne, G. M., 1983. Ediacaran (Precambrian) fossils from the Wernecke Mountains, Northwestern Canada. *Science* 221: 455-457.

Hofmann, H. J., Grey, K., Hickman, A. H. and Thorpe, R. I., 1999. Origin of 3.45 Ga-old stromatolites in Warrawoona Group, Western Australia. *Geological Society of America, Bulletin* 111: 1256–1262.

Hofmann, H. J. and Mountjoy, E. W., 2001. *Namacalathus-Cloudina* assemblage in Neoproterozoic Miette Group (Byng Formation), British Columbia: Canada's oldest shelly fossils. *Geology* 29: 1091-1094.

Hofmann, H. J., Mountjoy, E. W. and Teitz, M. W., 1985. Ediacaran fossils from the Miette Group, Rocky Mountains, British Columbia, Canada. *Science* 221: 455-457.

Hofmann, H. J., Narbonne, G. M. and Aitken, J. D., 1990. Ediacaran remains from intertillite beds in northwestern Canada. *Geology* 18: 1199-1202.

Hofmann, H. J., O'Brien, S. J. and King, A. F., 2005. Ediacaran fossils on Bonavista Peninsula, Newfoundland, Canada. *North American Paleontological Convention Abstract Volume*, 2005.

Holland, H. D., 1994: Early Proterozoic atmospheric change. In: *Early Life on Earth*. Edited by S. Bengtson. *Nobel Symposium No. 84*. New York: Columbia University Press: 237-244.

Holland, H. D., 1999. When did the Earth's atmosphere become oxic? A reply. *The Geochemical News* 100: 20-23.

Holland, H. D., 2002. Volcanic gases, black smokers, and the great oxidation event. *Geochimica et Cosmochimica Acta* 66: 3811-3826.

Holland, H. D., 2004. The geological history of seawater. In: *Treatise on Geochemistry*. Edited by H. D. Holland and K. K. Turekian. Oxford: Elsevier: 583-625.

Holland, H. D. and Kasting, J. F., 1991. The environment of the Archean Earth. In: *The Proterozoic Biosphere. A Multidisciplinary Study*. Edited by J.W. Schopf and C. Klein. Cambridge: Cambridge University Press: 21-24.

Holmes, J. and Maslin, M., 2007. Stable Isotopes in Paleoclimatology. Oxford: Blackwell Publishing.

Hoover, R. B. and Rozanov, A. Yu., 2002. Chemical biomarkers and microfossils in carbonaceous meteorites. Instruments, methods, and missions for astrobiology. *Proceedings of SPIE, the International Journal for Optical Engineering* 4495: 1-18.

Hoppert, M. and Mayer, F., 1999. Prokaryotes. *American Scientist* 87 (6): 518-525.

Horodyski, R. J., 1982. Problematic bedding-plane markings from the Middle Proterozoic Appekunny Argillite, Belt Supergroup, Northwestern Montana. *Journal of Paleontology* 56: 882-889.

Horodyski, R. J., 1983. Sedimentary geology and stromatolites of the Middle Proterozoic Belt Supergroup, Glacier National Park, Montana. *Precambrian Research* 20: 391-425.

Horodyski, R. J., 1989. Paleontology of the Middle Proterozoic Belt Supergroup. In: *Middle Proterozoic Belt Supergroup, Western Montana*. Edited by D. Winston, R. J. Horodyski and J. W. Whipple. *28th International Geological Congress, Field Trip Guidebook* T334: 7-26.

Horodyski, R. J., 1991. Late Proterozoic megafossils from southern Nevada. *Geological Society of America, Abstracts with Programs* 26 (5): 163.

Hou, S.-G., Aldridge, R. J., Bergstrom, J., Siveter, D. J., Siveter, D. J. and Feng, X.-H., 2004. *The Cambrian Fossils of Chengjiang, China. The Flowering of Early Animal Life*. London: Blackwell Publishing.

Hoyle, F., 1950. *The Nature of the Universe*. New York: Harper & Brothers.

Hoyle, F., 1960. On the origin of the solar system. *Quarterly Journal of the Royal Astronomical Society* 1: 28.

Hoyt, D. and Shatten, K. H., 1997. *The Role of the Sun in Climate Change*. Oxford: Oxford University Press.

Hua, H., Chen, A., Yuan, X. and Xiao, S., 2005. The first biomineralizing *Cloudina* reconsidered. *North American Paleontological Convention, Abstract Volume*.

Hua, H., Chen, Z., Yuan, X., Zhang, L. and Xiao, S., 2005. Skeletogenesis and asexual reproduction in the earliest biomineralizing animal *Cloudina*. *Geology* 33: 277-280.

Hua, H., Pratt, B. R. & Zhang, L.-Y., 2003. Borings in *Cloudina* shells: Complex predator-prey dynamics in the terminal Neoproterozoic. *Palaios* 18: 454-459.

Huang, D.-Y., Chen, J.-Y., Vannier, J. and Salinas, J. I. S., 2004. Early Cambrian sipunculan worms from southwest China. *Proceedings of the Royal Society of London* B 271 (1549): 1671-1676.

Hunter, D. R., ed., 1984. *Precambrian of the Southern Hemisphere*. Amsterdam: Elsevier.

Hutchison, R., 1983. *The Search for Our Beginning*. Oxford: Oxford University Press.

Hyde, S. T., Carnerup, A. M., Larsson, A. K. and Christy, A. G., 2004. Self-assembly of carbonate-silica colloids: Between living and non living form. *Physica* 339: 24-33.

Hyde, W. T., Crowley, T. J., Baum, S. K. and Peltier, W. R., 2000. Neoproterozoic 'Snowball Earth' simulations with a coupled climate/ice-sheet model. *Nature* 405: 425-429.

Igolkina, N. S., 1956. Age of sandy-clayey deposits at the "Zimnii Coast" of the White Sea. *Materialy po geologii Evropeiskoi territorii SSSR Leningrad. Vses. Geological Institute, Leningrad,* 14: 169-173.

Igolkina, N. S., 1959. A possibility to distinguish the Baltic Complex in the northern Russian Platform. *Geologiya Evropeiskoi territorii SSSR (Geology of the USSR European Territory). Geological Institute, Leningrad* 17-23.

Isley, A. E. and Abbot, D. H., 1999. Plume-related mafic volcanism and the deposition of banded iron formations. *Journal of Geophysical Research* 104: 15461-15477.

Ivantsov, A. Yu., 1999. A new Dickinsonid from the Upper Vendian of the White Sea Winter Coast (Russia, Arkhangelsk Region). *Paleontological Journal* 33 (3): 211-221.

Ivantsov, A. Yu., 2001. Dependence of reconstructions on the preservation character of the Ediacaran organisms. In: *Ecosystem Restructure and the Evolution of the Biosphere* 4. Edited by A. G. Ponomarenko, A. Yu. Rozanov and M. A. Fedonkin. Paleontological Institute of the Russian Academy of Science, Moscow: 64-66. (in Russian)

Ivantsov, A. Yu., 2001. Traces of active moving of the large late Vendian Metazoa over the sediment surface. In: Ponomarenko, A. G., Rozanov, A. Yu. and Fedonkin, M. A.., eds., *Ecosystem Restructure and the Evolution of the Biosphere* 4. Paleontological Institute of the Russian Academy of Sciences, Moscow: 119-120.

Ivantsov, A. Yu., 2003. Vendian organism recognized by the impressions. *Priroda* 10: 3-9. (in Russian)

Ivantsov, A. Yu., 2004. *Vendia* and other Precambrian "arthropods." *Paleontological Journal* 35 (4): 335-343.

Ivantsov, A. Yu., 2004. Vendian animals in the Phyum: Protoarticulata. In: *The Rise and Fall of the Ediacara (Vendian) Biota. Symposium IGCP493*. Edited by P. Komarower and P. Vickers-Rich. Prato: Monash University: 52.

Ivantsov, A. Y. and Fedonkin, M. A., 2001. Locomotion trails of the Ediacara-type organisms preserved with the producer's body fossils, White Sea, Russia. Abstract. *Geological Association of Canada Joint Annual Meeting – Late Neoproterozoic Evolution of Life and Earth*: 69.

Ivantsov, A. Y. and Fedonkin, M. A., 2001. Locomotion trails of the Vendian invertebrates preserved with the producer's body fossils, White Sea, Russia. North American Paleontological Convention Programs & Abstracts, *PaleoBios* 21: 72.

Ivantsov, A. Y. and Fedonkin, M. A., 2001.Traces of self-dependent movement: Final evidence for the animal nature of Ediacaran organisms. In: *Proceedings of the II International Symposium "Evolution of Life on the Earth*. Edited by V.M. Podobina. Tomsk: NTL: 133-137. (in Russian)

Ivantsov, A. Y. and Fedonkin, M. A., 2002. Conulariid-like fossil from the Vendian of Russia: A metazoan clade across the Proterozoic/Palaeozoic boundary. *Palaeontology* 45 (6): 1219-1229.

Ivantsov, A. Yu. and Grazhdankin, D. V., 1997. A new representative of the Petalonamae from the Upper Vendian of the Arkhangelsk region. *Paleontological Journal* 31 (1): 1-16.

Ivantsov, A. Yu. and Malakhovskaya, Y. E., 2002. Giant traces of Vendian animals. *Doklady Akademii Nauk* 385 (6): 618-622.

Ivantsov, A. Yu., Malakhovskaya, Y. E. and Serezhnikova, E. A., 2004. Some problematic fossils from the Vendian of the southeastern White Sea Region. *Paleontological Journal* 38 (1): 1-9.

Jackson, M. S., Muir, M. P. and Plumb, K. A., 1987. Geology of the southern McArthur Basin, Northern Territory. *Bulletin of the Bureau of Mineral Resources, Geology and Geophysics* 200: vii, 1-173.

James, N. P., Narbonne, G. M. and Kyser, T. K., 2001. Late Neoproterozoic cap carbonates: Mackenzie Mountains, northwestern Canada: Precipitation and global glacial meltdown. *Canadian Journal of Earth Sciences* 38: 1229-1262.

James, N.P., Narbonne, G.M., Kyser, T. K. and Dalrymple, R.W., 2005. Glendonites in Neoproterozoic low latitude, interglacial sedimentary rocks, NW Canada: Insights into the Cryogenian ocean and Precambrian cold-water carbonates. *Geology* 33: 9-12.

James, N. P. Narbonne, G. M. and Sherman, A. G., 1998. Molar-tooth carbonates: Shallow subtidal facies of the mid-to late Proterozoic. *Journal of Sedimentary Research* 68: 716-722.

Jankauskas, T. V., Mikhailova, N. S. and German, T. N., eds, 1989. *Mikrofossilii Dokembriya SSSR. Precambrian Microfossils of the USSR.* Lenningrad: Trudy Institut Geologii i Geokhronologii [Proceedings of the Institute of Geology and Geochronology] Akademiya Nauk SSSR. (in Russian)

Jankauskas, T. V. and Sarjeant, W. A. S., 2001. Boris V. Timofeyev (1916–1982): Pioneer of Precambrian and Early Paleozoic palynology. *Earth Sciences History* 20: 178–192.

Javaux, E., Knoll, A. H. and Walter, M. R., 2001. Ecological and morphological complexity in early eukaryotic ecosystems. *Nature* 412: 66-69.

Javaux, E. J., Knoll, A. H. and Walter, M. R., 2003. Recognising and interpreting the fossils of early eukaryotes. *Origins of Life and Evolution of the Biosphere* 33: 75–94.

Jell, P. A., Jago, J. B. and Gehling, J. G., 1992. A new conocoryphid trilobite from the Lower Cambrian of the Flinders Ranges, South Australia. *Alcheringa* 16: 189-200.

Jenkins, R. J. F., 1981. Concept of an 'Ediacaran period' and its stratigraphic significance in Australia. *Transactions of the Royal Society of South Australia* 105: 179-194.

Jenkins, R. J. F., 1983. Interpreting the oldest fossil cnidarians. Proceedings of the 4th International Symposium on Fossil Cnidaria. *Palaeontographica Americana* 54: 95-104.

Jenkins, R. J. F., 1984. Ediacaran events: Boundary relationships and correlation of key sections, especially in 'Armorica.' *Geological Magazine* 121: 635-643.

Jenkins, R. J. F., 1984. Interpreting the oldest fossil Cnidaria. *Palaeontographica Americana* 54: 95-10.

Jenkins, R. J. F., 1985. The enigmatic Ediacaran (late Precambrian) genus *Rangea* and related forms. *Paleobiology* 11 (3): 336-355.

Jenkins, R. J. F., 1989. The "supposed terminal Precambrian extinction event" in relation to the Cnidaria. *Memoir of the Association of Australasian Palaeontologists* 8: 307-317.

Jenkins, R. J. F., 1992. Functional and ecological aspects of Ediacaran Assemblages. In: *Origin and Early evolution of the Metazoa.* Edited by J.H. Lipps and P.W. Signor. New York: Plenum Press: 131-176.

Jenkins, R. J. F. 1995. The problems and potential of using animal fossils and trace fossils in terminal Proterozoic biostratigraphy. *Precambrian Research* 73: 51-69.

Jenkins, R. J. F., 1996. Aspects of the geological setting and palaeobiology of the Ediacara assemblage. In: *Natural History of the Flinders Ranges.* Edited by M. Davies, C.R.Twidale and M. J. Tyler. *Royal Society of South Australia:* 33-45.

Jenkins, R. J. F., Ford, C. H. and Gehling, J. G., 1983. The Ediacara Member of the Rawnsley Quartzite: The context of the Ediacara assemblage (late Precambrian, Flinders Ranges). *Journal of the Geological Society of Australia* 30: 101-119.

Jenkins, R. J. F. and Gehling, J. G., 1978. A review of frond-like fossils of the Ediacara assemblage. *Records of the South Australian Museum* 17: 347-359.

Jenkins, R. J. F., McKirdy, D. M. and Nedin, C., compilers, 1998. *The Ediacaran I South Australia: Proposal and Field Guide Supporting GSSP Position 'C' at Wearing Dolomite, Flinders Ranges.* IUGS Working Group on the Terminal Proterozoic System, Flinders Ranges Field Trip, 16-22 June 1998.

Jenkins, R. J. F., Plummer, P. S. and Moriarty, K., 1981. Late Precambrian pseudofossils from the Flinders Ranges, South Australia. *Transactions of the Royal Society of South Australia* 105: 67-83.

Jensen, S., Droser, M. L. and Gehling, J. G., 2005. Trace fossil preservation and the early evolution of animals. *Palaeogeography, Palaeoclimatology, Palaeoecology* 220: 19-29.

Jensen, S., Droser, M. L. and Gehling, J. G., 2006. A critical look at the Ediacaran trace fossil record. In: *Neoproterozoic Geobiology and Paleobiology.* Edited by S. Xiao and A. J. Kaufman. Dordrecht: Springer: Chap. 5.

Jensen, S., Gehling, J. G., and Droser, M. L., 1998. Ediacaran-type fossils in Cambrian sediments. *Nature* 393: 567-569.

Jensen, S., Gehling, J. G., Droser, M. L. and Grant, S. W. F., 2002. A scratch circle origin for the medusoid fossil *Kullingia.* *Lethaia* 35 (15): 291-299.

Jensen, S., Palacios, T. and Mus, M., 2005. Megascopic filamentous organisms preserved as grooves and ridges in Ediacaran siliciclastics. *North American Paleontological Convention 2005, Abstracts Volume.*

Jensen, S., Saylor, B. Z., Gehling, J. G. and Germs, G. J. B., 2000. Complex trace fossils from the terminal Proterozoic of Namibia. *Geology* 28: 143-146.

Jezek, P., Willner, A. P., Aceñolaza, F. G. and Miller, H., 1985. The Puncoviscana Trough – a large basin of Late Precambrian to Early Cambrian age on the Pacific edge of the Brazilian shield. *Geologische Rundschau* 74 (3): 573-584.

Kalberg E. A. 1940. Geological description of Onega Peninsula. *Transaction of Northern Geological Department.* Leningrad: Gostoptechizdat.

Kalkowsky, E., 1908. Oolith und stromatolith im norddeutschen Bundsandstein. *Zeitschrift der Deutschen Geologischen Gesellschlat* 60: 68-125.

Kanuth, L. P., 1998. Salinity history of the Earth's early ocean. *Nature* 395: 554-555.

Kanuth, L. P., 2005. Temperature and salinity history of the Precambrian ocean: implication for the course of microbial evolution. In: *Geobiology: Objectives, concepts, Perspectives.* Edited by N. Noffke. New York: Elsevier: 53-69.

Karhu. J. A. and Holland, H. D., 1996. Carbon isotopes and the rise of atmospheric oxygen. *Geology* 24 (10): 867-870.

Kasting, J., 1993. Earth`s early atmosphere. *Science* 259: 920-926.

Kasting, J. and Siefert, J. L., 2002. Life and the evolution of Earth`s atmosphere. *Science* 296: 1066-1068.

Kasting, J. F., 1988. Runaway and moist greenhouse atmospheres and the evolution of Earth and Venus. *Icarus* 74: 472.

Kasting, J. F., Whitmire, D. P. and Reynolds, R. T., 1993. Habitable zones around main sequence stars. *Icarus* 101: 108-128.

Kaufman, A. J., Knoll, A. H. and Narbonne, G. M., 1997: Isotopes, ice ages, and terminal Proterozoic earth history. *Proceedings of the National Academy of Sciences of the United States of America* 94: 6600-6605.

Kaufman, A. J. and Xiao, S., 2003. High CO_2 levels in the Proterozoic atmosphere estimated from analyses of individual microfossils. *Nature* 425: 279-282.

Kearey, P., Klepeis, K. A. and Vine, F. J., 2008. Global Tectonics. Oxford: Blackwell Publishing.

Keller, B. M., 1969. Imprint of an unknown animal from the Valdai Series of the Russian Platform In: The Tommotian Stage and the Problem of the Lower Boundary of the Cambrian. Edited by A. Yu. Rozanov. *Transactions of the Geological Institute, SSSR,* 206: 175-176. (in Russian; English translation edited by M. E. Raaben, published in 1981 for the U. S. Department of the Interior and the National Science Foundation (Washington, D.C.) by Amerind, New Dehli.

Keller, B. M., 1974. Stratotype of the Vendomian in the South Urals: In: *Essays of Stratigraphy.* Edited by A. L. Yanshin. Moscow: Nauka: 97-101. (in Russian)

Keller, B. M. and Fedonkin, M. A., 1977. New organic fossil finds in the Precambrian Valdai Series along the Syuz'ma River. *International Geological Review* 19 (8): 924-930.

Keller, B. M., Menner, V. V., Stepanov, V. A. and Chumakov, N. M.. 1974. New finds of fossils in the Precambrian Valday Series along the Syuzma River. *Izvestia Akademii Nauk SSSR, Seriya Geologicheskaya* 12: 130-134.

Kendall, B., Creaser, R. A., Ross, G. M. and Selby, D., 2004. Constraints on the timing of Marinoan "Snowball Earth" glaciation by 187 Re – 187 Os dating o f a Neoproterozoic post-glacial black shale in western Canada. *Earth and Planetary Science Letters* 22: 729-740.

Kennedy, M. J., Christie-Blick, N. and Prave, A. R., 2001. Carbon isotopic composition of Neoproterozoic glacial carbonates as a test of paleoceanographic models for Snowball Earth phenomena. *Geology* 29 (12): 1135-1138.

Kennedy, M. J., Christie-Blick, N. and Sohl, L. E., 2001. Are Proterozoic cap carbonates and isotopic excursions a record of gas hydrate destabilization following Earth's coldest intervals? *Geology* 29 (5): 443-446.

Kennedy, M. J., Droser, M., Mayer, L. M., Pevear, D. and Mrofka, D., 2006. Late Precambrian oxygenation; Inception of the clay mineral factory. *Science*, 311: 1446-1449.

Kennedy, M. J., Runegar, B., Prave, A. R., Hoffman, K.-H. and Arthur, M. A., 1998. Two or four Neoproterozoic glaciations? *Geology* 26: 1059-1063.

Keppie, J. D. and Dostal, J., 1998. Birth of the Avalon arc in Nova Scotia, Canada: Geochemical evidence for ~700-630 Ma back-arc rift volcanism off Gondwana. *Geological Magazine* 135: 171-181.

Khomentovsky V. V. and Karlova G.A., 2002. Boundary between the Nemakit-Daldyn and Tommotian stages (Vendian-Cambrian) in Siberia. *Stratigraphy and Geological Correlation* 10 (3): 13-34. (in Russian and English).

Khomentovsky V. V., 1990. Vendian of the Siberian Platform. In: Sokolov, B. S. and Fedonkin, M. A. (eds). *The Vendian System, Vol. 2 Regional Geology.* Berlin: Springer-Verlag.

Khomentovsky V. V. and Karlova G. A. 2002. Boundary between the Nemakit-Daldyn and Tommotian stages (Vendian-Cambrian) in Siberia. *Stratigraphy and Geological Correlation* 10 (3): 13-34. (in Russian and English)

King, A. F., 1988. Geology of the Avalon Peninsula, Newfoundland. *Newfoundland Department of Mines and Energy, Geological Survey,* Map 88-01, scale 1:250 000.

King, A. F., 1990. Geology of the St John's area. *Newfoundland Department of Mines and Energy, Geological Survey Branch Report* 90-2: 1-88.

King, A. F., Anderson, M. M. and Benus, A. P., 1988. Late Precambrian sedimentation and related orogenesis of the Avalon Peninsula, eastern Avalon Zone. *Field Trip Guidebook Trip 4A, Geological Association of Canada, Annual Meeting,* St John's, Newfoundland.

Kippenhahn, R. and Weigert, A., 1990. *Stellar Structure and Evolution.* Berlin: Springer.

Kirschvink, J. L., 1992. Late Proterozoic low latitude glaciation: The Snowball Earth. In: *The Proterozoic Biosphere: A Multidisciplinary Study.* Edited by J.W. Schopf and C. Klein. Cambridge: Cambridge University Press: 51-52.

Kirschvink, J. L., Gaidos, E. J., Bertani, L. E., *et al.,* 2000. Paleoproterozoic Snowball Earth: Extreme climatic and geochemical global change and its biological consequences. *Proceedings of the National Academy of Science Sciences of the United States of America* 97 (4): 1400-1405.

Kirschvink, J. L. and Raub, T. D., 2003. A methane fuse for the Cambrian explosion. *C. R. Geoscience, Science Direct* 335: 65-78.

Kirschvink, J. I., Ripperdan, R. L. and Evans, D. A., 1997. Evidence for a large-scale reorganization of Early Cambrian continental masses by inertial interchange true polar wander. *Science* 277: 541-545.

Klein, C. and Beukes, N. J., 1992. Proterozoic iron-formations. In: *Proterozoic Crustal Evolution.* Edited by K. C. Condie. Amsterdam: Elsevier: 383-418.

Knauth, L. P., 2004. Temperature and salinity history of the Precambrian ocean: Implication for the course of microbial evolution. *Palaeogeography, Palaeoclimatology, Palaeoecology* 219 (2005): 53-69.

Knoll, A. H., 1991. End of the Proterozoic Eon. *Scientific American* 265 (4): 64-73.

Knoll, A. H., 1992. The early evolution of eukaryotes: A geological perspective. *Science* 256: 622-627.

Knoll, A. H., 1994. Proterozoic and early Cambrian protists: Evidence for accelerating evolutionary tempo. *Proceedings of the National Academy of Sciences of the United States of America* 91: 6743-6750.

Knoll, A. H., 1996. Archaean and Proterozoic palaeontology. In: *Palynology: Principles and Applications.* Edited by J. Jansonius and D. C. McGregor. American Association of Stratigraphic Palynologists Foundation 1. Salt Lake City: Publishers Press: 51-80.

Knoll, A. H., 1996: Breathing room for early animals. *Nature* 382: 111-112.

Knoll, A. H., 2000. Learning to tell Neoproterozoic time. *Precambrian Research* 100: 3–20.

Knoll, A.H., 2003. *Life on a Young Planet. The First Three Billion Years of Evolution on Earth.* Princeton and Oxford: Princeton University Press.

Knoll, A. H., Barghoorn, E. S. and Golubic, S., 1975. *Palaeopleurocapsa wopfneri* gen. et sp. nov.: A late Precambrian alga and its modern counterpart. *Proceedings of the National Academy of Science USA* 72: 2488-2492.

Knoll, A. H. and Butterfield, N. J., 1989. New window on Proterozoic life. *Nature* 337: 602-603.

Knoll, A. H. and Carroll, S. B., 1999. Early animal evolution: Emerging views from comparative biology and geology. *Science* 284: 2129-2137.

Knoll, A. H. and Golubic, S., 1979. Anatomy and taphonomy of a Precambrian algal stromatolite. *Precambrian Research* 10: 115-151.

Knoll, A. H., Grotzinger, J. P., Kaufman, A. J. and Kolosov, P., 1995. Integrated approaches to terminal Proterozoic stratigraphy: An example from the Olenek Uplift, northeastern Siberia. *Precambrian gas Research* 73: 251-270.

Knoll, A. H. and Walter, M. R., 1992. Latest Proterozoic stratigraphy and Earth history. *Nature* 356: 673-678.

Kooijman, S.A.L.M, Auger, P., Poggiale, J.C. and Kooi, B.W., *in press.* Quantitative steps in symbiogensesis and the evolution of homeostasis. *Biological Review.*

Kooijman, S. A. L. M. and Hengeveld, R., *in press.* The symbiotic nature of metabolic evolution. *Acta Biotheoretica.*

Kopal, Z., 1973. *The Solar System.* Oxford: Oxford University Press.

Kopal, Z., 1979. *The Realm of the Terrestrial Planets.* Toronto : Halstead Press.

Kopp, R. E., Kirschvink, J. L., Hilburn, I. A. and Nash, C. Z., 2005. The Paleoproterozoic Snowball Earth: A climate disaster by the evolution of oxygenic phytosynthesis. *Proceedings of the National Academy of Sciences of the United States of America* 102: 11131-11136.

Krogh, T. E., Strong, D. F., O'Brien, S. J. and Papezik, V., 1988. Precise U-Pb dates from zircons in the Avalon Terrane of Newfoundland. *Canadian Journal of Earth Sciences* 25: 442-453.

Laflamme, M., Narbonne, G. M. and Anderson, M. M., 2004. Morphometric analysis of the Ediacaran frond *Charniodiscus* from the Mistaken Point Formation, Newfoundland. *Journal of Paleontology* 78 (5): 827-837.

Laflamme, M., Narbonne, G. M., Greentree, C. and Anderson, M. M., 2007. Morphology and Taphonomy of an Ediacaran Frond: *Charnia* from the Avalon Peninsula of Newfoundland. In: *The Rise and Fall of the Ediacaran Biota.* Edited by P. Vickers-Rich and P. Komarower. *Geological Society of London Special Publication* 286..

Lahov, N., 1999. *Biogenesis: Theories for Life's Origin.* Oxford: Oxford University Press.

Landing, E., 1994. Precambrian-Cambrian boundary ratified and a new perspective of Cambrian time. *Geology* 22: 179-182.

Landing, E., 1996. Avalon: Insular continent by the latest Precambrian. In: *Avalonian and Related Peri-Gondwanan Terrances of the Circum-North Atlantic.* Edited by R.D. Nance and M.D.Thomson. *Geological Society of America Special Paper* 304: 29-63.

Landing, E., Narbonne, G. M. and Myrow, P. M., eds , 1988. Trace Fossils, Small Shelly Fossils, and the Precambrian-Cambrian Boundary. *New York State Museum and Geological Survey Bulletin* 463.

Landing, E., Narbonne, G. M., Myrow, P. M., Benus, A. P. and Anderson, M. M., 1988. Faunas and depositional environments of the Upper Precambrian through Lower Cambrian, southeastern Newfoundland. In: *Trace Fossils, Small Shelly Fossils, and the Precambrian-Cambrian Boundary.* Edited by E. Landing, G.M. Narbonne and P.M. Myrow. *New York State Museum and Geological Survey Bulletin* 463: 27-32.

Lane, N., 2002. *Oxygen. The Molecule that Made the World.* Oxford: Oxford University Press.

Lasaga, A. C. and Ohmoto, H., 2002. The oxygen geochemical cycle: Dynamics and stability. *Geochimica and Cosmochimica Acta* 66: 361-381.

Lazcano, A., 1994. The RNA world, its predecessors and descendants. In: *Early Life on Earth.* Edited by S. Bengtson. Nobel Symposium 84. New York: Columbia University Press: 70-80.

Lee, J. S., 1924. Geology of the Gorge District of the Yangtse from Ichang to Tzehui, with special reference to the development of the Gorges. *Bulletin Geological Society of China* 3: 351-391.

Legouta, A. and Seilacher, A., 2001. Ediacaran trace fossils reinterpreted as xenophyophoran protests. Abstracts. *GAC/MAC Annual Meeting.* St John's, Newfoundland.

Le Grand, H. E., 1994. *Drifting Continents and Shifting Theories.* Cambridge: Cambridge University Press.

Leiming, Y., Zhu, M., Knoll, A. H., Yuan, X., Zhang, J. and Hu, J., 2007. Doushantuo embryos preserved inside diapause egg cysts. *Nature* 446: 661-663.

Leitch, A. M., 2001. Archaean plate tectonics and MOM. In: *International Archaean Symposium.* Edited by K.F. Cassidy, J.M. Dunphy and M.J. Van Kranendonk. *Extended Abstracts. AGSO – Geoscience Australia Record* 2001/37: 61-63.

Lemonick, M., 1998. *Other Worlds: The Search for Life in the Universe.* New York: Simon and Schuster.

Leonov, M. V., 2004. Comparative taphonomy of the Vendian genera *Beltanelloides* and *Nemiana* as a key to their true nature. In: *The Rise and Fall of the Ediacara (Vendian) Biota. Symposium IGCP493.* Edited by P. Komarower and P. Vickers-Rich. Prato: Monash University: 66-67.

Lerman. L. and Teng, J., 2004. In the beginning. A functional First Principles Approach to chemical evolution. In: *Origins. Genesis, Evolution and Diversity of Life.* Edited by J. Seckbach. Dordrecht: Kluwer Academic Publishers: 37-55.

Leslie, J., 1998. *Modern Cosmology and Philosophy.* New York: Prometheus.

Li. C.-W., 2006. Diversity and taphonomy of early metazoan embryos. *Abstract Volume,* IGCP 493 Symposium 27-31 January 2006, Kyoto, Japan. Kyoto: Kyoto University: 17-18.

Li, C.-W., Chen, J.-Y. and Hua, T.-E., 1998. Interpreting Late Precambrian microfossils: Response. *Science* 282: 1783a.

Li, C.-W., Chen, J.-Y. and Hua, T.-E., 1998. Precambrian sponges with cellular structures. *Science* 279: 879-882.

Li, Z. X., Powell, C. McA., 2001. An outline of the palaeogeographic evolution of the Australasian region since the beginning of the Neoproterozoic. *Earth-Science Reviews* 53 (2001): 237-277.

Lichtenberg. D., 2007. *The Universe and the Atom.* London: Imperial College.

Lidsey, J. E., 2000. *The Bigger Bang.* Cambridge: Cambridge University Press.

Lightman, A., 1991. *Ancient Light. Our Changing View of the Universe.* Cambridge: Harvard University Press.

Lin, J.-P., Gon III, S. M., Gehling, J. G., Babcock, L. E., Zhao, Y.-L., Zhang, X.-L., Hu, S.-X., Yuan, J.-L., Yu, M. Y. and Peng, J., 2006. A *Parvancorina*-like arthropod from the Cambrian of South China. *Historical Biology* 18 (1): 33-45.

Lin, J.-P., Scott, A. C., Li, C.-W., Wu, H.-J., Ausich, W. I., Zhao, Y.-L. and Swu, Y.-K., 2006. Silicified egg clusters from a Middle Cambrian Burgess Shale-type deposit, Guizhou, south China. *Geology* 34 (12): 1037-1040.

Lipps, P. S., 1975. Chemistry of the sea surface microlayer. In: *Chemical Oceanography.* Edited by J. P. Ridley and G. Skirrow. London: Academic Press: 193-243.

Lipps, J. H., Collins, A. G. and Fedonkin, M. A. 1998. Evolution of biological complexity: Evidence from geology, paleontology and molecular biology. In *Instruments, Methods, and Missions for Astrobiology.* Edited by R.B. Hoover. *Proceedings of The International Society for Optical Engineering* 3441: 138-148.

Lipps, J. H. and Signor. P. W., eds., 1992. *Origin and Early Evolution of the Metazoa.* New York: Plenum Press.

Liu, S. V., Zhou, J., Zhang, C., Cole, D. R., Gajdarziska-Josifovska, M. and Phelps, T. J., 1997. Thermophilic Fe(III)-reducing bacteria from the deep subsurface: The evolutionary implications. *Science* 277: 1106-1109.

Lodish, H., Berk, A., Matsudaira, P., Daiser, C. A., Krieger, M., Scott, M. P., Zipursky, S. L. and Darnell, J., 2003. *Molecular Cell Biology.* New York: W. H. Freeman and Company, 5th Edition.

Logan, B. W., 1961. *Cryptozoon* and associated stromatolites from the Recent, Shark Bay, Western Australia. *Journal of Geology* 95: 329–338.

Logan, G. A., Hayes, J.M., Hieshima, G.B. and Summons, R. E., 1995. Terminal Proterozoic reorganization of biogeochemical cycles. *Nature* 376: 53-56.

Love, G. D., Fike, D. A., Grosjean, E., Stalvies, C., Grotzinger, J., Bradley, A. S., Bowring, S., Condon, D. and Summons, R. E., 2006. Constraining the timeing of basal metazoan radiation using molecular biomarkers and U-Pb isotope dating. *Geochimica et Cosmochimica Acta* 70 (18): A371.

Luminet, J. P., *Weeks, F. R., Riazvelo, A., Lehovcq, R. and Uzan, J. P.,* 2003. Dodecahedral space topology as an explanation for weak wide-angle temperature correlations in the cosmic microwave background. *Nature* 425: 593-595.

Luo Qiling, 1991. New data on the microplants from the Changlongshan Formation of Upper Precambrian in western Yanshan Range. *Bulletin of the Tanjin Institute of Geology and Mineral Resources Chinese Academy of Geological Sciences* 25: 107–118 (in Chinese with English summary)

Lurquin, P. F., 2003. *The Origin of Life and the Universe.* New York: Columbia University Press.

MacGabhann, B., 2007. *Ediacaria booleyi* - weeded from the Garden ot Ediacara? In: *The Rise and Fall of the Ediacara (Vendian) Biota.* Edited by P. Vickers-Rich and P. Komarower. *Geological Society of London Special Publication* 286.

MacIntyre, F., 1974. Chemical fractionation and sea-surface microlayer processes. In: *The Sea.* Edited by E. D. Goldberg. New York: John Wiley & Sons: 245-299.

MacIntyre, F., 1974. The top millimeter of the ocean. *Scientific American* 230 (5): 66-77.

MacIntyre, F. and Winchester, J. W., 1969. Phosphate ion enrichment in drops from breaking bubbles. *Journal of Physical Chemistry* 73: 2163-2169.

Macleod, N., 2007. PaleoBase: Macrofossils. Oxford: Blackwell Publishing.

MacNaughton, R. B., Narbonne, G. M. and Dalrymple, R. W., 2000. Neoproterozoic slope deposits, Mackenzie Mountains, northwestern Canada: Implication for passive margin development and Ediacaran faunal ecology. *Canadian Journal of Earth Sciences* 37: 997-1020.

Madigan, C. T., 1932. The geology of the western MacDonnell Ranges, central Australia. *Quarterly Journal of the Geological Society of London* 88: 672-711.

Madigan, C. T., 1932. The geology of the eastern MacDonnell Ranges, central Australia. *Transactions of the Royal Society of South Australia* 56: 71-117.

Madigan, C. T., 1935. The geology of the MacDonnell Ranges, Central Australia. *Report of the Australian and New Zealand Association for the Advancement of Science* 21: 75-86.

Magnum, C., 1991. Precambrian oxygen levels, the sulfide biosystem, and the origin of the Metazoa. *The Journal of Experimental Zoology* 260: 33-42.

Maithy, P. K. and Kumar, G., 2004. Biota in the terminal Proterozoic successions on Indian subcontinent: A review. In: *Abstracts of The Rise and Fall of the Vendian Biota. International Geological Correlation Project 493 Symposium*, Prato, Italy 30-31 August 2004. Edited by P. Komarower and P. Vickers-Rich. Monash University, Melbourne: 77-82.

Maithy, P. K., Narain, K. and Sarkar, A., 1986. Body and trace fossils from the Rohtas Formation (Vindhyan Super group) exposed around Akbarpur Rohtas District. *Current Science (India)* 55 (20): 1029-1030.

Mangano, M. G. and Buatois, L. A., 2004. Reconstructing early Phanerozoic intertidal ecosystems: Ichnology of the Cambrian Campanario Formation in northwest Argentina. *Fossils and Strata* 51: 17-38.

Margulis, L, 1981. *Symbiosis in Cell Evolution.* New York: W. H. Freeman and Co.

Margulis, L. and Schwartz, K. V., 1988. *Five Kingdoms. An Illustrated Guide to the Phyla of Life on Earth.* 2nd ed. New York: W. H. Freeman and Company.

Mapstone, N. B. & McIlry, D., 2006. Ediacaran fossil preservation: Taphonomy and diagenesis of a discoid biota from the Amadeus Basin, central Australia. *Precambrian Research* 149 (3-4): 126-148.

Margulis, L and Schwartz, K. V., 1998. *Five Kingdoms – An Illustrated Guide to the Phyla of Life on Earth.* 3rd ed. New York: Freeman.

Marshall, C. R., 2006. Explaining the Cambrian "Explosion" of animals. *Annual Review of Earth and Planetary Science* 34: 355-384.

Martí Mus, M. and Moczydowska, M., 2000. Internal morphology and taphonomic history of the Neoproterozic vase-shaped microfossils from the Visingsö Group, Sweden. *Norsk Geologisk Tidsskrift* 80: 213-228.

Martin, D. McB., 2004. Depositional environment and taphonomy of the 'strings of beads': Mesoproterozoic multicellular fossils in the Bangemall Supergroup, Western Australia. *Australian Journal of Earth Sciences* 51: 555-561.

Martin, D. McB. and Thorne, A..M., 2002. Revised lithostratigraphy of the Mesoproterozoic Bangemall Supergroup on the Edmund and Turee Creek 1:250 000 sheets, Western Australia. *Western Australia Geological Survey, Record 2002* 15: 1-27.

Martin, D. McB., Thorne, A. M. and Copp, I. A., 1999. A provisional revised stratigraphy for the Bangemall Group on the Edmund 1:250 000 sheet: *Western Australia Geological Survey, Annual Review* 1998-1999: 51-55.

Martin, K. A., *et al.*, 2003. Cyanobacterial signature genes. *Photosynthetic Research* 75: 211-221.

Martin, M. W., Grazhdankin, D. V., Bowring, S. A., Evans, D. A. D., Fedonkin, M. A. and Kirschvink, J. L., 2000. Age of Neoproterozoic bilaterian body and trace fossils, White Sea, Russia: Implications for metazoan evolution. *Science* 288: 841-845.

Martin, W., 2003. Gene transfers from organelles to the nucleus: Frequent and in big chunks. *Proceedings of the National Academy of Science* 100: 8612-8614.

Martin, W. and Muller, M., 1998. The hydrogen hypothesis for the first eukaryote. *Nature* 392: 37-41.

Martin, W., Rotte, C., Hoffmeister, M., *et al.*, 2003. Early cell evolution, eukaryotes, anoxia, sulfide, oxygen, fungi first (?), and a tree of genomes revisited. *IUBMB Life* 55 (4-5): 193-204.

Martin, W. and Russell, M. J., 2003. On the origins of cells: An hypothesis for the evolutionary transitions from abiotic geochemistry to chemoautotrophic prokaryotes, and from prokaryotes to nucleated cells. *Philosophical Transactions of the Royal Society of London* B 358: 59-85.

Martin, W., *et al.*, 2001. An overview of endosymbiotic models for origins of eukaryotes, their ATP-producing organelles, mitochondria and hydrogenosomes, and their heterotrophic lifestyle. *Biological Chemistry* 382: 1521-1539.

Martindale, M. Q. and Henry, J. Q., 1998. The development of radial and bilateral symmetry: The evolution of bilaterality. *American Zoologist* 38 (4): 672-684.

Maslov, A. V., *et. al.*, 1997. The main tectonic events, deposition history, and the palaeogeography of the southern Urals during the Riphean-early Palaeozoic. *Tectonophysics* 276: 313-335.

Mathur, V. K. and Srivastava, D. K., 2004. Record of tissue grade colonial eukaryote and microbial mat associated with Ediacaran fossils in Krol Group, Garhwal Syncline, Lesser Himalaya, Uttaranchai. *Journal of the Geological Society of India* 63: 100-102.

Mathur, V. K. & Srivastava, D. K., 2004. Record of Terminal Neoproterozoic Ediacaran fossils from Krol Group, Nigalidhar Syncline, Sirmaur District, Mimachal Pradesh, India. *Journal of the Geological Society of India* 64: 231-233.

Mawson, D., 1925. Evidence and indications of algal contributions in the Cambrian and pre-Cambrian limestones of South Australia. *Transactions of the Royal Society of South Australia* 49: 186–190.

Mawson, D., 1938. Cambrian and Subcambrian Formations at Parachilna Gorge. *Transactions of the Royal Society of South Australia* 62 (2): 255–261.

Mawson, D., 1942. The structural character of the Flinders Ranges. *Transactions of the Royal Society of South Australia* 66 (2): 262–272.

McBride, N. and Gilmour, I., 2003. *An Introduction to the Solar System*. Cambridge: The Open University and Cambridge University Press.

McCaffrey, M. A., Moldowan, J. M., Lipton, P. A., Summons, R. E., Peters, K. E., *et al.*, 1994. Paleoenvironmental implications of novel C30 steranes in Precambrian to Cenozoic age petroleum and bitumens. *Geochimica et Cosmochimica Acta* 58: 529-532.

McCall, G. J. H., 2006. The Vendian (Ediacaran) in the geological record: Enigmas in geology's prelude to the Cambrian explosion. *Earth-Science Reviews* 77 (1-3): 1-229.

McIlroy, D. and Walter, M. R., 1997. A reconsideration of the biogenicity of *Arumberia banksi* Glaessner & Walter. *Alcheringa* 21: 79-80.

McKay, D. S. E. *et al.* 1996. Search for past life on Mars: Possible relic biogenic activity in Martian meteorite ALH84001. *Science* 273: 924-930.

McKerrow, W. S., Scortese, C. R. and Brasier, M. D., 1992. Early Cambrian continental reconstructions. *Journal of the Geological Society London* 149: 599-606.

McKirdy, D. M., Webster, L. J., Arouri, K. R. and Gostin, V. A., 2003. Contrasting sterane signatures in Neoproterozoic marine sediments of the Centralian Superbasin before and after the Acraman bolide impact. *21st International Meeting on Organic Geochemistry, Krakow, Abstracts*: 132-133.

McKirdy, D. M., Webster, L. J., Arouri, K. R., Gostin, V. A. and Grey, K., 2006. Contrasting sterane signatures in Neoproterozoic marine sediments of the Centralian Superbasin before and after the Acraman bolide impact. *Organic Geochemistry* 37: 189-207.

McMenamin, M. A. S., 1986. The garden of Ediacara. *Palaios* 1: 178-182.

McMenamin, M. A. S., 1996. Ediacaran biota from Sonora, Mexico. *Proceedings of the National Academy of Science,* 93: 4990-4993.

McMenamin, M. A. S., 1998. *The Garden of Ediacara: Discovering the First Complex Life*. New York: Columbia University Press.

McMenamin, M. A. and McMenamin, D. L. S., 1989. *The Emergence of Animals. The Cambrian Breakthrough*. New York: Columbia University Press.

McSween, H. Y., Jr., 1999. *Meteorites and Their Parent Planets*. New York: Cambridge University Press, 2nd ed.

McSween, H. Y., Jr., and Murchie, S. L., 1999. Rocks at the Mars Pathfinder landing site. *American Scientist* 87 (1): 36-45.

Meert, J. G. and van der Voo, R., 1994. The Neoproterozoic (1000-540 Ma) glacial intervals: No more snowball earth. *Earth and Planetary Science Letters* 123: 1-13.

Melezhik, V. A., *et al.*, 2005. Emergence of the aerobic biosphere during the Archean-Proterozoic transition: Challenges for future research. *GSA Today* 15 (11): 4-11.

Mendelsohn, J., Jarvis, A., Roberts, C. and Robertson, T., 2002. *Atlas of Namibia. A Portrait of the Land and Its People*. Cape Town: New Africa Books.

Microsoft Encarta Encyclopedia. 2002. Microsoft Corporation. http://encarta.msn.com

Milton, L. W., 1985. Chimney fossils. *Earth Science* 38 (3): 24-27.

Mincham, H. 1958. The oldest fossils found in South Australia. *Education Gazette,* 17 June 1958.

Misra, S. B., 1969. Late Precambrian (?) fossils from southeastern Newfoundland. *Geological Society of America Bulletin* 80: 2133-2140.

Misra, S. B., 1971. Stratigraphy and deposition history of late Precambrian coelenterate-bearing rocks, southeastern Newfoundland. *Geological Society of America Bulletin* 82: 979-988.

Misra, S. B., 1981. Depositional environment of the late Precambrian fossil-bearing rocks of southeastern Newfoundland, Canada. *Journal of the Geological Society of India* 22: 375-382.

Moczydlowska, M., 2004. Taxonomic review of some Ediacaran acritarchs from the Siberian Platform. *Precambrian Research* 136: 283-307.

Mojzsis, S. J., Arrhenius, G., McKeegan, K. D., Harrison, T. M., Nutman, A. P. and Friend, C. R. L., 1996. Evidence for life on Earth before 3,800 million years ago. *Nature* 384: 55-59.

Moore, P., 1990. *The Universe for the Under Tens*. London: George Philip.

Morbath, S., 2005. Dating earliest life. *Nature* 434: 155.

Mossman, D. J., 2001. Hydrocarbon habitat of the Paleoproterozoic Franceville Series, Republic of Gabon. *Energy Sources* 23 (1): 45-53.

Mulder, T., Syvitski, J. P. M. and Skene, K. I., 1998. Modeling of erosion and deposition by turbidity currents generated at river mouths. *Journal of Sedimentological Research* 68 (1): 124-137.

Müller, M., 1998. Enzymes and compartmentation of core energy metabolism of anaerobic protists – a special case in eukaryotic evolution of special interest . In: *Evolutionary Relationships Among Protozoa*. Edited by G.H. Coombs, K. Vickermann, M.A. Sleigh and A. Waren. Dordrecht: Kluwer Academic Publishers: 109-132.

Murchison R. I., Verneuil, E. de, and Keiserling, A. von, 1849. *Geological Description of the European Russia and Ural Range. Part 1. Geological Description of the European Russia*. Sankt-Petersburg. (in Russian)

Murphy, J. B., Keppie, J. D., Dostal, J. and Nance, R. D., 1999. Neoproterozoic-early Palaeozoic evolution of Avalonia. In: *Laurentia-Gondwana Connections before Pangea*. Edited by V. A. Ramos and J. D. Keppie. *Geological Society of America, Special Paper* 336: 253-266.

Murphy, J. B. and Nance, R. D., 2004. How do supercontinents assemble. *American Scientist* 92: 324-333.

Murphy, J. B., Pisarevsky, S. A., Nance, R. D. and Keppie, J. D., 2001. Animated history of Avalonia in Neoproterozoic-Early Palaeozoic. In: *General Contributions, 2001*. Edited by M.J. Jessell. *Journal of the Virtual Explorer* 3: 45-58.

Murphy, J. B., Strachan, R. A., Nance, R. D., Parker, K. D. and Fowler, M. B., 2000. Proto-Avalonia: A 1.2-1.0 Ga tectonothermal event and constraints for the evolution of Rodinia. *Geology* 28: 1071-1074.

Murray, A., 1868. *Report of the Geological Survey of Newfoundland for the year 1868*. St John's Newfoundland: Robert Winton: 1-65.

Murray, A., 1873. *Report upon the Geological Survey for the year 1872*. St John's Newfoundland: 1-34.

Mutch, T. A., *et al.,* 1976. *The Geology of Mars*. Princeton: Princeton University Press.

Myrow, P. M., 1995. Neoproterozoic rocks of the Newfoundland Avalon Zone. *Precambrian Research* 73: 123-136.

Myrow, P. M. and Kaufman, A. J., 1999. A newly discovered cap carbonate above Varanger-age glacial deposits in Newfoundland, Canada. *Journal of Sedimentary Research* 69: 784-793.

Nakashima, K., Yamamda, L., Satou, Y., Azuma, J.-I. and Satoh, N., 2004. The evolutionary origin of animal cellulose synthase. *Development Genes and Evolution* 214: 81-88.

Nance, R. D., Murphy, J. B. and Keppie, J. D., 2002. A Cordilleran model for the evolution of Avalonia. *Tectonophysics* 352: 11-31.

Nance, R. D. and Thompson, M. C., 1996. Avalonian and related peri-Gondwanan terranes of the Circum-North Atlantic. *Geological Society of America, Special Paper* 304.

Narbonne, G. M., 1994. New Ediacaran fossils from the Mackenzie Mountains, northwestern Canada. *Journal of Paleontology* 68: 411-416.

Narbonne, G. M., 1998. The Ediacara biota: A terminal Neoproterozoic experiment in the evolution of life. *GSA Today* 8 (2): 1-6.

Narbonne, G. M., 2004. Modular construction of early Ediacaran complex life forms. *Science* 305: 1141-1144.

Narbonne, G. M., 2005. Earth's earliest Ediacarans. *The Palaeontological Association, 49th Annual Meeting, Abstracts Newsletter*, Oxford: 12.

Narbonne, G. M., 2005. The Ediacara biota. Neoproterozoic origin of animals and their ecosystems. *Annual Review of Earth and Planetary Science* 33: 421-442.

Narbonne, G. M. and Aitken, J. D., 1985. Precambrian-Cambrian boundary sequence, Wernecke Mountains, Yukon Territory. *Geological Survey of Canada Paper* 85-1A: 603-608.

Narbonne, G. M. and Aitken, J. D., 1990. Ediacaran fossils from the Sekwi Brook Area, Mackenzie Mountains, Northwestern Canada. *Palaeontology* 33 (4): 945-980.

Narbonne, G. M. and Aitken, J. D., 1995. Neoproterozoic of the Mackenzie Mountains, northwestern Canada. *Precambrian Research* 73: 101-121.

Narbonne, G.M., Dalrymple, R. W., Gehling, J. G., Wood, D. A., Clapham, M. E. and Sala, R. A., 2001. *Field Trip B5. Neoproterozoic Fossils and Environments of the Avalon Peninsula, Newfoundland.* Geological Association of Canada, Mineralogical Association of Canada, Joint Annual Meeting, Memorial University, St John's, Newfoundland.

Narbonne, G., Dalrymple, R. W., La Flamme, M., Gehling, J. and Boyce, W. D., 2005. Life After Snowball. Mistaken Point Biota and the Cambrian of the Avalon. *NAPC 2005, Halifax, Nova Scotia, North American Paleontological Convention, Field Trip Guidebook*.

Narbonne, G.M. and Gehling, J.G., 2003. Life after snowball: The oldest Ediacaran fossils. *Geology* 31: 27-30.

Narbonne, G.M. and Hofmann, H.J., 1987. Ediacaran biota of the Wernecke Mountains, Yukon, Canada. *Palaeontology* 30: 647-676.

Narbonne, G.M., James, N.P., Rainbird, R.H. and Morin, J., 2000. Early Neoproterozoic (Tonian) patch reef complexes, Victoria Island, Arctic Canada. In: *Carbonate Sedimentation and Diagenesis in the Evolving Precambrian World*. Edited by J. P. Grotzinger and N. P. James. *SEPM Special Publication* 67: 163-178.

Narbonne, G. M., Kaufman, A. J. and Knoll, A. H., 1994. Integrated chemostratigraphy and biostratigraphy of the Windermere Supergroup, northwestern Canada: Implications for Neoproterozoic correlations and the early evolution of animals. *Geological Society of America Bulletin* 106: 1281-1292.

Narbonne, G.M., Myrow, P.M., Landing, E. and Anderson, M.M., 1987. A candidate stratotype for the Precambrian-Cambrian boundary, Fortune Head, Burin Peninsula, southeastern Newfoundland. *Canadian Journal of Earth Sciences* 24: 1277-1293.

Narbonne, G. M., Myrow, P.M., Landing, E. and Anderson, M.M., 1991. A Chondrophorine (medusoid hydrozoan) from the basal Cambrian (Placentian) of Newfoundland. *Journal of Paleontology* 65 (2): 186-191.

Narbonne, G. M., Saylor, B. Z. and Grotzinger, J. P., 1997. The youngest Ediacaran fossils from southern Africa. *Journal of Paleontology* 71 (6): 953-967.

Narlikar, J., 1977. *The Structure of the Universe*. London: Oxford University Press.

Naumova, S. N., 1951. Spores in ancient formations on the western slope of the South Urals. *Trudy Moscovskogo Obchestva Ispytatelei Prirody (Transactions of Moscow Society of Naturalists), Section Geology*: 183-187. (in Russian)

Navarro-Gonzalez, R., *et al.*, 2001. A possible nitrogen crisis for Archaean life due to reduced nitrogen fixation by lightning. *Nature* 412: 61-64.

Nedin, C. and Jenkins, R. J. F., 1998. The first occurrence of the Ediacaran fossil *Charnia* from the southern hemisphere. *Alcheringa* 22: 315-316.

Nelson, R P., *et al.*, 2000. The migration and growth of protoplanets in protostellar disks. *Monographs Not. Royal Astronomical Society* 318: 18.

Nevesskaya, L.A. and Kurochkin, E.N., 2000. Past and present studies at the Paleontological Institute of the Russian Academy of Sciences (the 70th Anniversary of the Paleontological Institute). *Paleontological Journal* 34 (5): 475-485.

Nikishin, A. M., *et al.*, 1996. Late Precambrian to Triassic history of the East European craton; dynamics of sedimentary basin evolution. *Tectonophysics* 268: 23-63.

Nisbet, E.G., 1991. *Living Earth. A Short History of Life and Its Home*. London: HarperCollins Academic.

Nisbet, E.G., 1995. Archaean ecology: A review of evidence for the early development of bacterial biomes, and speculations on the development of a global-scale biosphere. In: *Early Precambrian Processes*. Edited by M. P. Coward and A. C. Ries. *Geological Society Special Publication* 95: 27-51.

Nisbet, E.G., 2000. The realms of Archaean life. *Nature* 405: 625-626.

Nisbet, E.G. and Fowler, D.M.R., 1999. Archaean metabolic evolution of microbial mats. *Proceedings of the Royal Society of London* 266: 2375-2382.

Nisbet, E.G. and Sleep, N.H., 2001. The habitat and nature of early life. *Nature* 409: 1083-1109.

Nursall, J. R., 1959. Oxygen as a prerequisite to the origin of the Metazoa. *Nature* 183: 1170-1172.

Nutman, A. P., *et al.*, 1997. Recognition of ~3850 Ma water-lain sediments in W. Greenland and their significance for the early Archaen Earth. *Geochemica et Cosmochimica Acta* 61: 2475-2484.

O'Brien, S. J. and King, A. F., 2004. Ediacaran fossils from the Bonavista Peninsula (Avalon Zone), Newfoundland: Preliminary descriptions and implications for regional correlation. *Current Research (2004) Newfoundland and Labrador Department of Natural Resources. Geological Survey Report* 04-1: 203-212.

O'Brien, S. J. and King, A. F., 2005. Late Neoproterozoic (Ediacaran) stratigraphy of Avalon Zone sedimentary rocks, Bonavista Peninsula, Newfoundland. *Current Research (2005) Newfoundland and Labrador Department of Natural Resources. Geological Survey Report* 05-1: 101-113.

O'Brien, S. J., King, A. F. and Hofmann, H. J. 2006. Lithostratigraphic and biostratigraphic studies on the eastern Bonavista Peninsula: An update. *Current Research (2005) Newfoundland and Labrador Department of Natural Resources. Geological Survey Report* 06-1: 257-263.

Oehler, J. H. and Logan, R. G., 1977. Microfossils, cherts, and associated mineralization in the Proterozoic McArthur (H. Y. C.) Lead-Zinc-Silver deposit. *Economic Geology* 72 (8): 1393-1409.

Ohmoto, H., 1997. When did the Earth's atmosphere become oxic? *The Geochemical News* 93: 12-13.

Ohmoto, H. and Felder, R. P., 1987. Bacterial activity in the warmer, sulphate-bearing Archaean oceans. *Nature* 328: 244-246.

Ohmoto. H., Watanabe, Y., Yamaguchi, K.E., Ono, S., Bau, M., Kakegawa, T., Naraoka, H., Nedachi, M. and Lasaga, A. C., 2001. The Archaean atmosphere, oceans, continents and life. In: *International Archaean Symposium. Extended Abstracts*. Edited by K.F. Cassidy, J. M. Dunphy and M. J. Van Kranendonk. AGSO – *Geoscience Australia Record* 2001/37: 19-21.

Omer, A. D., *et al.*, 2000. Homologs of small nucleolar RNA's in Archaea. *Science* 288: 517-522.

Oparin, A. I., 1936. *Origin of Life*. New York: Dover Reprint Edition, 1953.

Oro, J., 2004. Comets and the origin of life on primitive earth. In: *Origins. Genesis, Evolution and Diversity of Life*. Edited by J. Seckbach. Dordrecht: Kluwer Academic Publishers: 553-567.

Palij, V. M., 1969. On a new species of cyclomedusae from the Vendian of Podolia. *Paleontologie Sbornik Luniversiteta* 6: 1. (in Russian)

Palij, V. M., 1974. Bilobate traces from the deposits of the Baltic series in the Dniester Region. *Doklady AN URSR*, B (1). (in Ukrainian)

Palij, V. M., 1976. Remains of soft-bodied animals and trace fossils from the Upper Precambrian and Lower Cambrian of Podolia. In: *Palaeontology and Stratigraphy of the Upper Precambrian and Lower Paleozoic of the Southwestern Part of the East European Platform*. Edited by V. A. Ryabenko. Kiev: Naukov Dumka: 63-76. (in Russian)

Palij, V. M., Posti, E. and Fedonkin, M. A., 1979. Soft-bodied Metazoa and trace fossils of the Vendian and Lower Cambrian. In: *Paleontology of Upper Precambrian and Cambrian Deposits of East-European Platform* Edited by. B.M. Keller and A.Yu Rozanov. Moscow: Nauka: 49-82. (in Russian)

Parker, A., 2003. *In the Blink of an Eye: The Cause of the Most Dramatic Event in the History of Life*. London: The Free Press.

Parkhaev, P. Yu., 1999. Siphonoconcha – A new class of Early Cambrian bivalved organisms. *Paleontological Journal* 32 (1): 1-15.

Parkhaev, P. Yu., 2001. Muscle scars of the Cambrian univalved molluscs and their significance for systematics. *Paleontological Journal*, 36 (5): 453-459.

Pavlov, A. A., Hurtgen, M. T., Kasting, J.F., *et al.*, 2003. Methane-rich Proterozoic atmosphere? *Geology* 31(1): 87-90.

Pearce, F., 2003. Doomsday Scenario. *New Scientist* 2422: 40-43.

Pearse, J. S., McClintock, J. B. and Bosch, I., 1991. Reproduction of Antarctic benthic invertebrates: Tempos, modes, and timing. *American Zoologist* 31: 65-80.

Pederson, K., 1993. The deep subsurface biosphere. *Earth Science Reviews* 34: 243-260.

Pedley, H. M. and Carannante, G., 2006. *Cool-water Carbonates: Deposition Systems and Palaeoenvironmental Controls. Geological Society of London, Special Publication* 255.

Penny, D., 2005. Relativity for molecular clocks. *Nature* 436: 183-184.

Pesonen, L. J., Elming, S.-A., Mertanen, S., Pisarevsky, S., D'Ggrella-Filho, M. S., Meert, J. G., Schmidt, P. W., Abrahamsen, N. and Bylund, G., 2003. Palaeomagnetic configuration of continents during the Proterozoic. *Tectonophysics* 375 (2003): 289-324.

Peterson, K. J. and Butterfield, N. J., 2005. Origin of the Eumetazoa: Testing ecological predictions of molecular clocks against the Proterozoic fossil record. *PNAS*, 102 (27): 9547-9552.

Peterson, K. J. and Davidson, E. H., 2000. Regulatory evolution and the origin of the bilaterians. *Pinas* 97 (9): 4430-4433.

Peterson, K. J., Waggoner, B. and Hagadorn, J. W., 2003. A fungal analog for Newfoundland Ediacaran fosslls? *Integrative Comparative Biology* 43: 127-136.

Pflug, H. D., 1966. Neue Fossilreste aus den Nama-Schichten in Sudwest-Afrika. *Palaeontologische Zeitschrift* 40: 14-25.

Pflug, H. D., 1970. Zur Fauna der Nama-Schichten in Sudwest-Afrika. I. Pteridinia, Bau und Systematische Zugehorigkeit. *Sonder-Abdruk, Palaeontographica. Beitrage zur Naturgeschichte der Vorzeit* Bd. 134, Abt. A: 226-262.

Pflug, H. D., 1970. Zur Fauna der Nama-Schichten in Sudwest-Afrika. II. Rangeidae, Bau und Systematische Zugehorigkeit. *Sonder-Abdruk, Palaeontographica. Beitrage zur Naturgeschichte der Vorzeit* 135, Abt. A: 198-231.

Pflug, H. D., 1972. Systematik der jung-prakambrischen Petalonamae Pflug 1970. *Palaeontologische Zeitschrift* 46: 56-67.

Pflug, H. D., 1972. Zur Fauna der Nama-Schichten in Sudwest-Afrika. III. Erniettomorpha, Bau und Systematik. *Sonder-Abdruk, Palaeontographica. Beitrage zur Naturgeschichte der Vorzeit* Bd. 139, Abt. A: 134-170.

Pflug, H. D., 1973. Zur Fauna der Nama-Schichten in Sudwest-Afrika. IV. Mikroskopische Anatomie der Petal-Organismen. Palaeontographica *Palaeontographica. Beitrage zur Naturgeschichte der Vorzeit* Bd. 144, Abt. A: 166-202.

Pflug, H. D., 1974. Vor- und Fruhgeschichte der Metazoen. Precambrian history of the Metazoa. *Neues Jahrbuch Geologie und Palaeontologie* 145: 328-374.

Pfluger, F., 1999. Matground structures and redox facies. *Palaios* 14: 25-39.

Philippe, H., *et al.*, 2000. The new phylogeny of eukaryotes. *Current Opinion of Genetic Development* 10: 596-601.

Pickering, K. T., 1982. The shape of deep-water siliciclastic systems: A discussion. *Geo-Marine Letters* 2: 41-46.

Pickering, K. T., Hiscott, R. N. and Hein, F. J., 1989. *Deep-marine Environments: Clastic Sedimentation and Tectonics*. London: Unwin Hyman.

Pickford, M. H. L., 1995. Review of the Riphean, Vendian and early Cambrian palaeontology of the Otavi and Nama Groups, Namibia. *Communications of the Geological Survey of Namibia. Special Issue. Proterozoic Crustal and Metallogenic Evolution* 10 (1995): 57-81.

Pickford, M. and Senut, B., 2002. *The Fossil Record of Namibia*. Namibia: Geological Survey.

Pidgeon, R. T, Nemchin, A. A. and Williams, I. S., 2001. Internal structure of >3.9 Ga detrital zircons from the Jack Hills, Western Australia. In: *International Archaean Symposium Extended Abstracts*. Edited by K.F. Cassidy, J.M. Dunphy and M.J. Van Kranendonk. *AGSO – Geoscience Australia Record* 2001/37: 75-77.

Pirrus, E. A., 1992. Freshening of the Late Vendian Basin on the East European Craton. *Proceedings of the Estonian Academy of Sciences, Geology* 41: 115-123.

Polkinghorne, J. C., 1979. *The Particle Play. An Account of the Ultimate Constituents of Matter*. Oxford: W. H. Freeman and Co.

Poole, A., *et al.*, 1998. The path from the RNA world. *Journal of Molecular Evolution* 46: 1-17.

Poole, A., *et al.*, 1999. Procaryotes, the new kids on the block. *Bio Essays* 21: 880-889.

Porter, S. M., 2004. The Fossil Record of Early Eukaryotic Diversification. In: *Neoproterozoic-Cambrian Bilogical Revolution*. Edited by J. H. Lipps and B. M. Waggoner. *The Paleontological Papers* 10: 35-50.

Porter, S.M. and Knoll, A.H., 2000. Testate amoebae in the Neoproterozoic Era: Evidence from vase-shaped microfossils in the Chuar Group, Grand Canyon. *Paleobiology* 26: 360-385.

Prave, A. R., 2002. Life on land in the Proterozoic: Evidence from the Torridonian rocks of northwest Scotland. *Geology* 30 (9): 811-814.

Preiss, W. V., compiler, 1987. The Adelaide Geosyncline - late Proterozoic stratigraphy, sedimentation, palaeontology and tectonics. *Bulletin of the Geological Survey of South Australia* 33.

Preiss, W. V. 1993. The Precambrian. In: T*he Geology of South Australia, Vol. 1*. Edited by J. F. Drexel, W. V. Preiss and A. J. Parker. *South Australian Geological Survey Bulletin* 54: 171-204.

Preiss, W. V., 2000. The Adelaide Geosyncline of South Australia and its significance in Neoproterozoic continental reconstruction. *Precambrian Research* 100: 21–63.

Puchkov. V. N., 1997. *Structure and Geodynamics of the Uralian Orogen. Geological Society Special Publication* 121: 201-236.

Pullman, B., 1972. Electronic factors in biochemical evolution. In: *Exobiology*. Edited by C. Ponnamperuma. Amsterdam: North Holland Publishing Company: 136-169.

Purves, W. K., Sadava, D., Orians, G. H. and Heller, C., 2003. *Life: The Science of Biology*, 7th ed. Gordensville: W. H. Freeman & Co.

Pyle, L. J., Narbonne, G. M., James, N. P., Dalrymple, R. W., and Kaufman, A. J., 2004. Integrated Ediacaran chronostratigraphy, Wernecke Mountains, northwestern Canada. *Precambrian Research* 132: 1-27.

Raff, R. A., 1996. *The Shape of Life. Genes, Development, and the Evolution of Animal Form*. 7th ed. Chicago: University of Chicago Press.

Raghav, K. S., De, C, and Jain, R. L., 2005. The first record of Vendian medusoids and trace fossil-bearing algal mat grounds from the basal part of the Marwar Supergroup of Rajasthan, India. *Indian Minerals* 59 (1-2): 23-30.

Rainbird, R. H., Stern, R. A., Khudoley, A. K., Kropachev, A. P., Heaman, L. M. and Sukhorukov, V. I., 1988. U–Pb geochronology of Riphean sandstone and gabbro from southeast Siberia and its bearing on the Laurentia-Siberia connection. *Earth and Planetary Science Letters* 164: 409-420.

Rasmussen, B., Bengtson, S., Fletcher, I.R. and McNaughton, N.J., 2002. Discoidal impressions and trace-like fossils more than 1200 million years old. *Science* 296: 1112-1115.

Rasmussen, B. and Buick, R., 1999. Redox state of the Archean atmosphere: Evidence from detrital heavy minerals in ca. 3250-2750 Ma sandstones from the Pilbara Craton, Australia. *Geology* 27 (2): 115-118.

Raven, P.H. and Johnson, G.B., 1986. *Biology*. Toronto and St Louis: Times Mirror/Mosby College Publishing.

Raymond, J. and Blankenship, R. E., 2003. Horizontal gene transfer in eukaryotic algal evolution. *Proceedings of the National Academy of Science USA* 100 (13): 7419-7420.

Raymond, J., *et al.*, 2000. Whole-genome analysis of photosynthetic prokaryotes. *Science* 298: 1616-1620.

Rees, M. J., 1997. *Before the Beginning: Our Universe and Others*. Reading: Addison-Wesley.

Reich, E. S., 2005. Earth plunged into dust and came out icy white. *New Scientist* 2487: 9.

Reinking, G. F., 2003. *Cosmic Legacy. Space, Time, and the Human Mind*. New York: Vantage Press.

Retallack, G. J., 1994. Were the Ediacaran fossils lichens? *Paleobiology* 20: 523-544.

Richter, R., 1955. Die altesten Fossilien Sud-Afrikas. *Senckenberg Lethia* 36 (3/4): 243-289.

Riding, R., 2006. Cyanobacterial calcification, carbon dioxide concentrating mechanisms and Proterozoic-Cambrian changes in atmospherelc composition. *Geobiology* 4 (4): 299.

Riva, M. C. and Lake, J. A., 2004. The ring of life provides evidence for a genome fusion origin of eukaryotes. *Nature* 431: 152-155.

Rizzotti, M., 1996. *Defining Life: The Central Problem in Theoretical Biology*. Padua: University of Padua.

Rizzotti, M., 2000. *Early Evolution. From the Appearance of the First Cell to the First Modern Organisms*. Basel: Birkhauser.

Roberts, D. and Siedlecka, A., 2002. Timanian orogenic deformation along the northeastern margin of Baltica, northwest Russia and northeast Norway, and Avalonian-Cadomian connections. *Tectonophysics* 352: 169-184.

Robison, R. A. and Teichert, C., 1979. *Treatise on Invertebrate Paleontology, Part A. Introduction*. Boulder (Colorado) and Lawrence (Kansas): Geological Society of America and the University of Kansas.

Roger, A. J., 1999. Reconstructing early events in eukaryotic evolution. *American Naturalist* 154 (S4): 146-163.

Ross, G. M., 1991. Tectonic setting of the Windermere Supergroup revisited. *Geology* 19: 1125–1128.

Rollinson, H., 2006. Early Earth Systems. Oxford: Blackwell Publishing.

Rothschild, L. J. and Mancinelli, R. L., 2001. Life in extreme environments. *Nature* 409: 1092-1101.

Rouse, G. W. and Pleijel, F., 2001. *Polychaetes*. Oxford: Oxford University Press.

Roy, S., 2000. Late Archean initiation of manganese metallogenesis: Its significance and environmental controls. *Ore Geology Review* 17 (3): 179-198.

Rozanov, A. Yu. and Fedonkin, M. A., 1982. Skeletal growth of aquatic organisms: Biological record of environmental change. A review. *Journal of Paleontology* 56 (5): 1313-1314.

Rozanov, A. Yu. and Kurochkin, E. N., 2000. The Paleontological Institute has turned seventy. *Herald of the Russian Acadmey of Sciences* 70 (5): 499-507.

Rozanov, A. Yu. and Zhurarter, A. Yu., 1992. The Lower Cambrian Fossil Record of the Soviet Union. In: *Origin and Evolution of the Metazoa*. Edited by J. H. Dipps and P. W. Signor. New York: Plenum Press.

Rubey, W. W., 1955. Development of the hydrosphere and atmosphere, with special reference to probable composition of the early atmosphere. In: *Crust of the Earth*. Edited by A. Poldervaart. New York: Geological Society of America: 631-650.

Runnegar, B., 1982. Oxygen requirements, biology and phylogenetic significance of the late Precambrian worm, *Dickinsonia*, and the evolution of the burrowing habit. *Alcheringa*, 6: 223-239.

Runnegar, B., 1991. Oxygen and the early life of metazoans. In: *Metazoan Life without Oxygen*. Edited by C. Bryant. London: Chapman & Hall: 65-87.

Runnegar, B., 1992. Proterozoic Fossils of Soft-bodied Metazoans (Ediacara Faunas). In: J. W. Schopf and C. Klein. *The Proterozoic Biosphere. A Multidisciplinary Study*. Cambridge: Cambridge University Press: 999-1007.

Runnegar, B., 1995. Vendobionta or Metazoa? Developments in understanding the Ediacara "Fauna." *Neues Jahrbuch Geologie und Palaeontologie Abhandlung* 195: 303-318.

Runnegar, B. N. and Fedonkin, M. A. 1992. Proterozoic metazoan body fossils. In: *The Proterozoic Biosphere. A Multidisciplinary Study*. Edited by J. W. Schopf and C. Klein. New York: Cambridge University Press: 369-388.

Russell, M.J. and Hall, A.J., 1997. The emergence of life from iron monosulphide bubbles at a submarine hydrothermal redox and pH front. *Journal of the Geological Society of London* 154 (3): 377.

Rye, R. and Holland, H.D., 1998. Paleosols and the evolution of atmospheric oxygen: A critical review. *American Journal of Science* 298 (8): 621-672.

Sabehi, G. *et al.*, 2003. Novel proteorhodopsin variants from the Mediteranean and Red Seas. *Environmental Microbiology* 5: 842-849.

Sagan, C., 1979. *Broca's Brain. Reflections on the Romance of Science*. New York: Random House.

Sagan, C. 1997. *Billions & Billions. Thoughts on Life and Death at the Brink of the Millennium*. New York: Random House.

St Jean, J., 1973. A new Cambrian trilobite from the Piedmont of North Carolina. *American Journal of Science* 273A: 196-216.

Samuelsson, J. and Butterfield, N. J., 2001. Neoproterozoic fossils from the Franklin Mtns., northwestern Canada: Stratigraphic and palaeobiological implication. *Precambrian Research* 107: 235-251.

Sando, W. J., 1972. Bee-nest pseudofossils from Montana, Wyoming and South-west Africa. *Journal of Paleontology* 46 (3): 421-425.

Savazzi, E., 2003. Pattern formation and function in palaeobiology. In: *Morphogenesis and Pattern Formation in Biological Systems: Experiments and Models.* Edited by T. Sekimura, S. Noji, N. Ueno and P. K. Maini. Berlin: Springer-Verlag: 329-343.

Savazzi, E., 2006. Reflection on staggered bilateral symmetry in *Dickinsonia* and other Ediacarans. *Abstract Volume, IGCP 493 Symposium, Japan, 27-31 January 2006.* Edited by P. Vickers-Rich and T. Ohno. Kyoto: Kyoto University: 7-8.

Savazzi, E., 2006. Were efficient burrowers responsible for the disappearance of Ediacarans? *Abstract Volume, IGCP 493 Symposium Japan, 27-31 January 2006.* Edited by P. Vickers-Rich and T. Ohno. Kyoto: Kyoto University: 15-16.

Saylor, B. Z., Grotzinger, J. P. and Germs, G. J. B., 1995. Sequence stratigraphy and sedimentology of the Neoproterozoic Kuibis and Schwarzrand Subgroups (Nama Group), southwest Namibia. *Precambrian Research* 73: 153-171.

Sciama, D. W., 1972. *Modern Cosmology.* Cambridge: Cambridge University Press.

Schmidt, P. W. and Williams, G. E., 1995. The Neoproterozoic climatic paradox: Equatorial palaeolatitude for Marinoan glaciation near sea level in South Australia. *Earth and Planetary Science Letters* 134.

Schneider, G., 2004. The Roadside Geology of Namibia. *Sammlung Geologischer Fuhrer* 9: Berlin/Stuttgart.

Schopf, J. W., 1968. Microflora of the Bitter Springs Formation, Late Precambrian, Central Australia. *Journal of Paleontology* 42: 651–688.

Schopf, J., ed., 1983. *Earth's Earliest Biosphere, Its Origin and Evolution.* Princeton: Princeton University Press.

Schopf, J., ed., 1992. *Major Events in the History of Life.* Boston: Jones and Bartlett.

Schopf, J., 1992. Microfossils of the early Archean Apex Chert: New evidence of the antiquity of life. *Science* 260: 640-646.

Schopf, J. W., 1992. Paleobiology of the Archean. In: *The Proterozoic Biosphere. A Multidisciplinary Study.* Edited by J.W. Schopf and C. Klein. New York: Cambridge University Press: 25-39.

Schopf, J. W., 1999. *Cradle of Life.* Princeton: Princeton University Press.

Schopf, J. W., ed., 2002. *Life's Origin.* Berkeley: University of California Press.

Schopf, J. W. and Barghoorn, E. S., 1969. Microorganisms from the Late Precambrian of South Australia. *Journal of Paleontology* 43: 111–118.

Schopf, J. W. and Klein, C., eds., 1992. *The Proterozoic Biosphere. A Multidisciplinary Study.* Cambridge: Cambridge University Press.

Schopf, W. A. and Packer, B. M., 1987. Early Archean microfossils from Warrawoona Group, Australia. *Science* 237: 70-73.

Schopf, J. W., *et al.,* 2002. Laser-Raman imagery of Earth's earliest fossils. *Nature* 416: 73-76.

Schrodinger, E., 1977. *What is Life? Mind and Matter.* London: Cambridge University Press.

Sciama, D. W., 1972. *Modern Cosmology.* Cambridge: Cambridge University Press.

Seckback. J., ed., 2004. *Origins. Genesis, Evolution and Diversity of Life.* Dordrecht: Kluwer Academic Publishers.

Seilacher, A., 1964. Sedimentological classification and nomenclature of trace fossils. *Sedimentology* 3: 253-256.

Seilacher, A., 1984. Late Precambrian and Early Cambrian Metazoa: Preservational or real extinctions? In: *Patterns of Change in Earth Evolution* (Dahlem Konferenzen). Berlin: Springer-Verlag: 159-168.

Seilacher, A. 1989. Vendozoa: Organismic construction in the Proterozoic biosphere. *Lethaia* 22: 229-239.

Seilacher, A., 1992. Vendobiota and Psammocorallia: Lost constructions of Precambrian evolution. *The Journal of the Geological Society of London* 149: 607-614.

Seilacher, A., 1999. Biomat-related lifestyles in the Precambrian. *Palaios* 14: 86-93.

Seilacher, A, Bose, P.K. and Pfluger, F., 1998. Triploblastic animals more than 1 billion years ago: Trace fossil evidence from India. *Science* 282 (5386): 80-83.

Seilacher, A., Buatois, L. A. and Mangano, M. G., 2005. Trace fossils in the Ediacaran-Cambrian transition: Behavioral diversification, ecological turnover and environmental shift. *Palaeogeography, Palaeoclimatology, Paleoecology* 227: 323-356..

Seilacher, A., Grazhdankin, D. and Legouta, A., 2003. Ediacaran biota: The dawn of animal life in the shadow of giant protests. *Paleontological Research* 7 (1): 43-54.

Seilacher, D. and Meschede, M., 1998. Precambrian fossil turns into tectonic diary. *Geological Society of America Annual Meeting. Abstract Volume.*

Seilacher, A. and Pfluger, F., 1994. From biomats to benthic agriculture: A biohistorical revolution. In: *Biostabilization of Sediments.* Edited by W. E. Krumbein, D. M. Paterson and L. J. Stal. Oldenburg: Bibliotheks und Informationssystem der Carl von Ossietzky Universität Oldenburg: 97-105.

Semikhatov, M. A., *et al.,* 1970. Yudomskij kompleks stratipicheskaj mestnosti. Akademiya Nouk SSSR, Trudy: 210.

Semikhatov, M. A., Kuznetsov, A. B., Podkovyrov, V. N., Bartley, J. K. and Davydov, Yu. V., 2004. The Yudoma Group stratotype area: C-isotope chemostratigraphic correlations and Yudomian-Vencian relation. *Stratigrafia I Geologicheskaia Korreliatsia* 12 (5): 435-459.

Semikhatov, M. A., Ovchinnikova, G. V., Gorokhov, I. M., Kuznetsov, A. B., Kaurova, O. K. and Petrov, P. Yu., 2003. Pb-Pb isochronic age and Sr isotope characteristics of the Upper Yudoma carbonate deposits (Vendian of the Uydoma-Maya Basin, Eastern Siberia). *Doklady Akedemii Nauk* 393 (1): 83-87.

Seong-Joo, L. and Golubic, S., 1998. Multi-trichomous cyanobacterial microfossils from the Mesoproterozoic Gaoyuzhuang Formation, China: Paleoecological and taxonomic implications. *Lethaia* 31 (3): 169-184.

Sepkoski, J. J., Jr., 1978. A kinetic model of Phanerozoic taxonomic diversity, I. Analysis of marine orders. *Paleobiology* 4 (3): 223-251.

Sepkoski, J. J., Jr., 1979. A kinetic model of Phanerozoic taxonomic diversity, II. Early Phanerozoic families and multiple equilibria. *Paleobiology* 5: 222-252.

Sepkoski, J. J., Jr., 1981. A factor analytic description of the Phanerozoic marine fossil record. *Paleobiology* 7: 36-53.

Sepkoski, J. J., Jr., 1984. A kinetic model of Phanerozoic taxonomic diversity, III. Post-Paleozoic families and mass extinction. *Paleobiolology* 10 (2): 246-267.

Sepkoski, J. J., Jr., 1986. Phanerozoic overview of mass extinction. In: *Patterns and Processes in the History of Life.* Edited by D.M. Raup and D. Jablonski. Berlin: Springer-Verlag: 277-295.

Sepkoski, J. J., Jr., 1989. Periodicity in extinction and the problem of catastrophism in the history of life. *Journal of the Geological Society of London* 146: 7-19.

Sepkoski, J. J., Jr., 1992. Proterozoic-Early Cambrian diversification of metazoans and metaphytes. In: *The Proterozoic Biosphere. A Multidisciplinary Approach.* Edited by J. W. Schopf and C. Klein. Cambridge: Cambridge University Press: 553-561.

Sepkoski, J. J., Jr., Bambach, R .K., Raup, D. M. and Valentine, J. W., 1981. Phanerozoic marine diversity and the fossil record. *Nature* 293: 435-437.

Serezhnikova, E. A., 2005. New explanation of *Hiemalora* fossil prints from Vendian of Olenek Uplife (northeastern Siberian Platform). *Bulletin of the Moscow Society of Naturalists, Division of Geology* 80 (3): 26-32. (in Russian)

Serezhnikova, E. A., 2007. Vendian *Hiemalora* from Arctic Siberia reinterpreted as holdfasts of benthic organisms. In: *The Rise and Fall of the Ediacaran Biota. Environmental and Biotic Dynamics of the Neoproterozoic and Early Phanerozoic.* Edited by P. Vickers-Rich and P. Komarower. *Geological Society of London, Special Publication* 286.

Sergeev, V. N., 1992. Okremnennyye mikrofossilii avzyanskoy svity Yuzhnogo Urala [Silicified microfossils from the Avzyan Formation of the Southern Urals]. *Paleontologicheskii Zhurnal* 2, 103–122. (in Russian with English translation in *Paleontological Journal, Scripta Technica Inc.* 1993: 128–141).

Sergeev, V. N., 1994. Microfossils in cherts from the Middle Riphean (Mesoproterozoic) Avzyan Formation, Southern Ural Mountains, Russian Federation. *Precambrian Research* 65: 231–254.

Seward, A. C., 1931. *Plant Life through the Ages.* Cambridge: Cambridge University Press.

Shanker, R., Bhattacharya, D. D., Pande, A. C. & Mathur, V. K., 2004. Ediacaran biota from the Jarashi (Middle Krol) and Mahi (Lower Krol) Formations, Drol Group, Lesser Himalaya, India. *Journal of the Geologcial Society of India* 63: 649-654.

Shanker, R. and Mathur, V. K., 1992. Precambrian-Cambrian sequence in Krol belt and additional Ediacaran fossils. Proceedings of the Birbal Sahni Birth Centenary Palaeobotanical Conference. *Geophytology* 22: 27-29.

Shapiro, R., 1988. Prebiotic ribose synthesis: A critical analysis. *Origins of Life and Evolution of the Biosphere* 18: 71-85.

Shapiro, R., 1995. The prebiotic role of adenine: A critical analysis. *Origins of Life and Evolution of the Biosphere* 25: 83-98.

Shen, Y., Canfield, D.E. and Knoll, A.H., 2002. Middle Proterozoic ocean chemistry: Evidence from the McArthur Basin, Northern Australia. *American Journal of Science* 302: 81-109.

Shen, Y., Knoll, A.H. and Walter, M..R. , 2003. Evidence for low sulphate and anoxia in a mid-Proterozoic marine basin. *Nature* 423 (6940): 632-635.

Shen, B., Xiao, S., Zhou, C., Yuan, X. and Sie, G., 2004. A possible frondose Ediacaran fossil from Neoproterozoic bituminous limestone of the Dengying Formation: Its body plan, lifestyle, and taphonomy. *Geological Society of America, Annual Meeting, Denver, Abstracts with Programs* 36 (5): 521.

Shixing, Z. and Huineng, C., 1995. Megascopic multicellular organisms from the 1700-million-year-old Tuanshanzi Formation in the Jixian area, North China. *Science* 270: 620-622.

Shu, D.-G., Conway Morris, S., Han, J., Chen, L., Zhang, X.-F., Liu, H. Q. and Liu, J.-N., 2001. Primitive deuterostomes from the Chengjiang Lagerstatte (Lower Cambrian, China). *Nature* 414: 419-424.

Shu, D., Conway Morris, S., Han, J., Zhang, Z.-F., Yasuis, K., Janvier, P., Chen, L., Zhang, S.-L., Liu, J.-N., Li., Y. and Liu, H.-Q., 2003. Head and backbone of the Early Cambrian vertebrate *Haikouichthys*. *Nature* 421: 526-529.

Shu, D., Conway Morris, S. and Zhang, S.-L., 1996. A *Pikaia*-like chordate from the Lower Cambrian of China. *Nature* 384: 157-158.

Shu, D., Conway Morris, S., Zhang, S.-L., Chen, L., Li, Y. and Han, J., 1999. A pipiscid-like fossil from the Lower Cambrian of south China. *Nature* 400: 746-749.

Shu, D.-G., Luo, H.-L., Conway Morris, S., Zhang, X.-L., Hu, S.-H., Han, J. U., Zhu, M., Li, Y. and Chen, L.-Z., 1999. Lower Cambrian vertebrates from south China. *Nature* 402: 42-46.

Shukla, M., Tewari, V. C. Babu, R. and Kumar, P., 2004. Vendian non-mineralized sponges from the Buxa Dolomite, NE Lesser Himalaya, India. In: *Abstracts Volume, The Rise and Fall of the Vendian Biota. IGCP493 Symposium*. Prato, Italy 30-31 August 2004. Edited by P. Komarower and P. Vickers-Rich. Prato: Monash University: 94-95.

Simonson, B. M. and Hassler, S., 1996. Was the deposition of large Precambrian iron formations linked to major marine transgressions? *Journal of Geology* 104: 665-676.

Smolin, L., 1997. *Life of the Cosmos*. New York : Oxford University Press.

Smoot, G. and Davidson, K., 1995. *Wrinkles in Time. The Imprint of Creation*. London: Abacus.

Sohl, L. E., Christie-Blick, N. and Kent, D. V., 1999. Paleomagnetic polarity reversals in Marinoan (ca. 600 Ma) glacial deposits of Australia: Implication for the duration of low-latitude glaciation in Neoproterozoic time. *Geological Society of America Bulletin* 111 (8): 1120-1139.

Sokolov, B. S., 1952. On the age of the oldest sedimentary cover of the Russian Platform. *Izvestiya Akademii Nauk SSSR. Ser. Geol.* 5: 21-31.

Sokolov, B. S., ed., 1965. *All-Union Symposium on the Paleontology of the Precambrian and Early Cambrian. Abstracts*. Novosibirsk, Institute of Geology and Geophysics, Siberian Branch of the USSR Academy of Sciences. (in Russian).

Sokolov, B. S. 1971. Vendian of northern Euroasia. *Geologia i Geofizika* 6: 13-22. (in Russian)

Sokolov, B. S., 1972. The Vendian stage in Earth history. In: *XXIV IGC Session. Sect 1: Precambrian Geology*: 78-84.

Sokolov, B. S., 1972. Vendian and Early Cambrian Sabellitida (Pogonophora) of the USSR. In: Proceedings of IPU. XXIII IGC (Prague, 1968 W-wa: 79-84.

Sokolov, B. S., 1973. Vendian of northern Eurasia. Arctic Geology. In: *Proceedings Second Interntional Symposium on Arctic Geology*. Edited by M. G.Pitcher. San Francisco, California USA: 204-218.

Sokolov, B. S., 1976. Organic world of the Earth on its way to Phanerozoic differentiation. *Vestnik Akademii, Nauk SSSR* 1: 126-143. (in Russian)

Sokolov, B. S., 1997. *Essays on the Advent of the Vendian System*. Moscow: KMK Scientific Press Ltd. (in Russian with English summary)

Sokolov, B. S. and Fedonkin, M. A. 1984. The Vendian as the terminal system of the Precambrian. *Episodes* 7 (1): 12-19.

Sokolov, B. S. and Fedonkin, M. A., eds., 1985. *Vendskaja Sistema 2, Istoriko-Geologiceskoe i Paleontologiceskoe Obosnovanie Paleontologija*. Moscow: Nauka.

Sokolov, B. S. and Fedonkin, M. A., 1986. Global biological events in the Late Precambrian. *5th A. Wegener Conference "Global Bio-events," University of Goettingen*: 124-129.

Sokolov, B. S. and Fedonkin, M. A., 1986. Global biological events in the Late Precambrian. In: *Global Bio-events. Lecture Notes in Earth Sciences*. Edited by O. Walliser. Berlin and Heidelberg: Springer-Verlag: 105-108.

Sokolov, B. S. and Fedonkin, M. A., eds., 1990. *The Vendian System*. Vol. 2. *Regional Geology*. Berlin: Springer-Verlag.

Sokolov, B. S. and Iwanovsky, A. B., eds., 1985. *Vendskaya Sistema 1, Istoriko-Geologiceskoe i Paleontologiceskoe Obosnovanie Paleontologija*. Moscow: Nauka.

Sokolov, B. S. and Iwanowski, A. B., eds., 1990. *The Vendian System*. 1. *Paleontology*. Berlin: Springer-Verlag

Sonnet, C. P., Kvale, E. P., Zakharian, A., Chen, M. A. and Demko, T. M., 1996. Late Proterozoic and Paleozoic tides, retreat of the Moon, and rotation of the Earth. *Science* 273: 100-104.

Sorokhtin, O.G. and Ushakov, S.A., 2002. *Development of Earth*. Moscow: Moscow State University Press.

Spirin, A.S., 2003. Ribonucleic acids as the central link of living matter. *Vestnik rossiiskoi Akademii Nauk* 73(2): 117-127.

Sprigg, R. G., 1947. Early Cambrian (?) jellyfishes from the Flinders Ranges, South Australia. *Transactions of the Royal Society of South Australia* 71 (2): 212-224.

Sprigg, R. C., 1949. Early Cambrian "jellyfishes" of Ediacara, South Australia and Mount John, Kimberley District, Western Australia. *Transactions of the Royal Society of South Australia* 73 (1): 72-99.

Sprigg, R. C., 1984. *Arkaroola – Mount Painter in the Northern Flinders Ranges, S. A.: The Last Billion Years*. Adelaide: Published by the Author and produced by Gilligham Printers Pty. Ltd.

Sprigg, R., 1989. *Geology is Fun (Recollections) or The Anatomy and Confessions of a Geological Addict*. Adelaide: Published by the Author and produced by Gilligham Printers Pty Ltd.

Sprigg, R. C., 1991. Martin F. Glaessner Palaeontologist Extraordinalre. *Geological Society of India Memoir* 20: 13-20.

Squire, R. J., Campbell, Charlotte, M. & Wilson, C. J. L., 2006. Did the Transgondwanan Supermountain trigger the explosive radiation of animals on Earth? *Earth and Planetary Sciences Letters* 250 (1-2): 116-133.

Squire, R. J., Campbell, A. H., Allen, C. M. and Wilson, C. J. L., 2006. Did the Transgondwanan Supermountain trigger the explosive radiation of animals on Earth. *Earth and Planetary Science Letters* 250.

Stankovsky, A.F., Sinitsin, A.B. and Shinkarev, N.F. 1972. Buried traps of Onega Peninsula of the White Sea. *Vestnik Leningradskogo Universiteta* 18: 12-20.

Stankovsky, A.F., Verichev, E.M. and Dobeiko, I.P., 1990. The Vendian of the South-Eastern White Sea Region. In: *The Vendian System. Vol. 2. Stratigraphy and Geological Processes*. Edited by B. S. Sokolov and M. A. Fedonkin. Berlin: Springer-Verlag: 67-76.

Star, J., 1999. *The Inner Treasure. An Introduction to the World's Sacred and Mystical Writings*. New York: Jeremy P. Tarcher/Putnam.

Steiner, M., 1994. Die Neoproterozoischen Megaalgen Südchinas. *Berliner geowissenschaftliche Abhandlungen (E)* 15: 1-146.

Steiner, M. and Reitner, J., 2001. Evidence of organic structures in Ediacara-type fossils and associated microbial mats. *Geology* 29: 1119-1122.

Stern, R. J., 2005. Evidence from ophiolites, blueschists, and ultrahigh-pressure metamorphic terranes that the modern episode of subduction tectonics began in Neoproterozoic time. *Bulletin of the Geological Society of America* 33 (7): 557-560.

Stetter, K.O., 1994. The lesson of Archaeobacteria. In: *Early Life on Earth*. Edited by S. Bengtson. Nobel Symposium No. 84. New York: Columbia University Press: 143-151.

Stow, D. A. V. and Lovell, J. P. B., 1979. Contourites: Their recognition in modern and ancient sediments. *Earth-Science Reviews* 14: 251-291.

Summons, R.E., *et al.*, 1999. 2-methylhopanoids as biomarkers for cyanobacterial oxygenic photosynthesis. *Nature* 400: 554-557.

Sun, W. G., 1986. Late Precambrian pennatulids (sea pens) from the Eastern Yangtze gorge, China: *Paracharnia* gen. nov. *Precambrian Research* 31: 361-375.

Sun, W. G., 1994. Early multicellular animals. In: *Early Life on Earth*. Nobel Symposium, No. 84. Edited by S. Bengston. New York: Columbia University Press: 359-362.

Sun, W., Wang, G.-X. and Zhou, B., 1986. Macroscopic worm-like body fossils from the Upper Precambrian (900-700 Ma), Huainan district, Anhui, China and their stratigraphic and evolutionary significance. *Precambrian Research* 31: 377-403.

Swift, D. J. P., Hudelson, P. M., Brenner, R. L. and Thompson, P., 1987. Shelf construction in a foreland basin: Storm beds, shelf sandbodies and shelf-slope depositional sequences in the Upper Cretaceous Mesaverde Group, Book Cliffs, Utah. *Sedimentology* 34: 423-457.

Swift, D. J. P., Phillips, S. and Thorne, J. A., 1991. Sedimentation on continental margins: Lithofacies and depositional systems. In: *Shelf Sand and Sandstone Bodies: Geometry, Facies, and Sequence Stratigraphy*. Edited by D. J. P. Seift, G. F. Tillman, R. W. Oertel and J. A Thorne. Oxford: Blackwell: 89-152.

Tang, F., Yin, C., Liu, Y., Wang, Z. and Gao, L., 2005. Discovery of macroscopic carbonaceous compression fossils from the Doushantuo Formation in eastern Yangtze Gorges. *Chinese Science Bulletin*, in press.

Tappan, H., 1980. *The Paleobiology of Plant Protists*. San Francisco: W.H. Freeman & Co.

Tegmark, M., 2003. Parallel universes. *Scientific American* 288 (5): 30-41.

Teske, A., Dhillon, A. and Sogin, M. L., 2003. Genomic markers of ancient anaerobic microbial pathways: Sulfate reduction, methanogenesis, and methane oxidation. *The Biological Bulletin* 204 (2): 186-191.

Tevesz, M. J. S. and McCalll, P. L., eds., 1983. *Biotic Interactions in Recent and Fossil Benthic Communities. Topics in Geobiology*, Vol. 3. New York and London: Plenum.

Tian, F., Toon, O. B., Pavlov, A. A. and Sterck, H. De, 2005. A hydrogen-rich early Earth atmosphere. *Science* 308: 1014-1017.

Timofeev, B. V., 1966. Micropaleophytological studies of ancient formations. Moscow and Leningrad: Nauka. (in Russian)

Timofeev, B. V., Herman, T. N. and Mikhailova, N. S., 1976. *Microphytofossils of the Precambrain, Cambrian and Ordovician*. Leningrad: Nauka (in Russian)

Trifonov, E. N., 2000. Leap into life's beginnings. Tracking the chronology of amino acids. *Science Spectra* 20: 62-71.

Trindade, R. I. F. and Macouin, M., 2007. Paleolatitude of glacial deposits and palaeogeography of Neoproterozoic ice ages. *Compte Rendus Geoscience* 339 (3-4): 181-288.

Trompette, R., 1996. Temporal relationship between cratonization and glaciation: The Vendian-early Cambrian glaciation in Western Gondwana. *Palaeogeography, Palaeoclimatology, Palaeoecology* 123: 373-383.

Truswell, J. F., 1977. *The Geological Evolution of Africa*. Cape Town: Purnell.

Tucker, M. E., 1992. The Precambrian-Cambrian boundary: Seawater chemistry, ocean circulation and nutrient supply in metazoan evolution, extinction and biomineralization. *Journal of the Geological Society* 149: 655-668.

Turbeville, C., 1986. An ultrastructural analysis of coelomogenesis in the hoplonemertin *Prosorhochmus americanus* and the plychaete *Magelona* sp. *Journal of Morphology* 187: 51-60.

Turner, E. C., James, N. P. and Narbonne, G. M., 1997. Growth dynamics of Neoproterozoic calcimicrobial reefs, Mackenzie Mountains, northwest Canada. *Journal of Sedimentary Research* 67: 437-450.

Turner, E. C., Narbonne, G. M. and James, N. P., 1993. Neoproterozoic reef microstructures from the Little Dal Group, northwestern Canada. *Geology* 21: 259-262.

Turner, E. C., Narbonne, G. M. and James, N. P., 2000. Framework composition of Early Neoproterozoic calcimicrobial reefs and associated microbialites, Mackenzie Mountains, N. W. T., Canada. In: *Carbonate Sedimentation and Diagenesis in the Evolving Precambrian World*. Edited by J. Grotzinger and N. James. *SEPM Special Publication* 67: 179-205.

Tyler, S. A. and Barghoorn, E. S., 1954. Occurrence of structurally preserved plants in Pre-Cambrian rocks of the Canadian Shield. *Science*, 119: 606–608. University of California Glossary. (http://www.ucmp.Berkeley.edu/glossary)

Valentine, J. W., 1986. Fossil record of the origin of Bauplans and its implication. In: *Patterns and Processes in the History of Life*. Edited by D. M. Raup and D. Jablonski. Berlin: Springer-Verlag: 209-222.

Valentine, J. W., 1992. The macroevolution of phyla. In: *Origin and Early Evolution of the Metazoa*. Edited by J. J. Lipps and P. W. Signor. New York: Plenum Press: 525-553.

Valentine, J. W., 1994. The Cambrian explosion. In: *Early Life on Earth*. Edited by S. Bengtson. Nobel Symposium 84. New York: Columbia University Press: 401-411.

Valentine, J. W., 2004. *On the Origin of Phyla*. Chicago: University of Chicago Press.

Valentine, J. W., 2006. Ghosts of bilaterians past. In: Abstract volume, IGCP 493 Symposium, Kyoto University, Japan, 27-31 January 2006. Edited by P. Vickers-Rich and T. Ohno. Kyoto: Kyoto University: 10.

Valentine, J. W., Erwin, D. H. and Jablonski, D., 1996. Developmental evolution of metazoan Body Plans: The fossil evidence. *Developmental Biology* 173: 373-381.

Valentine, J. W., Jablonski, D. and Erwin, D. H., 1999. Fossils, molecules and embryos: New perspectives on the Cambrian. *Development* 126: 851-859.

Valley, J. W., 2005. A cool early Earth? *Scientific American* 293 (4): 58-65.

Van Dover, C.L., 2000. *The Ecology of Deep-Sea Hydrothermal Vents*. Princeton: Princeton University Press.

van Iten, H., de Moraes Leme, J., Rodrigues, S. C. and Simones, M. G., 2005. Reinterpretation of a conularild-like fossil from the Vendian of Russia. *Palaeontology* 48 (3): 619-622.

Van Straden, A. and Zimmermann, U., 2004. Tillites or ordinary conglomerates? Provenance studies on diamictites of the Neoproterozoic Puncoviscana in NW Argentina. In: *Geoscience Africa 2004. Abstract Volume*. Edited by L. D. Ashwal. Johannesburg: University of the Witwatersrand: 668-669.

Veizer, J., 1983. Geologic evolution of the Archean-Early Proterozoic Earth. In *Earth's Earliest Biosphere*. Edited by J. W. Schopf. Princeton: Princeton University Press: 240-259.

Veis, A. F., Vorob'eva, N. G. & Golubkova, E. Yu., 2006. The Early Vendian microfossils first found in the Russian Plate: Taxonomic composition and biostratigraphic significance. *Stratigraphy and Geological Correlation* 14 (4): 368-38.

Velikanov, V. A., 1990. Key section of the Vendian in Podolia. In: *The Vendian System*. Vol. 2. *Regional Geology*. Edited by B.S. Sokolov, and M.A. Fedonkin. Berlin and London: Springer-Verlag: 38-75.

Velikonov, V. A., Asseva, E. A. & Fedonkin, M. A., 1983. *The Vendian of the Ukraine*. Akademia Nauk Ukranskoi SSR. Institute Geologicheskich Nauk: 162 pp.

Velikanov, V. A., Aseeva, Ye. A. and Fedonkin, M. A., 1983. The Vendian of the Ukraine. Kiev: Naukova Dumka. (in Russian).

Vellai, T. and Vida, G., 1999. The origin of eukaryotes: The difference between prokaryotic and eukaryotic cells. *Proceedings of the Royal Society of London, Biological Sciences* 1266(1428): 1571-1577.

Vickers-Rich, P., 2006. "Saline giants," cold cradles and the cold playgrounds of Neoproterozoic Earth: The origin of Animalia. *Abstract Volume, IGCP 493 Symposium*, Kyoto University, Japan 27-31 January 2006. Edited by P. Vickers-Rich and T. Ohno. Kyoto: Kyoto University: 11-14.

Vickers-Rich, P., 2007. Opportunities in the weedy "cold playgrounds" of the Neoproterozoic and the rise of Metazoa. In: *The Rise and Fall of the Ediacaran Biota*. Edited by P. Vickers-Rich and P. Komarower. *Geological Society of London Special Publication* 286.

Vickers-Rich, P. and Komarower, P., eds., 2007. *The Rise and Fall of the Ediacaran Biota. Environmental and Biotic Dynamics of the Neoproterozoic and Early Phanerozoic*. Geological Society of London, Special Publication 286.

Vickers-Rich, P. and Rich, T. H., 1999. *Wildlife of Gondwana. Dinosaurs & Other Vertebrates from the Ancient Supercontinent*. Bloomington: Indiana University Press.

Vidal, G. and Ford, T. D., 1985. Microbiotas from the late Proterozoic Chuar Group (northern Arizona) and Uinta Mountain Group (Utah) and their chronostratigraphic implications. *Precambrian Research* 28: 349-389.

Vidal, G. and Jensen, S., 1994. Neoproterozoic (Vendian) ichnofossils from Lower Alcudian strata in central Spain. *Geological Magazine* 131: 169-179.

Vidal, G. and Maczydlowska-Vidal, M., 1997. The Neoproterozoic of Baltica – stratigraphy, palaeobiology and general geological evolution. *Precambrian Research* 73: 197-216.

Vincent, W. F., Mueller, D., van Hove, P. and Howard-Williams, C., 2004. Glacial periods on early Earth and implications for the evolution of life. In: *Origins. Genesis, Evolution and Diversity of Life*. Edited by J. Seckbach. Berlin: Kluwer Academic Publishers: 483-501.

Vodanjuk, S. A. 1989. Ostatki besskeletnykh metazoa iz khatyspytskoj svity Olenekskogo podnjatija. In: *Pozdnij Dokembrij i Rannij Paleozoj Sibiri*. Edited by V. V. Khomentovskij and Ju. K. Sovetov. Akademija Nauk SSSR, Sibirskoe Otdelenie, Institut Geologii i Geofiziki, Novosibirsk: 61-74. (in Russian).

Volkova, N. A., Ragozina, A. L., Sivertseva, I. A., Jankauskas, T. V., Hermann, T. N., Pyatiletov, V. G., Rudavskaya, V. V. and Yakshin, M. S., 1990. Vendian microfossils. *The Vendian System*. 1. *Paleontology*. Edited by B.S. Sokolov, and A.B.Iwanowski. Berlin & New York: Springer-Verlag: 154-192.

Vologdin, A. G. and Maslov, A. B., 1960. A new group of fossil extinct organisms from the lowermost Yodoma Group of the Siberian Platform. *Doklady Akademii Nauk SSSR* 135 (3): 691-693.

Wade, M., 1968. Preservation of soft bodied animals in Precambrian sandstones at Ediacara, South Australia. *Lethaia* 1: 238-267.

Wade, M., 1969. Medusae from uppermost Precambrian or Cambrian sandstones, central Australia. *Palaeontology* 12: 351-365.

Wade, M. J., 1970. The stratigraphic distribution of the Ediacara fauna in Australia. *Transactions of the Royal Society of South Australia* 94: 87-104.

Wade, M. J., 1971. Bilateral Precambrian chondrophores from the Ediacara fauna, South Australia. *Proceedings of the Royal Society of Victoria* I84: 183-188.

Wade, M. J., 1972. *Dickinsonia*: Polychaete worms from the late Precambrian Ediacara fauna, South Australia. *Memoirs of the Queensland Museum* 16: 171-190.

Wade, M. J., 1972. Hydrozoa and Scyphozoa and other medusoids from the Precambrian Ediacara fauna, South Australia. *Palaeontology* 15: 197-255.

Waggoner, B. M., 1995. Ediacaran lichens: A critique. *Paleobiology* 21 (3): 393-397.

Waggoner, B. M., 1999. Interpreting the earliest Metazoan fossils: What can we learn? *American Zoologist* 38: 9756-982.

Waggoner, B. M., 2003, The Ediacaran biotas in space and time. *Integrative and Comparative Biology* 43: 104-113.

Waggoner, B. and Hagadorn, J. W., 2002. New fossils from terminal Neoproterozoic strata of southern Nye County, Nevada. In: *Proterozoic-Cambrian of the Great Basin and Beyond*. Edited by F. A. Corsetti. Field Trip Guidebook and Volume Prepared for the Annual Pacific Section SEPM Fall Field Trip: 87-96.

Walcott, C. D., 1883. Pre-Carboniferous strata in the Grand Canyon of the Colorado, Arizona. *American Journal of Science* 26: 437–442.

Walcott, C. D., 1895. Algonkian rocks of the Grand Canyon of the Colorado. *Journal of Geology* 3: 312–330.

Walcott, J.D., 1899. Pre-Cambrian fossiliferous formations. *Bulletin of the Geological Society of America* 10: 199-244.

Walker, G., 2003. *Snowball Earth: The Story of the Great Global Catastrophe that Spawned Life as We Know It*. New York: Crown Publishers.

Walliser, O. H., ed., 1996. Global Events and Event Stratigraphy in the Phanerozoic. *Results of the International Interdisciplinary Cooperation in the IGCP-Project 216 "Global Biological Events in Earth History*. New York: Springer.

Walter, M., 2001. *To Mars and Beyond. Search for the Origins of Life*. Sydney: Art Exhibitions Australia.

Walter, M. R., 1972. Stromatolites and the Biostratigraphy of the Australian Precambrian and Cambrian. Special Paper Palaeontology, 11.

Walter, M. R., Buick, R. and Dunlop, J. S. R., 1980. Stromatolites 3,400-3,500 Myr old from the North Pole area, Western Australia. *Nature* 248: 443-445.

Walter, M. R. and Cloud, P., 1983. Microfossils from the Albinia Formation. *Bureau of Mineral Resources Australia Bulletin* 212: 82–83.

Walter, M. R., Elphinstone, R. and Heys, G. R., 1989. Proterozoic and Early Cambrian trace fossils from the Amadeus and Georgina Basins, Central Australia. *Alcheringa* 13: 209-256.

Walter, M. R. and Heys, G. R., 1985. Links between the rise of the Metazoa and the decline of stromatolites. *Precambrian Research* 29: 149-174.

Walter, M. R. and Hofmann, H. J., 1983. The palaeontology and palaeoecology of Precambrian iron-formations. In: *Iron-formation: Facts and Problems*. Edited by A. F. Trendall and R. C. Morris. Amsterdam: Elsevier: 373-400.

Wang, G.-X., 1982. Late Precambrian Annelida and Pogonophora from the Huanian of Anhui Province. *Bulletin of the Tianjin Institute of Geology and Mineral Resources* 6: 9-22. (in Chinese)

Ward, P. D., 2006. Impact from the Deep. Scientific American, Oct. 2006: 42-49.

Ward, P. D. and Brownlee, D., 2000. *Rare Earth: Why Complex Life Is Rare in the Universe*. New York: Copernicus.

Waren, A., Bengston, S., Goffredi, S. K. and van Dover, G. O. L., 2003. A hot-vent gastropod with iron sulfide dermal sclerites. *Science* 302: 1007.

Washburn, M., 1982. *Distant Encounters. The Exploration of Jupiter and Saturn*. London: Harcourt Brace Jovanovich.

Watts, W. W., 1947. *The Geology of the Ancient Rocks of Charnwood Forest*. Leicestershire: Leicester.

Weaver, P. G., McMenamin, M. A. S. and Tacker, C., 2006. Paleoenviornmental and paleobiogeogrphic implication of a new Ediacaran body fossil from the Neoproterozoic Carolina Terrane, Stanley County, North Carolina. *Precambrian Research* 150 (3-4): 123-135.

Wei-Guo, S., 1986. Late Precambrian scyphozoan medusa *Mawsonites randellensis* sp. nov. and its significance in the Ediacara metazoan assemblage, South Australia. *Alcheringa* 10: 169-181.

Weil, A.B., Van der Voo, R., MacNiocaill, C. and Meert, J.G., 1998. The Proterozoic Supercontinent Rodinia: Paleomagnetically derived reconstructions for 1100 to 800 Ma. *Earth and Planetary Science Letters* 154: 13-24.

Weinberg, S., 1977. *The First Three Minutes. A Modern View of the Origin of the Universe*. Glasgow: Fontana/Collins.

Weinberg, S., 1992. *Dreams of a Final Theory*. New York: Pantheon Books.

Weistheimer, F. H., 1987. Why nature chose phosphates. *Science* 235: 1173-1178.

Westall, F., 2005. Life on the Early Earth: A sedimentary view. *Science* 308: 366-367.

Wetherall, D., 2003. Evolving with the enemy. *New Scientist* 2422: 44-47.

White, M. G., 1984. Marine Benthos. In: *Antarctic Ecology* 2. Edited by R. M. Laws. London: Academic Press: 421-461.

Whitman, W. B., Coleman, D. C. and Wiebe, W. J., 1998. Procaryotes: The unseen majority. *Proceedings of the National Academy of Sciences* 95: 6578-6583.

Whittington, H. B., 1985. *The Burgess Shale*. New Haven: Yale University Press.

Wiggins, A. W. and Wynn, C. M., 2003. *The Five Biggest Unsolved Problems in Science*. New York: John Wiley & Sons.

Wikipedia. The Free Encyclopedia. http://en.wikipedia.org/wiki/

Wilde, S. A., Valley, J. W., Peck, W. H. and Graham, C. M., 2001. Evidence from detrital zircons for the existence of continental crust and oceans on the Earth 4.4 Gyr ago. *Nature* 409: 175-178.

Williams, G. C., 1995. Living genera of sea pens (Coelenterata: Octocorallia: Pennatulacea): Illustrated key and synopses. *Zoological Journal of the Linnean Society* 113: 93-140.

Williams, G. C., 1997. Preliminary assessment of the phylogeny of Pennatulacea (Anthozoa: Octocorallia), with a reevaluation of Ediacaran frond-like fossils, and a synopsis of the history of evolutionary thought regarding the sea pens. *Proceedings of the 6th International Conference on Coelenterate Biology* 1995: 497-509.

Williams, G. E. and Schmidt, P. W., 2003. Possible fossil impression in sandstone from the late Palaeoproterozoic-early Mesoporterozoic Semri Group (lower Vendhyan Supergroup), central India. *Alcheringa*: 75-76.

Williams, G. E. and Wallace, M. W., 2003. The Acraman asteroid impact, South Australia: Magnitude and implication for the late Vendian environment. *Journal of the Geological Society, London* 160: 545-554.

Williams, H. and King, A. F., 1979. Trepassey map area, Newfoundland. *Geological Survey of Canada Memoir* 389: 1-24.

Williams, R.J.P. and Frausto da Silva, J.J.R., 1996. *The Natural Selection of the Chemical Elements. The Environment and Life's Chemistry*. Oxford : Clarendon Press.

Williams, R.J.P. and Frausto da Silva, J.J.R., 2001. *Bringing Chemistry to Life: From Matter to Man*. Oxford: Oxford University Press.

Wills, M. A. and Fortey, R. A., 2000. The shape of life: How much is written in stone? *BioEssays* 22: 1142-1152.

Wilson, E. O., 2003. *The Future of Life*. New York: Vintage Books.

Woese, C. R., Kandler, O. and Wheelis, M. L., 1990. Towards a natural system of organisms: Proposal for the domains Archaea, Bacteria, and Eukarya. *Proceedings of the National Academy of Sciences of the United States of America* 87: 4576-4579.

Wood, D. A., Dalrymple, R. W., Narbonne, G. M., Gehling, J. G. and Clapham, M. E., 2003. Paleoenvironmental analysis of the Late Neoproterozoic Mistaken Point and Trepassey Formations, southeastern Newfoundland. *Canadian Journal of Earth Sciences* 40: 1375-1391.

Wood, F., *et al.,* 2002. The genome sequence of *Schizosaccharomyces pombe*. *Nature* 415: 871-880.

Wood, R. A., Grotzinger, J. P. and Dickson, J. A. D., 2002. Proterozoic modular biomineralized metazoan from the Nama Group, Namibia. *Science* 296: 2383-2386.

Wray, G.A., Levinton, J.S. and Shapiro, L.H., 1996. Precambrian divergences among metazoan phyla. *Science* 274: 568-573.

Wright, L. D. and Coleman, J. M., 1974. Mississippi River Mouth processes: Effluent dynamics and morphologic development. *Journal of Geology* 82: 751-778.

Xian-Guang, H., Aldridge, R., Bergstrom, J., Siveter, D. J., Siveter, D. and Xiang-Hong, F., 2006. The Cambrian Fossils of Chengjiang, China. Oxford: Blackwell Publishing.

Xiao, S., 2002. Mitotic topologies and mechanics of Neoproterozoic algae and animal embryos. *Paleobiology* 28 (2): 244-250.

Xiao, S., 2004. New multicellular algal fossils and acritarch in Doushantou chert nodules (Neoproterozoic, Yangtze Gorges, South China). *Journal of Paleontology* 78 (2): 393-401.

Xiao, S. and Kaufman, A. J., 2006, eds. *Neoproterozoic Geobiology and Paleobiology*. Dorsrecht: Springer.

Xiao, S. and Knoll, A. H., 2000. Phosphatized animal embryos from the Neoproterozoic Doushantuo Formation at Weng'an, Guizhou, South China. *Journal of Paleontology* 74 (5): 767-788.

Xiao, S., Knoll, A. H., Yuan, X. and Pueschel, C. M., 2004. Phosphatized multicellular algae in the Neoproterozoic Doushantuo Formation, China and the early evolution of florideophyte red algae. *American Journal of Botany* 91: 214-227.

Xiao, S., Shen, B., Zhou, C., Xie, G. and Yuan, X., 2005. A uniquely preserved Ediacaran fossil with direct evidence for a quilted bodyplan. *Proceedings National Academy of Sciences USA* 102:10227-10232.

Xiao, S., Yuan, X. and Knoll, A. H., 2000. Eumetazoan fossils in terminal Proterozoic phosphorites? *Proceedings of the National Academy of Sciences* 97: 13684-13689.

Xiao, S., Yuan, X., Steiner, M. and Knoll, A. H., 2002. Macrospcopic carbonaceous compressions in a terminal Proterozoic shale: A systematic reassessment of the Miaohe biota, South China. *Journal of Paleontology* 76: 345-374.

Xiao, S., Zhang, Y. and Knoll, A., 1998. Three-dimensional preservation of algae and animal embryos in a Neoproterozoic phosphorite. *Nature* 391: 553-558

Xiong, J., *et al.,* 2000. Molecular evidence for the early evolution of photosynthesis. *Science* 289: 1724-1730.

Xiong, J. and Bauer, C. E., 2002. Complex evolution of photosynthesis. *Annual Review of Plant Biology* 53: 503-521.

Yakshin, M. S., 1987. Vendian of the Olenek Uplift. In: *Pozdnii dokembrii I ranni paleozoi Sibiri. Sibirskaia platforma I ee iuzhnoe skladchatoe obramlenie*. Edited by V. V. Khomentovskiy and V. Yu. Shenfil. Novosibirsk: IGG SO AN SSSR (Institute of Geology and Geophysics, Siberian Branch, Academy of Science, USSR): 18-30. (in Russian).

Yahkshin, M. S. and Vodanjuk, S. A., 1986. Khorbusuonka Group in the Khorbusuonka River Basin (Olenek Uplift). In: *Pozdnii dokembrii I ranni paleozoi Sibiri Stratigrafia I paleontologia*. Edited by V. V. Khomentovskiy and V. Yu. Shenfil. Novosibirsk: IGG SO AN SSSR (Institute of Geology and Geophysics, Siberian Branch, Academy of Science, USSR): 21-39. (in Russian)

Yakobson, K. E., Kuznetsova, M. Yu., Stankovsky, A. F., *et al.,* 1991. Riphean of the Winter Coast of the White Sea. *Sovetskaya Geologia* 11: 44-48. (in Russian).

Yang, W. and Holland, H. D., 2003. The Hekport paleosol profile in strata 1 at Gaborone Botswana: Soil formation during the great oxidation event. *American Journal of Science* 303: 187-220.

Yanichevsky, M. E., 1924. Ob ostatkah trubchatyh chervei iz kembriiskih sinih glin (On the remains of the tubular worms from the Cambrian blue clays). *Ezhegodnik Russkogo Paleontologicheskogo Obchestva* 4: 99-112. (in Russian)

Yin, C., Bengtson, S. and Yue, Z., 2004. Silicified and phosphatized *Tainzhushania*, spheroidal microfossils of possible animal origin from the Neoproterozoic of South China. *Acta Palaeontologica Polonica* 49 (1): 1–12.

Yin, C., Feng, T., Liu, Y., Gao, L., Liu, P., Xing, Y., Yang, Z., Wan, Y. and Wang, Z., 2005. U-Pb zircon age from the base of the Ediacaran Doushantuo Formation in the Yangtze Gorges, South China: Constraint on the age of Marinoan glaciation. *Episodes* 28 (1): 48-49.

Yin, C. and Gao, L., 1996. The early evolution of the acanthomorphic acritarchs in China and their biostratigraphical implication. *Acta Geologica Sinica* 9: 193–206.

Yin, L., Xiao, S. and Yuan, Y., 2001. New observations on spicule-like structures from Doushantuo phosphorites at Weng'an, Guizhou Province. *Chinese Science Bulletin* 46: 1031-1036.

Yochelson, E. L., 1996. Discovery, collection, and description of the Middle Cambrian Burgess Shale Biota by Charles Doolittle Walcott. *Proceedings of the American Philosophical Society* 140 (4): 469-545.

Yochelson, E. L., 1998. *Charles Doolittle Walcott 1850-1927*. National Academy of Sciences Biographical Memoirs 39: 471-540.

Yochelson, E. L. and Fedonkin, M. A., 1991. Paleozoic trail. *National Geographic Research and Exploration* 7(4): 453-455.

Yochelson, E. L. and Fedonkin, M. A., 1993. Paleobiology of *Climactichnites*, an enigmatic Late Cambrian fossil. *Smithsonian Contribution to Paleobiology* 74: 1-74.

Yochelson, E. L. and Fedonkin, M. A., 2000. A new tissue-grade organism 1.5 billion years old from Montana. *Proceedings of the Biological Society of Washington* 113(3): 843-847.

Yoshida, M., Windley, B.F. and Dasgupta, S., eds., 2003. Proterozoic East Gondwana: Supercontinent Assembly and Breakup. *Geological Society Special Publication* 206. London: The Geological Society.

Yuan, X., Li, J. and Cao, R., 1999. A diverse metaphyte assemblage from the Neoproterozoic black shales of South China. *Lethaia* 32: 143-155.

Yuan, X., Xiao, S., Li, J., Yin, L. and Cao, R., 2001. Pyritized chuarids with excystment structures from the late Neoproterozoic Lantian formation of Anhui, South China. *Precambrian Research* 107: 253-263.

Yuan, X., Xiao, S., Li, J., Yin, L., Knoll, A., Zhou, C. and Mu, S. I, 2002. *Doushantou Fossils. Life on the Eve of Animal Radiation*. Hefei: China University of Science and Technology Press: 1-171.

Yuan, X., Xiao, S. and Taylor, T. N., 2005. Lichen-like symbiosis 600 million years ago. *Science* 308: 1017-1020.

Yushkin, N. P., 2002. Biomineral homologies, abiotic biomorphs, mineral organismobiosis, and the problem of the genetic indication of geo- and astrobioproblematics. *Syktyvkar, Geoprint*: 1-44.

Zahnle, K. L., 2001. Mega-impacts and the late heavy bombardment: Effects on the Archaean geological and biological record. In: *International Archaean symposium Extended Abstracts*. Edited by K. F. Cassidy, J. M. Dunphy and M. J. Van Kranendonk. *AGSO – Geoscience Australia Record* 2001/37: 28-30.

Zaika-Novatsky, V. S., 1965. New problematics from the Upper Precambrian on the Dniester River Basin. In: *All-Union Symposium on Paleontology of the Precambrian and Early Cambrian*. Abstracts. Edited by B. A. Sokolov. Nauka, Novosibirsk: 98-99. (in Russian)

Zaika-Novatski, V. S., Velikanov, V. A. and Koval, A. P., 1968. First representative of Ediacara fauna in the Vendian of the Russian Platform. *Paleontological Journal* 2: 131-134. (in Russian)

Zang, W., 1995. Early Neoproterozoic sequence stratigraphy and acritarch biostratigraphy, eastern Officer Basin, South Australia. *Precambrian Research* 74: 119–175.

Zang, W. and Walter, M. R., 1992. Late Proterozoic and Early Cambrian microfossils and biostratigraphy, Amadeus Basin, central Australia. *Memoirs of the Association of Australasian Palaeontologists* 12: 1–132.

Zavarzin, G. A., 2002. Rol' kombinatornykh sobytii v razvitii bioraznoobraziya (Role of the coincident events in the development of biodiversity). *Priroda* 1: 12-19. (in Russian)

Zehnder, A. J. B., 1988. *Biology of Anaerobic Microorganisms*. New York: John Wiley & Son.

Zhang, S., Jiang, G., Zhang, J., Song, B., Kennedy, M. J. and Christie-Blick, N., 2005. U-Pb sensitive high-resolution ion microprobe ages from the Doushantuo Formation in south China; Constraints on late Neoproterozoic glaciations. *Geology* 33: 473-476.

Zhang, X., Hua, H. and Reitner, J., 2005. A new type of Precambrian megascopic fossils: The Jinxian biota from northeastern China. *North American Paleontological Convention 2005, Abstract Volume*.

Zhang, Y., 1989. Multicellular thallophytes with differentiated tissues from late Proterozoic phosphate rocks of South China. *Lethaia* 22: 113-132.

Zhang, Y., Yin, L., Xiao, S. and Knoll, A. H., 1998. Permineralized fossils from the terminal Proterozoic Doushantuo Formation, South China. *The Paleontological Society Memoir* 50: 1-52.

Zhang, Y. and Yuan, X., 1992. New data on multicellular thallophytes and fragments of cellular tissues from late Proterozoic phosphate rocks, South China. *Lethaia* 25: 1-18.

Zhang, Z. Y., 1997. A new Palaeoproterozoic clastic-facies microbiota from the Changzhougou Formation, Changcheng Group, Jixian, North China. *Geological Magazine* 134 (2): 145-150.

Zhao, Y., He, M., Chen, M. E., Peng, J., Yu, M., Wang, Y., Yang, R., Wang, P. and Zhang, Z., 2005. The discovery of Miaohe-type biota from the Neoproterozoic Doushantuo Formation in Jiangkou County, Guizhou Province, China. *Chinese Science Bulletin* 49: 1916-1918.

Zhao, Y., Zhu, M., Babcock. L. E., Yuan, J., Parsley, R. L., Peng, J., Yang, X and Wang, Y., 2005. Kalli biota. A taphonomic window on diversification of metazoans from the basal Middle Cambrian: Guizhou, China. *Acta Geologica Sinica* 79: 751-765.

Zheng, Wenwu, 1980. A new occurrence of fossil group of *Chuaria* from the Sinian system in north Anhui and its geological meaning. *Bulletin of the Tianjin Institute of Geology and Mineral Resources* 1: 49-69.

Zhou, C. and Xiao, S., 2006. Ediacaran δ^{13} chemostratography of South China. *Chemical Geology* (doi:10.1016/j.chemgeo.2006.06.021).

Zhou, C., Xie, G., McFadden, K., Xiao, S. and Yuan, X., in press. The diversification and extinction of Doushantuo-Pertatataka acritarchs in South China: Causes and biostratigraphic significance. *Geological Journal*.

Zhu, W. and Chen, M., 1984. On the discovery of macrofossil algae from the Late Sinian in the eastern Yangtze Gorges, south China. *Acta Botanica Sinica* 26 (5): 558-560.

Zhuravlev, A. Yu., 1993. Were Ediacaran Vendobionta multicellulars? *Neues Jahrbuch für Geologie und Paläontologie Abhandlungen*, 190 (2/3): 299-314.

Zhuravlev, A. Yu. and Riding, R., 2001. *The Ecology of the Cambrian Radiation*. New York: Columbia University Press.

Zimmer, C., 2000. *Parasite Rex. Inside the Bizarre World of Nature's Most Dangerous Creatures*. London: Arrow Books.

Zoricheva, A. I., 1963. The northern Russian Platform. In: *Geology of the USSR*. Vol. 2. *Arkhangel'sk to Vologda*. Edited by A. I. Zoricheva. Oblast and Komi ASSR: 79-99.

Zuckerman, B. and Malkan, M. A., eds., 1996. *The Origin and Evolution of the Universe*. Boston: Jones and Bartlett.

INDEX

Page numbers in **bold** type refer to illustrations.

Chengjiang fauna, 75, 112, **252–5**

Chernov, 131

Cherny Kamen (Chenokamen) Formation, **169–73**

"Chimney" Fossils, 189

China, 36, **49**, 51, 75, 104, 117

China University of Geosciences, 104

Chiquitos Line, **194**

chirodropid cubozoans, 139

Chistyakov, V.G., 130

chitinozoans, 220

chondrophores, 188

Chordata, **43**, 74, **236**, 239, **250–1**

chordates, 44, 51, 74, **113**, 150

Choromoro River, 192

chromatin, **23**

chromosomes, 22, 32, 43–4

Chuaria, 36, 176, 197

Chuaria circularis, 196, 224, 226

Chumakov, N., 120, **159**

Chung, M., 196

ciliates, **23**, 32

circulatory system, 45

Circulichnus (Circulichnis), **162**, 208, **212**, 213

clade, 237

Clapham, M., **59**

Clarke, Arthur C., xi

classification of metazoans, 239–42

Clemente Formation, 200

Climactichnites, **211**

Cloud, P., 186

Cloud-Walker-Holland-Kasting Model, 16, **46**

Cloudina, 39–40, 45, 51, **64**, 69, 72, 83–4, **85–6**, 183, **188–9**,192, 244, **245–6**

Cloudina hartmannae, 199

Cnidaria, 42, **43**, 45, 51, 58, 109, **236**, 239, **256**

cnidarian polyps, 59

cnidarian tubes, 198

coal, 18

Coates Lake Formation, **177**

Cochleatina, 154

Cochlichnus, **154**

Coelenterata, 74

"coelenterates," 58

coelom, 35, **42–3**

coelomates, 239

cold cradles, 38–9

cold playgrounds, 39

collagen, 40

Collins, A., 145–**6**

combjellies, 42, 45, 51, 75, 239, 242

combs, **58**

comets, 11, 15

compartmentalization, 24

complex organic compounds, 10

Conception Bay, 54, 62

Conception Group, 50, 56–**7**, 66

Conception-style preservation, 50

Condon, D., 39, 194

Congo Craton, 38, 72

*Conomedusite*s, 109, 137, 151

*Conomedusite*s *lobatus,* **200**

Conophyton, **177, 223**

Conotubus, 199

continental crust, 8, 11, 13–4

"contourites," 179

conularids, 74

Conulata, 74

convection, 11

Conway Morris, S., 107

Cooper, A., 247

Copley, **208**

coralomorphs, 85

corals, 42, 45, 51, 99, 107

Core, 8, 9, 11

Corumba Group, 192

Corumba-Ladario region, 193

Corumbella, 188–**9**

Corumbella werneri, 193–**4**

cosmic rays, 3

Crassicorium pendjariensis, 224

cratons, 11, 19

crown group, 237

Crust, 11

crustacean, **254**

crustal rocks, 11–2

Cryogenian, 38, 177, 224–6

Ctenophora, 42–**3**, 45, 75, **236,** 238–9, 242

Cubichnia, **208**

Cuckold Formation, **57**

Curvolithos, **85**

cyanobacteria, 15, 22–**3**, **27, 32,** 39, 94, 161, 167

cyanobacterial mats, **26**

Cycliophora, **43**

Cyclomedusa, 72, 85, 111, 130, 133, 141, 150, 157, **165,** 167,

170, **171,** 172, **177,**179, **182,** 188, 198, **200**

Cyclomedusa davidi, 200

Cyclomedusa delicata, 121

Cyclomedusa plana, **151, 154, 182,** 200

Cyclomedusa serebrina, **154**

Cyclozoa, 237

Cylindrichnus, 157

Cymatiosphaeroides sullingii, 224

cytoplasm, 24

cytosine, 43

cytoskeleton, 31, 32

Dabis Formation, 70, 73, 75, 78, **85**

Daily, B., 96, **97**

Dalrymple, R.W., 179

Daltaenia, **177**

Darwin, C., 41, 49, 67, 186, **219**

Darwin's dilemma, 41, 67, 90

David, E., 94

Davis Ice Shelf, 36

Dawson, Sir W., 185–6

Dead Sea Scrolls, 104

"death masks," 41, 50–1, 58, 110–**1,** 135

deep water, 54, 56

deep water occurrences, **57**

Denying Formation, **195**–6, 198–9, 243

deoxyribose, 43

Desulforococcus, **23**

deuterostomes, 43–5, 239

Deuterostomia, **43, 236**

Devil's Peak, 99, 132

Diabaig Formation, 199

Diamantino Formation, 192

diamicites, 36, 37, **57,** 69, 71, 73, 90–1, **225**

diamondoids, 21

Diapsida, **45**

diatoms, 220

dickinsomiomorphs, **64**

Dickinsonia, 50, 54, **64**, 79, **95–6, 102–3,** 108–10, **111,** 112–**3, 121,** 124–**5,** 131, 133, 135, 137, 141, 151–2, 154, **169,** 170, 172, **200,** 206, 209, 213–**4, 238–**9, 242–3, **250–1**

Dickinsonia costata, 93, 96, 98, 103, **109, 112, 214**

Dickinsonia elongata, 100

dickinsoniid, **180**

Didymaulichnus, **85**

Didymaulichnus tirasensis, **154**

Digermul Peninsula, 199

Dimoth/Ohmoto Model,16, **46**

dinoflagellates, 220

Dipleurozoa, **238**

Diplichnites, **85**

diploblastic, **42**, 45, 243

Diplocraterion, **250–1**

diploid, 42

diplomonads, **23**

Distosphaera australica, **220, 226**

diversity gradients, **244**

DNA, 19, **23,** 32, 41, 43–4, 236–7

Dniester Member, **154**

Dniester River, **148,** 151, 152

Dniester River Basin, **49, 149–55**

dolomites, 40–1

Domichnia, **208**

Domodedovo Airport, 158

Doppler, J.C., 5

Doppler Effect, 5

double helix, 43

Doushantuo Assemblage, **195**

Doushantuo-Denying boundary, 194

Doushantuo embryos, 198, 243, **245**

Doushantuo Formation, 194, **195, 196–9, 203,** 243

Doushantuo phosphorites, 243

Doushantuo/Shuram/Kuibis anomaly, **246**

Dozy, J.J., 200

Dresser Formation, 222

Drook Formation, **57, 60,** 62–3, **64,** 65

dropstones, 36–**8,** 90

Droser, M., 104, 129, 210, 212

"dubiofossils," 83

Dunning, G., 56, 66

Durham, J.W., 42

Dyfed, 188

Dzhurzhev Member, **154**

Dzik, J., 75–6, 79, 107, 138, 211, 239

Earaheedia kuleliensis, **20, 223**

Earaheedy Group, 19, **20**

early atmosphere, 16

Earth, 3, **12**

earthquakes, **14**

earthworms, 45

ABOUT THE AUTHORS

Mikhail A. Fedonkin, Paleontological Institute, Russian Academy of Science, Moscow 117997, Russia; Honorary Research Fellow, School of Geosciences, Monash University, Melbourne, Victoria 3800 Australia.

James G. Gehling, South Australian Museum, Adelaide, 5000; Honorary Research Associate, School of Geosciences, Monash University, Melbourne, Victoria, 3800 Australia.

Kathleen Grey, Geological Survey of Western Australia, 100 Plain Street, East Perth, Western Australia 6004; Honorary Research Associate, School of Geosciences, Monash University, Melbourne, Victoria 3800 Australia.

Guy M. Narbonne, Department of Geological Sciences and Geological Engineering, Queen's University, Kingston, Ontario K7L 3N6 and Research Associate, Royal Ontario Museum, Toronto, Ontario M5S 2C6, Canada; Department of Earth and Planetary Sciences, Macquaire University, Sydney, New South Wales 2109 Australia.

Patricia Vickers-Rich, School of Geosciences and the Monash Science Centre, Monash University, Melbourne, Victoria, 3800; Research Associate, Museum Victoria, Melbourne, Victoria, Australia; Research Associate, Laboratory of Precambrian Organisms, Paleontological Institute, Russian Academy of Sciences, Moscow, 117997, Russia.